Progress in Mathematics
Volume 138

Analytic Number Theory

Volume 1

Proceedings of a Conference In Honor of Heini Halberstam

Bruce C. Berndt
Harold G. Diamond
Adolf J. Hildebrand
Editors

Birkhäuser
Boston • Basel • Berlin

Bruce C. Berndt
Harold G. Diamond
Adolf J. Hildebrand
Department of Mathematics
University of Illinois
Urbana, IL 61801

Printed on acid-free paper

Softcover reprint of the hardcover 1st edition 1996

Birkhäuser

ISBN-13: 978-1-4612-8645-5 e-ISBN-13: 978-1-4612-4086-0
DOI: 10.107/978-1-4612-4086-0

Typesetting by the editors in TEX

9 8 7 6 5 4 3 2

Contents of Volume 1

Contents of Volume 2

Preface

On May 16 - 20, 1995, approximately 150 mathematicians gathered at the Conference Center of the University of Illinois at Allerton Park for an International Conference on Analytic Number Theory. The meeting marked the approaching official retirement of Heini Halberstam from the mathematics faculty of the University of Illinois at Urbana-Champaign. Professor Halberstam has been at the University since 1980, for 8 years as head of the Department of Mathematics, and has been a leading researcher and teacher in number theory for over forty years.

The program included invited one hour lectures by G. Andrews, J. Bourgain, J. M. Deshouillers, H. Halberstam, D. R. Heath-Brown, H. Iwaniec, H. L. Montgomery, R. Murty, C. Pomerance, and R. C. Vaughan, and almost one hundred other talks of varying lengths. These volumes comprise contributions from most of the principal speakers and from many of the other participants, as well as some papers from mathematicians who were unable to attend. The contents span a broad range of themes from contemporary number theory, with the majority having an analytic flavor.

The conference was made possible by grants from the National Science Foundation, the National Security Agency, the Institute for Mathematics and Applications, the College of Liberal Arts and Sciences, and the Mathematics Department of the University of Illinois. We are grateful to these agencies and institutions for their support. We also extend our thanks to the director of the Allerton Park Conference Center, Beverly Becker, and her staff for their help in preparing the conference and for ensuring a pleasant stay for the participants. Finally, we thank the staff and several of the graduate students of the UIUC Mathematics Department for the work they did toward the conference and in the production of these proceedings.

Urbana, Illinois
March, 1996

B. C. B.
H. G. D.
A. J. H.

Heini Halberstam

List of Participants

S. Ahlgren (Univ. of Colorado)
K. Alladi (Univ. of Florida)
J. Andersson (Stockholm Univ.)
G. Andrews (Penn State Univ.)
S. Ashford (Southern Univ.)
G. Bachman (Univ. of Nevada)
J. Bae (Univ. of Illinois)
M. Baica (Univ. of Wisc. at Whitew.)
R. Baker (Brigham Young Univ.)
R. Balasubramanian (Univ. of Madras)
A. Balog (Univ. of Illinois)
P. Bateman (Univ. of Illinois)
P. Bauer (Universität Frankurt)
B. Berndt (Univ. of Illinois)
P. Berrizbeitia (Univ. Simon Bolivar)
P. Bialek (Westmont College)
R. Blecksmith (North. Ill. Univ.)
K. Boklan (Univ. of Michigan)
J. Borwein (Simon Fraser Univ.)
P. Borwein (Simon Fraser Univ.)
N. Boston (Univ. of Illinois)
J. Bourgain (Inst. for Adv. Study)
D. Bowman (Univ. of Illinois)
J. Brüdern (Univ. Stuttgart)
D. Bradley (Univ. of Illinois)
J. Brillhart (Univ. of Arizona)
J. Brueggeman (Univ. of Illinois)
R. Burthe (Univ. of Georgia)
H. H. Chan (Univ. of Illinois)
F. Cheng (Univ. of Georgia)
W. Chen (Macquarie Univ.)
Y. Choi (Univ. of Illinois)
T. Cochrane (Kansas State Univ.)
B. Conrey (Oklahoma State Univ.)
M. Conroy (Univ. of Colorado)
H. Daboussi (Univ. de Paris XI)
M. Davidson (Univ. of Michigan)
H. Delange (Univ. de Paris Sud)
J. M. Deshouillers (Unv. Bordeaux)
H. Diamond (Univ. of Illinois)
K. Dilcher (Dalhousie Univ.)
D. Doud (Univ. of Illinois)
D. Eichhorn (Univ. of Illinois)
P. Elliott (Univ. of Colorado)
G. Embertson (Northern Illinois Univ.)
P. Erdös (Hung. Academy of Sci.)
C. van den Eynden (Ill. State Univ.)
D. Farmer (MSRI)
M. Filaseta (Univ. of South Carolina)
K. Ford (Inst. for Advanced Study)
G. Freiman (Tel Aviv Univ.)
J. Friedlander (Univ. of Toronto)
W. Galway (Univ. of Illinois)
F. Garvan (Univ. of Florida)
E. Gethner (MSRI)
A. Ghosh (Oklahoma State Univ.)
D. Goldston (San Jose State Univ.)
S. Gonek (Univ. of Rochester)
S. Graham (Michigan Tech. Univ.)
J. Grantham (Univ. of Georgia)
A. Granville (Univ. of Georgia)
G. Greaves (Univ. of Wales)
J. Haglund (Kennesaw State Univ.)
H. Halberstam (Univ. of Illinois)
R. Heath-Brown (Oxford Univ.)
M. Helm (CUNY Graduate College)
D. Hensley (Texas A&M Univ.)
A. Hildebrand (Univ. of Illinois)
C. Hill (Univ. of Illinois)
A. Hoit (Univ. of Illinois)
S. Huang (Univ. of Illinois)
M. Huxley (Univ. of Cardiff)
A. Ivić (Univ. of Belgrade)
H. Iwaniec (Rutgers Univ.)
K. James (Univ. of Georgia)
X. Jia (Southwest Texas St. Univ.)
M. Jutila (Univ. of Turku)
D. Kim (Univ. of Illinois)
G. Kolesnik (Cal. St. Univ. at L.A.)
M. Kolountzakis (Inst. for Adv. Study)

K. Kovacs (Univ. of Illinois)
M. Laporta (Univ. of Toronto)
X. Li (Purdue Univ.)
S. Lou (Halifax)
W. Luo (MSRI)
G. Martin (Univ. of Michigan)
J. Meyer (Univ. of Illinois)
P. Morée (Macquarie Univ.)
Y. Motohashi (Nihon Univ.)
R. Murty (McGill Univ.)
M. Nathanson (Lehman College)
J. M. Nicolas (Université Lyon I)
J. Peral (Univ. del Pais Vasco)
W. Philipp (Univ. of Illinois)
C. Pomerance (Univ. of Georgia)
J. Prather (Lexington, KY)
R. Ream (Univ. of Colorado)
Z. Reti (Univ. of Florida)
S. Robins (Univ. of North. Colorado)
E. Scourfield (Royal Holloway Coll.)
P. Shiu (Loughborouh Univ.)
C. Skinner (Princeton Univ.)
A. Srinivasan (Univ. of Georgia)
H. Stark (MSRI)
K. Stolarsky (Univ. of Illinois)
J. Thunder (Northern Ill. Univ.)
R. C. Vaughan (Imperial College)
L. Walling (Univ. of Colorado)
D. Wan (Univ. of Nevada)
M. Wijsmuller (La Salle Univ.)
D. Wolke (Univ. Freiburg)
T. Wooley (Univ. of Michigan)
L. C. Zhang (Southw. Miss. St. Univ.)
Z. Zheng (Zhongshan Univ.)

P. Kwok (Univ. of Illinois)
W. Liaw (Univ. of Illinois)
S. Li (Univ. of Georgia)
L. Lucht (TU Clausthal)
H. Maier (Univ. of Ulm)
C. Methfessel (Univ. of Waterloo)
H. Montgomery (Univ. of Michigan)
M. Mossinghoff (Univ. of Texas)
J. Mueller (Fordham Univ.)
M. Nair (Univ. of Glasgow)
A. M. Nicolas (Université Lyon I)
K. Ono (Univ. of Illinois)
A. Perelli (Univ. di Genova)
A. Pollington (Brigham Young Univ.)
A. van der Poorten (Macquarie Univ.)
R. Rankin (Univ. of Glasgow)
D. Redmond (Southern Illinois Univ.)
B. Reznick (Univ. of Illinois)
W. Schmidt (Univ. of Colorado)
J. Selfridge (Northern Illinois Univ.)
V. Skarda (Brigham Young Univ.)
K. Soundararajan (Univ. of Michigan)
P. Stanica (SUNY at Buffalo)
C. Starr (Lexington, KY)
G. Tenenbaum (Univ. de Nancy)
K. Tsang (Univ. of Hong Kong)
S. Wagstaff (Purdue Univ.)
P. van Wamelen (Louisiana St. Univ.)
F. Wheeler (Supercomp. Res. Center)
E. Wirsing (Univ. of Ulm)
S. Wong (Univ. of Colorado)
T. Zhan (Shandong Univ.)
W. B. Zhang (Univ. of Jamaica)

EVENING CONCERT

Thursday, May 18, 1995, 7:30 p.m.
Allerton House Library

Program

Trio in C major, K. 548 (1788) **W. A. Mozart**

Allegro
Andante cantabile
Allegro

Quartet #3 for piano, violin, viola and cello, in C minor, Opus 60 (1875) **J. Brahms**

Allegro non troppo
Scherzo: Allegro
Andante
Finale: Allegro comodo

Intermission

Trio #1 in B-flat, Opus 99 (D. 898, 1827) **F. Schubert**

Allegro moderato
Andanta un poco mosso
Scherzo: Allegro
Rondo: Allegro vivace

William Heiles	**Piano**
Howard Osborn	**Violin**
Anne Mischakoff	**Viola**
Barbara Hedlund	**Cello**

EVENING PROGRAM

Friday, May 19, 1995, 7:30 p.m.
Allerton House Library

Piano Recital

Sonata in E flat **J. Haydn**

Variations on a Derbyshire folk song. A musical diary from 1970/1971

Variations on the Russian dance from the ballet " The woodmaiden of Wranizsky " **L. v. Beethoven**

Peter Elliott, Piano

Readings

The Ballad of Reading Gaol **Oscar Wilde**

The History of The World (According to Student Bloopers) **Richard Lederer**

All the World's a Stage **William Shakespeare**

They Tell Me You Are Big **Todd McEwan**

Albert and the Lion **Marriott Edgar**

Michael Halberstam

Writers' Theater of Chicago

Weighted partition identities and applications

Krishnaswami Alladi[1]

Dedicated to Heini Halberstam on the occasion of his retirement

Abstract. We obtain interesting new identities for the classical partition functions of Euler, Gauss and Rogers-Ramanujan by attaching multiplicative weights given by the gaps between parts. In some cases the weights are powers of 2 and yield information about the underlying partition function modulo powers of 2. Such weighted identities have important applications including a new interpretation for Jacobi's Triple Product Identity and a combinatorial correspondence for a deep partition theorem of Göllnitz.

1. Introduction

The theory of partitions is rich in examples of q-series identities whose combinatorial interpretation yields the equality of two partition functions $C(n)$ and $D(n)$. In many instances $C(n)$ enumerates partitions of n whose parts satisfy congruence conditions while $D(n)$ counts partitions of n into parts satisfying certain difference conditions. Among the most extensively studied partition functions are

(i) $p(n)$ = the number of (unrestricted) partitions of n,

(ii) $Q(n)$ = the number of partitions of n into distinct parts, and

(iii) $\rho(n)$ = the number of partitions of n into parts differing by ≥ 2.

For a systematic treatment of these and other partition functions, see Andrews [6].

Recently [1] we began a study of relationships between certain partition functions which are unequal but where equality is achieved by attaching weights to the parts being counted. More precisely, let S and T be two sets of partitions with $S \subseteq T$. Let $p_S(n)$ (resp. $p_T(n)$) denote the number of partitions π of n with $\pi \in S$ (resp. $\pi \in T$). Clearly $p_S(n) \leq p_T(n)$. The general problem is to determine positive integral weights $w(\pi) = w_{S,T}(\pi)$ such that

$$p_T(n) = \sum_{\pi \in S, \sigma(\pi)=n} w(\pi), \tag{1.1}$$

[1]Research supported in part by the National Science Foundation Grant DMS 9400191

where $\sigma(\pi)$ is the sum of the parts of π. In many instances the equality in (1.1) can be refined by replacing $w(\pi)$ by a function in parameters $a, b, c, \ldots$, which reduces to $w(\pi)$ when the parameters are set equal to 1.

Such weighted identities yield new connections among the fundamental functions $p(n), Q(n)$ and $\rho(n)$. We have chosen for presentation in this paper the most appealing among such results, and the ones that are proved here are those which are established using what we have called *the sliding operation.* The proof of Theorem 3 in §4 is simpler compared to the proof in [1] which makes use of an identity of Sylvester [10].

In all examples considered here, the weights $w(\pi)$ are defined multiplicatively in terms of the gaps between parts. Recently Bowman [7] has obtained some new results for certain classical partition functions by inserting numbers in gaps. Bowman's notions and results are different from ours but there are connections which might be worth exploring. Weighted identities like (1.1) and their refinements have several important applications a few of which are discussed briefly here. For a more complete treatment, see [1], [2] and [3].

Perhaps the most striking among our results is Theorem 3 of §4 connecting $Q(n)$ to partitions with difference ≥ 3 between parts. In this case the weights $w(\pi)$ are powers of 2 and so (1.1) becomes a 2-adic identity for $Q(n)$ which suggests the remarkable result that for any integer $k > 0$

$$Q(n) \equiv 0 \pmod{2^k}, \text{ for almost all } n. \tag{1.2}$$

It is customary to approach congruences for partition functions via the theory of modular forms and (1.2) has recently been established by Gordon and Ono [9] using modular forms. But Theorem 3 offers an elementary combinatorial approach to this problem.

One of the deepest results in the theory of partitions is a theorem proved by Göllnitz [8] in 1967. It had remained an open problem to find a combinatorial proof/approach to Göllnitz' theorem. It turns out that a substantial refinement of Theorem 3 involving three free parameters a, b, c, can be obtained (see Theorem 6, §5) and that this refinement is an important reformulation of Göllnitz' theorem (for a complete description, see [2]). As a consequence, a combinatorial approach to Göllnitz' theorem is now possible. From this we get a new interpretation for Schur's partition theorem and Jacobi's Triple Product Identity (see §5).

The paper concludes with §6 where two weighted partition identities are presented.

We will use standard notation. For a positive integer n, let

$$(a)_n = (a;q)_n = \prod_{j=0}^{n-1} (1 - aq^j)$$

and for $|q| < 1$,

$$(a)_\infty = \lim_{n\to\infty} (a)_n = \prod_{j=0}^{\infty} (1 - aq^j).$$

Let $\pi : b_1 + b_2 + \cdots + b_k$, with $b_1 \geq b_i \geq \cdots \geq b_k$ be a partition of a positive integer n. We let $\sigma(\pi) = \Sigma b_i = n$, the number being partitioned, $\nu(\pi) = k$, the number of parts of π, and $\lambda(\pi) = b_1$, the largest part of π. The partition π can be represented by a Ferrers graph where the i-th row consists of b_i nodes equally spaced. We make no distinction between the partition π and its Ferrers graph. Since the main idea used here to prove weighted identities is the sliding operation on Ferrers graphs, we begin by discussing this operation.

2. The sliding operation

The Ferrers graph of each partition π contains a largest square of nodes starting from the upper left hand corner. This is the *Durfee square* of the partition (named after W. P. Durfee, a student of Sylvester) and is denoted by $D(\pi)$. The number of nodes on one side of the Durfee square is denoted by $|D(\pi)|$ and is called the dimension of the square. Thus

$$|D(\pi)| = \max \{i | b_i \geq i\}.$$

By considering Durfee squares, it is possible to give nice combinatorial proofs of many partition identities. For example, consider the identity

$$\sum_{\nu=0}^{\infty} \frac{q^{\nu^2}}{(q)_\nu^2} = \frac{1}{(q)_\infty} = \sum_{n=0}^{\infty} p(n)q^n \tag{2.1}$$

due to Euler. This is somewhat deeper than the more commonly known identity

$$\sum_{\nu=0}^{\infty} \frac{q^{\nu}}{(q)_\nu} = \frac{1}{(q)_\infty} = \sum_{n=0}^{\infty} p(n)q^n \tag{2.2}$$

also due to Euler. Whereas the combinatorial proof of (2.2) is trivial (the term $\frac{q^\nu}{(q)_\nu}$ on the left being the generating function of partitions with largest part ν) the combinatorial interpretation of the left side of (2.1) is not obvious. The best way to prove (2.1) is to consider Ferrers graphs of all partitions π whose Durfee squares have fixed dimension $|D(\pi)| = \nu$. For such partitions π (see figure (2.3) below) there is a partition π_r of nodes to the right of $D(\pi)$ and a partition π_b below $D(\pi)$. The Durfee square has ν^2 nodes and is represented by the term q^{ν^2}. The partitions π_r have $\leq \nu$ parts and so their generating function is $\frac{1}{(q)_\nu}$. Similarly, the partitions π_b have parts $\leq \nu$ and their generating function is also

$\frac{1}{(q)_\nu}$. Thus the generating function of all partitions π for which $|D(\pi)| = \nu$ is fixed is

$$\frac{q^{\nu^2}}{(q)_\nu^2}.$$

Summing this over all ν yields (2.1).

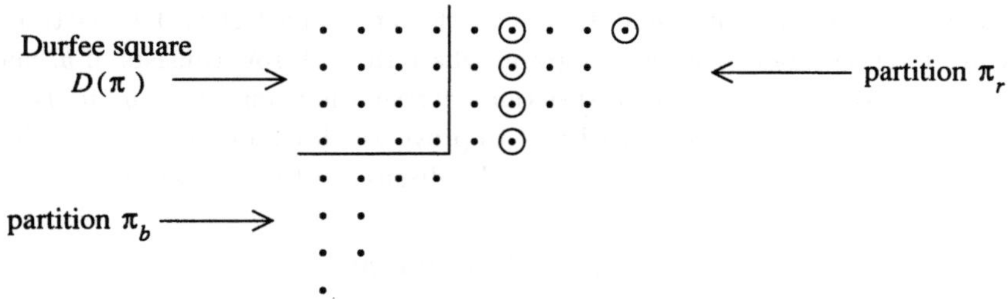

(2.3)

We will make crucial use of Durfee squares and in this connection we introduce the sliding operation.

Given a Ferrers graph π, by a *sliding operation* Ψ on π we mean the removal of certain columns to the right of $D(\pi)$ and the placement of these columns as rows below $D(\pi)$ to form a new Ferrers graph $\pi' = \Psi(\pi)$. On a given Ferrers graph π, several sliding operations can be performed to form new graphs π'. In (2.3) the columns to be moved by a specific sliding operation Ψ are circled and the resulting Ferrers graph $\pi' = \Psi(\pi)$ is shown below.

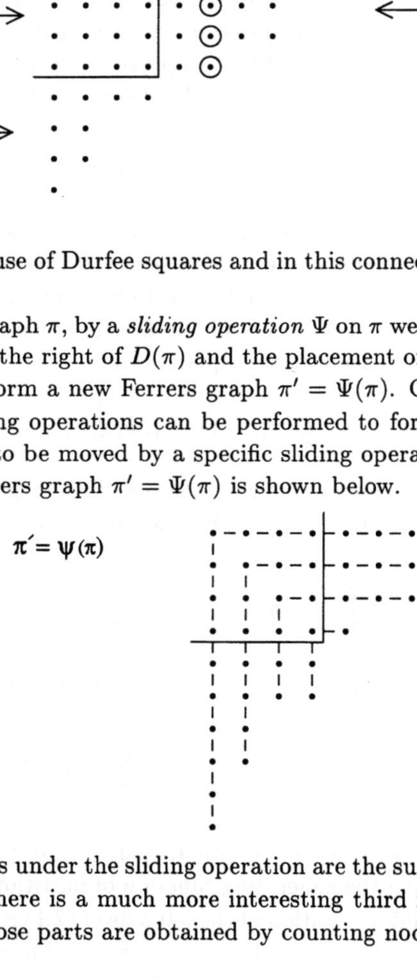

(2.4)

Two obvious invariants under the sliding operation are the sum of the parts and the Durfee square. There is a much more interesting third invariant, namely, the partition $\rho(\pi)$ whose parts are obtained by counting nodes along hooks of π. That is,

$$\rho(\pi) = \rho\big(\Psi(\pi)\big). \tag{2.5}$$

The hooks of the partition $\pi' = \Psi(\pi)$ are indicated in the figure above. For example, in (2.3) and (2.4) we see that $\rho(\pi) = \rho\big(\Psi(\pi)\big)$ is the partition

$16+12+8+4$. The importance of the third invariant will be seen in the next two sections.

3. Primary partitions

The celebrated first Rogers-Ramanujan identity is

$$\sum_{\nu=0}^{\infty}\frac{q^{\nu^2}}{(q)_\nu}=\frac{1}{(q;q^5)_\infty(q^4;q^5)_\infty}. \tag{3.1}$$

Notice that the series in (3.1) is obtained by blending the numerators of the series on the left in (2.1) with denominators of the series in (2.2). This suggests that there ought to be a fundamental link between partitions counted by the series in (3.1) and $p(n)$. We obtain such a connection in this section.

Clearly the right hand side of (3.1) is the generating function of partitions into parts $\equiv \pm 1 \pmod 5$. If we write the left hand side as

$$\sum_{n=0}^{\infty}\rho(n)q^n,$$

then $\rho(n)$ has two partition interpretations. The most well known interpretation is

(a) $\rho(n)=$ the number of partitions of n into parts differing by ≥ 2.

A less well known interpretation is

(b) $\rho(n)=$ the number of partitions of n where each part is at least as large as the number of parts.

From the point of view of Durfee squares, interpretation (b) is more natural. Indeed, from the analysis of §2, $\frac{q^{\nu^2}}{(q)_\nu}$ is the generating function for all partitions π for which $|D(\pi)|=\nu$ and such that there are no nodes below $D(\pi)$ as shown in the figure below.

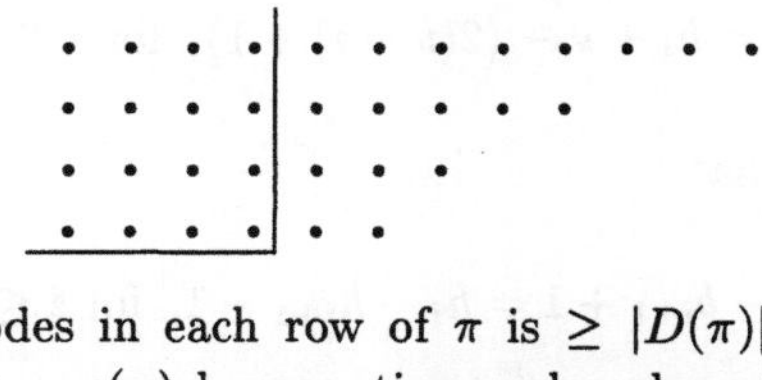

(3.2)

Thus the number of nodes in each row of π is $\geq |D(\pi)|$, as in (b). If we consider the partition $\tilde{\pi}=\rho(\pi)$ by counting nodes along hooks of π, we get a partition with difference ≥ 2 between parts. Conversely, given a partition $\tilde{\pi}: h_1+h_2+\cdots+h_\nu$ with $h_i-h_{i+1}\geq 2$ for $1\leq i<\nu$, consider a $\nu\times\nu$ Durfee square and add nodes to the right so that the hooks correspond to the parts $h_1, h_2, ..., h_\nu$. In view of the role of hooks in the interpretation of $\rho(n)$ as in (a), we introduced the notation $\rho(\pi)$ for the partition obtained from π by counting nodes along hooks. We call partitions π which have no nodes below $D(\pi)$ as *primary partitions.*

Now let $\tilde{\pi} : h_1 + h_2 + \cdots + h_\nu$ be given with $h_i - h_{i+1} \geq 2$, for $1 \leq i < \nu$. Let $\pi : b_1 + b_2 + \cdots + b_\nu$ be the primary Ferrers graph for which $\tilde{\pi} = \rho(\pi)$. We make the convention that $b_{\nu+1} = \nu$ and $h_{\nu+1} = -1$. Thus $b_1 \geq b_2 \geq \ldots b_\nu \geq \nu = b_{\nu+1}$. Given i such that $1 \leq i < \nu$, note that up to $h_i - h_{i+1} - 2$ columns of length i could be moved under the sliding operation and placed below $D(\pi)$ to form new Ferrers graphs. This gives $h_i - h_{i+1} - 1$ choices. Therefore define the weight of h_i to be $w_R(h_i) = h_i - h_{i+1} - 1$. Observe also that up to $h_\nu - 1$ columns of length ν could be moved below the Durfee square under the sliding operation. This gives h_ν choices and because of our convention $h_{\nu+1} = -1$, we have $h_\nu = h_\nu - h_{\nu+1} - 1$. Hence define $w_R(h_\nu) = h_\nu - h_{\nu+1} - 1$. Finally define $w_R(\tilde{\pi})$, the weight of $\tilde{\pi}$, multiplicatively by

$$w_R(\tilde{\pi}) = \prod_{i=1}^{\nu} w_R(h_i) = \prod_{i=1}^{\nu} (h_i - h_{i+1} - 1). \tag{3.3}$$

Thus every partition $\tilde{\pi}$ of n determines a unique primary partition π of n, and under sliding operations performed on π, each $\tilde{\pi}$ spawns $w_R(\tilde{\pi})$ unrestricted partitions of n. Since every partition of n can be generated in this fashion, this yields

Theorem 1. *Let $\mathcal{R}$ denote the set of all partitions into parts differing by ≥ 2. For $\tilde{\pi} \in R$, let $w_R(\tilde{\pi})$ be defined by (3.3). Then*

$$p(n) = \sum_{\tilde{\pi} \in \mathcal{R}, \sigma(\tilde{\pi}) = n} w_R(\tilde{\pi}).$$

It is natural to ask for a weight formula connecting primary partitions of n with $p(n)$. If $\tilde{\pi} = \rho(\pi)$ as above, then the parts h_i of $\tilde{\pi}$ and the parts b_i of π are connected via the relation

$$b_i = h_i + \nu - \big(2(\nu - i) + 1\big), \text{ for } i \leq i \leq \nu. \tag{3.4}$$

Thus (3.4) implies that

$$b_i - b_{i+1} + 1 = h_i - h_{i+1} - 1, \text{ for } i \leq i \leq \nu \tag{3.5}$$

with the conventions $b_{\nu+1} = \nu$ and $h_{\nu+1} = -1$. From (3.3) and (3.5) it follows that the weight $w_1(\pi)$ of the partition π should be defined as

$$w_1(\pi) = \prod_{i=1}^{\nu} (b_i - b_{i+1} + 1), \tag{3.6}$$

which would give $w_1(\pi) = w_R(\tilde{\pi})$. So we get

Theorem 2. *Let $\mathcal{D}_1$ denote the set of all primary partitions. For $\pi \in \mathcal{D}_1$, let $w_1(\pi)$ be as in (3.6). Then*

$$p(n) = \sum_{\pi \in \mathcal{D}_1, \sigma(\pi)=n} w_1(\pi).$$

Theorems 1 and 2 are combinatorial interpretations of the decomposition

$$\frac{q^{\nu^2}}{(q)_\nu^2} = \frac{q^{\nu^2}}{(q)_\nu} \circ \frac{1}{(q)_\nu}$$

connecting the series in (2.1) and (3.1) but these elegant results seem to have escaped attention. Theorems 1 and 2 can be refined and extended in many ways. One refinement is to introduce a parameter z whose power would represent the number of parts. Since the dimension $|D(\pi)|$ and the quantity $\lambda(\pi) + \nu(\pi)$ remain invariant under the sliding operation, refinements utilizing these partition statistics can also be obtained. In addition, Theorem 2 can be extended to yield a weighted identity connecting $p(n)$ with partitions into k successive Durfee squares, with Theorem 2 being the case $k = 1$. See [1] for a discussion of such refinements and extensions.

We now consider the effect of the sliding operation on partitions into distinct parts.

4. Primary partitions into distinct parts

The Ferrers graph of a partition π' into distinct parts contains in addition to the Durfee square of dimension say, $|D(\pi')| = \nu$, an isosceles triangle of $\nu(\nu-1)/2$ nodes to the right of $D(\pi')$ as shown in the figure below.

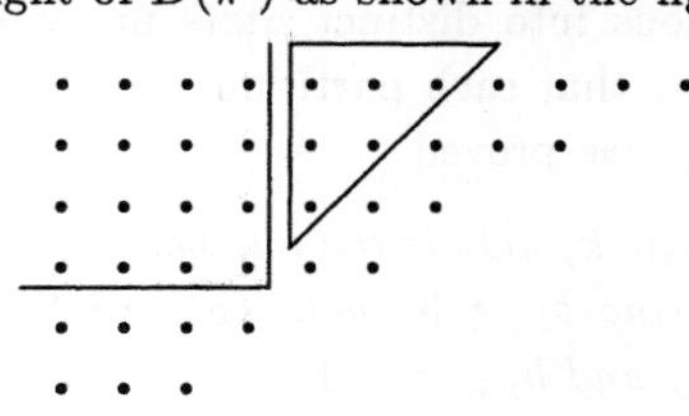

(4.1)

Thus the partition $\tilde{\pi} = \rho(\pi')$ obtained by counting nodes along hooks of π' will be a partition with difference ≥ 3 between parts. Note that several partitions π' may yield the same $\tilde{\pi} = \rho(\pi')$. Given $\tilde{\pi}$, in order to determine the number of π' which yield $\tilde{\pi}$ as $\rho(\pi')$, we first observe

Lemma. *To every partition $\tilde{\pi}$ into ν parts differing by ≥ 3, there is associated a unique primary partition π into ν distinct parts such that $\tilde{\pi} = \rho(\pi)$.*

It is to be noted that the Lemma is the next level result beyond the equivalence of the two interpretations (a) and (b) for $\rho(n)$.

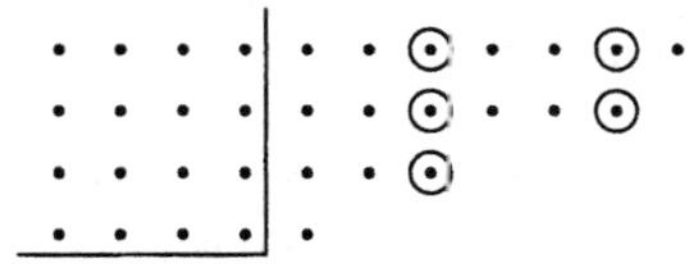

(4.2)

Given a partition $\tilde{\pi} : h_1 + h_2 + \cdots + h_\nu$ into parts differing by ≥ 3, let $\pi : b_1 + b_2 + \cdots + b_\nu$ be the primary partition into distinct parts as above with $\rho(\pi) = \tilde{\pi}$. Let us make the convention $h_{\nu+1} = -1$ and $b_{\nu+1} = \nu$ as before. Suppose for a certain i we have $h_i - h_{i+1} \geq 4$. Then in the Ferrers graph of π we could move exactly one column of length i under the sliding operation to form a Ferrers graph π' into distinct parts. This gives us two choices of either to move or not to move, when $h_i - h_{i+1} \geq 4$. If $h_i - h_{i+1} \leq 3$ for a certain i, then we cannot move a column of length i under the sliding operation. The number of choices here is 1, that is not to move. In (4.2) the columns that could be moved are circled and they correspond to the gaps ≥ 4 in the partition $\tilde{\pi}$ which is $14 + 11 + 6 + 2$. Thus the definition of weights $w(h_i)$ for each part h_i is

$$w(h_i) = \begin{cases} 1, \text{ if } h_i - h_{i+1} \leq 3, \\ 2, \text{ if } h_i - h_{i+1} \geq 4. \end{cases} \tag{4.3}$$

Now define the weight of the partition $\tilde{\pi}$ multiplicatively by

$$w(\tilde{\pi}) = \prod_{i=1}^{\nu} w(h_i). \tag{4.4}$$

Thus $w(\tilde{\pi}) = 2^k$ if there are exactly k gaps between the integers $h_1, h_2, \ldots, h_\nu$ and $h_{\nu+1} = -1$. Since every partition π' into distinct parts can be obtained from primary partitions into distinct parts under the sliding operation, the above argument shows that each partition $\tilde{\pi}$ spawns $w(\tilde{\pi})$ partitions into distinct parts. Thus we have proved

Theorem 3. *Let $g_3(n;k)$ denote the number of partitions $\tilde{\pi} : h_1+h_2+\cdots+h_\nu$ of n into parts differing by ≥ 3 such that exactly k of the gaps among the integers $h_1, h_2, \ldots, h_\nu$ and $h_{\nu+1} = -1$ are ≥ 4. Then*

$$Q(n) = \sum_k g_3(n;k)2^k.$$

The relationship $\tilde{\pi} = \rho(\pi)$ shows that the gaps between the parts h_i of $\tilde{\pi}$ and the gaps between the parts b_i of the primary partition π are connected by

$$h_i - h_{i+1} - 2 = b_i - b_{i+1}, \text{ for } 1 \leq i \leq \nu. \tag{4.5}$$

Hence the definition of the weights $w_p(b_i)$ of a part b_i of π is

$$w_p(b_i) = \begin{cases} 1, \text{ if } b_i - b_{i+1} \le 1, \\ 2, \text{ if } b_i - b_{i+1} \ge 2, \end{cases} \tag{4.6}$$

as given by (4.5) and (4.3). The weight of π is defined multiplicatively by

$$w_p(\pi) = \prod_{i=1}^{\nu} w_p(b_i). \tag{4.7}$$

Thus we have the following companion to Theorem 3:

Theorem 4. *Let $Q^*(n;k)$ denote the number of primary partitions $\pi : b_1 + b_2 + \cdots + b_\nu$ of n into distinct parts such that there are precisely k gaps ≥ 2 among the integers $b_1, b_2, \ldots, b_\nu$ and $b_{\nu+1} = \nu$. Then*

$$Q^*(n;k) = g_3(n;k)$$

and so

$$Q(n) = \sum_k Q^*(n;k) 2^k.$$

Theorem 3 is to be compared with Theorem 1 both of which have the convention $h_{\nu+1} = -1$. Similarly Theorem 4 is to be compared to Theorem 2 both of which have the convention $b_{\nu+1} = \nu$.

Using the sliding operation Theorems 3 and 4 can be refined as follows by introducing a parameter z which keeps track of the number of parts.

Theorem 5. *Let $Q(n;k)$ denote the number of partitions π' of n into distinct parts for which $\nu(\pi') = k$.*

Let $g_3(n;\nu,k)$ denote the number of partitions $\tilde{\pi} : h_1 + h_2 + \cdots + h_\nu$ of n into ν parts differing by ≥ 3 such that there are exactly k gaps ≥ 4 among the integers $h_1, h_2, \ldots, h_\nu$ and $h_{\nu+1} = -1$.

Let $Q^*(n;\nu,k)$ denote the number of primary partitions $\pi : b_1 + b_2 + \cdots + b_\nu$ of n into ν distinct parts such that there are exactly k gaps ≥ 2 among the integers $b_1, b_2, \ldots, b_\nu$ and $b_{\nu+1} = \nu$. Then

$$\sum_k z^k Q(n;k) = \sum_\nu \sum_k z^\nu (1+z)^k g_3(n;k) = \sum_\nu \sum_k z^\nu (1+z)^k Q^*(n;k).$$

In 1990 I showed that Theorem 5 is a new combinatorial interpretation of and equivalent to the following identity of Sylvester [10] in his classic paper of 1882:

$$(-zq)_\infty = 1 + \sum_{n=1}^{\infty} \frac{z^n q^{(3n^2-n)/2}(1+zq^{2n})(-zq)_{n-1}}{(q)_n}. \tag{4.8}$$

The proof of Theorem 5 via (4.8) utilizes the concept of imbeddings and is given in [1]. The proof given here using the sliding operation is simpler.

Theorem 3 and its refinements have many important consequences. For example, Theorem 3 immediately suggests the remarkable result that for every integer $k \geq 1$, the congruence

$$Q(n) \equiv 0 (\text{mod } 2^k) \tag{4.9}$$

holds for almost all n; that is the exceptional set where the congruence fails has density 0. Gordon and Ono [9] have recently established this result as well as a generalization to p-regular partitions using the theory of modular forms. But Theorem 3 can be used to study (4.9) combinatorially for small values of k.

For example, to decide when $Q(n)$ is odd, we need only determine when $g_3(n;0)$ is odd. Now $g_3(n;0)$ counts partitions $\tilde{\pi} : h_1 + h_2 + \cdots + h_\nu$ of n for which $h_i - h_{i+1} \geq 3$ for $1 \leq i < \nu$ and there are no gaps ≥ 4 among $h_1, h_2, \ldots, h_\nu$ and $h_{\nu+1} = -1$. So there are only two cases to consider.

<u>Case 1:</u> $h_\nu = 1$ and $h_i - h_{i+1} = 3$, for $1 \leq i < \nu$.

In this case $\tilde{\pi} : 1 + 4 + 7 + \cdots + 3\nu - 2 = (3\nu^2 - \nu)/2$.

<u>Case 2:</u> $h_\nu = 2$ and $h_i - h_{i+1} = 3$, for $1 \leq i < \nu$.

In this case $\tilde{\pi} : 2 + 5 + 8 + \cdots + 3\nu - 1 = (3\nu^2 + \nu)/2$.
Thus

$$Q(n) \equiv 0 (\text{mod } 2) \Leftrightarrow n \neq \frac{3\nu^2 \pm \nu}{2}. \tag{4.10}$$

Next, to determine when $Q(n) \not\equiv 0$ (mod 4), we need to see when $g_3(n;1)$ makes a contribution. By analyzing $g_3(n;1)$ it can be shown that

$$Q(n) \equiv 0 (\text{mod } 4) \text{ if } n \neq P_\nu + k^2; \tag{4.11}$$

that is if n is not a pentagonal number plus a square. The exceptional set is actually thinner and can be determined precisely. As k gets larger, the combinatorial analysis of (4.9) via Theorem 3 complicated and so it is best to treat (4.9) for all $k \geq 1$ using modular forms. For a combinatorial study of (4.9) for $k \leq 4$, see [3].

5. A three parameter refinement

One of the deepest results in the theory of partitions was proved by Göllnitz [8] in 1967. Using a new technique called *the method of weighted words*, Andrews, Gordon and I [4] recently obtained substantial refinements and generalizations of Göllnitz' theorem. It is often the case that refined partition identities can be proved combinatorially, because the set up for the bijection is naturally suggested by the refinement. Although Göllnitz' theorem was proved in a refined form in 1967, no combinatorial proof was known. Recently I realized [2] that a three parameter refinement of Theorem 3 could be obtained (see Theorem 6 below) and that this refinement is an important reformulation of Göllnitz' theorem. In [2] a natural bijection is constructed to explain this reformulation. In order to state this refinement of Theorem 3, let $\tilde{\pi} : h_1 + h_2 + \cdots + h_\nu$ be a partition into parts differing by ≥ 3 and let $h_{\nu+1} = -1$. Define the weight $w(h_i)$ of each part as follows:

Congruence	Gap	Weight $w(h_i)$
$h_i \equiv 1 \pmod 3$	$h_i - h_{i+1} \leq 3$	a
$h_i \equiv 2 \pmod 3$	$h_i - h_{i+1} \leq 3$	b
$h_i \equiv 0 \pmod 3$	$h_i - h_{i+1} \leq 3$	c
$h_i \equiv 1 \pmod 3$	$h_i - h_{i+1} \geq 4$	$a + ac$
$h_i \equiv 2 \pmod 3$	$h_i - h_{i+1} \geq 4$	$b + bc$
$h_i \equiv 0 \pmod 3$	$h_i - h_{i+1} \geq 4$	$c + ab$

The weight $w(\tilde{\pi})$ of the partition $\tilde{\pi}$ is defined multiplicatively by

$$w(\tilde{\pi}) = \prod_{i=1}^{\nu} w(h_i). \tag{5.1}$$

With this notion of weight, the following result is established in [2]:

Theorem 6. *Let $\mathcal{D}$ denote the set of all partitions into distinct parts. For $\pi' \in \mathcal{D}$, let $\nu_i(\pi')$ denote the number of parts of π' which are $\equiv i \pmod 3$, $i = 1, 2, 3$.*

Let $\mathcal{D}_3$ denote the set of partitions into parts differing by ≥ 3. For $\tilde{\pi} \in \mathcal{D}_3$ let $w(\tilde{\pi})$ be given by (5.1). Then

$$\sum_{\pi' \in \mathcal{D}, \sigma(\pi')=n} a^{\nu_1(\pi')} b^{\nu_2(\pi')} c^{\nu_3(\pi')} = \sum_{\tilde{\pi} \in \mathcal{D}_3, \sigma(\tilde{\pi})=n} w(\tilde{\pi}).$$

Theorem 6 reduces to Theorem 3 when $a = b = c = 1$. It is to be noted that although Theorems 3 and 4 are companions and are equivalent, it is only Theorem 3 which has this three parameter refinement.

We now discuss certain interesting special cases of Theorem 6.

First let $c = 0$ in Theorem 6. This means that on the left we are summing over partitions into distinct parts $\equiv \pm 1 \pmod 3$. On the right hand side we have $w(\tilde{\pi}) = 0$ whenever $h_i - h_{i+1} = 3$ and $h_i \equiv 0 \pmod 3$. Thus the partitions $\tilde{\pi}$ for which $w(\tilde{\pi}) \neq 0$ are those whose parts differ by ≥ 3, and such that consecutive multiples of 3 cannot occur as parts. Thus Theorem 6 yields the following well known refined form of Schur's celebrated partition theorem of 1926:

Refined Schur's Theorem. *Let $S(n;l,m)$ denote the number of partitions π' of n into distinct parts $\equiv 1$ or $2 \pmod 3$ such that $\nu_1(\pi') = l$ and $\nu_2(\pi') = m$. Let $S_1(n;i,j,k)$ denote the number of partitions $\tilde{\pi}$ of n into parts differing by ≥ 3 and with no consecutive multiples of 3 such that $\nu_1(\tilde{\pi}) = i, \nu_2(\tilde{\pi}) = j$ and $\nu_3(\tilde{\pi}) = k$. Then*

$$S(n;l,m) = \sum_{\substack{i+k=l\\ j+k=m}} S_1(n;i,j,k).$$

Next let $c = -1$ and $ab = 1$ in Theorem 6. Thus

$$c + ab = a + ac = b + bc = 0.$$

This means that $w(\tilde{\pi}) = 0$ if $\tilde{\pi}$ has at least one gap ≥ 4 among the integers $h_1, h_2, \ldots, h_\nu$ and $h_{\nu+1} = -1$. Thus the only partitions $\tilde{\pi}$ for which $w(\tilde{\pi}) \neq 0$ are those covered by Cases 1 and 2 in §4, namely the obvious partitions of the pentagonal numbers $(3\nu^2 \pm \nu)/2$. Thus the choices $c = -1$ and $ab = 1$ in Theorem 6 yield

$$(5.2) \qquad (-aq;q^3)_\infty(-a^{-1}q^2;q^3)_\infty(q^3;q^3)_\infty = \sum_{\nu=-\infty}^{\infty} a^\nu q^{(3\nu^2-\nu)/2}$$

which is equivalent to the famous Jacobi Triple Product Identity.

6. Two weighted identities

We conclude by mentioning two more weighted identities which are of the 2-adic type. Both identities can be refined by introducing parameters. The first of these connects Rogers-Ramanujan partitions with partitions into distinct parts.

Given a partition $\tilde{\pi} : h_1 + h_2 + \cdots + h_\nu$ into parts differing by ≥ 2, put $h_{\nu+1} = -1$. Define the weight of each part h_i by

$$(6.1) \qquad w(h_i) = \begin{cases} 1, & \text{if } h_i \text{ is even,} \\ 1, & \text{if } h_i \text{ is odd and } h_i - h_{i+1} = 2, \\ 2, & \text{if } h_i \text{ is odd and } h_i - h_{i+1} > 2. \end{cases}$$

As in the earlier examples, define $w(\tilde{\pi})$ multiplicatively by

$$w(\tilde{\pi}) = \prod_{i=1}^{\nu} w(h_i). \tag{6.2}$$

Thus $w(\tilde{\pi})$ is a power of 2. We have

Theorem 7. *Let $\mathcal{R}$ denote the set of all partitions with difference ≥ 2 between parts. For $\tilde{\pi} \in \mathcal{R}$, let $w(\tilde{\pi})$ be defined as in (6.2). Then*

$$Q(n) = \sum_{\tilde{\pi}\in\mathcal{R}, \sigma(\tilde{\pi})=n} w(\tilde{\pi}).$$

Equivalently, let $g_2(n;k)$ denote the number of partitions $\tilde{\pi} : h_1 + h_2 + \cdots + h_\nu$ into parts differing by ≥ 2 such that there are exactly k gaps ≥ 4 among the odd numbers in the list $h_1, h_2, \ldots, h_\nu$ and -1. Then

$$Q(n) = \sum_k g_2(n;k) 2^k.$$

Theorem 7 as well as a refinement involving two parameters a and b is proved in [1]. Compare Theorem 7 with Theorem 3 and note that Theorem 7 is another reason for the validity of the congruence (4.9) for almost all n. From Theorems 1 and 7 we see that weights can be attached to Rogers-Ramanujan partitions to get both $p(n)$ and $Q(n)$. For example, $9, 8+1, 7+2, 6+3$, and $5+3+1$ are the Rogers-Ramanujan partitions of 9. Their weights according to Theorem 7 are 2, 1, 2, 2 and 1. Summing these weights gives $Q(9) = 8$. On the other hand, the weights of these Rogers-Ramanujan partitions according to Theorem 1 are 9, 6, 8, 6 and 1. Summing these weights yields $p(9) = 30$.

The final weight formula we present connects $Q(n)$ with the function $g(n)$ given by

$$\prod_{m=1}^{\infty} \left(\frac{1+q^{2m}}{1-q^{2m-1}}\right) = \sum_{n=0}^{\infty} g(n) q^n. \tag{6.3}$$

Thus $g(n)$ is the number of partitions of n where the even parts do not repeat. By a combinatorial analysis of the well known Lebesgue identity

$$\sum_{\nu=0}^{\infty} \frac{q^{\nu(\nu+1)/2}(-bq)_\nu}{(q)_\nu} = \prod_{m=1}^{\infty} \left(\frac{1+bq^{2m}}{1-q^{2m-1}}\right), \tag{6.4}$$

Gordon and I [5] obtained a refined form of the following result:

Theorem 8. *Let $g_1(n;k)$ denote the number of partitions $\pi' : b_1+b_2+\cdots+b_\nu$ into distinct parts such that there are exactly k gaps ≥ 2 among the integers*

$b_1, b_2, \ldots, b_\nu$ and 0. Then

$$g(n) = \sum_k g_1(n;k)2^k.$$

Theorem 8 is interesting for a variety of reasons. First it bears a great similarity to Theorem 4. Although the weights in both theorems are powers of 2, there is a difference between the weights in Theorem 4 and those in Theorem 8 even if restricted to primary partitions. Theorem 4 is a 2-adic formula connecting primary partitions into distinct parts with $Q(n)$, while Theorem 8 is a 2-adic formula connecting partitions into distinct parts with $g(n)$.

Another reason that Theorem 8 is interesting is because it suggests that for each $k \geq 1$,

$$g(n) \equiv 0 (\text{mod } 2^k) \tag{6.5}$$

is valid for almost all n and indeed provides a combinatorial approach to this problem. For example to decide when $g(n)$ is odd, we need only determine when $g_1(n;0)$ is odd. Observe that the only partition counted by $g_1(n;0)$ is $1 + 2 + \cdots + \nu = \frac{\nu(\nu+1)}{2} = n$, when n is a triangular number. Thus

$$g(n) \text{ is even } \Leftrightarrow n \neq \text{ triangular}.$$

Compare this with (4.10). Next by analyzing $g_1(n;1)$ it can be shown that

$$g(n) \equiv 0 \ (\text{mod } 4) \text{ if } n \neq \text{ triangular } + \text{ square}$$

This is to be compared to (4.11). It turns out that $g(n)$ and $Q(n)$ have a similar behavior with triangular numbers replacing pentagonal numbers in the discussion of $g(n)$. In a forthcoming paper [3], the distribution of $g(n)$ and $Q(n)$ modulo 2^k for small values of k will be discussed combinatorially.

Acknowledgement: I would like to thank the referee for suggestions concerning §5 and Frank Garvan for help with the diagrams.

References

[1] K. Alladi, *Partition identities involving gaps and weights*, 1995 preprint.

[2] K. Alladi, *A combinatorial correspondence related to Göllnitz' (Big) partition theorem and applications*, 1995 preprint.

[3] K. Alladi, *On the divisibility of certain partition functions by small powers of 2*, in preparation.

[4] K. Alladi, G. E. Andrews and B. Gordon, *Generalizations and refinements of a partition theorem of Göllnitz*, J. Reine Angew. Math. **460** (1995), 165–188.

[5] K. Alladi and B. Gordon, *Partition identities and a continued fraction of Ramanujan*, J. Comb. Th. Ser. A **63** (1993), 275-300.

[6] G. E. Andrews, *The theory of partitions*, Encyclopedia of Mathematics and its applications, vol. 2, Addison-Wesley, Reading, 1976.

[7] D. Bowman, *Partitions with numbers in their gaps*, Acta. Arith (to appear).

[8] H. Göllnitz, *Partitionen mit Differenzenbedingungen*, J. Reine Angew. Math. **225** (1967), 154–190.

[9] B. Gordon and K. Ono, *Divisibility of certain partition functions by powers of primes*, The Ramanujan Journal **1** (1997), to appear.

[10] J. J. Sylvester, *A constructive theory of partitions arranged in three Acts, an Interact and an Exodion*, Amer. J. Math. **5** (1882), 251–330.

Krishnaswami Alladi
Department of Mathematics
University of Florida
Gainesville, FL 32611

Rogers-Ramanujan polynomials for modulus 6

George E. Andrews[1]

Dedicated to my friend, Heini Halberstam

Abstract. In 1967, it was proved that the partitions of n into parts $\not\equiv 0$, $\pm 1 \pmod 6$ are equinumerous with partitions wherein the difference between parts is never 1, no part appears more than twice and no 1's appear. Polynomial generating functions related to this theorem have useful and interesting representations involving Gaussian polynomials and q-trinomial coefficients.

1. Introduction

In 1967, the following partition identity was proved in [2; p. 432].

Theorem 1. *The number $A(n)$, of partitions of any positive integer, n, into parts of the forms $6m+2$, $6m+3$, or $6m+4$ equals $B(n)$, the number of partitions of n into parts each $\geqq 2$ and which neither contain sequences nor allow any part to appear more than twice.*

Examination of the proof in [2] provides an equivalent formulation of Theorem 1.

Theorem 2. *Given $B(n)$ in Theorem 1,*

$$\sum_{n=0}^{\infty} B(n)q^n = \frac{1}{\prod_{n=0}^{\infty}(1-q^n)} \sum_{j=-\infty}^{\infty} (-1)^j q^{3j^2+2j}, \tag{1.1}$$

where $|q| < 1$.

In a subsequent paper [3], further results related to these theorems were derived. Of particular interest for us here is the following result which follows immediately from the work in [3]:

Theorem 3. *If $|q| < 1$,*

$$\sum_{n=0}^{\infty} C(n)q^n = \frac{1}{\prod_{n=0}^{\infty}(1-q^n)} \sum_{j=-\infty}^{\infty} (-1)^j q^{3j^2}, \tag{1.2}$$

[1]Partially supported by National Science Foundation Grant DMS 8702695-04.

where $C(n)$ is the number of partitions of n which neither contain sequences nor allow any part to appear more than twice.

In 1990, in a seemingly completely unrelated paper [6], an extensive study of trinomial and q-trinomial coefficients was given. In the notation of [6], the trinomial coefficient $\binom{n}{j}_2$ is given by

$$\sum_{j=-n}^{n} \binom{n}{j}_2 x^j = (1+x+x^{-1})^n. \tag{1.3}$$

Theorem 4. [6; eq. (3.2), p. 659]

$$\begin{aligned}&\mathcal{E}_{m+1}(6;0,1) = \mathcal{E}_{m+1}(6;2,3)\\ &= \mathcal{E}_m(6;1,2) = \mathcal{E}_{m-1}(6;0,3)\\ &= (2^m - (-1)^m)/3,\end{aligned}$$

where

$$\mathcal{E}_n(k;a,b) = \sum_{\lambda=-\infty}^{\infty} \left(\binom{n}{k\lambda+a}_2 - \binom{n}{k\lambda+b}_2\right). \tag{1.4}$$

The primary object of [6] was to examine a classical discovery of Euler and its q-analogs. This required focusing on $\mathcal{E}_n(10;a,b)$. The numbers $\mathcal{E}_n(6;a,b)$ were viewed as elementary prototypes, and only q-analogs of $\mathcal{E}_n(6;0,2)$ and $\mathcal{E}_n(6;1,3)$ were given.

Recently in connection with his joint work with Barry McCoy in statistical mechanics, Alexander Berkovich asked me whether q-analogs of Theorem 4 above were known.

Our object in this paper is to provide full q-analogs for Theorem 4. In doing so we must study two appealing sequences of polynomials.

$$r_n(q) = \begin{cases} 0 & \text{if} \quad n=0\\ 1 & \text{if} \quad n=1\\ r_{n-1}(q) + (q^{n-2}+q^{2n-4})r_{n-2}(q), & \text{if} \quad n>1. \end{cases} \tag{1.5}$$

and

$$s_n(q) = \begin{cases} 0 & \text{if} \quad n=0\\ 1 & \text{if} \quad n=1\\ s_{n-1}(q) + (q^{n-1}+q^{2n-2})s_{n-2}(q), & \text{if} \quad n>1. \end{cases} \tag{1.6}$$

Our primary representations of these polynomials involve the q-binomial coefficients [5; p. 35]

$$\begin{bmatrix} A \\ B \end{bmatrix}_q = \begin{cases} 0 & \text{if} \quad B < 0 \text{ or } B > A \\ \frac{(1-q^A)(1-q^{A-1})\cdots(1-q^{A-B+1})}{(1-q^B)(1-q^{B-1})\cdots(1-q)}, & \text{otherwise}, \end{cases} \tag{1.7}$$

and the q-trinomial coefficients [7; eq. (2.15)]

$$\binom{m;B;q}{A}_2 = \sum_{j=0}^{m} q^{j(j+B)} \begin{bmatrix} m \\ j \end{bmatrix}_q \begin{bmatrix} m-j \\ j+A \end{bmatrix}_q . \tag{1.8}$$

We shall prove the following results:

Theorem 5. *For $n \geqq 1$,*

$$r_n(q) = \sum_{j=-\infty}^{\infty} q^{j(3j+1)/2} \begin{bmatrix} n-1 \\ \lfloor \frac{n+3j}{2} \rfloor \end{bmatrix}_{q^2}, \tag{1.9}$$

$$s_n(q) = \sum_{j=-\infty}^{\infty} q^{6j^2+3j} \begin{bmatrix} n \\ \lfloor \frac{n+6j+2}{2} \rfloor \end{bmatrix}_{q^2}. \tag{1.10}$$

Theorem 6.

$$r_n(q) = \sum_{j=-\infty}^{\infty} (-1)^j q^{3j^2} \binom{n-1;3j;q}{3j}_2 \tag{1.11}$$

$$s_n(q) = \sum_{j=-\infty}^{\infty} (-1)^j q^{3j^2+2j} \binom{n;3j+1;q}{3j+1}_2 \tag{1.12}$$

$$q^n s_{n-1}(q) = \sum_{j=-\infty}^{\infty} q^{12j^2+2j} \left(\binom{n;6j;q}{6j}_2 - \binom{n;6j+1;q}{6j+1}_2 \right) \tag{1.13}$$

$$q^{n-2} s_{n-1}(q) = \sum_{j=-\infty}^{\infty} q^{12j^2+10j} \left(\binom{n;6j+2;q}{6j+2}_2 - \binom{n;6j+3;q}{6j+3} \right). \tag{1.14}$$

We note in passing that inspection of (1.5) reveals that $r_m(q)$ is the generating function for partitions of the type enumerated by $C(n)$ with the added condition that each part is $\leqq m-2$. Also by (1.6), $s_m(q)$ is the generating function for partitions of the type enumerated by $B(n)$ with the added condition that each part is $\leqq m-1$.

Furthermore it follows immediately from (1.5) and (1.6) that

$$r_n(1) = s_n(1) = (2^n - (-1)^n)/3. \tag{1.7}$$

Therefore Theorem 6 is the q-analog of Theorem 4. Theorem 2 (and consequently Theorem 1) follows by letting $n \to \infty$ in either (1.10) or (1.12) while Theorem 3 follows by letting $n \to \infty$ in either (1.9) or (1.11).

In Section 2 we assemble the necessary background. In Section 3 we prove Theorem 5 and in Section 4 we prove Theorem 6. Section 5 is devoted to the limits of the reciprocal polynomials $q^{2N^2-2N}r_{2N}(q^{-1})$, $q^{2N^2}r_{2N+1}(q^{-1})$, $q^{2N^2-2}s_{2N}(q^{-1})$, and $q^{2N^2+2N}s_{2N+1}(q^{-1})$. We conclude with a discussion of open questions.

2. Background

In this section we shall collect the several recurrences necessary to prove Theorem 5 and 6.

First there are the classical recurrences for Gaussian polynomials [5; p. 35, eqs. (3.3.4) and (3.3.3)]

$$\begin{bmatrix} A \\ B \end{bmatrix}_q = \begin{bmatrix} A-1 \\ B-1 \end{bmatrix}_q + q^B \begin{bmatrix} A-1 \\ B \end{bmatrix}_q, \tag{2.1}$$

and

$$\begin{bmatrix} A \\ B \end{bmatrix}_q = \begin{bmatrix} A-1 \\ B \end{bmatrix}_q + q^{A-B} \begin{bmatrix} A-1 \\ B-1 \end{bmatrix}_q. \tag{2.2}$$

These are, of course, equivalent recurrences once one observes that [5; p. 35, eq. (3.3.2)]

$$\begin{bmatrix} A \\ B \end{bmatrix}_q = \begin{bmatrix} A \\ A-B \end{bmatrix}_q. \tag{2.3}$$

The recurrences we shall require for the q-trinomial coefficients are as follows [7; eqs. (2.25), (2.28), (2.29) and (2.27) corrected]:

$$\begin{aligned} \binom{m; A-1; q}{A}_2 &= q^{m-1}\binom{m-1; A-1; q}{A}_2 \\ &\quad + q^A \binom{m-1; A+1; q}{A+1}_2 + \binom{m-1; A-1; q}{A-1}_2; \end{aligned} \tag{2.4}$$

$$\binom{m;B;q}{A}_2 = \binom{m-1;B;q}{A}_2 + q^{m-A-1+B}\binom{m-1;B;q}{A+1}_2 + q^{m-A}\binom{m-1;B-1;q}{A-1}_2; \tag{2.5}$$

$$\binom{m;B;q}{A}_2 = \binom{m-1;B;q}{A}_2 + q^{m-A}\binom{m-1;B-2;q}{A-1}_2 + q^{m+B}\binom{m-1;B+1;q}{A+1}_2; \tag{2.6}$$

$$\binom{m;A;q}{A} + q^m\binom{m;A;q}{A+1}_2 - \binom{m;A+1;q}{A+1}_2 - q^{m-A}\binom{m;A-1;q}{A}_2 = 0. \tag{2.7}$$

Finally there is a symmetry relation [7; eq. (2.15)]:

$$\binom{m;B;q}{-A}_2 = q^{A(A+B)}\binom{m;B+2A;q}{A}_2. \tag{2.8}$$

3. The proof of Theorem 5

To facilitate matters we define $\rho_n(q)$ to be the righthand side of (1.9) and $\sigma_n(q)$ to be the righthand side of (1.10).

Inspection shows immediately that $\rho_1(q) = \rho_2(q) = \sigma_1(q) = \sigma_2(q) = 1$. Hence to establish Theorem 5 we need only show $\rho_n(q)$ and $\sigma_n(q)$ satisfy the recurrences given in (1.5) and (1.6) respectively.

To this end we note that

$$\rho_{2N}(q) = \sum_{j=-\infty}^{\infty} q^{6j^2+j}\begin{bmatrix}2N-1\\N+3j\end{bmatrix}_{q^2} + \sum_{j=-\infty}^{\infty} q^{6j^2-5j+1}\begin{bmatrix}2N-1\\N+3j-2\end{bmatrix}_{q^2}; \tag{3.1}$$

$$\rho_{2N+1}(q) = \sum_{j=-\infty}^{\infty} q^{6j^2+j}\begin{bmatrix}2N\\N+3j\end{bmatrix}_{q^2} + \sum_{j=-\infty}^{\infty} q^{6j^2-5j+1}\begin{bmatrix}2N\\N+3j-1\end{bmatrix}_{q^2}; \tag{3.2}$$

$$\sigma_{2N}(q) = \sum_{j=-\infty}^{\infty} q^{6j^2+3j}\begin{bmatrix}2N\\N-3j-1\end{bmatrix}_{q^2} \tag{3.3}$$

$$\sigma_{2N+1}(q) = \sum_{j=-\infty}^{\infty} q^{6j^2+3j} \begin{bmatrix} 2N+1 \\ N-3j \end{bmatrix}_{q^2}. \tag{3.4}$$

Hence by (3.1), (3.2), (2.1) and (2.2)

$$\begin{aligned}
&\rho_{2N+1}(q) - \rho_{2N}(q) \\
&= q^{2N-1}\left(\sum_{j=-\infty}^{\infty} q^{6j^2-5j+1}\begin{bmatrix} 2N-1 \\ N+3j-1 \end{bmatrix}_{q^2} + \sum_{j=-\infty}^{\infty} q^{6j^2+j}\begin{bmatrix} 2N-1 \\ N+3j-1 \end{bmatrix}_{q^2}\right) \\
&= q^{2N-1}\left\{\sum_{j=-\infty}^{\infty} q^{6j^2-5j+1}\left(\begin{bmatrix} 2N-2 \\ N+3j-2 \end{bmatrix}_{q^2} + q^{2N+6j-2}\begin{bmatrix} 2N-2 \\ N+3j-1 \end{bmatrix}_{q^2}\right)\right. \\
&\qquad \left. + \sum_{j=-\infty}^{\infty} q^{6j^2+j}\left(\begin{bmatrix} 2N-2 \\ N+3j-1 \end{bmatrix}_{q^2} + q^{2N-6j}\begin{bmatrix} 2N-2 \\ N+3j-2 \end{bmatrix}_{q^2}\right)\right\} \\
&= q^{2N-1}\rho_{2N-1}(q) + q^{4N-2}\rho_{2N-1}(q).
\end{aligned} \tag{3.5}$$

Again by (3.1), (3.2), (2.1) and (2.2)

$$\begin{aligned}
&\rho_{2N+2}(q) - \rho_{2N+1}(q) \\
&= q^{2N+2}\left(\sum_{j=-\infty}^{\infty} q^{6j^2+7j}\begin{bmatrix} 2N \\ N+3j+1 \end{bmatrix}_{q^2}\right. \\
&\qquad \left. + \sum_{j=-\infty}^{\infty} q^{6j^2-11j+3}\begin{bmatrix} 2N \\ N+3j-2 \end{bmatrix}_{q^2}\right) \\
&= q^{2N}\left\{\sum_{j=-\infty}^{\infty} q^{6j^2-5j+1}\begin{bmatrix} 2N \\ N+3j-2 \end{bmatrix}_{q^2}\right. \\
&\qquad \left. + \sum_{j=-\infty}^{\infty} q^{6j^2+j}\begin{bmatrix} 2N \\ N+3j+1 \end{bmatrix}_{q^2}\right\} \\
&= q^{2N}\left\{\sum_{j=-\infty}^{\infty} q^{6j^2-5j+1}\left(\begin{bmatrix} 2N-1 \\ N+3j-2 \end{bmatrix}_{q^2} + q^{2N-6j+4}\begin{bmatrix} 2N-1 \\ N+3j-3 \end{bmatrix}_{q^2}\right)\right. \\
&\qquad \left. + \sum_{j=-\infty}^{\infty} q^{6j^2+j}\left(\begin{bmatrix} 2N-1 \\ N+3j \end{bmatrix}_{q^2} + q^{2N+6j+2}\begin{bmatrix} 2N-1 \\ N+3j+1 \end{bmatrix}_{q^2}\right)\right\} \\
&= q^{2N}\rho_{2N}(q) + q^{4N+2}\left(\sum_{j=-\infty}^{\infty} q^{6j^2+7j}\begin{bmatrix} 2N-1 \\ N+3j+1 \end{bmatrix}_{q^2}\right. \\
&\qquad \left. + \sum_{j=-\infty}^{\infty} q^{6j^2-11j+3}\begin{bmatrix} 2N-1 \\ N+3j-3 \end{bmatrix}_{q^2}\right)
\end{aligned} \tag{3.6}$$

$$= q^{2N}\rho_{2N}(q) + q^{4N}\left(\sum_{j=-\infty}^{\infty} q^{6j^2-5j+1}\begin{bmatrix} 2N-1 \\ N+3j-2 \end{bmatrix}_{q^2}\right.$$

$$\left. + \sum_{j=-\infty}^{\infty} q^{6j^2+j}\begin{bmatrix} 2N-1 \\ N+3j \end{bmatrix}_{q^2}\right)$$

$$= q^{2N}\rho_{2N}(q) + q^{4N}\rho_{2N}(q).$$

Therefore for each $n > 1$,

$$\rho_n(q) = \rho_{n-1}(q) + (q^{n-2} + q^{2n-4})\rho_{n-2}(q), \tag{3.7}$$

and so

$$\rho_n(q) = r_n(q) \tag{3.8}$$

which establishes (1.9).

Next we see by (3.3), (3.4) and (2.1)

$$\sigma_{2N+1}(q) - \sigma_{2N}(q) \tag{3.9}$$

$$= \sum_{j=-\infty}^{\infty} q^{6j^2+3j} q^{2N-6j}\begin{bmatrix} 2N \\ N-3j \end{bmatrix}_{q^2}$$

$$= q^{2N}\sum_{j=-\infty}^{\infty} q^{6j^2+3j}\left(\begin{bmatrix} 2N-1 \\ N-3j-1 \end{bmatrix}_{q^2} + q^{2N-6j}\begin{bmatrix} 2N-1 \\ N-3j \end{bmatrix}_{q^2}\right)$$

$$= q^{2N}\sigma_{2N-1}(q) + q^{4N}\sum_{j=-\infty}^{\infty} q^{6j^2-3j}\begin{bmatrix} 2N-1 \\ N-3j \end{bmatrix}_{q^2}$$

$$= q^{2N}\sigma_{2N-1}(q) + q^{4N}\sum_{j=-\infty}^{\infty} q^{6j^2+3j}\begin{bmatrix} 2N-1 \\ N-1-3j \end{bmatrix}_{q^2}$$

$$= q^{2N}(1+q^{2N})\sigma_{2N-1}(q).$$

Finally by (3.3), (3.4) and (2.2)

$$\sigma_{2N+2}(q) - \sigma_{2N+1}(q) \tag{3.10}$$

$$= q^{2N+4} \sum_{j=-\infty}^{\infty} q^{6j^2+9j} \begin{bmatrix} 2N+1 \\ N-3j-1 \end{bmatrix}_{q^2}$$
$$= q^{2N+1} \sum_{j=-\infty}^{\infty} q^{6j^2-3j} \begin{bmatrix} 2N+1 \\ N-3j+2 \end{bmatrix}_{q^2}$$
$$= q^{2N+1} \sum_{j=-\infty}^{\infty} q^{6j^2+3j} \begin{bmatrix} 2N+1 \\ N-3j-1 \end{bmatrix}_{q^2}$$
$$= q^{2N+1} \sum_{j=-\infty}^{\infty} q^{6j^2+3j} \left(\begin{bmatrix} 2N \\ N-3j-1 \end{bmatrix}_{q^2} + q^{2N+6j+4} \begin{bmatrix} 2N \\ N-3j-2 \end{bmatrix}_{q^2} \right)$$
$$= q^{2N+1}\sigma_{2N}(q) + q^{4N+5} \sum_{j=-\infty}^{\infty} q^{6j^2+9j} \begin{bmatrix} 2N \\ N-3j-2 \end{bmatrix}_{q^2}$$
$$= q^{2N+1}\sigma_{2N}(q) + q^{4N+2} \sum_{j=-\infty}^{\infty} q^{6j^2-3j} \begin{bmatrix} 2N \\ N-3j+1 \end{bmatrix}_{q^2}$$
$$= q^{2N+1}\sigma_{2N}(q) + q^{4N+2} \sum_{j=-\infty}^{\infty} q^{6j^2+3j} \begin{bmatrix} 2N \\ N-3j-1 \end{bmatrix}_{q^2}$$
$$= q^{2N+1}(1+q^{2N+1})\sigma_{2N}(q).$$

Therefore for each $n > 1$,

$$\sigma_n(q) = \sigma_{n-1}(q) + (q^{n-1} + q^{2n-2})\sigma_{n-2}(q), \tag{3.11}$$

and so

$$\sigma_n(q) = s_n(q). \tag{3.12}$$

Consequently Theorem 5 is proved.

4. The proof of Theorem 6

Equations (1.11) and (1.12) may be deduced directly from the following identity:

$$\begin{aligned}
&\binom{m; A; q}{A}_2 - \binom{m-1; A; q}{A}_2 - (q^{m-1} + q^{2m-2})\binom{m-2; A; q}{A}_2 \\
&= q^{2m-3}\binom{m-2; A; q}{A+1}_2 + q^{m+A}\binom{m-2; A+2; q}{A+2}_2 \\
&+q^{m-A}\binom{m-2; A-1; q}{A-1}_2 + q^{2m-2A}\binom{m-2; A-3; q}{A-2}_2
\end{aligned} \tag{4.1}$$

Equation (4.1) is proved as follows:

$$
\begin{aligned}
(4.2)\qquad & \binom{m;A;q}{A}_2 - \binom{m-1;A;q}{A}_2 \\
&= q^{m-1}\binom{m-1;A;q}{A+1}_2 + q^{m-A}\binom{m-1;A-1;q}{A-1}_2 \\
&\qquad \text{(by (2.5))} \\
&= q^{m-1}\left(q^{m-2}\binom{m-2;A;q}{A+1}_2 + \binom{m-2;A;q}{A}_2 + q^{A+1}\binom{m-2;A+2;q}{A+2}_2\right) \\
&\quad + q^{m-A}\left(q^{m+A-2}\binom{m-2;A;q}{A}_2\right. \\
&\quad \left. + \binom{m-2;A-1;q}{A-1}_2 + q^{m-A}\binom{m-2;A-3;q}{A-2}_2\right), \\
&\qquad \text{(by (2.4) and (2.6))}
\end{aligned}
$$

and this is just (4.1) rewritten.

We now let $\mathcal{R}_n(q)$ and $\mathcal{S}_n(q)$ denote the righthand sides of (1.11) and (1.12) respectively.

Clearly $\mathcal{R}_1(q) = \mathcal{S}_1(q) = \mathcal{R}_2(q) = \mathcal{S}_2(q) = 1$. By (4.1),

$$
\begin{aligned}
(4.3)\qquad & \mathcal{R}_{n+1}(q) - \mathcal{R}_n(q) - (q^{n-1} + q^{2n-2})\mathcal{R}_{n-1}(q) \\
&= \sum(-1)^j q^{3j^2}\left\{q^{2n-3}\binom{n-2;3j;q}{3j+1}_2 + q^{n+3j}\binom{n-2;3j+2;q}{3j+2}_2\right. \\
&\quad \left. + q^{n-3j}\binom{n-2;3j-1;q}{3j-1}_2 q^{2n-3j}\binom{n-2;3j-3;q}{3j-2}_2\right. \\
&= (I) + (II) + (III) + (IV) \\
&= 0,
\end{aligned}
$$

since replacing j by $j+1$ in (III) and (IV) reveals that (IV) = - (I) and (III) = - (II). Hence (1.11) is established.

Next by (4.1),

$$
(4.4)\qquad \mathcal{S}_n(q) - \mathcal{S}_{n-1}(q) - (q^{n-1} + q^{2n-2})\mathcal{S}_{n-2}(q)
$$

$$= \sum_{j=-\infty}^{\infty} (-1)^j q^{3j^2+2j} \left\{ q^{2n-3} \begin{pmatrix} n-2; 3j+1; q \\ 3j+2 \end{pmatrix}_2 \right.$$

$$+ q^{n+3j+1} \begin{pmatrix} n-2; 3j+3; q \\ 3j+3 \end{pmatrix}_2 + q^{n-3j-1} \begin{pmatrix} n-2; 3j; q \\ 3j \end{pmatrix}_2$$

$$\left. + q^{2n-6j-2} \begin{pmatrix} n-2; 3j-2; q \\ 3j-1 \end{pmatrix}_2 \right\}$$

$$= (i) + (ii) + (iii) + (iv)$$

$$= 0,$$

since replacing j by $j+1$ in (iii) and (iv) reveals (iv) = - (i) and (iii) = - (ii). Hence (1.12) is established.

We now treat (1.13).

$$\text{(4.5)} \qquad \sum_{j=-\infty}^{\infty} q^{12j^2+2j} \left(\begin{pmatrix} n; 6j; q \\ 6j \end{pmatrix}_2 - \begin{pmatrix} n; 6j+1; q \\ 6j+1 \end{pmatrix}_2 \right)$$

$$= \sum_{j=-\infty}^{\infty} q^{12j^2-2j} \left(\begin{pmatrix} n; 6j; q \\ 6j \end{pmatrix}_2 - \begin{pmatrix} n; 6j-1; q \\ 6j-1 \end{pmatrix}_2 \right) \qquad \text{(by (2.8))}$$

$$= \sum_{j=-\infty}^{\infty} q^{12j^2+2j} \left(\begin{pmatrix} n-1; 6j; q \\ 6j \end{pmatrix}_2 + q^{n-1} \begin{pmatrix} n-1; 6j; q \\ 6j+1 \end{pmatrix}_2 \right.$$

$$\left. + q^{n-6j} \begin{pmatrix} n-1; 6j-1; q \\ 6j-1 \end{pmatrix}_2 \right) - \sum_{j=-\infty}^{\infty} q^{12j^2-2j} \left(\begin{pmatrix} n-1; 6j-1; q \\ 6j-1 \end{pmatrix}_2 \right.$$

$$\left. + q^{n-1} \begin{pmatrix} n-1; 6j-1; q \\ 6j \end{pmatrix}_2 + q^{n-6j+1} \begin{pmatrix} n-1; 6j-2; q \\ 6j-2 \end{pmatrix}_2 \right) \qquad \text{(by (2.5))}$$

$$= q^n s_{n-1}(q) + \sum_{j=-\infty}^{\infty} q^{12j^2+2j} \left(\begin{pmatrix} n-1; 6j; q \\ 6j \end{pmatrix}_2 + q^{n-1} \begin{pmatrix} n-1; 6j; q \\ 6j+1 \end{pmatrix}_2 \right.$$

$$\left. - q^{n-1} \begin{pmatrix} n-1; -6j-1; q \\ -6j \end{pmatrix}_2 - \begin{pmatrix} n-1; 6j+1; q \\ 6j+1 \end{pmatrix}_2 \right)$$

$$= q^n s_{n-1}(q) + \sum_{j=-\infty}^{\infty} q^{12j^2+2j} \left(\begin{pmatrix} n-1; 6j; q \\ 6j \end{pmatrix}_2 + q^{n-1} \begin{pmatrix} n-1; 6j; q \\ 6j+1 \end{pmatrix}_2 \right.$$

$$\left. + q^{n-6j-1} \begin{pmatrix} n-1; 6j-1; q \\ 6j \end{pmatrix}_2 - \begin{pmatrix} n-1; 6j+1; q \\ 6j+1 \end{pmatrix}_2 \right) \qquad \text{(by (2.8))}$$

$$= q^n s_{n-1}(q)$$

by (2.7). So (1.13) is proved.

Finally we prove (1.14).

$$
\begin{aligned}
(4.6)\qquad &\sum_{j=-\infty}^{\infty} q^{12j^2+10j}\left(\binom{n;6j+2;q}{6j+2}_2 - \binom{n;6j+3;q}{6j+3}_2\right)\\
&= \sum_{j=-\infty}^{\infty} q^{12j^2+10j}\left(\binom{n-1;6j+2;q}{6j+2}_2 + q^{n-1}\binom{n-1;6j+2;q}{6j+3}_2\right.\\
&\qquad\left. +q^{n-6j-2}\binom{n-1;6j+1;q}{6j+1}_2\right)\\
&\quad - \sum_{j=-\infty}^{\infty} q^{12j^2-10j}\left(\binom{n-1;6j-3;q}{6j-3}_2 + q^{n-1}\binom{n-1;6j-3;q}{6j-2}_2\right.\\
&\qquad\left. +q^{n-6j+3}\binom{n-1;6j-4;q}{6j-4}_2\right)\\
&= q^{n-2}s_{n-1}(q) + \sum_{j=-\infty}^{\infty} q^{12j^2+10j}\left(\binom{n-1;6j+2;q}{6j+2}_2\right.\\
&\qquad + q^{n-1}\binom{n-1;6j+2;q}{6j+3}_2\\
&\qquad\left. - \binom{n-1;6j+3;q}{6j+3}_2 - q^{n-1}\binom{n-1;-6j-3;q}{-6j-2}_2\right)\\
&= q^{n-2}s_{n-1}(q) + \sum_{j=-\infty}^{\infty} q^{12j^2+10j}\left(\binom{n-1;6j+2;q}{6j+2}_2\right.\\
&\qquad + q^{n-1}\binom{n-1;6j+2;q}{6j+3}_2 \binom{n-1;6j+3;q}{6j+3}_2\\
&\qquad\left. -q^{n-6j-3}\binom{n-1;6j+1;q}{6j+2}_2\right) \qquad \text{(by (2.8))}\\
&= q^{n-2}s_{n-1}(q)
\end{aligned}
$$

by (2.7). So (1.14) is proved.

5. The reciprocal polynomials

In various applications of results like Theorems 5 and 6, it is useful to know the limiting behavior of the reciprocal polynomials.

Theorem 7.

$$
\begin{aligned}
(5.1)\qquad &\lim_{N\to\infty} q^{2N^2-2N} r_{2N}(q^{-1}) = \frac{1}{(q^2;q^2)_\infty}\sum_{j=-\infty}^{\infty} q^{12j^2+5j}\left(1+q^{6j+2}\right)\\
&= \frac{(-q^7;q^{24})_\infty(-q^{17};q^{24})_\infty(q^{24};q^{24})_\infty + q^2(-q;q^{24})_\infty(-q^{23};q^{24})_\infty(q^{24};q^{24})_\infty}{(q^2;q^2)_\infty};
\end{aligned}
$$

(5.2) $$\lim_{N\to\infty} q^{2N^2} r_{2N+1}(q^{-1}) = \frac{1}{(q^2;q^2)_\infty} \sum_{j=-\infty}^{\infty} q^{12j^2+j}(1+q^{6j+1}) =$$
$$\frac{(-q^{11};q^{24})_\infty(-q^{13};q^{24})_\infty(q^{24};q^{24})_\infty + q(-q^5;q^{24})_\infty(-q^{19};q^{24})_\infty(q^{24};q^{24})_\infty}{(q^2;q^2)_\infty};$$

(5.3) $$\lim_{N\to\infty} q^{2N^2-2} s_{2N}(q^{-1})$$
$$= \frac{1}{(q^2;q^2)_\infty} \sum_{j=-\infty}^{\infty} q^{12j^2+9j} = \frac{(-q^3;q^{24})_\infty(-q^{21};q^{24})_\infty(q^{24};q^{24})_\infty}{(q^2;q^2)_\infty};$$

(5.4) $$\lim_{N\to\infty} q^{2N^2+2N} s_{2N+1}(q^{-1})$$
$$= \frac{1}{(q^2;q^2)_\infty} \sum_{j=-\infty}^{\infty} q^{12j^2+3j} = \frac{(-q^9;q^{24})_\infty(-q^{15};q^{24})_\infty(q^{24};q^{24})_\infty}{(q^2;q^2)_\infty};$$

Proof. Each of these formulas follows immediately from Theorem 5 once one makes the following observations:

(5.5) $$\begin{bmatrix} A \\ B \end{bmatrix}_{q^{-1}} = q^{-B(A-B)} \begin{bmatrix} A \\ B \end{bmatrix}_q ,$$

and

(5.6) $$\lim_{N\to\infty} \begin{bmatrix} 2N+\alpha \\ N+\beta j \end{bmatrix}_q = \frac{1}{(q;q)_\infty},$$

where

(5.7) $$(A;q)_n = \prod_{j=0}^{n-1} (1 - Aq^j).$$

Since the proofs are routine, we only carry out the details for (5.1). By (3.1), (5.5) and (5.6)

(5.8) $$\lim_{N\to\infty} q^{2N^2-2N} r_{2N}(q^{-1})$$

$$= \lim_{N\to\infty} q^{2N^2-2N} \left\{ \sum_{j=-\infty}^{\infty} q^{-6j^2-j-2(N+3j)(N-1-3j)} \begin{bmatrix} 2N-1 \\ N+3j \end{bmatrix}_{q^2} \right.$$

$$\left. + \sum_{j=-\infty}^{\infty} q^{-6j^2+5j-1-2(N+3j-2)(N-3j+1)} \begin{bmatrix} 2N-1 \\ N+3j-2 \end{bmatrix}_{q^2} \right\}$$

$$= \frac{1}{(q^2;q^2)_\infty} \left\{ \sum_{j=-\infty}^{\infty} q^{12j^2+5j} + \sum_{j=-\infty}^{\infty} q^{12j^2-13j+3} \right\}$$

$$= \frac{1}{(q^2;q^2)_\infty} \left\{ \sum_{j=-\infty}^{\infty} q^{12j^2+5j} + \sum_{j=-\infty}^{\infty} q^{12j^2+11j+2} \right\}$$

$$= \frac{1}{(q^2;q^2)_\infty} \sum_{j=-\infty}^{\infty} q^{12j^2+5j}(1+q^{6j+2}),$$

which is (5.1).

The remainder follow in the same way.

6. Conclusion

It is possible to carry out a more refined analysis than the one presented here by considering the two variable polynomials

$$(6.1) \qquad t_n(x,q) = \begin{cases} 0 & \text{if } n=0 \\ 1 & \text{if } n=1 \\ t_{n-1}(x,q) + (xq^{n-2} + x^2q^{2n-4})t_{n-2}(x,q), & \text{if } n>1. \end{cases}$$

Clearly $t_n(1,q) = r_n(q)$ and $t_n(q,q) = s_n(q)$. Furthermore one can show from (6.1) that

$$(6.2) \qquad F(z) := \sum_{n=0}^{\infty} \frac{t_n(x,q)}{(-x;q)_n} z^n$$

satisfies a first order inhomogeneous q-difference equation in z, namely

$$(6.3) \qquad F(z) = \frac{z}{1-z} + \frac{z^2x - xq^{-1}}{1-z} F(zq).$$

Consequently by the standard iteration [8; pp. 3–4]

$$(6.4) \qquad F(z) = z\sum_{n=0}^{\infty} \frac{(-x)^n (z^2q;q^2)_n}{(z;q)_{n+1}}.$$

Identity (6.4) can then be utilized to find new representations for $r_n(q)$ and $s_n(q)$, and can be utilized to lead to the proof of Theorems 1–3 alluded to in [4; p. 267, lines 6 and 7].

Also the methods of [1] could be used to provide partition theoretic interpretations of the reciprocal polynomials considered in Section 5.

Finally I would like to acknowledge the valuable assistance provided by the symbolic algebra package AXIOM. It was essential in the discovery of (1.13) and (1.14).

References

[1] A. K. Agarwal and G. E. Andrews, *Hook differences and lattice paths*, J. Planning and Stat. Inf. **14** (1986), 5–14.

[2] G. E. Andrews, *Some new partition theorems*, J. Comb. Th. **2** (1967), 431–436.

[3] ———, *Some new partition theorems, II*, J. Comb. Th. **7** (1969), 262–263.

[4] ———, *A new property of partitions with applications to the Rogers-Ramanujan identities*, J. Comb. Th. **10** (1971), 266–270.

[5] ———, The Theory of Partitions, Encyclopedia of Math. and Its Appl., Vol. 2, G.-C. Rota, ed., Addison-Wesley, Reading, 1976 (Reprinted: Cambridge University Press, London, 1985).

[6] ———, *Euler's "Exemplum memorabile inductionis fallacis" and q-trinomial coefficients*, J. Amer. Math. Soc. **3** (1990), 653–669.

[7] G. E. Andrews and R. J. Baxter, *Lattice gas generalizations of the hard hexagon model, III. q-trinomial coefficients*, J. of Stat. Physics **47** (1987), 297–330.

[8] N. J. Fine, Basic Hypergeometric Series and Applications, Math Surveys and Monographs, No. 27, Amer. Math. Soc., Providence 1988.

George E. Andrews
Department of Mathematics
The Pennsylvania State University
University Park, PA 16802
andrews@math.psu.edu

On subset-sum-distinct sequences

Jaegug Bae

Dedicated to Professor Halberstam on the occasion of his retirement

Abstract. We use Tomić's inequality to show some interesting properties of subset-sum-distinct sequences including an elementary proof of a theorem of Steele, Hanson, and Stenger. Also, we comment on a theorem of L. Moser.

1. Introduction

In this paper, by a *sequence* we mean a *strictly increasing sequence of positive integers.* We start with a definition.

Definition 1.1. Let A be a set of real numbers. We say that A is a subset-sum-distinct set (briefly, A is an SSD-set or A is SSD) if for any two finite subsets X, Y of A,

$$\sum_{x\in X} x = \sum_{y \in Y} y \qquad \Longrightarrow \qquad X = Y.$$

Also, we say that a sequence $\{a_n\}_{n=1}^{\infty}$ is an SSD-sequence if it is SSD.

One of the most interesting and natural SSD-sequences is

$$\mathbf{t} \;=\; \{1, 2, 2^2, 2^3, \cdots\}.$$

Now, for a given SSD-sequence $\{a_n\}_{n=1}^{\infty}$, how can one compare the size of this sequence with $\mathbf{t}$? The following basic lemma, which is a starting point of the next section, gives some insight.

Lemma 1.2. *Let* $\{a_n\}_{n=1}^{\infty}$ *be an SSD-sequence. Then*

$$a_1 + a_2 + \cdots + a_n \;\geq\; 2^n - 1$$

for every $n \geq 1$.

Proof. Let

$$A = \{a_1, a_2, \cdots, a_n\} \quad \text{and} \quad J = \left\{ \sum_{b \in B} b \,:\, \phi \neq B \subset A \right\}.$$

Note that all the elements of J are positive integers. Since A has the SSD-property,

$$B, B' \subset A \quad \text{and} \quad B \neq B' \quad \Longrightarrow \quad \sum_{b \in B} b \neq \sum_{b' \in B'} b'.$$

Hence $|J| = 2^n - 1$. Because $a_1 + a_2 + \cdots + a_n$ is the greatest element in J, we have

$$a_1 + a_2 + \cdots + a_n \geq 2^n - 1. \quad \blacksquare$$

As $\{1, 2, 2^2, 2^3, \cdots\}$ suggests, SSD-sequences are quite sparse. It seems very natural to ask how dense they can be. We will consider a question of this flavor in this paper.

As a way of obtaining finite "dense" SSD-sets, one can use the Conway-Guy sequence (see [8], [10]). Here we explain the construction of the Conway-Guy sequence. First, define an auxiliary sequence u_n by

$$u_0 = 0, \quad u_1 = 1 \qquad \text{and} \qquad u_{n+1} = 2\,u_n - u_{n-r}, \quad n \geq 1,$$

where $r = \langle \sqrt{2n} \rangle$, the nearest integer to $\sqrt{2n}$. Now, for a given positive integer n, we define

$$a_i = u_n - u_{n-i}, \qquad 1 \leq i \leq n.$$

The well-known Conway-Guy conjecture is that $\{a_i : 1 \leq i \leq n\}$ is SSD for any positive integer n. This is still open. It is known that it does yield SSD-sets for $n < 80$ (see [10, p.307, Theorem4.6]).

Throughout the paper, we assume that $\{a_n\}_{n=1}^{\infty}$ is an SSD-sequence.

(i) In the next section, we consider the Dirichlet series of an SSD-sequence, namely,

$$\sum_{i=1}^{m} a_i{}^s \text{ for real } s.$$

(ii) In the third section, we consider a lower bound on a_n. Here a result of L. Moser asserts that

$$\sum_{i=1}^{m} a_i{}^2 \geq \sum_{i=1}^{m} 2^{2(i-1)}.$$

We give a comment regarding this theorem.

2. Tomić 's Inequality

First of all, we introduce Tomić's inequality. It will turn out that his result is extremely useful for the estimation of $\sum_{i=1}^{m} {a_i}^s$, especially when s is not an integer.

Theorem 2.1.(M. Tomić, 1949) *Let*

$$u_1 \geq u_2 \geq \cdots \geq u_m\,, \qquad v_1 \geq v_2 \geq \cdots \geq v_m\,,$$

where the u_i's and v_i's are real numbers. Then

$$\sum_{i=1}^{k} u_i \leq \sum_{i=1}^{k} v_i \qquad for \;\; k = 1, 2, \cdots, m$$

if and only if

$$\sum_{i=1}^{m} f(u_i) \leq \sum_{i=1}^{m} f(v_i)$$

for every convex increasing function f.

Proof. See [11]. ■

Remark. Originally, Tomić gave a geometric proof based on Gauss' theorem on the centroid. But it can be proved easily by using convexity and summation by parts (See [11], [13], and [14]).

In order to apply Tomić's inequality to SSD-sequences, we give the following direct corollary of the theorem.

Corollary 2.2. *Let* $x_1 \leq x_2 \leq \cdots \leq x_m$ *and* $y_1 \leq y_2 \leq \cdots \leq y_m$, *where x_i's and y_i's are real numbers. Then the followings are equivalent.*

(i) $\sum_{i=1}^{k} x_i \leq \sum_{i=1}^{k} y_i \qquad for \;\; k = 1, 2, \cdots, m$

(ii) $\sum_{i=1}^{m} g(x_i) \geq \sum_{i=1}^{m} g(y_i)$ *for every convex decreasing function g,*

(iii) $\sum_{i=1}^{m} h(x_i) \leq \sum_{i=1}^{m} h(y_i)$ *for every concave increasing function h.*

The following theorem, which was first proved by F. Hanson, J. M. Steele and F. Stenger (see [9]) using a variation of Ryavec's generating function method, is immediate from the above corollary.

Theorem 2.3. *Let* $\{a_n\}_{n=1}^{\infty}$ *be an SSD-sequence. Then*

$$\sum_{i=1}^{m} a_i{}^s \le \sum_{i=1}^{m} 2^{(i-1)s} = \frac{1-2^{ms}}{1-2^s} \tag{2.1}$$

for all positive integers m *and all real numbers* $s \le 0$.

Proof. By Lemma 1.2, for all positive integers $k \le m$ we have

$$a_1 + a_2 + \cdots + a_k \ \ge \ 1 + 2 + 2^2 + \cdots + 2^{k-1}. \tag{2.2}$$

Now, the conclusion follows from Corollary 2.2 when we set $x_i = 2^{i-1}$, $y_i = a_i$ and $g(x) = x^s$. ■

At this point, one may ask whether one can have the reverse inequality of (2.1) for non-negative s. In other words, is

$$\sum_{i=1}^{m} a_i{}^s \ge \sum_{i=1}^{m} 2^{(i-1)s} = \frac{1-2^{ms}}{1-2^s} \tag{2.3}$$

for all positive integers m and all $s \ge 0$?
In view of (2.2), inequality (2.3) is true for $s = 1$. What about other positive values of s? In connection with this question we have the following theorem.

Theorem 2.4. *Let* $\{a_n\}_{n=1}^{\infty}$ *be an SSD-sequence and* β *a fixed positive real number. If*

$$\sum_{i=1}^{m} a_i{}^\beta \ge \sum_{i=1}^{m} 2^{(i-1)\beta} = \frac{1-2^{m\beta}}{1-2^\beta}$$

for all positive integers m, *then (2.3) is true for all positive integers* m, *and all* $0 \le s \le \beta$.

Proof. Starting from inequality (2.2), put $x_i = 2^{i-1}$, $y_i = a_i$ and $h(x) = x^{s/\beta}$ in Corollary 2.2. Note that $h(x)$ is concave and increasing on $x \ge 0$, whence the result follows. ■

By the previous theorem, we may define λ to be the supremum of all s that satisfy (2.3) for all SSD-sequences $\{a_n\}_{n=1}^{\infty}$ and for all positive integers m. Here we give lower and upper bounds for λ.

Theorem 2.5. $2 \le \lambda \le 3.6906742\cdots$.

Proof. By Moser's Theorem (Theorem 3.1), we have $\lambda \ge 2$. Let

$$\{a_i\}_{i=1}^{7} \ = \ \{20, 31, 37, 40, 42, 43, 44\}. \tag{2.4}$$

By routine calculations or by using the construction from the Conway-Guy sequence (see [8], [10]), one can show the set of (2.4) is SSD. Then

$$\sum_{i=1}^{7} a_i{}^s - \sum_{i=1}^{7} 2^{(i-1)s}$$

changes sign from + to − at $s = 3.6906742\cdots$. Note that it is obvious from Theorem 2.6 that it cannot change sign in the opposite direction. Hence we have the theorem. ■

Remark. Note that the method used by F. Hanson, J. M. Steele and F. Stenger in [9] has no immediate extension to $s \geq 0$ because of convergence problems.

3. A comment on a lower bound for a_n

Let $\{a_n\}_{n=1}^{\infty}$ be an SSD-sequence. It is an old problem to find a lower bound for a_n (see [1, pp.47-48], [2], [4], [5, p.467], [6, pp.59-60], [8] and [12]). This problem can be stated inversely: Under the condition $a_n \leq x$, find an upper bound on n. In his paper [3], N. Elkies mentioned the inter-relation between a lower bound on a_n and an upper bound on n in terms of x. A lower bound of the form

$$a_n \geq C n^{-s} 2^n (1 + o(1)) \tag{3.1}$$

corresponds to an upper bound of the form

$$n \leq \log_2 x + s \log_2 \log_2 x + \log_2 \frac{1}{C} + o(1). \tag{3.2}$$

The famous conjecture of Erdös is that (3.1) and (3.2) hold with $s = 0$ (see [7, p.114, problem C8]). Also, Elkies showed that

$$a_n \geq 2^{-n} \binom{2n}{n} \sim \frac{1}{\sqrt{\pi}} n^{-\frac{1}{2}} 2^n$$

by an analytic method. But we point out that one can derive a better bound (Theorem 3.2) immediately from the following theorem of Moser.

Theorem 3.1. *For any integer $n \geq 1$, we have*

$$\sum_{i=1}^{n} a_i{}^2 \geq \sum_{i=1}^{n} 2^{2(i-1)}.$$

Proof. See [4, p.137] or [8, p.142, Theorem 2]. ■

Theorem 3.2. *For any integer* $n \geq 2$, *we have*

$$a_n > \frac{1}{\sqrt{3}} n^{-\frac{1}{2}} 2^n$$

Proof. Theorem 3.1 says that

$$\sum_{i=1}^{n} {a_i}^2 \geq \sum_{i=1}^{n} 2^{2(i-1)} = \frac{1}{3}(4^n - 1).$$

Hence, for $n \geq 2$, we have $n\, {a_n}^2 > \frac{1}{3} 4^n$ which proves the theorem. ■

Acknowledgement The author wishes to thank Professor Kenneth B. Stolarsky for help and encouragement.

References

[1] N. Alon, J. H. Spencer, and P. Erdös, *The Probabilistic Method*, John Wiley & Sons, Inc., New York, 1992.

[2] J. H. Conway and R. K. Guy, *Solution of a problem of P. Erdös*, Colloquium Mathematicum **20** (1969), 307.

[3] N. D. Elkies, *An improved lower bound on the greatest element of a sum-distinct set of fixed order*, Journal of Combinatorial Theory, Series A **41** (1986), 89-94.

[4] P. Erdös, *Problems and results in additive number theory*, Colloque sur la Théorie des Nombres (1955), Bruxelles, 127-137.

[5] P. Erdös, *Some of my favorite unsolved problems*, A tribute to Paul Erdös (A. Baker, B. Bollobás and A. Hajnal, eds.), Cambridge University Press, Cambridge, 1990, pp. 467-478.

[6] P. Erdös and R. L. Graham, *Old and new problems and results in combinatorial number theory*, Monographies de L'Enseignement Mathématique **28** (1980), Université de Genève, L'Enseignement Mathématique, Geneva.

[7] R. K. Guy, *Unsolved Problems in Intuitive Mathematics, Vol.I, Number Theory*, Springer-Verlag, New York, 1994.

[8] R. K. Guy, *Sets of integers whose subsets have distinct sums*, Annals of Discrete Mathematics **12** (1982), 141-154.

[9] F. Hanson, J. M. Steele, and F. Stenger, *Distinct sums over subsets*, Proc. Amer. Math. Soc. **66** (Sep. 1977), no. 1, 179-180.

[10] W. F. Lunnon, *Integer sets with distinct subset-sums*, Mathematics of Computation **50** (Jan. 1988), no. 181, 297-320.

[11] G. Pólya, *Remark on Weyl's note "Inequalities between the two kinds of eigenvalues of a linear transformation"*, Proc. Nat. Acad. Sci. U.S.A. **36** (1950), 49-51.

[12] P. Smith, *Problem E2526*, Amer. Math. Monthly **82** (1975), 300; **83** (1976), 484; **88** (1981), 538-539.

[13] M. Tomić, *Gauss' theorem on the centroid and its application*, Bull. Soc. Math. Phys. Serbie 1 (1949), no. 1, 31-40. (Serbian)

[14] H. Weyl, *Inequalities between the two kinds of eigenvalues of a linear transformation*, Proc. Nat. Acad. Sci. U.S.A. **35** (1949), 408-411.

Jaegug Bae
Department of Mathematics
University of Illinois
Urbana, IL 61801
bae@math.uiuc.edu

The Brun-Titchmarsh Theorem on average

R. C. Baker[1] *and G. Harman*

To Professor Heini Halberstam on the occasion of his retirement

1. Introduction

Throughout this paper a denotes a fixed non-zero integer and the letter p with or without subscript denotes a prime variable. As usual, for $(q, a) = 1$ we write

$$\pi(x; q, a) = \sum_{\substack{p \le x \\ p \equiv a (\text{mod } q)}} 1.$$

By the form of the Brun-Titchmarsh inequality given by Montgomery and Vaughan [24] we know that

$$\pi(x; q, a) < \frac{2x}{\phi(q) \log(x/q)}, \quad \text{for } 1 \le q < x.$$

It follows from the Bombieri-Vinogradov theorem [6] that

$$\pi(x; q, a) = \frac{x}{\phi(q) \log x} \left(1 + O((\log x)^{-1})\right)$$

for most $q \le B$ (where $B \le x^{1/2-\epsilon}$). By 'most q' we mean the number of exceptional q is at most $B(\log x)^{-\Theta}$ for every $\Theta > 0$, provided that $x \ge x_0(a, \Theta)$. Hooley [19] was the first to obtain upper bounds for $\pi(x, q, a)$ for most $q \le B$ with B exceeding $x^{1/2}$. This result has an interesting application to the greatest prime factor of $p + a$ which we will discuss later in this section.

The best upper bounds on $\pi(x; q, a)$ for most $q \le B$ are due to Fouvry [11]. We must also note the work of Rousselet [26] (both upper and lower bounds) in the interval

$$x^{1/2-\epsilon} < B < x^{1/2+\rho}, \rho = 10^{-100}.$$

The method we present here will improve all these results up to $x^{0.56}$. Before stating our main result, we fix an arbitrarily small positive absolute constant ϵ.

---

[1]Research supported in part by a grant from the National Science Foundation

Theorem 1. *There are functions* $C_1(\theta), C_2(\theta)(0.5 \le \theta \le 0.6)$ *such that, for most* $q \le x^\theta$,

$$\frac{C_1(\theta)x}{\phi(q)\log x} < \pi(x;q,a) < \frac{C_2(\theta)x}{\phi(q)\log x}. \tag{1.1}$$

These functions satisfy

(i) $C_2(\theta)$ *monotonic increasing,* $C_2(\theta)$ *monotonic decreasing,*
(ii) $C_2(0.5) = 1+\epsilon, C_2(0.51) \le 1.015, C_2(0.533) < 2$,
(iii) $C_1(0.5) = 1-\epsilon, C_1(0.52) > 0.16$,
(iv) *we have*

$$\int_{1/2}^{3/5} C_2(\theta)d\theta < 0.241. \tag{1.2}$$

More details on the value of C_1 and C_2 will be supplied in section 9, where we shall see that

$$C_2(\theta) = 1 + O((\theta - 1/2+\epsilon)^2), C_1(\theta) = 1 + O(\theta - 1/2 + \epsilon).$$

Rousselet [26] obtained a result like our Theorem 1 with $C_2 = 1.48$ and $C_1 = 0.85$, for the interval

$$\frac{1}{2} - \epsilon < \theta < \frac{1}{2} + 10^{-100}.$$

Our theorem may also be compared to the following result of Bombieri, Friedlander and Iwaniec [5].

Theorem 2. *Let* $a \ne 0$ *be an integer and* $A > 0, 2 \le Q \le x^{3/4}$ *be reals. Let* $\mathcal{Q}$ *be the set of all integers* q, *prime to* a, *from an interval* $Q' < q \le Q$. *Then*

$$\sum_{q\in Q} |\pi(x;q,a) - \frac{\pi(x)}{\phi(q)}|$$
$$\le \{K(\theta-1/2)^2 x\mathcal{L}^{-1} + O(x\mathcal{L}^{-3}(\log\mathcal{L})^2)\} \sum_{q\in\mathcal{Q}} \frac{1}{\phi(q)} + O(x\mathcal{L}^{-A})$$

where $\theta = \log Q/\log x, \mathcal{L} = \log x, K$ *is absolute, the first implied constant depends only on* A, *and the second one on* a *and* A.

Because $\mathcal{Q}$ consists of *all* integers in an interval, it is not possible to deduce a result of our form, where only $Q(\log x)^{-\Theta}$ exceptions are allowed. Our result does imply their theorem with $(\theta-1/2)^2$ weakened to $\theta - 1/2+\epsilon$ (the ϵ could probably be removed with more work) but with Q now allowed to be *any* set of integers coprime to a from an interval $Q' < q \le Q$. The reason $(\theta - 1/2)^2$ appears in [5], while we require $\theta - 1/2$ (coming from $C_1(\theta)$) is that we are unable to appeal to Theorem 2 of [4] in our context. It is the appeal to that result which forces $\mathcal{Q}$ to consist of all integers coprime to a in an interval. This condition may be relaxed so that $\mathcal{Q}$ consists of all primes in an interval, for example, as shown by Theorem 2 of [9], but at the expense of making the constant K larger.

The lower bound in (1.1) has an interesting application for Linnik's constant on average, which shows that, in a certain sense, one can go beyond the Generalized Riemann Hypothesis on average. See [18] for a detailed discussion of Linnik's constant.

Corollary. *Let $p(a,q)$ be the smallest prime congruent to $a(mod\ q)$ and put*

$$C(a,q) = \frac{\log p(a,q)}{\log q}.$$

Then, for most $q > q_o(a)$ we have $C(a,q) < (0.52)^{-1}$.

The point of interest here is that we have gone below 2. It would be very desirable to replace $q > q_0(a)$ by $q > k$ for some absolute constant k.

The bound (1.2) when combined with Fouvry's result [11] that

$$\frac{1}{x} \sum_{x^{3/5} \le p < x^{\lambda}} \pi(x;p,a) \log p < 8 \log\left(\frac{12}{15-5\lambda}\right) + \epsilon \tag{1.3}$$

for $\lambda < 5/7$, leads to the following result.

Theorem 3. *For infinitely many primes p, we have*

$$P^{+}(p+a) > p^{0.676}, \tag{1.4}$$

where $P^{+}(n)$ denotes the greatest prime factor of n.

Goldfeld [14] was the first to observe that one could obtain $P^{+}(p+a) > p^{\delta}$ with $\delta > 1/2$. This result was improved by several authors whose results we tabulate as follows.

Motohashi [25]	$\delta = 0.6105\ldots$
Hooley [19]	$\delta = 0.6199\ldots$
Hooley [20]	$\delta = 0.625 - \epsilon$ for any $\epsilon > 0$
Iwaniec [22]	$\delta = 0.6381$
Deshouillers and Iwaniec [8]	$\delta = 0.6563$
Fouvry [10]	$\delta = 0.6578$
Fouvry [11]	$\delta = 0.6687$

The last entry in the table is particularly interesting because a slight variant of the argument, given by Fouvry [11], yields the following consequence (Adleman and Heath-Brown [1]):

The first case of Fermat's theorem is true for an infinity of prime exponents.

This is now only of historical interest since A. Wiles and R. Taylor (work in preparation) have proved Fermat's theorem.

Unfortunately, one of Fouvry's Lemmas (Lemma 5 of [10]), which is quoted from [3], is stated incorrectly; there should be an extra condition

$$N^4 L^3 \leq X^{2-\epsilon};$$

see Lemma 2, below. When this error is corrected we obtain $\delta = 0.6683$, which still suffices (as $\delta > 2/3$) to prove the Adleman-Fouvry-Heath-Brown theorem. The corrections were provided by Fouvry in a letter to the first named author, and we reproduce them here in Section 4 with Fouvry's permission.

The method and new ideas introduced into the bounding of $\pi(x; q, a)$ are explained in Section 5. Sections 2 and 3 contain material which is necessary for Fouvry's modified argument (Section 4) or the method presented from Section 5 onwards. Here we present the deduction of Theorem 2 from Theorem 1. The argument goes back to Chebyshev; see Hooley [21], Chapter 2. To prove our theorem it suffices to show that

$$\sum_{x^{0.676} \leq p \leq x+a} \pi(x; p, -a) \log p > \epsilon x \tag{1.5}$$

for some $\epsilon > 0$ and $x \geq x_0(a, \epsilon)$. Now, writing $\Lambda(n)$ for the Von Mangoldt function,

$$\sum_{p\le x}\log(p+a)=\sum_{p\le x}\sum_{t|p+a}\Lambda(t)$$
$$=\sum_{p_1\le x}\sum_{p_2|p_1+a}\log p_2+O(x).$$

Assuming x is large, it follows that

$$\sum_{p\le x+a}(\log p)\pi(x;p,-a)=\sum_{p_1\le x}\sum_{p_2|p_1+a}\log p_2>(1-\epsilon)x.$$

It is an easy consequence of the Bombieri-Vinogradov theorem [6] and the Brun-Titchmarsh theorem in simplest form [24] that

$$\sum_{p\le x^{1/2}}(\log p)\pi(x;p,-a)<\left(\frac{1}{2}+\epsilon\right)x$$

for $x\ge x_0(a,\epsilon)$. It remains to show that

$$\sum_{x^{1/2}<p\le x^{0.676}}\pi(x;p,-a)(\log p)<\left(\frac{1}{2}-3\epsilon\right)x$$

in order to establish (1.5). Combining Theorem 1 with (1.3) gives

$$\sum_{x^{1/2}<p\le x^{0.676}}\pi(x;p,-a)(\log p)<\left(0.241+8\log\left(\frac{12}{11.62}\right)\right)x$$
$$<0.4985x$$

as required. This completes the deduction of Theorem 2.

The following list of definitions should help the reader to navigate the remainder of the paper.

2. The sieve machinery

In this section we assemble the results we need from sieve theory. The linear sieve with flexible error term (Lemma 1) is needed for Fouvry's modified argument. The fundamental Lemma (Lemma 2) is needed for Section 5 onwards. A novelty of our approach, shared with the method of [5], is that we will use the one, two and three dimensional versions of this result. In the rest of this paper we suppose $x \geq x_0(a, \epsilon)$ and $x^{1/2} \leq q \leq x^{3/5}$. Let $\mathcal{A}^q$ denote the arithmetic progression $\{qk + a : (x - a)/q < k \leq (2x - a)/q\}$. We shall have occasion to consider other finite sequences $\mathcal{E}$ besides $\mathcal{A}^q$. We write $|\mathcal{E}|$ for the number of terms in the sequence $\mathcal{E}$,

$$\mathcal{E}_d = \{m : dm \in \mathcal{E}\}.$$

We set

$$r(\mathcal{A}^q, d) = |\mathcal{A}^q_d| - \frac{x}{qd} \tag{2.1}$$

for square free d coprime to q.

Henceforth, p with or without subscript will be tacitly assumed coprime to q and we write

$$P(z) = \prod_{p<z} p,\ V(z) = \prod_{p<z}\left(1 - \frac{1}{p}\right)$$

$$S(\mathcal{E}_d, z) = \prod_{\substack{n\in\mathcal{E}_d \\ (n,P(z))=1}} 1.$$

Lemma 1. *Let $2 \le z \le D \le z^3$ be given, together with $\eta > 0$. Then we have*

$$S(\mathcal{A}^q, z) \le \frac{xV(z)}{q}\left(\frac{2e^\gamma}{s} + E\right) + R_q^*. \tag{2.2}$$

Here γ is Euler's constant, $s = (\log D)/(\log z)$, $E = O\left(\eta + \eta^{-8}(\log D)^{-1/3}\right)$ and R_q^* is a remainder term of the form

$$R_q^* = \sum_{(D)} \sum_{\substack{\nu<D^\eta \\ \nu|P(z)}} c_{(D)}(\nu, \eta) \sum_{D_i \le p_i < D_i^{1+\eta^9}} r(\mathcal{A}^q, \nu p_1 \dots p_t) \tag{2.3}$$

where (D) runs through the sequences (including the empty sequence) $D_1 \ge D_2 \ge \cdots \ge D_\ell$ with

$$D_i = D^{\eta^2(1+\eta^9)^n} \text{ for some } n \ge 0,$$

$$D_1 \dots D_{2j} D_{2j+1}^3 < D \qquad (0 \le j \le (\ell - 1)/2),$$

and

$$|c_{(D)}(\nu, \eta)| \le 1.$$

Proof. See [22], Theorem 4.

Lemma 2 (The Fundamental Lemma). *Let $z \ge 2$, $s \ge 2$ and $y = z^s$. There exist two sequences $\{\lambda_d^\pm\}_{d|P(z)}$ such that*

(i) $\lambda_1^\pm = 1,\qquad |\lambda_d^\pm| \le 1$ *and* $\lambda_d^\pm = 0$ *for* $d \ge y$;

(ii) *for all* $k|P(z)$,

$$\sum_{d|k} \lambda_d^+ \geq 0, \qquad \sum_{d|k} \lambda_d^- \leq 0; \tag{2.4}$$

(iii) *for all multiplicative functions* $f(d)$ *satisfying the condition*

$$\prod_{w<p<z} \left(1 - \frac{1}{f(p)}\right) \geq \frac{1}{K}\left(\frac{\log w}{\log z}\right)^{\kappa}, \tag{2.5}$$

for some $K \geq 1$, *some* $\kappa > 0$, *and all* z, w *with* $z > w > 1$, *we have*

$$\sum_{d|P(z)} \frac{\lambda_d^{\pm}}{f(d)} = \prod_{p|P(z)} \left(1 - \frac{1}{f(d)}\right)\left(1 + O(s^{-s})\right) \tag{2.6}$$

where the implied constant depends only on K *and* κ.

Proof. This is essentially Lemma 5 of [13], with the error e^{-s} improved, as intimated there, to s^{-s}.

Remarks. The value κ in Lemma 2 is usually referred to as the *dimension* of the sieve. We shall take κ as 1, 2 or 3 later. The $\lambda_d^{\pm}$ are actually $\mu(d)\chi^{\pm}(d)$ where $\chi^{\pm}$ are characteristic functions of certain sets and $\mu(\,)$ denotes the Möbius function.

3. Remainders and asymptotic formulae

In order to employ Lemma 1 we need to estimate (2.3), at least on average over q, obtaining a bound $\ll x\mathcal{L}^{-\Theta}q^{-1}$ for any $\Theta > 0$. Here we have written $\mathcal{L}$ as an abbreviation for $\log x$. In effect we need asymptotic formulae for sums of the form

$$\sum_{q\sim x^{\theta}} \gamma_q \sum_{n\nu p_1\ldots p_t \equiv a (\mathrm{mod}\ q)} c^+(\nu, \eta) \tag{3.1}$$

where $p_1, \ldots, p_j$ are restricted as indicated in Lemma 1 and $n \sim N$ means $N \leq n < 2N$. We shall require similar estimates to utilize Lemma 2. We will make use of a variety of basic asymptotic formulae, eventually building up further formulae. These formulae constitute the number-theoretic input into the proof and for convenience we classify them as follows (using an elaboration of the usual type I/type II division):

$$\text{Type } I_j: \qquad \sum_q \gamma_q \sum_m a_m \sum_{mn_1\ldots n_j \equiv a (\mathrm{mod}\ q)} 1;$$

$$\text{Type } II_j: \qquad \sum_q \gamma_q \sum_{m_1 \ldots m_j \equiv a (\bmod q)} a(m_1, 1) \ldots a(m_j, j);$$

$$\text{Type } I/II: \qquad \sum_q \gamma_q \sum_{m,n} a_m b_n \sum_{\ell mn \equiv a (\bmod q)} 1.$$

We shall discuss Type $II_2, II_3, I/II$ sums here. We will require estimates for type I_2, I_3 sums in sections 7 and 8 respectively. In this section we appeal to the work of Bombieri, Friedlander and Iwaniec [3,4]; only the first of these was available to Fouvry when he wrote [11], and this is one factor which leads to our improved result. The further paper [5] is not of much direct help to us. All these depend in an essential way on the work of Deshouillers and Iwaniec [7] on Kloosterman sums, and so prevent our results from being uniform in a. In contrast, our bounds for type I_2 and I_3 sums are uniform in this parameter.

For sequences $\alpha_m (m \sim M)$ and $\beta_n (n \sim N)$ the convolution sequence $\gamma(k) = \alpha * \beta(k)$ is defined by

$$\gamma(k) = \sum_{mn=k} \alpha_m \beta_n \qquad (MN \le k < 4MN).$$

For an arithmetic function f we define

$$\Delta(f, q, a) = \sum_{\substack{n \equiv a(q) \\ n \sim x}} f(n) - \frac{1}{\phi(q)} \sum_{\substack{(n,q)=1 \\ n \sim x}} f(n). \tag{3.2}$$

Here $n \equiv a(q)$ is an abbreviation for $n \equiv a(\bmod q)$. In the remainder of the paper we suppose that ϵ is sufficiently small and write $\eta = \exp\left(\frac{-2}{\epsilon}\right)$. Let K, L, M be positive numbers, $KLM = X$,

$$\min(K, L, M) > X^\eta.$$

For sequences $\nu = (\nu_k), k \sim K, \lambda = (\lambda_\ell), \ell \sim L$ and $\sigma = (\sigma_m), (m \sim M)$, we write

$$\Delta(K, L, M; Q) = \sum_{\substack{q \sim Q \\ (a,q)=1}} |\Delta(\nu * \lambda * \sigma; q, a)|.$$

Similarly, if $\min(K, L) > x^\eta$ and $KL = X$, we write

$$\Delta(K, L, Q) = \sum_{\substack{q \sim Q \\ (a,q)=1}} |\Delta(\nu * \lambda; q, a)|.$$

The above notation has been taken from [4] with the modification of adding the condition $n \sim x$ in (3.2). This does not affect the validity of the results we quote from [4] in view of the argument presented on pages 388-389 of that paper.

We need to assume that one convolution factor is well distributed in arithmetic progressions in the following sense:

(A_1) *For any* $d \geq 1$, $k \geq 1$, $b \neq 0$, $(k,b)=1$ *we have*

$$\sum_{\substack{\ell \equiv b(k) \\ (n,d)=1}} \lambda_\ell = \frac{1}{\phi(k)} \sum_{(\ell,dk)=1} \lambda_\ell + O_\Theta\left(\|\lambda\| \tau^\Gamma(d) L^{1/2} (\log L)^{-\Theta}\right) \tag{3.3}$$

for any $\Theta > 1$, *where* $\|\lambda\| = \left(\sum_\ell |\lambda_\ell|^2\right)^{1/2}$ *and* $\tau(d)$ *denotes the number of divisors of* d.

In practice, we will have, at worst,

$$\lambda_\ell = \sum_{\substack{p_1 p_2 \dots p_t = \ell \\ p_j \sim D_j}} 1,$$

for some $D_j > x^\eta$, $t < \eta^{-1}$. The condition (A_1) will then follow from the Siegel-Walfisz theorem applied to any of $p_1, \dots, p_t$.

In (3.3) and below we will write the letter Γ, with or without subscript, to denote a constant which is $O_\epsilon(1)$, not necessarily the same at each occurrence. Similarly, Θ is not necessarily the same at each occurrence, but the conditions or results will work for any given Θ with suitable modification to the implied constants.

The sequences that we need to consider satisfy the condition

$$|\lambda_\ell| \leq \Gamma_1 \tau(\ell)^\Gamma. \tag{A_2}$$

Some sequences are supported on 'almost primes' in the sense that

(A_3) $\lambda_\ell = 0$ *whenever* ℓ *has a prime factor less than* $\exp\left(\frac{\log x}{(\log\log x)^2}\right)$.

In the next Lemma we need the hypothesis

$$L^{1-\epsilon} \sum_\ell |\lambda_\ell|^4 \ll \left(\sum_\ell |\lambda_\ell|^2\right)^2. \tag{A_4}$$

Here and below implied constants are absolute. This may seem surprising, but we accomplish this by supposing $x \geq x_0(a, \epsilon, \Theta)$ and, for instance, $L \geq x^\eta$.

Lemma 3. *Let $MN = x$, $\min(M,N) > x^\eta$, and $\nu = (\nu_m)$, $m \sim M$, $\lambda = (\lambda_n), n \sim N$. Suppose λ satisfies $(A_1 - A_4)$ and ν satisfies (A_2). Then, for any $\Theta > 0$,*

$$\Delta(M,N,Q) \ll x\mathcal{L}^{-\Theta}$$

provided that

$$x^{\epsilon-1}Q^2 < N < X^{5/6-\epsilon}Q^{-4/3}. \tag{3.4}$$

Proof. This follows from Theorem 3 of [3].

Lemma 4. *Let $\Theta > 0$, $KLM = x$, $\min(K,L,M) > x^\eta$, $\nu = (\nu_k)$, $k \sim K$, $\lambda = (\lambda_\ell), \ell \sim L, \sigma = (\sigma_m)$, $m \sim M$. Suppose that λ, σ, ν satisfy $(A_2), (A_3)$, and λ satisfies (A_1) as well. Suppose further that*

$$Q\mathcal{L}^{\Theta_0} < KL, \tag{3.5}$$

$$K^2L^3 < Qx^{1-\eta}, \tag{3.6}$$

$$K^4L^2(K+L) < x^{2-\eta}. \tag{3.7}$$

Here Θ_0 depends on Θ. Then

$$\Delta(K,L,M,Q) \ll x\mathcal{L}^{-\Theta}. \tag{3.8}$$

Proof. This follows from [4], Theorem 3.

Lemma 5. *Let $KLM = x$, $\min(K,L,M) > x^\eta$, $\nu = (\nu_k)$, $k \sim K$, $\lambda = (\lambda_\ell)$, $\ell \sim L$, $\sigma = (\sigma_m)$, $m \sim M$. Suppose that ν, λ, σ satisfy $(A_2), (A_3)$ and λ satisfies (A_1) as well. Let $\Theta > 0$. Then we have (3.8) provided that (3.5) holds and*

$$KL^2Q^2 < x^{2-3\eta}, \tag{3.9}$$

$$K^5L^2 < x^{2-3\eta}. \tag{3.10}$$

Proof. This is a slight variant of Theorem 4 of [4]. We have relaxed the condition $K^2 < Q$ stated there to (3.10). It is remarked in [4] that the condition on K could be relaxed, but the improvement is not stated there.

From [4], (2.2) we may suppose that

$$Q > x^{1/2}\mathcal{L}^{-\Theta}. \tag{3.11}$$

In the notation of [4] we must show that

$$\mathcal{B} = \mathcal{B}(2N, 2L, |a|R, H, 2K) \ll K^3L^3\mathcal{L}^{-\Theta},$$

where $N = KL$, $R = KLQ^{-1}$ and $H = M^{-1}Q\mathcal{L}^2$. Just as in the proof of [4], Theorem 4, we have

$$\mathcal{B} \ll \frac{KLQ}{M}\left(L + \frac{KL}{Q}\right)\left(\frac{Q}{M} + \frac{KL}{Q}\right)K\mathcal{L}^{\Gamma}$$
$$+ \frac{QK^2}{M}x^{\eta}\left[\left(\frac{KL}{Q}\right)^2 K^3L^2 + \left(\frac{KL}{Q}\right)^3\frac{QK}{M} + \left(\frac{KL}{Q}\right)^2(KL)^2 + \frac{KL^2}{Q}(KL)^2\right]^{1/2}$$
$$+ \left(L + \frac{KL}{Q}\right)^{1/2} K^2L\left(\frac{KL}{M}\right)^{3/2} x^{\eta}.$$

Since $K < Q$ and $QM^{-1} > (KLQ^{-1})\mathcal{L}^{-2\Theta}$ from (3.10) and (3.11) the upper bound for $\mathcal{B}$ can be simplified to

$$\frac{K^4L^4Q^2\mathcal{L}^{\Gamma}}{x^2} + x^{\eta}\left(\frac{K^{11/2}L^3}{x} + \frac{K^{9/2}L^3Q^{1/2}}{x} + \frac{K^5L^{9/2}}{x^{3/2}}\right),$$

where we have also used $KLM = x$. Hence

$$\frac{|\mathcal{B}|}{K^3L^2\mathcal{L}^{-\Theta}} \ll KL^2Q^2\mathcal{L}^{\Gamma} + x^{\eta}\left(\left(\frac{K^5L^2}{x^2}\right)^{1/2} + \left(\frac{KL^2Q^2}{x^2}\right)^{1/4}\left(\frac{K^5L^2}{x^2}\right)^{1/4}\right.$$
$$\left. + \frac{(KL^2)^{17/16}(K^5L^2)^{3/16}}{x^{3/2}}\right)$$
$$\ll 1,$$

using (3.9–11). This completes the proof of Lemma 5.

Lemma 6. *Suppose that $KLM = x$, that (ν_k) and (σ_m) satisfy (A_2) and*

$$\lambda_\ell = \begin{cases} 1 & (L \le \ell < L_1) \\ 0 & \text{otherwise} \end{cases} \tag{3.12}$$

where $L_1 \in [L, 2L)$. Then (3.8) holds provided that

$$Q\mathcal{L}^{\Theta_0} < KL, \tag{3.13}$$

(3.14) $$MK^4Q < x^{2-\eta},$$

(3.15) $$MK^2Q^2 < x^{2-\eta}.$$

Here Θ_0 depends on Θ.

Proof. This is Theorem 5 of [4].

We end this section by using Lemmas 3, 4 and 5 to obtain some asymptotic formula that will play a crucial role later. First we require some notation. We must obtain formulae of the form

$$\sum_{p_j} S(\mathcal{A}^q_{p_1\dots p_k}, p_k) = \frac{(1+O(\mu))}{\phi(q)} \sum_{p_j} S(\mathcal{B}^q_{p_1\dots p_k}, p_k)$$

where μ is a suitable small error, $\mathcal{B}^q = \{n \sim x : (n,q) = 1\}$, and the p_j are restricted in some way. We write

$$T = \{s \geq 0, t \geq 0, s+t \leq 1\},$$

and

$$A_j = A_j(\beta_0, \beta_1) = \{(\alpha_1, \dots, \alpha_j); \beta_0 \leq \alpha_j < \dots < \alpha_1 < \beta_1,$$

$$\alpha_1 + \dots + \alpha_j \leq 1\}.$$

We write $\boldsymbol{\alpha}_j$ for a point in $\mathbb{R}^j$, usually in A_j. We write

$$\sum_{\boldsymbol{\alpha}_j \in \mathcal{E}} S(\mathcal{A}^q_{p_1\dots p_j}, z)$$

to mean that we sum over $p_1, \dots, p_j$ with

$$\mathcal{L}^{-1}(\log p_1, \dots, \log p_j) \in \mathcal{E}$$

where $\mathcal{E} \subseteq A_j$. Given any set $\mathcal{H} \subset \mathbb{R}^j$ with $\boldsymbol{\alpha}_j \in \mathcal{H} \Rightarrow 0 \leq \alpha_i$, $\alpha_1 + \dots + \alpha_j < 1$, we say that $\mathcal{H}$ *partitions* into $\mathcal{D} \subseteq T$ if for every $\boldsymbol{\alpha}_j \in \mathcal{H}$, there exist $\mathcal{I}, \mathcal{J}$ with $\mathcal{I} \cup \mathcal{J} \subseteq \{1, \dots, j\}, \mathcal{I} \cap \mathcal{J} = \emptyset$, and

$$\left(\sum_{i \in \mathcal{I}} \alpha_i, \sum_{j \in \mathcal{J}} \alpha_j\right) \in \mathcal{D}.$$

If $\mathcal{I} \cup \mathcal{J} = \{1, \ldots, j\}$ for every $\boldsymbol{\alpha} \in \mathcal{H}$ we say that $\mathcal{H}$ *partitions exactly* into $\mathcal{D}$. We say that a subset of $\mathbb{R}^j$ is *polyhedral* if it is the union of $O(1)$ convex polytopes.

Lemma 7. *Let $\mathcal{G}^1$ be the set of (s,t) in T satisfying*

$$2\theta - 1 + \epsilon \le s \le (5 - 8\theta)/6 - \epsilon. \tag{3.16}$$

Let $\mathcal{G}^2$ be the set of (s,t) in T satisfying

$$s + t \ge \theta + \epsilon \tag{3.17}$$

$$2s + 3t \le 1 + \theta - \epsilon \tag{3.18}$$

$$5s + 2t \le 2 - \epsilon \tag{3.19}$$

$$4s + 3t \le 2 - \epsilon. \tag{3.20}$$

Let $\mathcal{G}^3$ be the set of (s,t) in T satisfying (3.17), (3.19) and

$$s + 2t \le 2 - 2\theta - \epsilon. \tag{3.21}$$

Let $\mathcal{G}$ be the set of (s,t) for which either (s,t) or $(s, 1-s-t)$ belongs to $\mathcal{G}^1 \cup \mathcal{G}^2 \cup \mathcal{G}^3$. Let $G_j = \{\boldsymbol{\alpha}_j \in \mathbb{R}^j : \alpha_j$ partitions into $\mathcal{G}\}$. Let C be a polyhedral subset of $\mathcal{G}_j$ and suppose $\min \alpha_i \ge \eta\,(\boldsymbol{\alpha}_j \in C)$. *Then*

$$\sum_{\boldsymbol{\alpha}_j \in C} S(\mathcal{A}^q_{p_1 \ldots p_j}, p_j) = \frac{(1 + O(\mathcal{L}^{-\Theta}))}{\phi(q)} \sum_{\boldsymbol{\alpha}_j \in C} S(\mathcal{B}^q_{p_1 \ldots p_j}, p_j) \tag{3.22}$$

for most $q \sim x^\theta$.

Proof. By arguing as on pp. 388–389 of [4] we may replace the condition $\boldsymbol{\alpha}_j \in C$ by $p_i \sim P_i\,(i = 1, \ldots, j)$ with $\mathcal{L}^{-1}(\log P_1, \ldots, \log P_j) \in \mathcal{C}$. The left hand side of (3.22) is then

$$\sum_{t=1}^{[1/\eta]} |V_t|$$

where

$$V_t = \{p_1 \ldots p_j p'_1 \ldots p'_t : p_i \sim P_i, p'_1 \ge p'_2 \ge \cdots \ge p'_t \ge p_j,$$

$$p_1 \dots p_j p'_1 \dots p'_t \sim x, p_1 \dots p_j p'_1 \dots p'_t \equiv a(q)\}.$$

Suppose (for example) that $\boldsymbol{\alpha}_j$ partitions into $\mathcal{G}^3$. Then there exist $\mathcal{I}, \mathcal{J}$ with

$$\left(\sum_{i \in \mathcal{I}} \alpha_i, \sum_{i \in \mathcal{J}} \alpha_i\right) \in \mathcal{G}^3.$$

Hence, if we write

$$K = \prod_{i \in \mathcal{I}} P_i, \; L = \prod_{i \in \mathcal{J}} P_i,$$

we have (3.5), (3.9) and (3.10). We write

$$\nu_k = \sum_{\prod_{i \in \mathcal{I}} p_i = k} 1, \; \lambda_\ell = \sum_{\prod_{i \in \mathcal{J}} p_i = \ell} 1,$$

$$\sigma_m = \sum_{p'_i \dots p'_k \prod_{i \notin \mathcal{I} \cup \mathcal{J}} p_i} 1.$$

The hypotheses of Lemma 5 are then satisfied, except for the condition $p'_t \geq p_j$ which can be removed using the argument given on pages 388–389 of [4]. We thus obtain, for most $q \sim x^\theta$,

$$|V_t| = \frac{1}{\phi(q)}(1 + O(\mathcal{L}^{-\mathcal{A}}))|U_t|,$$

where U_t is defined as V_t but with the condition

$$p_1 \dots p_j p'_1 \dots p'_t \equiv a(q)$$

removed. Thus

$$\sum_{\boldsymbol{\alpha}_j} S(\mathcal{A}^q_{p_1 \dots p_j}, p_j) = \sum_{t=1}^{[1/\eta]} |V_t| = \frac{(1 + O(\mathcal{L}^{-\Theta}))}{\phi(q)} \sum_{t=1}^{1/\eta} |U_t|$$

$$= \frac{(1 + O(\mathcal{L}^{-\Theta}))}{\phi(q)} \sum_{\boldsymbol{\alpha}_j} S(\mathcal{B}^q_{p_1 \dots p_j}, p_j)$$

as desired (since $p_i \geq x^\eta$ for each i the condition $(n, q) = 1$ in $\mathcal{B}$ has negligible effect). This completes the proof in this case; the proof for other cases follows similarly.

Remarks. For $\frac{1}{2} \le \theta \le 0.56$ it is convenient to define the following subsets of $\mathbb{R}^2$.

$$\begin{aligned}
A = &\ \{(\alpha_1, \alpha_2) : \kappa \le \alpha_2 < \alpha_1, \alpha_1 + \alpha_2 < \theta + \epsilon\},\\
B = &\ \{(\alpha_1, \alpha_2) : \kappa \le \alpha_2 < \alpha_1, 5\alpha_2 + 2\alpha_1 \le 2 - \epsilon, \alpha_1 + \alpha_2 \ge \theta + \epsilon;\\
&\quad (\alpha_2, \alpha_1) \text{ satisfies either } \{(3.18), (3.20)\} \text{ or } (3.21)\},\\
B' = &\ \{(\alpha_1, \alpha_2) : \alpha_2 > \alpha_1, \alpha_1 + \alpha_2 \ge \theta + \epsilon, 5\alpha_2 + 2\alpha_1 \le 2 - \epsilon\},\\
E = &\ \{(\alpha_1, \alpha_2) : \alpha_1 \le 1 - \theta - \epsilon, \alpha_1 + 2\alpha_2 < 1, 3\alpha_2 \le 2\alpha_1 - \epsilon;\\
&\quad (\alpha_2, 1 - \alpha_1 - \alpha_2) \text{ satisfies either } \{(3.18), (3.20)\} \text{ or } (3.21)\},\\
E' = &\ \{(\alpha_1, \alpha_2) : \alpha_1 \le 1 - \theta - \epsilon, \alpha_1 + 2\alpha_2 \ge 1, 3\alpha_2 \le 2\alpha_1 - \epsilon\},\\
C = &\ \{(\alpha_1, \alpha_2) : \kappa \le \alpha_2 < \alpha_1, \alpha_1 + 4\alpha_2 \le 3 - 3\theta - \epsilon, (\alpha_1, \alpha_2) \notin A \cup B \cup E\},\\
D = &\ \{(\alpha_1, \alpha_2) : \alpha_2 < \alpha_1, \alpha_1 + 2\alpha_2 < 1, (\alpha_1, \alpha_2) \notin A \cup B \cup C \cup E\}.
\end{aligned}$$

Some of these regions are mentioned in the following discussion; others will be referred to later.

For $\theta < 29/56 - \epsilon, \mathcal{G}^2$ lies within $\mathcal{G}^3$. The region $\mathcal{G}^1$ is empty for $\theta \ge 11/20$ and $\mathcal{G}^3$ is empty for $\theta \ge 4/7$. The region $\mathcal{G}^2$ is small, but non-empty, for $\theta \in [4/7, 3/5)$. Diagrams 1, 2 and 3 in the Appendix illustrate the cases $\theta = 0.5, 0.52, 0.54$ respectively. The strip $\mathcal{G}^1$ is only depicted in Diagram 3. The regions B, B', E, E' all belong to $\mathcal{G}$.

For $\theta = 0.5, B \cup B' = \mathcal{G}^3 \cap \{(s,t) : t \ge 1/6\}$. For $\theta = 0.52$, most of $B \cup B'$ is made up of $\mathcal{G}^3 \cap \{(s,t) : t \ge 0.14\}$ with a tiny part coming from $\mathcal{G}^2 \backslash \mathcal{G}^3$. On Diagram 3 ($\theta = 0.54$) it is clear that $B \cup B'$ and $E \cup E'$ are both made up of the two overlapping shapes of $\mathcal{G}^2$ and $\mathcal{G}^3$.

4. Fouvry's modified argument

The purpose of this section is to give Fouvry's corrections to [11]. We shall require Lemma 10 in our proof since we can do no better than Fouvry for $\theta \ge 0.56$. We will give a much stronger result than Lemma 8 in section 6.

Lemma 8. *Let $(\sigma_m)(m \sim M)$ satisfy (A_2) and suppose that $MN = X$,*

$$\lambda(n) = \begin{cases} \Lambda(n) & (N \le n < N_1) \\ 0 & \textit{otherwise}, \end{cases} \tag{4.1}$$

where $x^{1/2} \le Q < x^{4/7-\epsilon}, N_1 \le 2N$, and

$$Qx^{\epsilon} \le N \le x^{-\epsilon} \min\left((xQ)^{3/8}, x^{4/7}\right). \tag{4.2}$$

Then for every $\Theta > 0$,

$$\Delta(M, N; Q) \ll x\mathcal{L}^{-\Theta}.$$

Proof. This is Lemma 9 of [11]. As indicated earlier, the only problems arise from the omission of the term K^4L^3 from (3.7) here. Thus, on page 393 of [11], one requires

$$L_1^4 L_1'^3 < x^{2-\epsilon}, \tag{4.3}$$

in addition to

$$L_1^5 L_1'^2 < x^{2-\epsilon}.$$

Following the argument of [11], we know that

$$L_1 L_1' \le N,\ L_1 \le 2N^{1/2},$$

hence

$$L_1^4 L_1'^3 \ll N^{7/2} \ll x^{2-2\epsilon}$$

from (4.2). This completes the proof of Lemma 8.

Lemma 9. *Let $D_1 \ge \cdots \ge D_\ell \ge 1$ and suppose that*

$$D_1 \dots D_{t-1} D_t^2 < D \quad (1 \le t \le \ell).$$

Then whenever $D = D'D''$ with $D', D'' \ge 1$ we may partition $D_1 \dots D_\ell$ into two products Δ' and Δ'' with $\Delta' \le D'$ and $\Delta'' \le D''$.

If, further, no subproduct of $D_1, \dots, D_\ell$ lies in $[U, V]$, where $1 \le U \le V \le D$, we have $D_1 \dots D_\ell \le DUV^{-1}$.

Proof. The first assertion is established on p. 312 of [22]. For the second assertion, partition $D_1 \dots D_\ell$ into products $\Delta' \le V$ and $\Delta'' \le DV^{-1}$. By hypothesis, $\Delta' < U$ and the result follows.

Lemma 10. *Let $Q = x^\theta$ with $10/19 \le \theta \le 3/5 - \epsilon$. Let*

$$\chi(\theta) = \frac{12 - 13\theta}{7}, \tag{4.4}$$

$$D = x^{\chi(\theta) - 100\eta}. \tag{4.5}$$

Let γ_q be complex numbers whose modulus does not exceed 1 for $q \sim Q$. Then, in the notation of (2.3) we have, for any $\Theta > 0$,

$$\sum_{q \sim Q} \gamma_q R_q^* \ll x\mathcal{L}^{-\Theta}.$$

Remark. It is possible to increase $\chi(\theta)$, at least for $\theta < 13/24$, but we shall not need this improvement.

Proof. For given $D_1, \ldots, D_\ell$ satisfying $D_1 \geq \cdots \geq D_\ell$ and the conditions given in Lemma 1, it suffices to show that

$$\sum_q \gamma_q \left\{ \sum_{\substack{\nu p_1 \ldots p_t n \equiv a(q) \\ \nu | P(z)}} c_\nu - \frac{1}{\phi(q)} \sum_{\substack{(n,q)=1 \\ \nu|P(z) \\ p_1,\ldots,p_t}} c_\nu \right\} \ll x\mathcal{L}^{-\Theta}. \tag{4.6}$$

Here we leave implicit the conditions

$$|c_\nu| \leq 1, D_i \leq p_i < D_i^{1+\eta^9}, p_i | P(z), \text{ and } \nu < x^\eta; \tag{4.7}$$

$$\nu p_1 \ldots p_t n \sim x. \tag{4.8}$$

Here we have used, for any interval $\mathcal{I}$ of length I,

$$\frac{1}{\phi(q)} \sum_{\substack{(n,q)=1 \\ n \in \mathcal{I}}} 1 = \frac{I}{q} + O(q^\eta),$$

to obtain (4.6).

Now let $D_1 \ldots D_t = x^u$. By Lemma 6, we may partition $D_1 \ldots D_t$ into two products P' and P'' satisfying

$$P' \leq x^{1-\theta-3\eta}, P'' \leq x^{\theta+\chi(\theta)-1-97\eta}, P'P'' = x^u.$$

We consider two cases.

(a) $u \leq \dfrac{9-8\theta}{7} = \psi(\theta)$ say.

Since $\theta \geq 10/19$,

$$u + 4\theta + 3\chi(\theta) \leq 5 \text{ and } u + 3\theta + \chi(\theta) \leq 3.$$

Consequently,

$$\begin{aligned} P'(P'')^4 Q &= P'P''(P'')^3 Q \\ &\leq x^{u+3\theta+3\chi(\theta)-3-200\eta+\theta} \leq x^{2-200\eta} \end{aligned} \tag{4.9}$$

and

$$P'(P'')^2 Q^2 \leq x^{u+\theta+\chi(\theta)-1-97\eta+2\theta} \leq x^{2-97\eta}. \tag{4.10}$$

By allowing ℓ to run over products of ν and the p_i corresponding to P', and n to run over products of the p_i corresponding to P'', we bound the modulus of the sum in (4.6) by

$$\sum_{M,K,L} \Delta(M,K,L;Q) \tag{4.11}$$

where M, K, L run over $O(\mathcal{L}^2)$ powers of 2, satisfying

$$P' \leq M \leq (P')^{1+\eta^9} x^\eta, P'' \leq K \leq (P'')^{1+\eta^9}. \tag{4.12}$$

Moreover (A_2) holds for the coefficients corresponding to M and K, and (3.12) holds for the coefficients corresponding to L. We observe that

$$KL \gg x^{1-2\eta}(P')^{-1} \gg x^{\theta+\eta}. \tag{4.13}$$

From (4.9), (4.10), (4.12) and (4.13), the hypotheses of Lemma 6 are satisfied, giving

$$\Delta(M,K,L;Q) \ll x\mathcal{L}^{-\Theta}.$$

(b) $u \geq \psi(\theta)$. Let

$$U = x^{-1+u+\theta+4\eta}, V = x^\lambda$$

where

$$\lambda = \min\left(-1 + \frac{3u}{2} + \frac{\theta}{2} - 2\eta, \frac{2u}{5} - \frac{4\eta}{5}, -\frac{1}{4} + \frac{3u}{4} - \eta\right).$$

It is easy to see that

$$1 \leq U \leq V \leq D.$$

We now show that

$$VU^{-1} > x^{\chi(\theta)-u-100\eta} = D(D_1 \dots D_t)^{-1}. \tag{4.14}$$

It suffices to verify the inequalities

$$\begin{aligned} 3u &\geq 2\chi(\theta) + \theta \\ 2u + 5 &\geq 5\chi(\theta) + 5\theta \\ 3u + 3 &\geq 4\chi(\theta) + 4\theta, \end{aligned}$$

and we may do this since $u \geq \psi(\theta)$.

Combining (4.14) with Lemma 9, we find that there is a subproduct of the D_i in $[U, V]$, say

$$x^f = P' \in [U, V]; P'P'' = x^u.$$

We now aim to bound the sum (4.6) in the form (4.11), where Lemma 4 is applicable to each summand. We let ℓ be the product of the p_i corresponding to P' and set $k = n$. Then all convolution factors satisfy $(A_2), (A_3)$ and λ_ℓ satisfies (A_1). To verify the hypotheses of Lemma 4 we must check the following inequalities:

$$\begin{aligned} f + (1-u) &\geq \theta + 4\eta, \\ 2f + 3(1-u) &\leq 1 + \theta - 4\eta, \\ 5f + 2(1-u) &\leq 2 - 4\eta, \\ 4f + 3(1-u) &\leq 2 - 4\eta. \end{aligned}$$

The reader can check that these are equivalent to $x^f \in [U, V]$. Accordingly, Lemma 4 yields (4.6), and the proof of Lemma 10 is complete.

It is easy to combine Lemma 10 with Lemma 1 to obtain

$$S(\mathcal{A}^q, z) \lesssim \frac{2x}{\phi(q) \log D} \tag{4.15}$$

(here $U \lesssim V$ means $U \leq (1 + O(\epsilon))V$) for most $q \sim x^\theta$, with D as in (4.5) and $z = D^{1/3}$. We 'deduct' those p_1p_2 with

$$p_1 \in \left[x^{\theta+\epsilon}, x^{4/7-\epsilon}\right],$$

for which Lemma 8 gives an asymptotic formula. Just as in [11] we arrive at

$$\pi(x; q, a) \lesssim \frac{xC(\theta)}{\phi(q) \log x} \text{ for most } q \sim x^\theta \tag{4.16}$$

with

$$C(\theta) = \begin{cases} \frac{14}{12-13\theta} - \log\left(\frac{4(1-\theta)}{3\theta}\right) \left(\frac{17}{32} - \epsilon \leq \theta \leq \frac{4}{7} - \epsilon\right) \\ \frac{14}{12-13\theta} \left(\frac{4}{7} - \epsilon < \theta \leq \frac{3}{5} - \epsilon\right). \end{cases}$$

In estimating $\int_{1/2}^{3/5} C(\theta)d\theta$, we must replace

$$\int_{17/32}^{8/15} \frac{22d\theta}{15 - 13\theta} + \int_{8/15}^{3/5} \frac{10}{9 - 10\theta} d\theta \; (\leq 0.00568 + 0.20068)$$

on p. 406 of [11], by

$$\int_{17/32}^{3/5} \frac{14}{12-13\theta} d\theta \leq 0.20778.$$

Leaving unchanged the contribution from $\theta \in [1/2, 17/32]$ and the 'deductible'

$$\int_{17/32}^{4/7} \log\left(\frac{4(1-\theta)}{3\theta}\right) d\theta,$$

we arrive at

$$\int_{1/2}^{3/5} C(\theta) d\theta \leq 0.26887.$$

Since

$$\int_{3/5}^{0.6683} \frac{8}{3-\theta} d\theta < 0.231,$$

the argument given at the end of Section 1 shows that $\delta = 0.6683$ is admissible.

5. Outline of an alternative sieve approach

The method we use in the following sections is an elaboration of the technique used in [15]. Essentially the same basic method was discovered independently by Fouvry [11] and used for $\theta \leq 17/32$ in the current problem. The idea is to use the Buchstab identity:

$$S(\mathcal{A}, z_1) = S(\mathcal{A}, z_2) - \sum_{z_2 \leq p < z_1} S(\mathcal{A}_p, p), \tag{5.1}$$

(we henceforth omit the superscript q for clarity) in conjunction with the fundamental Lemma and the basic asymptotic formulae to obtain further formulae for sums

$$S(\mathcal{A}, z), \sum_p S(\mathcal{A}_p, z), \sum S(\mathcal{A}_{p_1 p_2}, z), \ldots,$$

for a certain z and $p, p_1, p_2, \ldots$ in suitable ranges. Using (5.1) again, and the fact that $S(\mathcal{A}_{p_1}, p_1), S(\mathcal{A}_{p_1p_2p_3}, p_3) \ldots$ are non-negative we obtain

$$\begin{aligned} S(\mathcal{A}, (2x)^{1/2}) \le S(\mathcal{A}, z) &- \sum_{\alpha_1 \in R_1} S(\mathcal{A}_{p_1}, z) + \sum_{\alpha_2 \in R_2} S(\mathcal{A}_{p_1p_2}, z) \\ &- \sum_{\alpha_3 \in R_3} S(\mathcal{A}_{p_1p_2p_3}, p_3) \\ &= S_0 - S_1 + S_2 - S_3 \text{ say,} \end{aligned}$$

where $z = x^\kappa$,

$$R_1 \text{ is a subset of } \left[\kappa, \frac{1}{2}\right] = U_1$$

$$R_2 = \left\{(\alpha_1, \alpha_2) \in T : \alpha_1 \in R_1, \kappa \le \alpha_2 < \min\left(\alpha_1, \frac{1-\alpha_1}{2}\right)\right\}$$

R_3 is a subset of

$$\left\{(\alpha_1, \alpha_2, \alpha_3) : (\alpha_1, \alpha_2) \in R_2, \kappa \le \alpha_3 < \min\left(\alpha_2, \frac{1-\alpha_1-\alpha_2}{2}\right)\right\} = U_3$$

and we can give asymptotic formulae for $S_j, 0 \le j \le 3$, of the form

$$S(\mathcal{A}, z) = \frac{1 + O(\epsilon)}{\phi(q)} S(\mathcal{B}, z), \ldots$$

Since (see [12])

$$S(\mathcal{B}, z) = \frac{(1 + O(\epsilon)) w\left(\frac{\log x}{\log z}\right) x}{\log z},$$

$$\sum_{\gamma \le \alpha_1 \le \delta} S(\mathcal{B}_{p_1}, p_1) = \frac{(1 + O(\epsilon))x}{\log x} \int_\gamma^\delta \frac{w\left(\frac{1-\alpha_1}{\alpha_1}\right)}{\alpha_1^2} d\alpha_1, \text{ etc. },$$

a corresponding application of Buchstab's identity to $S(\mathcal{B}, (2x)^{1/2})$ leads to the inequality

$$S(\mathcal{A}, (2x)^{1/2}) \lesssim \frac{x}{\phi(q) \log x} (1 + I_1 + I_3).$$

Here

$$I_1 = \int\limits_{U_1 \backslash R_1} \frac{w\left(\frac{1-\alpha_1}{\alpha_1}\right)}{\alpha_1^2} d\alpha_1,$$

$$I_3 = \int\limits_{U_3 \backslash R_3} \frac{w\left(\frac{1-\alpha_1-\alpha_2-\alpha_3}{\alpha_3}\right)}{\alpha_1\alpha_2\alpha_3^2} d\alpha_1 d\alpha_2 d\alpha_3.$$

We use the notation $w(u)$ to denote Buchstab's function, defined by

$$w(u) = \frac{1}{u}, \qquad 1 \le u \le 2,$$
$$(uw(u))' = w(u-1), \qquad u > 2.$$

To obtain a good upper bound, clearly I_1 and I_3 must be small, that is, the measure of $U_1 \backslash R_1$ and $U_3 \backslash R_3$ must be small. In other words we discard some negative sums we cannot evaluate, but strive to incorporate as many as possible. Of course we must be able to evaluate all of S_2 for this to work, and this adds a further constraint on R_1. Clearly we could continue getting more sums in extra variables by this process.

To obtain a lower bound we have, for example,

$$S(\mathcal{A}, (2x)^{1/2}) \ge S(\mathcal{A}, z) - \sum_{\alpha_1 \in U_1} S(\mathcal{A}_{p_1}, z) + \sum_{\alpha_2 \in V_2} S(\mathcal{A}_{p_1 p_2}, p_2)$$

where V_2 is a subset of $\left\{(\alpha_1, \alpha_2) \in T : \alpha_1 \in U_1, \kappa < \alpha_2 < \min\left(\alpha_1, \frac{1-\alpha_1}{2}\right)\right\}$. Here it is vital that we can evaluate the negative contribution and make V_2 as large as possible.

As we have discussed the method so far, we use only type II_2, II_3 and I/II sums. We shall discover that this is of no use for $1-\theta \le \alpha_1 \le 1/2$, for example. However, we then can use

$$\sum_{1-\theta \le \alpha_1 \le 1/2} S(\mathcal{A}_{p_1}, p) = S(\mathcal{F}, x^{(\theta+\epsilon)/2})$$

where

$$\mathcal{F} = \{mn \in \mathcal{A} : x^{1-\theta} \le M \le x^{1/2}\}.$$

We can treat this new expression using a two dimensional sieve (see section 7 below), where we need type I_2 arithmetical information. Similarly we can consider sums

$$\sum S(\mathcal{A}_{p_1 p_2}, p_2)$$

using a three-dimensional sieve approach if p_1 and p_2 are in the right ranges. This entails the application of type I_3 sums (see section 8).

In practice, for θ near 1/2, after applying (5.1) twice more to (5.2) and obtaining a further integral I_5, we find $I_1 + I_3 + I_5$ is small, and so we get a good upper bound for $S(\mathcal{A}, (2x)^{1/2})$. However, for larger θ we have to discard a large part of U_1 (the intervals $[\tau, 3/7], [1-\theta, 1/2]$, where τ is defined later). We are then effectively bounding $S(\mathcal{A}, x^\tau)$ but 'deducting' the sum

$$\sum_{3/7<\alpha_1<1-\theta} S(\mathcal{A}_{p_1}, p_1).$$

The sum $S(\mathcal{A}, x^\tau)$ will have expected size about twice $S(\mathcal{A}, (2x)^{1/2})$, but our approach still improves on Lemma 10 up to $\theta \approx 0.56$. In fact our approach theoretically improves on Lemma 10 up to $\theta = 0.6 - \epsilon$, that is so long as there are non-trivial bounds for type II_3 sums, but the calculations become very complicated and tedious.

In the following sections we say

$$\sum a_m S(\mathcal{A}_m, z)$$

'can be evaluated with error μ' if

$$\sum a_m S(\mathcal{A}_m, z) = \frac{1}{\phi(q)} \sum a_m S(\mathcal{B}_m, z)(1 + O(\mu)).$$

Also we write

$$d\mu_j = \frac{d\alpha_1 \ldots d\alpha_j}{\alpha_1 \ldots \alpha_j}.$$

It will become apparent that there are possible ways of producing further very small improvements to our results but the authors believe that these improvements would not justify the extra details and complications necessary. As an alternative to our approach the interested reader might like to begin with the Linnik identity as used by Heath-Brown [17] and Bombieri, Friedlander and Iwaniec [5].

6. The alternative sieve in one dimension

We use this approach alone for $\kappa \le \alpha_1 \le \tau$ and $3/7 \le \alpha_1 \le 1 - \theta$, where

$$\kappa = \begin{cases} (5-8\theta)/6 - \epsilon & (\theta \le 17/32 - \epsilon) \\ (5-8\theta)/12 - 3\epsilon & (17/32 - \epsilon < \theta \le 7/13 - \epsilon) \\ (3-5\theta)/7 - 2\epsilon & (7/13 - \epsilon < \theta \le 4/7 - \epsilon), \end{cases}$$

$$\tau = \begin{cases} 3(1-\theta)/5 - \epsilon, & \theta \leq 11/21 \\ 2/7 - \epsilon, & 11/21 < \theta \leq 6/11 - \epsilon \\ (5-6\theta)/7 - \epsilon, & \theta > 6/11 - \epsilon. \end{cases}$$

We also write

$$\kappa' = \begin{cases} (11-20\theta)/6 - 2\epsilon & \text{if } 7/13 - \epsilon < \theta \leq 0.55 - \epsilon \\ \kappa & \text{otherwise}, \end{cases}$$

$$\tau' = \begin{cases} (5-6\theta)/7 & \text{if } 7/13 - \epsilon < \theta \leq 0.55 - \epsilon \\ \tau & \text{otherwise}. \end{cases}$$

The significance of these values will become apparent as we proceed.

We write S for the set of (s,t) in T for which

(6.1) $$s \leq 1 - \theta - \epsilon,\ s + 2t \leq 2 - 2\theta - \epsilon,\ s + 4t \leq 2 - \theta - \epsilon.$$

We let A_j be as defined in Section 3 with $\beta_0 = \eta, \beta_1 = \tau$. Let S_j be the subset of $\mathbb{R}^j$ which partitions exactly into S. We now need some elementary Lemmas giving conditions for points to belong to G_j or S_j.

Lemma 11. *Let $\theta \leq 0.55 - \epsilon$. Let $\boldsymbol{\alpha}_k \in A_k$ and suppose that*

(6.2) $$\alpha_{j+1} + \cdots + \alpha_{k-1} < 2\theta - 1 + \epsilon \leq \alpha_{j+1} + \cdots + \ alpha_k,$$

(6.3) $$\alpha_{j+1} < \kappa',$$

for some $j, 0 \leq j \leq k-1$. Then $\boldsymbol{\alpha}_k \in G_k$.

Remark. Since $\kappa = \kappa'$ for $\theta < 7/13 - \epsilon$, the result holds with (6.3) replaced by

(6.4) $$\alpha_{j+1} < \kappa,\ \theta < 7/13 - \epsilon.$$

Proof. We consider three cases.

a) $\alpha_k < \frac{11-20\theta}{6} - 2\epsilon$ (the only case for $\theta > 7/13 - \epsilon$ in view of (6.3)). From (6.2),

$$2\theta - 1 + \epsilon \leq \alpha_{j+1} + \cdots + \alpha_k < 2\theta - 1 + \epsilon + \frac{11-20\theta}{6} - 2\epsilon = \frac{5-8\theta}{6} - \epsilon.$$

Hence $\boldsymbol{\alpha}_k \in G_k$ (see (3.16)).

b) $\alpha_k \geq \frac{11-20\theta}{6} - 2\epsilon$, $\theta \leq 17/32 - \epsilon$. In this case

$$\frac{11-20\theta}{6} - 2\epsilon \geq 2\theta - 1 + \epsilon.$$

So, in view of (6.3),

$$2\theta - 1 + \epsilon \leq \alpha_k \leq \frac{5-8\theta}{6} - \epsilon.$$

Thus $\boldsymbol{\alpha}_k \in G_k$.

c) $\alpha_k \geq \frac{11-20\theta}{6} - 2\epsilon$, $\frac{17}{32} - \epsilon < \theta \leq \frac{7}{13} - \epsilon$. We now have

$$2\alpha_k \geq \frac{11-20\theta}{3} - 4\epsilon > 2\theta - 1 + \epsilon. \tag{6.5}$$

Since $\alpha_{j+1} < (5-8\theta)/12 - \epsilon < 2\theta - 1 + \epsilon$, it follows from (6.2) that $k \geq j+2$. We then have

$$2\theta - 1 + \epsilon \leq 2\alpha_k \leq \alpha_{j+1} + \alpha_{j+2} < 2\kappa = (5-8\theta)/6 - 2\epsilon.$$

Thus $\boldsymbol{\alpha}_k \in G_k$, as required to complete the proof.

Lemma 12. *Let $\theta \in (7/13-\epsilon, 4/7-\epsilon]$. Let $\boldsymbol{\alpha}_k \in A_k$ where $k \geq 1$, and suppose that*

$$\alpha_1 + \cdots + \alpha_k < \theta + \epsilon. \tag{6.6}$$

Then $\boldsymbol{\alpha}_k \in S_k$.

Proof. If $\alpha_1 + \cdots + \alpha_k \leq 1 - \theta - \epsilon$ this follows immediately with $s = \alpha_1 + \cdots + \alpha_k, t = 0$. We may therefore suppose that

$$\alpha_1 + \cdots + \alpha_k \geq 1 - \theta - \epsilon > 2\theta - 1 + 2\epsilon \tag{6.7}$$

and let i be the least integer with $\alpha_1 + \cdots + \alpha_i > 2\theta - 1 + 2\epsilon$, so that

$$\alpha_1 + \cdots + \alpha_{i-1} < 2\theta - 1 + 2\epsilon \leq \alpha_1 + \cdots + \alpha_i. \tag{6.8}$$

Let $v = \alpha_1 + \cdots + \alpha_i, u = \alpha_1 + \cdots + \alpha_k - v$.

If $\alpha_1 > 2\theta - 1 + \epsilon$, then $v = \alpha_1$. Otherwise, $v \leq 4\theta - 2 + 2\epsilon$. Now

$$u < \theta + \epsilon - v \leq 1 - \theta - \epsilon$$

from (6.6), (6.8). Also

$$u + 2v < \theta + \epsilon + v \leq \max(5\theta - 2 + 3\epsilon, \theta + \epsilon + \alpha_1) \leq 2 - 2\theta - \epsilon$$

since $\alpha_1 \le \tau$; similarly

$$u + 4v < \theta + \epsilon + 3v \le \max(13\theta - 6 + 7\epsilon,\ \theta + \epsilon + 3\alpha_1) \le 2 - \theta - \epsilon.$$

This completes the proof of Lemma 12.

Now let

$$T^* = \{(s,t) = 0 \le s \le (8\theta - 2)/7,\ 0 \le t \le (5 - 6\theta)/7\},$$

$$U_j' = U_j'(\theta) = \{\boldsymbol{\alpha}_j \in A_j : (\alpha_1, \dots, \alpha_j, 2\theta - 1 + \epsilon) \in S_{j+1}\},$$

$$U_j = U_j(\theta) = \begin{cases} U_j'(\theta) & \text{if } \theta \le 7/13 - \epsilon \\ \{\boldsymbol{\alpha}_j \in A_j : \boldsymbol{\alpha}_j \text{ partitions exactly into } T^*\} & \text{if } \theta > 7/13 - \epsilon. \end{cases}$$

Lemma 13. *Let $\boldsymbol{\alpha}_k \in A_k$, suppose that $\theta > 7/13 - \epsilon$,*

$$\alpha_1 + \dots + \alpha_k \ge \theta + \epsilon > \alpha_1 + \dots + \alpha_{k-1}, \tag{6.9}$$

$$\alpha_{j+1} < \kappa \tag{6.10}$$

$$(\alpha_1, \dots, \alpha_j) \in U_j$$

where $0 \le j \le k-1$. Then $\boldsymbol{\alpha}_k \in G_k$.

Proof. From the definition of U_j we can combine $\alpha_1, \dots, \alpha_j$ into two sums s, t with $(s,t) \in T^*$. The condition on s can be rewritten

$$s \le 1 - \theta - 3\kappa - 6\epsilon,$$

and so

$$t + \alpha_{j+1} + \dots + \alpha_k \ge \theta + \epsilon - s \ge 2\theta - 1 + 3\kappa + 4\epsilon. \tag{6.11}$$

We claim that there is an $\ell, j \le \ell \le k$, such that

$$\beta_\ell := t + \sum_{j+1 \le t \le \ell} \alpha_t \in [2\theta - 1 + 3\kappa + 4\epsilon, 2 - 3\theta - 3\kappa - 4\epsilon]. \tag{6.12}$$

To see this, the interval has length exceeding κ (we recall $\kappa = (3 - 5\theta)/7 - 2\epsilon$ for $\theta > 7/13 - \epsilon$), the α_t $(j+1 \le t \le \ell)$ are all bounded above by κ from (6.10), and β_k is at least $2\theta - 1 + 3\kappa + 4\epsilon$ from (6.11). Moreover, since $t \leq (5 - 6\theta)/7$, we have

$$\beta_j = t \leq 2 - 3\theta - 3\kappa - 4\epsilon.$$

If we take ℓ as the first value for which $\beta_\ell \geq 2\theta - 1 + 3\kappa + 4\epsilon$ then (6.12) will be satisfied.

Combining (6.9) and (6.10), we get

$$\alpha_1 + \cdots + \alpha_k \leq \theta + \kappa + \epsilon. \tag{6.13}$$

We claim that in the notation of Lemma 7,

$$(\beta_\ell, \alpha_1 + \cdots + \alpha_k - \beta_\ell) \in \mathcal{G}^2. \tag{6.14}$$

We obtain (3.17) from (6.9). Next we combine (6.12) and (6.13) to derive (3.18) since

$$2\beta_\ell + 3(\alpha_1 + \cdots + \alpha_k - \beta_\ell) \leq 3(\theta + \kappa + \epsilon) - (2\theta - 1 + 3\kappa + 4\epsilon) = 1 + \theta - \epsilon.$$

Finally, to establish (3.19) and (3.20) we note that

$$\begin{aligned} 5\beta_\ell + 2(\alpha_1 + \cdots + \alpha_k - \beta_\ell) &\leq 4\beta_\ell + 3(\alpha_1 + \cdots + \alpha_k - \beta_\ell) \\ &\leq 2 - 3\theta - 3\kappa - 4\epsilon + 3(\theta + \kappa + \epsilon) = 2 - \epsilon. \end{aligned}$$

Hence (6.14) is verified and so $\boldsymbol{\alpha}_k \in G_k$.

We are now in a position to prove some asymptotic formulae using the fundamental Lemma. For $s \geq 1$ we write

$$g(s) = \exp(-s \log s).$$

Lemma 14. *Let E be a polyhedral subset of S_j. We can evaluate*

$$\sum_{\boldsymbol{\alpha}_j \in E} S(\mathcal{A}_{p_1 \ldots p_j}, x^\eta) \tag{6.15}$$

with error $g(\epsilon/\eta)$. We can also evaluate $S(\mathcal{A}, x^\eta)$ with error $g(\epsilon/\eta)$.

Proof. We apply Lemma 2 with $z = x^\eta, y = x^\epsilon, f(d) = d$ to obtain upper and lower bounds for (6.15). We establish an upper bound in the following; the lower bound is proved similarly. We have

$$S(\mathcal{A}_{p_1 \ldots p_j}, x^\eta) \leq \sum_{d \leq y} \lambda_d^+ |\mathcal{A}_{p_1 \ldots p_j d}|,$$

and

$$\sum_{\boldsymbol{\alpha}_j \in E} \sum_{d \le y} \lambda_d^+ |\mathcal{A}_{p_1 \dots p_j d}| = \sum_{d \le y} \lambda_d^+ \sum_{\substack{\boldsymbol{\alpha}_j \in E \\ p_1 \dots p_j dn \in \mathcal{A}}} 1$$

$$= \sum_{\substack{d \le y \\ \boldsymbol{\alpha}_j \in E}} \lambda_d^+ \frac{x}{q d p_1 \dots p_j} + O(xq^{-1}\mathcal{L}^{-\Theta})$$

for most $q \sim x^\theta$, working as in Lemma 7, but using Lemma 6,

$$= \sum_{\boldsymbol{\alpha}_j \in E} \frac{xV(x^\eta)}{q d p_1 \dots p_j}(1 + O(g(\epsilon/\eta)) + O(xq^{-1}\mathcal{L}^{-\Theta})$$

from (2.6),

$$= \frac{1}{\phi(q)} \sum_{\boldsymbol{\alpha}_j \in E} S(\mathcal{B}_{p_1 \dots p_j}, x^\eta)(1 + O(g(\epsilon/\eta)) \tag{6.16}$$

as required since $x > x_0(\epsilon, a, \Theta)$, so the $O(xq^{-1}\mathcal{L}^{-\Theta})$ term can be absorbed into the other error, and the formula for (6.16) is easy to obtain (see [12] for example).

The proof for $S(\mathcal{A}, x^\eta)$ is similar but simpler.

Lemma 15. *Let E be a polyhedral subset of U_j where $1/2 \le \theta \le 7/13 - \epsilon$. Then we can evaluate*

$$\sum\nolimits_j := \sum_{\boldsymbol{\alpha}_j \in E} S(\mathcal{A}_{p_1 \dots p_j}, x^\kappa), \tag{6.17}$$

with error ϵ. We can evaluate $S(\mathcal{A}, x^\kappa)$ with error ϵ. Also, we can increase θ to $0.55 - \epsilon$ if we replace U_j by U_j' and κ by κ'.

Proof. The evaluation of $S(\mathcal{A}, x^\kappa)$ is similar to, but simpler than, (6.17) so we deal with $\sum_j$. The proof with U_j replaced by U_j' and κ by κ' includes the proof of the assertion for $\theta \le 7/13 - \epsilon$, so we henceforth work with (6.17) with κ replaced by κ'. By Buchstab's identity,

$$\sum\nolimits_j = \sum_{\boldsymbol{\alpha}_j \in E} S(\mathcal{A}_{p_1 \dots p_j}, x^\eta) - \sum_{\substack{\boldsymbol{\alpha}_{j+1} \in A_{j+1} \\ \boldsymbol{\alpha}_j \in E \\ \alpha_{j+1} < \kappa'}} S(\mathcal{A}_{p_1 \dots p_{j+1}}, p_{j+1}), \tag{6.18}$$

where here, and henceforth, $\boldsymbol{\alpha}_j$ means the first j coordinates of $\boldsymbol{\alpha}_{j+\ell}$ for any ℓ. We write $\sum_{j+1}$ for the part of the second sum on the right of (6.18) with

$$\alpha_{j+1} < 2\theta - 1 + \epsilon \tag{6.19}$$

and $\sum''_{j+1}$ denotes the rest of that sum (which will be empty for $\theta > 17/32$). Thus

$$\sum_j = \sum'_j - \sum''_{j+1} - \sum_{j+1}, \text{ say .}$$

By Lemma 14 we can evaluate $\sum'_j$ with error $g(\epsilon/\eta)$. We can evaluate $\sum''_{j+1}$ with error $\mathcal{L}^{-\Theta} = O(g(\epsilon/\eta))$ by Lemma 7 (since $\alpha_{j+1} < \kappa'$).

We now apply Buchstab's identity to $\sum_{j+1}$. If we continue in this fashion we obtain a sequence of sums $\sum_{j+1}, \sum_{j+2}, \dots,$

$$\sum_k = \sum_{\boldsymbol{\alpha}_k} S(\mathcal{A}_{p_1 \dots p_k}, p_k)$$

with summation conditions

$$(6.20)_k \qquad \boldsymbol{\alpha}_k \in A_k,$$

$$(6.21) \qquad \boldsymbol{\alpha}_j \in E, \alpha_{j+1} < \kappa',$$

$$(6.22)_k \qquad \alpha_{j+1} + \dots + \alpha_k < 2\theta - 1 + \epsilon.$$

We have

$$\sum_k = \sum'_k - \sum''_{k+1} - \sum_{k+1},$$

where the conditions of summation are:

$\sum'_k$: same as $\sum_k$ ($S(\mathcal{A}_{p_1 \dots p_k}, p_k)$ replaced by $S(\mathcal{A}_{p_1 \dots p_k}, x^{\kappa'})$ in this sum);
$\sum''_{k+1}$: $(6.20)_k$, (6.21), $(6.22)_k$ and

$$\alpha_{j+1} \cdots + \alpha_{k+1} \geq 2\theta - 1 + \epsilon;$$

$\sum_{k+1}$: $(6.20)_{k+1}$, (6.21) and $(6.22)_{k+1}$.

We can evaluate $\sum'_k$ using Lemma 12 with error $g(\epsilon/\eta)$. This is possible since $(6.22)_k$ together with $\boldsymbol{\alpha}_j \in U'_j$ gives $\boldsymbol{\alpha}_k \in S_k$ (this explains the definition of U'_j). We can evaluate $\sum''_{k+1}$ with error $\mathcal{L}^{-\Theta} = O(g(\epsilon/\eta))$ by Lemma 7 using Lemma 11 to verify that $\boldsymbol{\alpha}_{k+1} \in G_{k+1}$. This leaves an unevaluated sum $\sum_{k+1}$ for the next step.

After $< [\eta^{-1}]$ steps, no unevaluated term remains. We can combine the main terms to form the sum

$$\sum_{\boldsymbol{\alpha}_j \in E} \phi(q)^{-1} S(\mathcal{B}_{p_1 \dots p_j}, x^{\kappa'})$$

by applying the Buchstab identity to this sum.

It remains to show that Y, the total of the moduli of the $O(\eta^{-1})$ errors, is (for most $q \sim x^{\theta}$),

$$O\left(\frac{\epsilon}{\phi(q)}\sum_{\alpha_j\in E} S(\mathcal{B}_{p_1\dots p_j}, x^{\kappa'})\right).$$

Now

$$Y = O\left(\frac{1}{\phi(q)}g(\epsilon/\eta)\sum_{\alpha_j\in E}\sum_{h=1}^{[\eta^{-1}]}\sum_{x^{\eta}\le p_{j+h}<\cdots<p_{j+1}} S(\mathcal{B}_{p_1\dots p_{j+h}}, x^{\eta})\right).$$

By Lemma 10 of [2] this is

$$O\left(2^{1/\eta}g(\epsilon/\eta)\frac{1}{\phi(q)}\sum_{\alpha_j\in E} S(\mathcal{B}_{p_1\dots p_j}, x^{\eta})\right)$$

$$= O\left(\frac{\eta^{-1}2^{1/\eta}g(\epsilon/\eta)}{\phi(q)}\sum_{\alpha_j\in E} S(\mathcal{B}_{p_1\dots p_j}, x^{\kappa'})\right).$$

Since ϵ is sufficiently small, and $\eta = \exp(-2/\epsilon)$, we have

$$\eta^{-1}2^{1/\eta}g(\epsilon/\eta) < \exp\left(\frac{1}{\eta} - \frac{\epsilon}{\eta}\log\left(\frac{\epsilon}{\eta}\right)\right) < \epsilon.$$

This completes the proof of Lemma 15.

Lemma 16. *Let* $7/13 - \epsilon < \theta \le 4/7 - \epsilon$. *We can evaluate* $S(\mathcal{A}, x^{\kappa})$ *with error* ϵ. *Let* E *be a polyhedral subset of* U_j. *Then we can evaluate (6.17) with error* ϵ.

Remark. In this result we have increased κ' to κ compared to the previous Lemma, but at the expense of deleting $U_j'\backslash U_j$.

Proof. It suffices to evaluate (6.17) and we follow the proof of Lemma 15 rather closely. The main difference is that κ' is replaced by κ, (6.19) is replaced by

$$\alpha_1 + \cdots + \alpha_{j+1} < \theta + \epsilon,$$

and, more generally, $(6.22)_k$ is replaced by

$$(6.23)_k \qquad \alpha_1 + \cdots + \alpha_k < \theta + \epsilon.$$

We evaluate $\sum_k'$ using Lemma 14 as before; this time $(6.23)_k$ together with Lemma 12 implies that the domain of summation is contained in S_k. Likewise we evaluate $\sum_{k+1}''$ using Lemma 7, since Lemma 13 yields a domain of summation contained in G_{k+1}. The remainder of the proof proceeds as before.

The work of this section so far is useful for (α_1, α_2) in region $A \cup C$. (Of course, for B and E, Lemma 7 is a better result.) We now prove some further results which are important when one variable (or product of variables) lies between $x^{3/7+\epsilon}$ and $x^{1-\theta-\epsilon}$. We write

$$T^{**} = \left\{(s,t) : 3/7 + \epsilon \le s \le 1 - \theta - \epsilon,\ 0 \le t < \frac{1-s}{2}\right\},$$
$$U_j^* = \left\{\boldsymbol{\alpha}_j \in \mathbb{R}^j : \boldsymbol{\alpha}_j \text{ partitions exactly into } T^{**}\right\},$$
$$\psi(\alpha) = \frac{2\alpha}{3} - \min(2 - \theta - 3\alpha, 2\theta - 2\alpha) - 2\epsilon.$$

In the following, $\psi(\alpha)$ plays the role of κ and U_j^* replaces U_j or U_j'.

Lemma 17. *Let $1/2 \le \theta \le 4/7 - \epsilon$. Let H be a polyhedral subset of U_j^*. Then we can evaluate*

$$\sum_{\boldsymbol{\alpha}_j \in H} S(\mathcal{A}_{p_1 \ldots p_j}, x^\lambda) \tag{6.24}$$

with error ϵ, for any $\lambda \in [\eta, \psi(s)]$. Here $s = s(\boldsymbol{\alpha})$ is a subsum of $\alpha_1 + \cdots + \alpha_j$ such that $(s,t) \in T^{**}, t = \alpha_1 + \cdots + \alpha_j - s$.

Proof. We follow the proof of Lemma 15 closely again. This time, (6.19) is replaced by

$$t + \alpha_{j+1} < \min(2 - \theta - 3s, 2\theta - 2s) + \epsilon$$

and, more generally, $(6.22)_k$ is replaced by

$$t + \alpha_{j+1} + \cdots + \alpha_k < \min(2 - \theta - 3s, 2\theta - 2s) + \epsilon. \tag{6.25}$$

We evaluate $\sum_k'$ using Lemma 14. For (6.25) implies

$$s + 2(t + \alpha_{j+1} + \cdots + \alpha_k) < s + 4 - 2\theta - 6s < 4 - 2\theta - \frac{15}{7} < 2 - 2\theta - \epsilon,$$

$$s + 4(t + \alpha_{j+1} + \cdots + \alpha_k) < s + 8 - 4\theta - 12s < 8 - 4\theta - \frac{33}{7} < 2 - \theta - \epsilon.$$

In the sum $\sum''_{k+1}$,

(6.26)
$$t' := t + \alpha_{j+1} \cdots + \alpha_k \leq \psi(s) + \min(2 - 3\theta - 2s, 2\theta - 2s) + \epsilon$$
$$\leq \frac{2s}{3} - \epsilon.$$

We evaluate $\sum''_{k+1}$ using Lemma 7. It suffices to show that $(t', 1 - s - t')$ satisfies one of the sets of inequalities {(3.17), (3.18), (3.19), (3.20)}, {(3.17), (3.19), (3.21)}.

We begin by observing that

$$t' + (1 - s - t') = 1 - s \geq \theta + \epsilon,$$

$$5t' + 2(1 - s - t') = 2 - 2s + 3t' \leq 2 - \epsilon$$

from (6.26). This deals with (3.17) and (3.19). For (3.21) we need

$$t' + 2(1 - s - t') = 2 - 2s - t' \leq 2 - 2\theta - \epsilon,$$

that is,

$$t' \geq 2\theta - 2s + \epsilon.$$

If this fails, then, from the definition of $\sum''_{k+1}$,

$$t' \geq 2 - \theta - 3s + \epsilon,$$

$$2t' + 3(1 - s - t') = 3 - (3s + t') \leq 1 + \theta - \epsilon.$$

Moreover, using (6.26),

$$4t' + 3(1 - s - t') = 3 - 3s + t' \leq 3 - \frac{7s}{3} \leq 2 - \epsilon.$$

Thus one of our two sets of inequalities is satisfied. We now complete the proof in a similar manner to Lemma 15.

We now use Lemma 17 to prove the following more powerful result.

Lemma 18. *Let $1/2 \leq \theta \leq 4/7 - \epsilon$. Let H be a polyhedral subset of U_j^* and let $\xi(\boldsymbol{\alpha}_j)$ be a continuous function on H with $\eta \leq \xi(\boldsymbol{\alpha}_j) \leq 1/2$. Then we can evaluate*

(6.27)
$$\sum_{\boldsymbol{\alpha}_j \in H} S(\mathcal{A}_{p_1 \ldots p_j}, x^{\xi(\boldsymbol{\alpha}_j)})$$

with error ϵ.

Corollary. *For* $1/2 \le \theta \le 4/7 - \epsilon$ *we can evaluate*

$$\sum_{x^{3/7+\epsilon} < p < x^{1-\theta-\epsilon}} S(\mathcal{A}_p, p)$$

with error ϵ.

Remark. This Lemma, and its immediate corollary, provide the promised strengthening of Lemma 8.

Proof. Define $s = s(\boldsymbol{\alpha})$ as in Lemma 17. In view of Lemma 17 we may suppose that $\xi(\alpha_j) > \psi(s)$. Also, there is no contribution from any terms with $\xi(\alpha_j) > 1 - \alpha_1 - \cdots - \alpha_j$. To simplify the proof we suppose henceforth that $\xi(\alpha_j)$ is fixed (so we write it just as ξ) with $\psi(s) < \xi \le 1 - \alpha_1 - \cdots - \alpha_j$.

We combine s, t into two variables as in Lemma 17. If $(s,t) \in \mathcal{G}$ for all $\boldsymbol{\alpha}_j$ there is nothing to prove, so we suppose $H \cap G_j = \emptyset$. The sum in (6.27) equals

$$\sum_{\boldsymbol{\alpha}_j \in H} S(\mathcal{A}_{p_1 \ldots p_j}, x^{\psi(s)}) - \sum_{\substack{\boldsymbol{\alpha}_j \in H \\ \psi(s) < \alpha_{j+1} \le \xi}} S(\mathcal{A}_{p_1 \ldots p_{j+1}}, p_{j+1}), \tag{6.28}$$

and the first sum in (6.28) can be estimated by Lemma 17.

Let the second sum be denoted by Σ_1. We can estimate any part of Σ_1 with $(s, \alpha_{j+1}) \in E$, so there remains a sum Σ_2 with

$$\alpha_{j+1} < \min(2 - \theta - 3s, 2\theta - 2s) + \epsilon = \rho(\theta, s), \text{ say}$$

(and t must also satisfy this inequality since $(s,t) \notin E$).

We can follow the previous proof, with t' replaced by $t + \alpha_{j+1}$, to estimate the part of S_2 with

$$\rho(\theta, s) \le \alpha_{j+1} + t \le \frac{2s}{3} - \epsilon.$$

There remain two subsums of S

(i) $\sum_3$, with

$$\alpha_{j+1} + t < \rho(\theta, s);$$

(ii) $\sum_4$, with

$$\alpha_{j+1} + t > \frac{2s}{3} - \epsilon.$$

Note that

$$2s/3 - \epsilon \geq (1-s)/2, \text{ so}$$

$$s + 2(\alpha_{j+1} + t) \geq 1$$

in Σ_4. It is the last inequality that we shall need.

In Σ_3, we apply Buchstab's inequality once more to obtain

$$\sum\nolimits_3 = \sum_{\substack{\alpha_j \in H \\ \psi(s) < \alpha_{j+1} \leq \xi \\ \alpha_{j+1} + t < \rho(\theta, s)}} S(\mathcal{A}_{p_1 \ldots p_{j+1}}, x^{\psi(s)}) - \sum_{\substack{\alpha_j \in H \\ \psi(s) < \alpha_{j+2} \leq \alpha_{j+1}}} S(\mathcal{A}_{p_1 \ldots p_{j+1} p_{j+2}}, p_{j+2}) \tag{6.29}$$

where we have suppressed some summation conditions in the interests of clarity. The first sum on the right of (6.29) can be estimated by Lemma 17.

In the second we note that

$$\alpha_{j+1} + \alpha_{j+2} \geq 2\psi(s) \geq \rho(\theta, s)$$

(at this point we require $s \geq 3/7 + \epsilon$; the worst case is at $\theta = 11/21$).

Hence $(s, \alpha_{j+1} + \alpha_{j+2}) \in E$ unless $2(\alpha_{j+1} + \alpha_{j+2}) + s \geq 1$. The working for this case thus reduces to the same as case (ii).

For Σ_4 we claim that

$$(t + \alpha_{j+1}, 1 - s - t - \alpha_{j+1}) \in \mathcal{G}^3.$$

To see this, we have $s \leq 1 - \theta - \epsilon$, so $t + \alpha_{j+1} + (1 - s - t - \alpha_{j+1}) \geq \theta$. Also

$$\begin{aligned}(1 - s - t - \alpha_{j+1}) + 2(t + \alpha_{j+1}) &= 1 + t + \alpha_{j+1} - s \\ &\leq 1 + 2(2 - \theta - 3s) - s = 5 - 2\theta - 7s \\ &< 2 - 2\theta - \epsilon \text{ as } s \geq 3/7 + \epsilon, \\ 5(1 - s - t - \alpha_{j+1}) + 2(t + \alpha_{j+1}) &= 5 - \frac{7}{2}s - \frac{3}{2}(s + 2(t + \alpha_{j+1})) \\ &\leq \frac{7}{2}(1 - s) < 2 - \epsilon \text{ as } s > 3/7 + \epsilon.\end{aligned}$$

This completes the proof of Lemma 18.

We remark that the above result is very useful when performing the calculations we need later, since it demonstrates that it is often possible to obtain an asymptotic formula if some of the variables can be grouped together to form a product lying between $x^{3/7+\epsilon}$ and $x^{1-\theta-\epsilon}$.

As might be expected, the method of Lemma 18 provides a useful lower bound for (6.27) for smaller values of α_1. We can obtain a result for $\alpha_1 \geq (6\theta-1)/5+\epsilon$ (which can be used for $\theta < 11/21$). However, it is more efficient to work as in section 7 (apart from a very small range of α_1, for certain θ).

7. The alternative sieve in two dimensions

We now deal with the region $1-\theta-\epsilon < \alpha_1 \leq 1/2$. If we were unable to deal with this region it would be impossible to give the lower bound on average. The upper bound would also be worse. In the following we suppose $1/2 < \theta < 17/32$. We write $\mathcal{C}$ for the finite sequence

$$\mathcal{C} = \{mn : x^{1-\theta-\epsilon} \leq m \leq x^{1/2}, mn \sim x, (mn, q) = 1\}$$

$$\mathcal{F} = \mathcal{C} \cap \mathcal{A}.$$

Lemma 19. *For most $q \sim x^\theta$ we have*

$$S(\mathcal{F}, x^\kappa) = \frac{1}{\phi(q)} S(\mathcal{C}, x^\kappa)(1+O(\epsilon)). \tag{7.1}$$

Remarks. The formula (7.1) is valid for $\theta < 11/20-\epsilon$, with κ replaced by κ', and one can generalize it to

$$\sum S(\mathcal{F}_{p_1\ldots p_j}, x^\kappa) = \frac{1}{\phi(q)} \sum S(\mathcal{C}_{p_1\ldots p_j}, x^\kappa)(1+O(\epsilon))$$

where $p_1 \ldots p_j < x^{2-7\theta/2-3\epsilon}$, but we will not need this result here.

Proof. We first show that we can give an asymptotic formula for

$$\sum_{\alpha_j \in H} S(\mathcal{F}_{p_1\ldots p_j}, x^\eta). \tag{7.2}$$

provided that $\alpha_1 + \cdots + \alpha_j < 1-\epsilon-3\theta/2$ for $\boldsymbol{\alpha}_j \in H$. Here $H \subset A_j(\eta, \kappa)$. To do this we need an expression for $|\mathcal{F}_d|$ for square-free d coprime to q. We have

$$|\mathcal{F}_d| = \sum_{e|d} \sum_{\substack{(n,d/e)=1 \\ emn\in\mathcal{F} \\ m\equiv 0(\mathrm{mod}\ d/e) \\ x^{1-\theta-\epsilon}\leq en \leq x^{1/2}}} 1 = \sum_{e|d} \sum_{f|d/e} \mu(f) \sum_{\substack{nfdm\equiv a(q) \\ x^{1-\theta-\epsilon}\leq nfe\leq x^{1/2} \\ mndf\sim x}} 1.$$

To evaluate this last sum, we first restrict nfe and md/e to the intervals $\mathcal{I} = [u, u(1+\mathcal{L}^{-\Theta}))$ and $\mathcal{J} = [v, v(1+\mathcal{L}^{-\Theta}))$ respectively for some u, v with $uv \sim x$. Denoting such a subsum by F_d we have

$$F_d = \sum_{f|d} \mu(f) \sum_{e|d/f} \left(\frac{\phi(q) I J}{q^2 df} + S(d,e,f) \right)$$

where $I = U\mathcal{L}^{-\Theta}$, $J = V\mathcal{L}^{-\Theta}$, and

$$S(d,e,f) = \sum\nolimits^{*} 1 - \frac{\phi(q) I J}{q^2 df}$$

with $*$ indicating the summation conditions:

$$nfdm \equiv a(q), nfe \in \mathcal{I}, md/e \in \mathcal{J}.$$

We note that

$$\sum_{f|d} \mu(f) \sum_{e|d/f} \frac{\phi(q)}{q^2 df} = \frac{\phi(q)}{q^2 d} \rho(d)$$

where $\rho(d) = \tau(d) \prod_{p|d} \left(1 - \frac{1}{2p}\right)$. Now

$$\sum_{\substack{f|d \\ f>w}} \sum\nolimits^{*} 1 < \frac{x^{1+\eta}}{dwq}$$

by a trivial bound. We thus have

$$F_d = \frac{\phi(q)\rho(d) I J}{q^2 d} + O\left(\frac{x^{1-\eta}}{qd}\right)$$

$$+ O\left(\max_{\substack{f < x^{2\eta} \\ e|d/f}} x^{\eta} |S(d,e,f)| \right).$$

We will bound $|S(d,e,f)|$ using the well known estimate

$$\sum_{\substack{A \le n \le B \\ (n,q)=1}} e\left(\frac{b\bar{n}}{q}\right) \ll q^{1/2+\eta} (b,q)^{1/2} \tag{7.3}$$

(see page 36 of [21]), which follows from Weil's estimate for the Kloosterman sum. Here $\bar{n}$ denotes inverse (mod q). We have

$$\begin{aligned}
{\sum}^{*} 1 &= \frac{1}{q}\sum_{r=0}^{q-1} \sum_{\frac{md}{e}\in\mathcal{J}} e\left(\frac{-rm}{q}\right) \sum_{\substack{(n,q)=1 \\ nfe\in\mathcal{I}}} e\left(\frac{a\bar{d}\bar{f}\bar{n}r}{q}\right) \\
&= \frac{1}{q}\left(\sum_{\frac{md}{e}\in\mathcal{J}} 1\right)\left(\sum_{\substack{(n,q)=1 \\ nfe\in\mathcal{I}}} 1\right) \\
&\quad + O\left(\frac{1}{q}\sum_{1\le|r|\le q/2} \min\left(J, \frac{q}{r}\right)\left|\sum_{\substack{(n,q)=1 \\ nfe\in\mathcal{I}}} e\left(\frac{\overline{adfnr}}{q}\right)\right|\right) \\
&= \frac{\phi(q)}{q^2 df} IJ + O\left(q^{\eta-1} + \frac{I}{feq} + \frac{Jeq^{\eta-1}}{d}\right) + O(q^{1/2+2\eta})
\end{aligned}$$

using (7.3) together with $|\{n : c \le n < d, (n,q) = 1\}| = (d-c)\phi(q)/q + O(q^{\eta})$. Thus

$$|F_d| = \frac{\phi(q)\rho(d)IJ}{q^2 d}(1 + O(\mathcal{L}^{-\Theta}))$$

provided that $d < x^{1-4\eta}q^{-3/2}$ and $\max(I, J) \le q^{3/2}$ (which certainly holds here). Using standard procedures (see pp. 388–389 of [4]) we obtain

$$|\mathcal{F}_d| = \frac{\phi(q)\rho(d)}{q^2 d} \sum_{\substack{mn\sim x \\ x^{1-\theta-\epsilon}\le m\le x^{1/2}}} (1 + O(\mathcal{L}^{-\Theta})).$$

We can now apply Lemma 2 with dimension 2 to (7.2) with $z = x^{\eta}, s = \epsilon/\eta$. This gives

$$\begin{aligned}
\sum_{\alpha_j\in H} S(\mathcal{F}_{p_1\dots p_j}, x^{\eta}) &= \sum_{\alpha_j\in H} \frac{\phi(q)}{q^2 p_1\dots p_j} \sum_{\substack{mn\sim x \\ x^{1-\theta-\epsilon}\le m\le x^{1/2}}} (1 + O(\mathcal{L}^{-\Theta}))\times \\
&\quad \times (1 + O(s^{-s})) \prod_{p<x^{\eta}} \left(1 - \frac{\rho(p)}{p}\right)
\end{aligned}$$

since (r denoting *any* prime number)

$$\prod_{p<x^{\eta}} \left(1 - \frac{\rho(p)}{p}\right) = \left(\frac{q}{\phi(q)}\right)^2 \prod_{r<x^{\eta}} \left(1 - \frac{1}{r}\right)^2 (1 + O(q^{-\eta})).$$

We can obtain the formula

$$\sum_{\alpha_j \in H} S(\mathcal{C}_{p_1 \dots p_j}, x^\eta) = \sum_{\alpha_j \in H} \frac{\phi^2(q)}{q^2 p_1 \dots p_j} \sum (1 + O(\mathcal{L}^{-\Theta})) \times$$
$$\times (1 + O(s^{-s})) \prod_{p < x^\eta} \left(1 - \frac{\rho(p)}{p}\right)$$

by a simplified version of the same argument. We conclude that

$$\sum_{\alpha_j \in H} S(\mathcal{F}_{p_1 \dots p_j}, x^\eta) = \frac{1}{\phi(q)} \sum_{\alpha_j \in H} S(\mathcal{C}_{p_1 \dots p_j}, x^\eta)(1 + O(g(\epsilon/\eta))).$$

Now, using Buchstab's identity,

$$S(\mathcal{F}, x^\kappa) = S(\mathcal{F}, x^\eta) - \sum_{x^\eta \le p < x^\kappa} S(\mathcal{F}_p, p)$$

$$= S_1 - S_2 \text{ say.}$$

By the above argument we can estimate S_1, while Lemma 7 (modified only slightly) deals with that part of S_2 for which $x^{2\theta - 1 + \epsilon} \le p < x^\kappa$. For the rest of S_2 we can use Buchstab again and continue as in the proof of Lemma 15. Eventually the sum is decomposed into sums for all of which either the two dimensional fundamental Lemma or Lemma 7 can be applied. This completes the proof since the accumulated errors can be dealt with as in Lemma 15.

For the rest of this section, s and r also denote prime numbers.

Lemma 20. *For most $q \sim x^\theta$ with $1/2 \le \theta \le 17/32 - \epsilon$ we can evaluate with error ϵ*

$$\sum_{x^{1-\theta-\epsilon} \le p \le x^{1/2}} S(\mathcal{A}_p, p) - \sum_{(*)} S(\mathcal{A}_{ptr}, r) \tag{7.4}$$

where $(*)$ indicates the summation conditions:

$$x^\kappa < r < p < x^{(\epsilon+\theta)/2}, s|t \Rightarrow s \ge p, \ x^{1-\theta-\epsilon} \le pt \le x^{\theta+\epsilon}.$$

Remark. One can increase θ above 17/32 but at the expense of an additional term in (7.4). In our context there is no point in pursuing this since the numerical advantage gained by Lemma 20 becomes negligible near 17/32.

Proof. We have

$$(7.5)\qquad \sum_{x^{1-\theta-\epsilon}\le p\le x^{1/2}} S(\mathcal{A}_p,p) = S(\mathcal{F},x^{(\theta+\epsilon)/2}) = S(\mathcal{F},x^{\kappa}) - \sum_{x^{\kappa}\le p<x^{(\theta+\epsilon)/2}} S(\mathcal{F}_p,p)$$

We can give an asymptotic formula for the first sum in (7.5) by Lemma 19, while the second sum (with error ϵ) is

$$(7.6)\qquad \sum_{\substack{x^{\kappa}\le p<t\\ s|t\Rightarrow s\ge p\\ x^{1-\theta-\epsilon}\le pt\le x^{\theta+\epsilon}}} S(\mathcal{A}_{pt},p) = \sum_{\substack{x^{\kappa}\le p<t\\ s|t\Rightarrow s\ge p\\ x^{1-\theta-\epsilon}\le pt\le x^{\theta+\epsilon}}} S(\mathcal{A}_{pt},x^{\kappa}) - \sum_{(*)} S(\mathcal{A}_{ptr},r).$$

We can evaluate the first sum on the right of (7.6) with error ϵ using Lemma 15. To see this, note that

$$tx^{2\theta-1+\epsilon} \le x^{3\theta-1+2\epsilon-\kappa} < x^{1-\theta-\epsilon} \text{ for } \theta\le 17/32-\epsilon,$$

$$tx^{2\theta-1+\epsilon}p^2 \le (tp)^{3/2}x^{2\theta-1+\epsilon} \le x^{3\theta/2+2\theta-1+5\epsilon/2} < x^{2-2\theta-\epsilon} \text{ as } \theta<6/11-\epsilon,$$

$$tx^{2\theta-1+\epsilon}p^4 \le (tp)^{5/2}x^{2\theta-1+\epsilon} \le x^{9\theta/2-1+11\epsilon/2} < x^{2-\theta-\epsilon} \text{ since } \theta<6/11-\epsilon.$$

It follows that we can evaluate (7.4) with error ϵ as desired.

We note that we can give an asymptotic formula for part of the second sum in (7.4) using Lemmas 7 or 18. We thus obtain

(7.7)

$$\sum_{x^{1-\theta-\epsilon}\le p\le x^{1/2}} S(\mathcal{A}_p,p) \le \frac{-1}{\phi(q)} \sum_{x^{1-\theta-\epsilon}\le p\le x^{1/2}} S(\mathcal{B}_p,p)(1+O(\epsilon)) + \frac{x}{\phi(q)\log x}\int_{\mathcal{D}} w(\alpha_2/\alpha_1)w\left(\frac{1-\alpha_1-\alpha_2-\alpha_3}{\alpha_3}\right)\frac{d\alpha_1 d\alpha_2 d\alpha_3}{\alpha_1^2\alpha_3^2}$$

where

$$\mathcal{D} = \{(\alpha_1,\alpha_2,\alpha_3)\notin G_3 : \kappa\le\alpha_1\le\theta/2, 1-\theta-\alpha_1\le\alpha_2\le\theta-\alpha_1, \kappa\le\alpha_3\le\min(\alpha_1,(1-\alpha_1-\alpha_2)/2)\}.$$

If we had discarded the left hand side of (7.7) we would suffer a loss $\log(\theta/(1-\theta))$, which is > 0.08 at $\theta = 0.52$ for example, whereas the integral over $\mathcal{D}$ is < 0.025 for this value of θ. It is clear that the integral has size $O((\theta-1/2+\epsilon)^2)$

since $(\alpha_2+\alpha_3,\alpha_1)$ belongs either to B, *or* $(\alpha_1, 1-\alpha_1-\alpha_2-\alpha_3)$ belongs to B, *or*

$$\theta+\epsilon \geq \alpha_2+\alpha_3 \geq 1-\theta-\epsilon, \text{ or}$$

$$2-2\theta-\epsilon \leq 2(\alpha_2+\alpha_3)+\alpha_1 \leq 2\theta+\epsilon.$$

Hence α_3 is also restricted to an interval of length $O(\theta-1/2+\epsilon)$ for all pairs (α_1,α_2).

We conclude this section by giving an upper bound for $S(\mathcal{A}_p,p)$ that comes immediately from stopping the proof of Lemma 20 at the left hand side of (7.6).

Lemma 21. *For most $q \sim x^\theta$ we have*

(7.8)
$$\sum_{x^{1-\theta-\epsilon}\leq p\leq x^{1/2}} S(\mathcal{A}_p,p) \leq \sum_{x^{1-\theta-\epsilon}\leq p\leq x^{1/2}} S(\mathcal{B}_p,p)(1+O(\epsilon))$$
$$+\frac{x}{\phi(q)\log x}\int_{\kappa}^{\theta/2}\int_{1-\theta-\alpha_1}^{\theta-\alpha_1} w\left(\frac{\alpha_2}{\alpha_1}\right) w\left(\frac{1-\alpha_1-\alpha_2}{\alpha_1}\right)\frac{d\alpha_2 d\alpha_1}{\alpha_1^3}.$$

Remarks. Clearly the integral on the right of (7.8) is a continuous function of θ of size $O(\theta-1/2+\epsilon)$. Indeed, for small $\theta-1/2$ the value is approximately $7.7(\theta-1/2+\epsilon)$. We have $(\alpha_1,\alpha_2)\notin G_2$ for any (α_1,α_2) in the domain of integration. However, a further saving can be made by writing (briefly)

(7.9)
$$\sum S(\mathcal{A}_{pt},p) = \sum S\left(\mathcal{A}_{pt},\left(\frac{x}{pt}\right)^{1/2}\right)$$
$$+\sum_{p_1\leq p_2<(x/p_1t)^{1/2}} S(\mathcal{A}_{p_1p_2t},p_2).$$

The first sum on the right in (7.9) must be discarded, but we are able to give an asymptotic formula for part of the second sum. A further decomposition of this second sum is possible for p_1 in a very short range, but this leads only to a very small additional saving.

8. The alternative sieve in three dimensions

In this section we consider the region D in Diagram 1. In this region $\alpha_1+4\alpha_2 \geq 3-3\theta$ and this prevents us from applying Buchstab's identity again since we cannot, as yet, evaluate

$$\sum_{\alpha_2 \in D} S(\mathcal{A}_{p_1 p_2}, x^{\kappa}). \tag{8.1}$$

Also, $3\alpha_2 \geq 2\alpha_1$, so we are not in the asymptotic formula region E. Now, let R be a polygonal subset of

$$\{(s,t) : 1/4 \leq t \leq s \leq \min(3/7, 4 - 7\vartheta - 3\epsilon; 7\theta - 3 + 3\epsilon \leq s + t\}. \tag{8.2}$$

Let

$$\mathcal{C}' = \{\ell mn : (\log \ell, \log m) \in (\log x)R, \ell mn \sim x\},$$

and

$$\mathcal{F}' = \mathcal{C}' \cap \mathcal{A}.$$

As the reader will by now expect, our next goal is the estimation of $|\mathcal{F}'_d|$ *en route* to evaluating $S(\mathcal{F}', x^{\kappa})$. The new arithmetical information we require (a type I_3 sum) is given by the following result due to Heath-Brown [16]. We here state a variant of his Lemma 5. It is of great importance that his result is uniform in the parameter b.

Lemma 22. *Suppose $\mathcal{I}, \mathcal{J}, \mathcal{K}$ are subintervals of $[1, y)$ such that $\ell \in \mathcal{I}, m \in \mathcal{J}, n \in \mathcal{K} \Rightarrow \ell mn \leq y$. Then for $(b,q) = 1$ we have*

$$\sum_{\substack{\ell mn \equiv b(q) \\ \ell \in \mathcal{I}, m \in \mathcal{J}, n \in \mathcal{K}}} 1 = \frac{\phi^2(q)}{q^3} IJK + O(E_1) + O(E_2) \tag{8.3}$$

where $\mathcal{I}, \mathcal{J}, \mathcal{K}$ have lengths I, J, K respectively, and

$$E_1 = \frac{y^{\eta}}{q} \max(IJ, IK, JK)$$
$$E_2 = y^{\eta} \left(q^{5/6} + q^{1/4} \{\min(IJ, KJ, IK)\}^{1/2} + q^{1/2} \{\max(K, I, J)\}^{1/2}\right).$$

Proof. Heath-Brown ([16], Lemma 5) shows that the left hand side of (8.3) is

$$N_1 + O(E_2)$$

for some N_1 independent of q. Summing over b with $(b,q) = 1$,

$$\sum_{\substack{(\ell mn, q) = 1 \\ \ell \in \mathcal{I}, m \in \mathcal{J}, n \in \mathcal{K}}} 1 = \phi(q) N_1 + O(\phi(q) E_2).$$

The left-hand side of the last equation is

$$\frac{\phi^3(q)}{q^3} IJK + O(\phi(q)E_1).$$

Lemma 22 now follows at once.

Lemma 23. *For most* $q \sim x^\theta$ *with* $\theta < 16/31 - \epsilon\,(= 0.516\dots)$ *we have*

$$S(\mathcal{F}', x^\kappa) = \frac{1}{\phi(q)} S(\mathcal{C}', x^\kappa)(1 + O(\epsilon)). \tag{8.4}$$

Proof. As in the proof of Lemma 19, we need only estimate

$$\sum_{d \le D} c_d S(\mathcal{F}'_d, x^\eta)$$

where $D = x^{2\theta - 1 + \epsilon}$, and $c_d = 0$ unless $p|d \Rightarrow p > x^\eta$. Using Lemma 2, now with dimension 3, this entails the estimation of

$$|\mathcal{F}'_d| \text{ for } d \le x^\epsilon D, (d, q) = 1.$$

Again working as before, we remove the restriction $\ell mn \sim x$ and instead confine the variables to intervals $\mathcal{I}, \mathcal{J}$ and $\mathcal{K}$ of the form $[V, V(1 + \mathcal{L}^{-\Theta}))$. Write F_d for this new set.

To calculate $|F_d|$, we classify the products ℓmn by taking h to be the largest factor of d that is coprime to n, and let e be the largest factor of h that is coprime to m. Let $ef = h$ and $efg = d$. Then

$$g|n, f|m, e|\ell.$$

Thus writing $n = n_1 g, m = m_1 f, \ell = \ell_1 e$,

$$\begin{aligned} |F_d| &= \sum_{efg=d} \sum_{\substack{(n_1, ef)=1 \\ n_1 g \in \mathcal{K}}} \sum_{\substack{(m_1, e)=1 \\ m_1 f \in \mathcal{J} \\ \ell_1 m_1 n_1 d \equiv a(q)}} \sum_{\ell_1 e \in \mathcal{I}} 1 \\ &= \sum_{efg=d} \sum_{\substack{k|e \\ j|fe}} \mu(k)\mu(j) \sum_{\substack{\ell mndkj \equiv a(q) \\ e\ell \in \mathcal{I}, mkf \in \mathcal{J}, njg \in \mathcal{K}}} 1 \\ &= \frac{\phi^2(q)}{q^3} \frac{\rho(d)}{d} IJK + O(E_3 x^{2\eta}) + O\left(\frac{x^{1-\eta}}{dq}\right). \end{aligned} \tag{8.5}$$

Here

$$\rho(d) = \sum_{efg=d} \sum_{\substack{k|e \\ j|fe}} \frac{\mu(k)\mu(j)}{kj} = \prod_{p|d} \left(3 - \frac{3}{p} + \frac{1}{p^2}\right),$$

and E_3 is the maximum error obtained from E_1 or E_2 in Lemma 22 with

$$y = 2x(dkj)^{-1}, b \equiv a\overline{dkj} (\bmod q), kj \le x^{2\eta}, d \le x^{2\theta-1+\epsilon},$$

and

$$\frac{x}{kjd}\mathcal{L}^{-3\Theta} \le IJK \le \frac{x}{kjd}, (\log(Ie), \log(Jfk)) \in R.$$

Now, since $s + t \le 6/7$ for $(s,t) \in R$ we have

$$E_1 \ll \frac{x^{1-\epsilon}}{qd},$$

with plenty of room to spare (since $\min(I,J,K) \ge x^{1/7}D^{-1}\mathcal{L}^{-\Theta}$). We therefore need to consider the terms in E_2. We have

$$y^{\eta}q^{5/6} < \frac{x^{1-\epsilon}}{dq}$$

since $16/31 < 12/23$.

For the next term in E_2 we note that $\min(IJ, IK, JK) \le (IJK)^{2/3}$. Hence

$$x^{\eta}q^{1/4}\min(IJ, IK, JK)^{1/2} \le x^{\eta}q^{1/4}\left(\frac{x}{dkj}\right)^{1/3} < \frac{x^{1-\epsilon}}{dq}$$

as $\theta \le 16/31 - \epsilon$.

For the final term in E_2 we note that, by the definition of R, if $(s,t) \in R$ then $\max(s, t, 1-s-t) \le 4 - 7\theta - 3\epsilon$. Hence

$$x^{\eta}q^{1/2}\max(I,J,K)^{1/2} \le x^{\eta-3\epsilon/2+2-3\theta} \le \frac{x^{1-\epsilon/2+\eta}}{dq}.$$

Hence

$$|F_d| = \frac{\phi^2(q)\rho(d)IJK}{q^3 d} + O\left(\frac{x^{1-2\eta}}{dq}\right),$$

and so

$$|\mathcal{F}'_d| = \frac{\phi^2(q)}{q^3}\frac{\rho(d)}{d}(1 + O(\mathcal{L}^{-\Theta})) \sum_{\ell mn \in \mathcal{C}'} 1.$$

From this we obtain the formula for $S(\mathcal{F}'_d, x^\eta)$, noting that

$$\prod_{p<x^\eta}\left(1-\frac{\rho(p)}{p}\right)=\left(\frac{q}{\phi(q)}\right)^3(1+O(q^{-\eta}))\prod_{r<x^\eta}\left(1-\frac{1}{r}\right)^3$$

(r here denoting a prime). We can then prove the formula for $S(\mathcal{F}', x^\kappa)$ by the procedure used in Lemmas 15 and 19.

We now use Lemma 23 to tackle (8.1). We tacitly assume $\theta < 16/31 - \eta$ in the following.

Lemma 24. *For most $q \sim x^\theta$ we have*

$$\sum_{\alpha_2\in R} S(\mathcal{A}_{p_1p_2}, x^\kappa) \lesssim \frac{1}{\phi(q)}\sum_{\alpha_2\in R} S(\mathcal{B}_{p_1p_2}, x^\kappa) + \frac{x}{\kappa\phi(q)\log x}(I_1+I_2) \tag{8.6}$$

where

$$I_1 = \int_{\alpha_3\in\mathcal{D}} w\left(\frac{1-\alpha_1-\alpha_2-\alpha_3}{\kappa}\right) d\mu_3,$$

$$I_2 = \int_{\substack{\alpha_4\notin G_4\\ \alpha_j\geq\kappa\\ (\alpha_1+\alpha_2,\alpha_3+\alpha_4)\in R}} w\left(\frac{1-\alpha_1-\alpha_2-\alpha_3-\alpha_4}{\kappa}\right) d\mu_4,$$

$$\mathcal{D} = \{\boldsymbol{\alpha}_3 \notin G_3 : (\alpha_1, \alpha_2+\alpha_3)\in R \textit{ with } \alpha_2\geq\alpha_3 \textit{ or } (\alpha_1+\alpha_2, \alpha_3)\in R \textit{ with } \alpha_1\geq\alpha_2, \alpha_j\geq\kappa\}.$$

Proof. We note that

$$\sum_{\alpha_2\in R} S(\mathcal{A}_{p_1p_2}, x^\kappa) = S(\mathcal{F}, x^\kappa) - \sum_{\substack{\ell mn\in\mathcal{F}\\ p\mid\ell mn\Rightarrow p\geq x^\kappa\\ \Omega(\ell m)\geq 3}} 1, \tag{8.7}$$

where $\Omega(n)$ denotes the total number of prime factors (counted with multiplicity) of n. Since $\theta < 29/56$, the value for which $\kappa = 1/7 - \epsilon$, and $m \leq \ell \leq x^{3/7}$, we have $\Omega(\ell) \leq 2$ and $\Omega(m) \leq 2$. Hence we get an upper bound for the right hand side of (8.7) which consists of the 'correct' answer (the first sum on the right of (8.6)) plus integrals arising from those parts of the second sum on the right of (8.7) which cannot be estimated by Lemma 7. These lead to I_1 ($\Omega(\ell m) = 3$) and I_2 ($\Omega(\ell m) = 4$), which completes the proof.

We now specify the region R. We must ensure that it lies in the set (8.2), and we may assume that the method of section 6, which requires $\boldsymbol{\alpha}_2 \in S$, fails.

Let

$$R = \{\boldsymbol{\alpha}_2 : \alpha_2 \le \alpha_1, \alpha_1 + 2\alpha_2 \le 1, \alpha_1 + 4\alpha_2 \ge 3 - 3\theta - \epsilon, 3\alpha_2 \ge 2\alpha_1,$$
$$\max\left(\tau, \frac{31\theta - 15}{3} + 5\epsilon\right) \le \alpha_1 \le \min(\frac{3}{7} + \epsilon, 4 - 7\theta - 3\epsilon)\}.$$

Then in R,

$$\alpha_1 + \alpha_2 = \frac{3\alpha_1}{4} + \frac{1}{4}(\alpha_1 + 4\alpha_2) \ge \frac{31\theta - 15}{4} + \frac{3 - 3\theta}{4} + 3\epsilon = 7\theta - 3 + 3\epsilon.$$

We remark that the range for α_1 is $\tau \le \alpha_1 \le 3/7$ for $\theta < 25/49$. By Buchstab's identity in conjunction with (8.6), we have

(8.8)
$$\sum_{\boldsymbol{\alpha}_2 \in R} S(\mathcal{A}_{p_1p_2}, p_2) \le \frac{1}{\phi(q)} \sum_{\boldsymbol{\alpha}_2 \in R} S(\mathcal{B}_{p_1p_2}, p_2) + \frac{x}{\phi(q)\log x}\left(\frac{1}{\kappa}(I_1 + I_2) + I_3\right)$$

where

$$I_3 = \int_{\boldsymbol{\alpha}_3 \in \mathcal{E}} w\left(\frac{1 - \alpha_1 - \alpha_2 - \alpha_3}{\alpha_3}\right)\frac{d\mu_3}{\alpha_3}$$

with

$$\mathcal{E} = \{\boldsymbol{\alpha}_3 \notin G_3 : (\alpha_1, \alpha_2) \in R, \kappa \le \alpha_3 \le \min(\alpha_2, (1 - \alpha_1 - \alpha_2)/2)\}.$$

The three integrals I_1, I_2 and I_3 are all continuous functions of θ and are $O((\theta - \frac{1}{2} + \epsilon)^2)$.

Now let R' be the subset of R which gives no contribution to I_1, I_2 or I_3. that is

$$R' = R \cap R_1 \cap R_2 \cap R_3 \cap R_4$$

with

$$R_1 = \{\boldsymbol{\alpha}_2 : \beta + \gamma = \alpha_2 \text{ with } \kappa \le \beta \le \gamma \Rightarrow (\alpha_1, \beta, \gamma) \in G_3\},$$
$$R_2 = \{\boldsymbol{\alpha}_2 : \beta + \gamma = \alpha_1 \text{ with } \kappa \le \beta \le \gamma \Rightarrow (\alpha_2, \beta, \gamma) \in G_3\},$$
$$R_3 = \{\boldsymbol{\alpha}_2 : \beta + \gamma = \alpha_1, \lambda + \mu = \alpha_2$$
$$\text{with } \kappa \le \beta \le \gamma, \kappa \le \lambda \le \mu \Rightarrow (\beta, \gamma, \lambda, \mu) \in G_4\},$$
$$R_4 = \{\boldsymbol{\alpha}_2 : \kappa \le \alpha_3 \le \min(\alpha_2, (1 - \alpha_1 - \alpha_2)/2) \Rightarrow (\alpha_1, \alpha_2, \alpha_3) \in G_3\}.$$

Then, working as above, we obtain

$$\sum_{\boldsymbol{\alpha}_2 \in R} S(\mathcal{A}_{p_1p_2}, p_2) \gtrsim \frac{1}{\phi(q)} \sum_{\boldsymbol{\alpha}_2 \in R} S(\mathcal{B}_{p_1p_2}, p_2) - \frac{x}{\phi(q)\log x} I_4 \tag{8.9}$$

where

$$I_4 = \int\limits_{R\backslash R'} w\left(\frac{1-\alpha_1-\alpha_2}{\alpha_2}\right)\frac{d\mu_2}{\alpha_2}.$$

We find that I_4 is again a continuous function of θ with $I_4 = O(\theta - 1/2 + \epsilon)$. By way of illustration we show that the measure of one part of $R\backslash R'$, namely

$$V = \{\boldsymbol{\alpha_2} \in R : \exists \beta, \gamma \text{ with } \beta + \gamma = \alpha_1, \kappa \leq \beta \leq \gamma, (\alpha_2, \beta, \gamma) \notin G_3\}$$

is $O(\theta - \frac{1}{2} + \epsilon)$. Let $\xi = 1 - \alpha_1 - \alpha_2$. Then we will show that $(\xi, \gamma) \in \mathcal{G}^3$ except for the small strip

$$W = \{(\alpha_1, \alpha_2) \in R : 2 - 2\theta - 2\epsilon \leq 2\alpha_2 + \alpha_1 \leq 1\},$$

which clearly has area $O(\theta - \frac{1}{2} + \epsilon)$.

Suppose $\boldsymbol{\alpha_2} \in V\backslash W$. Then

$$\xi + \gamma \geq 1 - \alpha_1 - \alpha_2 + \alpha_1/2 = 1 - \frac{2\alpha_2 + \alpha_1}{2} \geq \theta + \epsilon.$$

Also

$$\begin{aligned} 2\xi + \gamma &\leq 2(1 - \alpha_1 - \alpha_2) + (\alpha_1 - \kappa) \\ &= 2 - \kappa - \frac{1}{2}(4\alpha_2 + \alpha_1) - \frac{\alpha_1}{2} \\ &\leq (1 + 3\theta + \epsilon - \tau)/2 - \kappa \,(\text{as } (\alpha_1, \alpha_2) \in R) \\ &\leq 2 - 2\theta - \epsilon \text{ for } \theta < 79/154 - \epsilon. \end{aligned}$$

Moreover

$$\begin{aligned} 2\xi + 5\gamma &\leq 2(1 - \alpha_1 - \alpha_2) + 5(\alpha_1 - \kappa) \\ &= 2 - 5\kappa - \frac{1}{2}(4\alpha_2 + \alpha_1) + 7\alpha_1/2 \\ &\leq 2 - 5\kappa - \frac{1}{2}(3 - 3\theta - \epsilon) + 3/2 \\ &= 2 - \epsilon + (3\theta - 10\kappa + 3\epsilon)/2 \\ &\leq 2 - \epsilon \text{ for } \theta \leq 25/49 - \epsilon. \end{aligned}$$

Hence our claim that $(\xi, \gamma) \in \mathcal{G}^3$ is validated.

9. Conclusion

We begin with the upper bound in (1.1). We discuss a series of intervals for θ, beginning with

$$\theta \in [1/2, 25/49 - \epsilon],$$

moving to the right, and explaining how the method needs to be modified at each stage.

(i) $1/2 \leq \theta \leq 25/49 - \epsilon$. We use Buchstab's identity to write

(9.1)
$$\begin{aligned} S(\mathcal{A}, (2x)^{1/2}) &= S(\mathcal{A}, x^{\kappa}) - \sum_{\kappa \leq \alpha_1 \leq \frac{3}{7}+\epsilon} S(\mathcal{A}_{p_1}, p_1) \\ &\quad - \sum_{3/7+\epsilon \leq \alpha_1 \leq 1-\theta-\epsilon} S(\mathcal{A}_{p_1}, p_1) - \sum_{1-\theta-\epsilon < \alpha_1 \leq 1/2} S(\mathcal{A}_{p_1}, p_1) \\ &= S_{1,0} - S_{1,1} - S_{1,2} - S_{1,3}, \text{ say.} \end{aligned}$$

We can evaluate $S_{1,0}$ and $S_{1,2}$ with error ϵ by Lemmas 15 and 18 respectively. We get an upper bound for $-S_{1,3}$ by (7.7). We can apply Buchstab again to $S_{1,1}$ to obtain

(9.2)
$$\begin{aligned} S_{1,1} &= \sum_{\kappa \leq \alpha_1 \leq \frac{3}{7}+\epsilon} S(\mathcal{A}_{p_1}, x^{\kappa}) - \sum_{\alpha_2 \in A \cup C, \alpha_1 \leq 3/7+\epsilon} S(\mathcal{A}_{p_1 p_2}, p_2) \\ &\quad - \sum_{\alpha_2 \in B \cup E, \alpha_1 \leq 3/7+\epsilon} S(\mathcal{A}_{p_1 p_2}, p_2) - \sum_{\alpha_2 \in D} S(\mathcal{A}_{p_1 p_2}, p_2) \\ &= S_{2,0} - S_{2,1} - S_{2,2} - S_{2,3}, \text{ say.} \end{aligned}$$

The sums $S_{2,0}$ and $S_{2,2}$ can be evaluated with error ϵ. The sum $S_{2,3}$ can be bounded by (8.8). It is easily verified that $A \cup C \subset U_2$. We can apply Buchstab's identity again to $S_{2,1}$ to obtain

$$\begin{aligned} S_{2,1} &= \sum_{\alpha_2 \in A \cup C} S(\mathcal{A}_{p_1 p_2}, x^{\kappa}) - \sum_{\alpha_3} S(\mathcal{A}_{p_1 p_2 p_3}, p_3) - \sum_{\alpha_3} S(\mathcal{A}_{p_1 p_2 p_3}, p_3) \\ &= S_{3,0} - S_{3,1} - S_{3,2}, \end{aligned}$$

where Lemmas 7 or 18 furnish an asymptotic formula for $S_{3,2}$ but not for $S_{3,1}$ (so $\boldsymbol{\alpha}_3 \notin G_3$ in $S_{3,1}$ for example). Again we can evaluate $S_{3,0}$ and $S_{3,2}$. We write

$$S_{3,1} = S_{3,4} + S_{3,5}.$$

The definition of $S_{3,4}$ is as follows. In $S_{3,4}$, *either* $\boldsymbol{\alpha}_3 \notin U_3$ (so no further decomposition is possible) *or*, if $\boldsymbol{\alpha}_3 \in U_3$, the sum

$$\sum_{\substack{\boldsymbol{\alpha}_3 \\ \kappa \le \alpha_4 \le \min(\alpha_3, \frac{1-\alpha_1-\alpha_2-\alpha_3}{2}) \\ \boldsymbol{\alpha}_4 \notin U_4}} S(\mathcal{A}_{p_1 \ldots p_4}, x^\kappa)$$

cannot be evaluated with error ϵ. The sum $S_{3,4}$ splits into sums over sets of integers with (i) 4 prime factors, (ii) 5 prime factors, (iii) 6 or more prime factors:

$$\begin{aligned} S_{3,4} = & \sum_{\boldsymbol{\alpha}_3} S(\mathcal{A}_{p_1p_2p_3}, (x/p_1p_2p_3)^{1/2}) \\ & + \sum_{\substack{\boldsymbol{\alpha}_3 \\ \alpha_3 \le \alpha_4 \le \frac{1-\alpha_1-\alpha_2-\alpha_3}{2}}} S(\mathcal{A}_{p_1 \ldots p_4}, (x/p_1 \ldots p_4)^{1/2}) \\ & + \sum_{\substack{\boldsymbol{\alpha}_3 \\ \alpha_3 \le \alpha_4 \le \alpha_5 \le \frac{1-\alpha_1-\alpha_2-\alpha_3-\alpha_4}{2}}} S(\mathcal{A}_{p_1 \ldots p_5}, p_5) \\ = & \ S_{3,4,1} + S_{3,4,2} + S_{3,4,3} \text{ say.} \end{aligned}$$

No formula can be given for $S_{3,4,1}$, but it may be possible to give formulae for parts of $S_{3,4,2}$ or $S_{3,4,3}$ using Lemmas 7 and 19. The remainder of the sums must be discarded leading to an error

(9.3)

$$\frac{x}{\phi(q)\log x}\left(\int_{\boldsymbol{\alpha}_3} \frac{d\mu_3}{(1-\alpha_1-\alpha_2-\alpha_3)} + \int_{\boldsymbol{\alpha}_4} \frac{d\mu_4}{(1-\alpha_1-\alpha_2-\alpha_3-\alpha_4)} + \int_{\boldsymbol{\alpha}_5} w\left(\frac{1-\alpha_1-\alpha_2-\alpha_3-\alpha_4-\alpha_5}{\alpha_5}\right)\frac{d\mu_5}{\alpha_5}\right),$$

for $\boldsymbol{\alpha}_3, \boldsymbol{\alpha}_4, \boldsymbol{\alpha}_5$ in certain complicated regions.

By its definition, we can write $S_{3,5} = S_{4,0} - S_{4,1}$ where $S_{4,0}$ can be evaluated with error ϵ and $S_{4,1}$ can be further decomposed:

$$S_{4,1} = \sum S(\mathcal{A}_{p_1 \ldots p_4}, x^\kappa) - \sum S(\mathcal{A}_{p_1 \ldots p_5}, p_5) = S_{5,0} - S_{5,1},$$

and $S_{5,0}$ can be evaluated with error ϵ. For part (in fact *most*) of $S_{5,1}$, an asymptotic formula can be given and the rest must be discarded. This leads to an error

$$\frac{x}{\phi(q)\log x}\int_{\boldsymbol{\alpha}_5} w\left(\frac{1-\alpha_1-\alpha_2-\alpha_3-\alpha_4-\alpha_5}{\alpha_5}\right)\frac{d\mu_5}{\alpha_5}, \tag{9.4}$$

for $\boldsymbol{\alpha}_5$ in a certain region. Since $\kappa > 1/7$ for the values of θ under consideration, $S_{5,1}$ (and $S_{3,4,3}$ above) only counts products of 6 primes, so no further decomposition is possible (also $w(u) = 1/u$ in (9.3) and (9.4)).

Combining all our results we therefore obtain

$$S(\mathcal{A},(2x)^{1/2}) \leq \frac{x}{\phi(q)\log x}(1+O(\epsilon)+E_{1,3}+E_{2,3}+E_{3,4}+E_{5,1})$$

where $E_{j,k}$ denotes the 'error' in bounding $S_{j,k}$. Each of these errors is a continuous function of θ with $E_{j,k} = O(\theta - 1/2 + \epsilon)$.

We shall now give more details for $\theta \leq 0.51$. Our purpose for this is two-fold. We would like the reader to be confident that our numerical calculations are accurate–we shall give fewer details for the following cases. Also we wish to justify what might appear to be a surprisingly good value for $C_1(0.51)$. We shall need this value for a further application to be discussed elsewhere. We partition C into $\mathcal{C}_1, \mathcal{C}_2$ and $\mathcal{C}_3$, where (see Diagram 4):

$$\begin{aligned}
\mathcal{C}_1 &= \{(\alpha_1,\alpha_2) \in C : \alpha_1 < \theta - 1/5 + 2\epsilon\}\\
\mathcal{C}_2 &= \{(\alpha_1,\alpha_2) \in C : \alpha_1 \geq \theta - 1/5 + 2\epsilon, \alpha_2 \geq \theta/2\}\\
\mathcal{C}_3 &= \{(\alpha_1,\alpha_2) \in C : \alpha_2 < \theta/2\}.
\end{aligned}$$

We let $E'_{3,4}$ be the contribution to $E_{3,4}$ and $E_{5,1}$ from $\boldsymbol{\alpha}_2 \in A$.

There is no contribution to $E_{5,1}$ from C. To see this, we note that, for $\alpha_1+\alpha_2 > \theta, \alpha_3 > \alpha_4 \geq \kappa$,

$$\begin{aligned}
\alpha_1+\alpha_2+4(\alpha_3+\alpha_4) &> \theta+(20-32\theta)/3 > 3-3\theta\\
\alpha_2+\alpha_3+\alpha_4+4\alpha_1 &> \frac{5-8\theta}{2}+2\theta > 3-3\theta\\
\alpha_1+\alpha_3+\alpha_4 &> \frac{\theta}{2}+\frac{5-8\theta}{3} > 1-\theta
\end{aligned}$$

which, with a little thought, we find prevents $(\alpha_1,\alpha_2,\alpha_3,\alpha_4)$ from belonging to U_4.

Let EC_j be the contribution to $E_{3,4}$ from $\boldsymbol{\alpha}_2 \in \mathcal{C}_j$. We shall prove below that $EC_2 = 0$. We obtain the following bounds (generously rounded up) by calculations performed on a personal computer in BASIC.

θ	$E_{1,3}$	$E_{2,3}$	$E'_{3,4}$	EC_1	EC_3
0.505	10^{-3}	10^{-3}	10^{-3}	10^{-4}	2×10^{-4}
0.51	4×10^{-3}	3×10^{-3}	5.5×10^{-3}	4×10^{-4}	1.5×10^{-3}

Let $\delta = \theta - 1/2 + \epsilon$. All the values above have a main term $O(\delta^2)$ except EC_3 which is $O(\delta^3)$. Altogether, for $0.5 \le \theta \le 0.51$ we have $C_1(\theta) \le 1 + 150\delta^2$.

We now focus on $\boldsymbol{\alpha}_2 \in \mathcal{C}_1 \cup \mathcal{C}_2$. Let $\alpha_4 = 1 - \alpha_1 - \alpha_2 - \alpha_3$. Suppose that

$$(\alpha_1 - \alpha_2) + (\alpha_4 - \alpha_3) > 2\delta. \tag{9.5}$$

Then

$$\alpha_1 + \alpha_4 = \frac{1}{2}(\alpha_1 + \alpha_2 + \alpha_3 + \alpha_4) + \frac{1}{2}(\alpha_1 - \alpha_2) + \frac{1}{2}(\alpha_4 - \alpha_3) > \frac{1}{2} + \delta,$$

$$\begin{aligned} 2\alpha_1 + 5\alpha_4 &= 5 - 3\alpha_1 - 5\alpha_2 - 5\alpha_3 \\ &\le 5 - 5\alpha_3 - \alpha_1 - 2 + \epsilon \\ &\lesssim 3 - 5\kappa - 2/7 < 2 \text{ since } \theta \le 0.51 < 29/56. \end{aligned}$$

Thus either $(\alpha_1, \alpha_3) \in \mathcal{G}^3$ or $(\alpha_1, \alpha_4) \in \mathcal{G}^3$. To see this,

$$2\alpha_1 + 5\alpha_3 \le 2\alpha_1 + 5\alpha_4 < 2$$

as shown above. If

$$2\alpha_1 + \alpha_4 \ge 2 - 2\theta - \epsilon$$

then

$$\begin{aligned} \alpha_1 + \alpha_3 \ge \alpha_1 + \kappa &= 2\alpha_1 + \alpha_4 + \alpha_2 + \alpha_3 + \kappa - 1 \\ &\ge 1 - 2\theta - \epsilon + \alpha_2 + 2\kappa > \theta + \epsilon \text{ since } \alpha_2 \ge \theta/2, \end{aligned}$$

and moreover

$$\begin{aligned} 2\alpha_1 + \alpha_3 &\le 4\alpha_1 + \alpha_3 + \alpha_4 - (2 - 2\theta - \epsilon) \\ &= 2 - 2\theta + 3\alpha_1 - \alpha_2 + 4\theta - 3 + \epsilon \\ &= 2 - 2\theta + \frac{7}{11}(\alpha_1 + 4\alpha_2) - \frac{13}{11}(3\alpha_2 - 2\alpha_1) + 4\theta - 3 + \epsilon \\ &\le 2 - 2\theta + \frac{7}{11}(3 - 3\theta) + 4\theta - 3 + \epsilon < 2 - 2\theta - \epsilon. \end{aligned}$$

We conclude that for $\boldsymbol{\alpha}_2 \in \mathcal{C}_1 \cup \mathcal{C}_2$, (9.5) implies $(\alpha_1, \alpha_2, \alpha_3, \alpha_4) \in G_4$.

Now in $\mathcal{C}_2$,

$$\begin{aligned}\alpha_1-\alpha_2 &\geq \alpha_1-\frac{3(1-\theta)-\alpha_1}{4}\\ &=\frac{5\alpha_1+3\theta-3}{4}>\frac{5(\theta-1/5+2\epsilon)+3\theta-3}{4}>2\delta.\end{aligned}$$

This establishes (9.5) and proves that $E\mathcal{C}_2=0$.

Turning our attention to $\mathcal{C}_1$, we obtain likewise

$$E\mathcal{C}_1=\int_{2/7}^{\theta-1/5}\int_{\substack{\alpha_1-2\delta\\ \alpha_1+4\alpha_2\leq 3-3\theta}}^{\alpha_1}\int_{\frac{1}{2}-\alpha_2-\delta}^{\frac{1}{2}(1-\alpha_1-\alpha_2)}\frac{d\mu_3}{\alpha_4}+O(\delta^3)$$

since

$$\alpha_3<\frac{1}{2}(1-\alpha_1-\alpha_2)$$

and since if (9.5) fails then

$$\alpha_1-\alpha_2+1-\alpha_1-\alpha_2-2\alpha_3\leq 2\delta,\ \text{that is, } \alpha_3\geq\frac{1}{2}-\alpha_2-\delta.$$

(For the part of $E\mathcal{C}_1$ arising from products of 5 or more primes, we would be able to restrict α_4 as well as α_3 to an interval of length $O(\delta)$ by working with $(\alpha_1,\alpha_2,\alpha_4,1-\alpha_1-\alpha_2-\alpha_4)$.)

By elementary integration,

$$E\mathcal{C}_1=\delta^2\int_{2/7}^{3/10}\frac{d\alpha}{\alpha^2(\frac{1}{2}-\alpha)^2}+O(\delta^3)=(2+16\log 9/8)\delta^2+O(\delta^3).$$

This supports the bound $4\delta^2$ in the above table. (Actually, $\boldsymbol{\alpha}_2\in B$ for part of the integral which more than compensates for the $O(\delta^3)$ term.)

Turning our attention to $\mathcal{C}_3$, we find that

$$\alpha_1>\frac{2}{5}+\frac{6\delta}{5}\Rightarrow(\alpha_1,\alpha_4)\in E\text{ or }(\alpha_1,\alpha_3)\in B\left(\theta\leq\frac{59}{116}-\epsilon\right)\tag{9.6}$$

while

$$\alpha_1<\frac{2}{5}-\frac{6\delta}{5}\Rightarrow(\alpha_1,\alpha_3)\in B.\tag{9.7}$$

(Unfortunately we must introduce another term for $\theta\geq\frac{59}{116}-\epsilon$. It is very small, certainly less than $1200\delta(\delta-1/116)^2$ $\left(<2.3\times10^{-5}\text{ for }\delta=0.01\right)$.)

To prove (9.6), we begin with

$$\alpha_1 < \frac{2\theta + \epsilon - \kappa}{2} < 1 - \theta.$$

Next,

$$\begin{aligned}\alpha_4 + 2\alpha_1 &\geq \frac{1 - \alpha_1 - \alpha_2}{2} + 2\alpha_1 = \frac{3\alpha_1 - \alpha_2 + 1}{2}\\ &> \frac{3\alpha_1 - (2\theta + \epsilon - 2\alpha_1)}{2} + \frac{1}{2} > 2\theta + \epsilon\end{aligned}$$

since $\alpha_1 > \frac{2}{5} + \frac{6\delta}{5}$. Finally, if $(\alpha_1, \alpha_3) \notin B$ then it is easy to see that

$$\begin{aligned}&2\alpha_1 + \alpha_3 > 2 - 2\theta - \epsilon,\\ &3\alpha_4 - 2\alpha_1 = 3 - 5\alpha_1 - 3\alpha_2 - 3\alpha_3\\ &\qquad = 3 - \frac{5}{4}(2\alpha_1 + \alpha_2) - \frac{5}{4}(2\alpha_1 + \alpha_3) - \frac{7}{4}(\alpha_2 + \alpha_3)\\ &\qquad \leq 5\theta - 2 - \frac{7}{4}(2\kappa) + \frac{5\epsilon}{2} < -\epsilon\end{aligned}$$

for $\theta \leq \frac{59}{116} - \epsilon$.

To prove (9.7), we begin with

$$2\alpha_1 + \alpha_2 > 2 - 2\theta - \epsilon, \quad \alpha_1 > 1 - \theta - \frac{\alpha_2}{2} - \epsilon \geq 1 - \frac{5\theta}{4} - \epsilon,$$

and so

$$\alpha_1 + \alpha_3 > 1 - \frac{5\theta}{4} - \epsilon + \kappa > \theta + \epsilon.$$

Further,

$$\begin{aligned}5\alpha_3 + 2\alpha_1 &\leq \frac{5}{2}(1 - \alpha_1 - \alpha_2) + 2\alpha_1\\ &= \frac{5}{2} - \frac{5}{2}(\alpha_2 + 2\alpha_1) + \frac{9}{2}\alpha_1\\ &\leq \frac{5}{2} - \frac{5}{2}(2 - 2\theta - \epsilon) + \frac{9}{2}\left(\frac{2}{5}\right) < 2 - \epsilon,\\ \alpha_3 + 2\alpha_1 &\leq \frac{1 - \alpha_1 - \alpha_2}{2} + 2\alpha_1 = \frac{5\alpha_1 - (\alpha_2 + 2\alpha_1) + 1}{2}\\ &\leq \frac{2 - 6\delta - 1 + 2\theta + \epsilon}{2} < 2 - 2\theta - \epsilon.\end{aligned}$$

Consequently in the integral for EC_3, each variable is constrained to a range of length $O(\delta)$. Indeed, upon writing

$$\alpha_1 = \frac{2}{5} + x\delta, \quad \alpha_2 = \frac{1}{5} + y\delta, \quad \alpha_3 = \frac{1}{5} + z\delta,$$

we obtain

$$EC_3 = \frac{I}{(1/5)^3(2/5)}\delta^3 + O(\delta^4)$$

where I is the volume of the region given by

$$-\frac{6}{5} \le x \le \frac{6}{5}, y \le x, -2 \le 2x + y \le 2, -2 \le 2x + z \le 2,$$

$$-2 \le x - y - z < 2, \quad z \le \min\left(y, -\frac{(x+y)}{2}\right).$$

We thus obtain

$$EC_3 \approx 1334\delta^3 + O(\delta^4)$$

which supports the bounds claimed above.

We now consider $E'_{3,4}$ in detail. To simplify matters we will not split $S_{3,4}$ into subsums, nor will we continue decomposition to obtain a five-fold integral. We split A into the following subsets (see Diagram 4):

$$\begin{aligned}
A_1 &= \{\boldsymbol{\alpha}_2 \in A : \alpha_1 + \alpha_2 \le \theta - \kappa + \epsilon\}\\
A_2 &= \{\boldsymbol{\alpha}_2 \in A : 4\alpha_2 + \alpha_1 \ge 2 - 2\theta, \alpha_1 + \alpha_2 \le 3/7 + \epsilon\}\\
A_3 &= \{\boldsymbol{\alpha}_2 \in A : 1 - \theta - \epsilon \le \alpha_1 + \alpha_2 \le \theta + \epsilon\}\\
A_4 &= \{\boldsymbol{\alpha}_2 \in A : \frac{3}{7} + \epsilon < \alpha_1 + \alpha_2 < 1 - \theta - \epsilon\}\\
A_5 &= \{\boldsymbol{\alpha}_2 \in A : \theta - \kappa + \epsilon \le \alpha_1 + \alpha_2 < 3/7 + \epsilon, \alpha_1 + 4\alpha_2 < 2 - 2\theta - \epsilon\}.
\end{aligned}$$

We can give an asymptotic formula, for $\boldsymbol{\alpha}_2 \in A_4$, by Lemma 18.

We can give an asymptotic formula, for $\boldsymbol{\alpha}_2 \in A_5$, since $(\alpha_1, \alpha_2 + \alpha_3) \in B$ in this case. For the region A_3 it is possible to have $(\alpha_1 + \alpha_3, \alpha_2) \in E$ or $(\alpha_2 + \alpha_3, \alpha_1) \in E \cup E'$. This *always* happens if

$$\frac{3\theta - 1}{2} + \frac{3\epsilon}{2} \le \alpha_1 \le \frac{4(1-\theta)}{7} - \epsilon. \tag{9.8}$$

To see this, suppose that $\boldsymbol{\alpha}_2 \in A_3$ and $\alpha_1 + \alpha_3 < 1 - \theta - \epsilon$. Then

$$3\alpha_2 - 2(\alpha_1 + \alpha_3) \le \frac{1}{2}(\alpha_1 + \alpha_2) - 2\kappa < \frac{1}{2}(\theta + \epsilon) - 2\kappa < -\epsilon,$$

$$\alpha_2 + 2(\alpha_1 + \alpha_3) \geq \frac{3}{2}(1-\theta) + 2\kappa \geq 2\theta + \epsilon$$

giving $(\alpha_1 + \alpha_3, \alpha_2) \in E$. We therefore consider only triples with

(9.9) $$\boldsymbol{\alpha}_2 \in A_3, \alpha_1 + \alpha_3 \geq 1 - \theta - \epsilon.$$

Now if α_1 satisfies (9.8),

$$\alpha_2 + \alpha_3 \leq \alpha_2 + \frac{1-\alpha_1-\alpha_2}{2} = \frac{\alpha_1+\alpha_2}{2} - \alpha_1 + \frac{1}{2}$$
$$\leq \frac{\theta}{2} - \frac{1}{2}(3\theta - 1) + \frac{1}{2} - \epsilon = 1 - \theta - \epsilon,$$

$$\alpha_1 + 2(\alpha_2 + \alpha_3) \geq 4(\alpha_1 + \alpha_3) - 3\alpha_1 \geq 4 - 4\theta - \frac{12}{7}(1-\theta) - \epsilon > 2\theta + \epsilon,$$

$$3\alpha_1 - 2(\alpha_2 + \alpha_3) = 7\alpha_1 - 2(\alpha_1 + \alpha_3) - 2(\alpha_1 + \alpha_2)$$

$$< 4(1-\theta) - 7\epsilon - 4(1 - \theta - \epsilon) < -\epsilon.$$

For those triples with (9.9) and

(9.10) $$\alpha_1 < \frac{3\theta - 1}{2} + \frac{3\epsilon}{2},$$

$$\alpha_2 + \alpha_3 \leq 1 - \theta - \epsilon,$$

we have

$$(\alpha_2 + \alpha_3, \alpha_1) \in E \cup E'.$$

To see this,

$$3\alpha_1 - 2(\alpha_2 + \alpha_3) = 7\alpha_1 - 2(\alpha_1 + \alpha_3) - 2(\alpha_1 + \alpha_2)$$
$$\leq 7\left(\frac{3\theta - 1}{2} + \frac{3\epsilon}{2}\right) - 4(1 - \theta - \epsilon) < -\epsilon;$$
$$\alpha_1 + 2(\alpha_2 + \alpha_3) = 2(\alpha_1 + \alpha_3) + 2(\alpha_1 + \alpha_2) - 3\alpha_1$$
$$\geq 4 - 4\theta - 4\epsilon - 3\left(\frac{3\theta - 1}{2} + \frac{3\epsilon}{2}\right) \geq 2 - 2\theta + \epsilon.$$

Accordingly, the loss from triples with $\boldsymbol{\alpha}_2 \in A_3$ and (9.10) is

$$2/3 \cdot 4^4 \delta^3 + O(\delta^4),$$

since 2/3 is the volume of the region

$$-1/2 \le x \le 3/2, -1-x \le y \le \min(x, 1-x), -1-y \le z \le \min\left(y, \frac{-(x+y)}{2}\right).$$

For $\alpha_1 > 4(1-\theta)/7$ there is a loss of

$$\int_{2/7}^{1/3} \frac{d\alpha_1}{\alpha_1^2\left(\frac{1}{2}-\alpha_1\right)^2} \int_{1-\theta-\alpha_1}^{\theta-\alpha_1} d\alpha_2 \int_{1-\theta-\alpha_1}^{\alpha_2} d\alpha_3 + O(\delta^3)$$

$$= 2\delta^2 \int_{2/7}^{1/3} \frac{d\alpha_1}{\alpha_1^2\left(\frac{1}{2}-\alpha_1\right)^2} + O(\delta^3) = 8\delta^2\left(\frac{11}{6} + 4\log\frac{3}{2}\right) + O(\delta^3).$$

For $\boldsymbol{\alpha}_2 \in A_1$ one may use the fact that $(1-\alpha_1-\alpha_2-\alpha_3)/\alpha_3 \ge 3 + O(\delta)$ and the formula

$$\omega(3) = \frac{1+\log 2}{3}$$

to bound the loss from above by

$$6^4\delta^3\frac{(1+\log 2)}{3}I + O(\delta^4)$$

where I is the volume of the region

$$-1/3 \le x \le 11/3, -4/3 \le y \le \min(7/3-x, x), -4/3 \le z \le y,$$

$$-1 \le x+y+z \le 1.$$

This gives a contribution $\approx 2000\delta^3 + O(\delta^4)$.

For $\boldsymbol{\alpha}_2 \in A_2, (\alpha_1+\alpha_2, \alpha_3)$ and $(\alpha_1+\alpha_3, \alpha_2)$ belong to $B \cup C \cup E$. Hence we can give an asymptotic formula (for $\theta < 59/116$) unless $(\alpha_1+\alpha_2, \alpha_3)$ and $(\alpha_1+\alpha_3, \alpha_2)$ both belong to C_3', where

$$C_3' = \left\{(s,t) \in C_3 : \frac{2-6\delta}{5} < s \le \frac{2+6\delta}{5}, 1-2\delta \le t+2s \le 1+2\delta\right\}.$$

This leads to a contribution $< 700\delta^3 + O(\delta^4)$.

Of course, we cannot just add together the figures for A since we did not quantify all errors $O(\delta^3), O(\delta^4)$. Nevertheless the discussion would indicate that

$$E'_{3,4} \approx 22.3\delta^2 + 3000\delta^3.$$

This accords well with our calculated bounds.

(ii) $\frac{25}{49} - \epsilon < \theta \leq \frac{21}{41}$. In this interval the upper bound on α_1 in R is $4 - 7\theta + O(\epsilon)$. This means that in (9.1) we must introduce another sum

$$\sum_{4-7\theta+O(\epsilon)\leq\alpha_1\leq 3/7+\epsilon} S(\mathcal{A}_{p_1}, p_1).$$

The additional error introduced here is

$$\int_{4-7\theta}^{3/7} \frac{d\alpha}{\alpha(1-\alpha)} + O(\epsilon) = \log\left(\frac{3}{4}\cdot\frac{7\theta-3}{4-7\theta}\right)$$

since we discard this sum. We remark that the value for C_1 now starts to rise sharply compared to its previous behavior because of the introduction of this and like terms. The contribution from integrals of dimension 3 and above remains small until θ is considerably larger (> 0.53). Let us tabulate here what happens between 0.511 and 0.524 in intervals of length 10^{-3}. The values are discussed further in the subsequent cases.

	E^*	$E_{1,3}$	$E_{2,3}$	$E'_{3,4}$	EC_1	EC_3
0.511	2.28×10^{-2}	5×10^{-3}	3.5×10^{-3}	6×10^{-3}	4×10^{-4}	2×10^{-3}
0.512	5.16×10^{-2}	6×10^{-3}	5×10^{-3}	7×10^{-3}	4×10^{-4}	2.5×10^{-3}
0.513	0.136	7×10^{-3}	6×10^{-3}	8×10^{-3}	3×10^{-4}	3×10^{-3}
0.514	0.230	8×10^{-3}	8×10^{-3}	8×10^{-3}	2×10^{-4}	2.5×10^{-3}
0.515	0.318	9×10^{-3}	1×10^{-2}	7×10^{-3}	2×10^{-4}	1×10^{-3}
0.516	0.401	1×10^{-2}	1×10^{-2}	7×10^{-3}	1×10^{-4}	2×10^{-4}
0.517	0.648	1.2×10^{-2}		5×10^{-3}	8×10^{-4}	
0.518	0.652	1.5×10^{-2}		6×10^{-3}	8×10^{-5}	
0.519	0.656	1.8×10^{-2}		8×10^{-3}	7×10^{-5}	
0.520	0.660	2.2×10^{-2}		1.1×10^{-2}	5×10^{-5}	
0.521	0.664	2.6×10^{-2}		1.5×10^{-2}	3×10^{-5}	
0.522	0.668	3.1×10^{-2}		1.8×10^{-2}	2×10^{-5}	
0.523	0.672	3.6×10^{-2}		2.2×10^{-2}	1×10^{-5}	
0.524	0.676	4.1×10^{-2}		2.7×10^{-2}		

Here E^* corresponds to the 1-dimensional integrals.

(iii) $21/41 < \theta \le 16/31$. For θ in this interval, the lower bound on α_1 in R is $(31\theta - 15)/3 + O(\epsilon)$. This introduces a further sum in (9.1) with

$$\tau \le \alpha_1 \le (31\theta - 15)/3 + O(\epsilon).$$

The additional error introduced here is $O(\theta - 21/41 + \epsilon)$.

In this range, the regions of $A, \mathcal{C}_1$ and $\mathcal{C}_3$ which contribute actually decrease because of the excluded values for α_1. Note that, as happens elsewhere, for increasing θ we are always discarding larger sums. When one sum (e.g. EC_1) is decreasing, the amount lost in E^* increases more rapidly. The behavior of $C_1(\theta)$ thus remains monotonic.

(iv) $16/31 < \theta \le 29/56$. There is a discontinuity at 16/31 (contrast the behavior of E^* below and above 0.516 in the table). This arises because the middle term in E_2 in Lemma 22 becomes too large for α_1 across a range of values at this point. Hence the whole sum

$$\sum_{\tau < \alpha_1 < \frac{3}{7} + \epsilon} S(\mathcal{A}_p, p)$$

must be discarded because we are unable to deal with region D. Consequently $E_{2,3}, EC_3$ and parts of $E'_{3,4}$ no longer appear.

(v) $29/56 \le \theta \le 11/21$. In this region we note that it is now possible for products of 7 primes to be counted in our sums. This would lead to a very small improvement on the bound obtained, at the expense of more effort.

(vi) $11/21 \le \theta \le 17/32$. The choice of τ previously was governed by the point of intersection of the lines $\alpha_1 = \alpha_2, \alpha_1 + 4\alpha_2 = 3 - 3\theta$. However, for $\theta > 11/21$, we can evaluate

$$\sum_{\substack{\frac{3(1-\theta)}{5} \le \alpha_1 \le 2/7 \\ \alpha_2 < \alpha_1, \alpha_2 \ge \theta - \alpha_1 + \epsilon}} S(\mathcal{A}_{pq}, q)$$

via Lemma 7, while if $\alpha_2 < \theta - \alpha_1 + \epsilon$ we have $\alpha_1 + 4\alpha_2 \le 3 - 3\theta - \epsilon$, $\alpha_1 + 2\alpha_2 \le 3 - 4\theta - \epsilon$, so $\boldsymbol{\alpha}_2 \in U_2$.

There is now no contribution from region C and the value of E^* is fixed. Only $E_{1,3}$ and $E'_{3,4}$ are increasing here with θ. In this interval,

$$C_1(\theta) \le 1.725 + 13.125(\theta - 0.523).$$

(vii) $17/32 - \epsilon \le \theta \le 7/13 - \epsilon$. At 17/32 the value of κ is halved. This is not the problem which at first sight it would appear to be, since the new sums introduced are $O(\theta - 17/32 + \epsilon)$. These new sums come from regions like H on Diagram 3. The 'asymptotic formula region' $2\theta - 1 + \epsilon \le s \le (5 - 8\theta)/6 - \epsilon$, which has not been of explicit use in this section, now comes into play. Also the region $\alpha_1 \geq 1 - \theta - \epsilon$ must be discarded, that is, we begin with

$$S(\mathcal{A}, (2x)^{1/2}) \leq S(\mathcal{A}, x^{\tau}) - \sum_{x^{3/7+\epsilon} \leq p \leq x^{1-\theta-\epsilon}} S(\mathcal{A}_p, p), \tag{9.10}$$

where we are giving an upper bound for the first sum on the right of (9.10) and an asymptotic formula for the second.

Since κ has halved in value we are now counting products of ≥ 16 primes in some sums which arise. Clearly there are now many different ways variables can combine to lie in 'good' ranges, and it may be possible to apply Buchstab several more times. It may not be practical to carry out the calculations in the most efficient manner, but some integrals can be shown to be small using

$$\int_{\kappa}^{\lambda}\int_{\kappa}^{\alpha_1} \frac{d\mu_2}{\alpha_2} = \frac{1}{\kappa}\left(\log\left(\frac{\lambda}{\kappa}\right) - 1\right) + \frac{1}{\lambda},$$

and

$$\int_{\kappa}^{\lambda} \cdots \int_{\kappa}^{\alpha_{2n-1}} \frac{d\mu_{2n}}{\alpha_{2n}} \leq \frac{(\log(\lambda/\kappa))^{2n}}{\kappa(2n)!}.$$

(viii) $7/13 - \epsilon \leq \theta < 6/11 - \epsilon$. At 7/13 the value of κ reduces again. Indeed, the value reduces to κ' if we wish to continue carrying out decompositions in the same way. We begin by writing

(9.11)
$$S(\mathcal{A}, (2x)^{1/2}) \leq S(\mathcal{A}, x^{\kappa}) - \sum_{\kappa \leq \alpha_1 \leq \tau'} S(\mathcal{A}_{p_1}, p_1) - \sum_{\tau' < \alpha_1 \leq \tau} S(\mathcal{A}_{p_1}, p_1) -$$
$$\sum_{3/7+\epsilon \leq \alpha_1 \leq 1-\theta-\epsilon} S(\mathcal{A}_{p_1}, p_1) = S_1 - S_2 - S_3 - S_4, \text{ say.}$$

For S_2 we can decompose further:

$$S_2 = \sum S(\mathcal{A}_{p_1}, x^{\kappa}) - \sum S(\mathcal{A}_{p_1 p_2}, p_2),$$

and we can do the same for S_3 with κ replaced by κ'. Of course we can evaluate S_4 as before.

We note that the above approach works because the line $\alpha_1 + 4\alpha_2 = 3 - 3\theta - \epsilon$ lies in or above the region B. This fails for $\theta > 6/11 - \epsilon$.

(ix) $6/11 - \epsilon \le \theta < 0.56$. For θ near $6/11 - \epsilon$ it only requires a minor modification to (9.11) to make the method work. If the sum with

$$\tau'' = \frac{3 - 3\theta}{5} - \epsilon \le \alpha_1 \le \frac{7\theta - 3}{3} = \tau'''$$

is discarded, it is still possible to perform further decompositions with $\tau' \le \alpha_1 \le \tau''$ and $\tau''' \le \alpha_1 \le 2/7$. One runs into serious practical difficulties quickly since κ is decreasing to zero at $\theta = 0.55$. It is best therefore to begin with

$$S(\mathcal{A}, (2x)^{1/2}) \le S(\mathcal{A}, x^{\tau}) - \sum_{3/7 \le \alpha_1 \le 1-\theta-\epsilon} S(\mathcal{A}_{p_1}, p_1)$$

for $\theta \ge 0.546$.

By performing the calculation as explained in (vi)–(ix) we arrive at the following values for $C_1(\theta)$

θ	$C_1(\theta)$
0.525	1.75
0.53	1.82
0.533	1.96
0.535	2.09
0.54	2.25
0.545	2.47
0.55	2.66
0.555	2.76
0.56	2.88

We omit the detailed calculation of the integral (1.2), but the reader should find the result entirely plausible in view of the above tables and (4.16).

Lower bounds. We now suppose $1/2 < \theta < 25/49 - \epsilon$. We begin with (9.1), but now we must use (7.8) to supply a lower bound for $-S_{1,3}$. We can apply Buchstab again to $S_{1,1}$ to obtain (9.2) as previously. It follows immediately from this that we can produce a lower bound for $\pi(x; q, a)$ when

$$\int_{\kappa}^{\theta/2} \int_{1-\theta-\alpha_1}^{\theta-\alpha_1} w(\alpha_2/\alpha_1) w\left(\frac{1-\alpha_1-\alpha_2}{\alpha_1}\right) \frac{d\alpha_2 d\alpha_1}{\alpha_1^3} + \int_{\alpha_2 \in A \cup C \cup D} w\left(\frac{1-\alpha_1-\alpha_2}{\alpha_2}\right) \frac{d\mu_2}{\alpha_2} < 1.$$

At $\theta = 0.5$ the first integral is zero, of course, while the second is less than 0.61.

We can do much better, though, as we now consider the regions A, C, and D in turn.

Region A. We split into subregions as previously. In A_1 it is possible to decompose twice more, so we obtain an error $O(\delta^4)$. In A_2 we can also decompose twice more. Recalling our previous discussion, we can obtain an asymptotic formula unless both $(\alpha_1+\alpha_2, \alpha_3)$ and $(\alpha_1+\alpha_3, \alpha_2)$ belong to $\mathcal{C}_3'$. In view of our discussion of A_2 and $\mathcal{C}_3$ in the upper bound case, this leads to an error $O(\delta^4)$. We discard the parts of A_3 with $\alpha_1 > 4(1-\theta)/7$ or $\alpha_1 < (3\theta-1)/2$. This leads to a loss $O(\delta)$. For the rest of A_3, together with A_4 and A_5, we obtain an asymptotic formula as in the upper bound case. Hence we can estimate

$$\sum_{\boldsymbol{\alpha}_2 \in A, \alpha_1 \leq 3/7+\epsilon} S(\mathcal{A}_{p_1p_2}, p_2)$$

with error $O(\delta)$ as required.

Region C. As with region A above, although we cannot now discard three fold sums, we can still borrow much of the working from the upper bound case. We discard that part of $\mathcal{C}_1$, say $\mathcal{C}_1'$, determined by

$$\mathcal{C}_1' = \{\boldsymbol{\alpha}_2 \in \mathcal{C}_1 : \alpha_2 > \alpha_1 - 2\delta\}.$$

This leads to an error of size $O(\delta)$. For the remainder of $\mathcal{C}_1$ and $\mathcal{C}_2$ we obtain an asymptotic formula as previously. For $\mathcal{C}_3$ we can do likewise except in the region $\mathcal{C}_3'$ defined above. This leads to an error of size $O(\delta^2)$. We conclude that we can evaluate

$$\sum_{\boldsymbol{\alpha}_2 \in C, \alpha_1 \leq 3/7+\epsilon} S(\mathcal{A}_{p_1p_2}, p_2)$$

with error $O(\delta)$ as required.

Region D. This has already been evaluated in section 8 (see (8.9)) with error $O(\theta-1/2+\epsilon)$, for $\theta < 16/31-\epsilon$. We note that one advantage of the lower bound case is that we can discard those parts of D which cannot be evaluated. The problems encountered for the upper bound case for $0.51 < \theta \leq 0.52$ are thus not so severe in the lower bound case.

We have already gone into the bounds obtained for $C_2(\theta)$ in detail. We give some values of the errors in the lower bound case for illustration. For simplicity, the region D was discarded for all calculations here. The entries are upper bounds, except for the values of C_1 which are, of course, lower bounds.

Region	0.51	0.515	0.52
A	0.135	0.21	0.29
D	0.115	0.126	0.138
C	0.045	0.085	0.15
$1-\theta \le \alpha_1 \le 1/2$	0.1	0.18	0.26
C_1	0.6	0.39	0.16

With the final entry in the table, we have completed the proof of Theorem 1.

Diagrams

Diagram 1

θ = 0.5

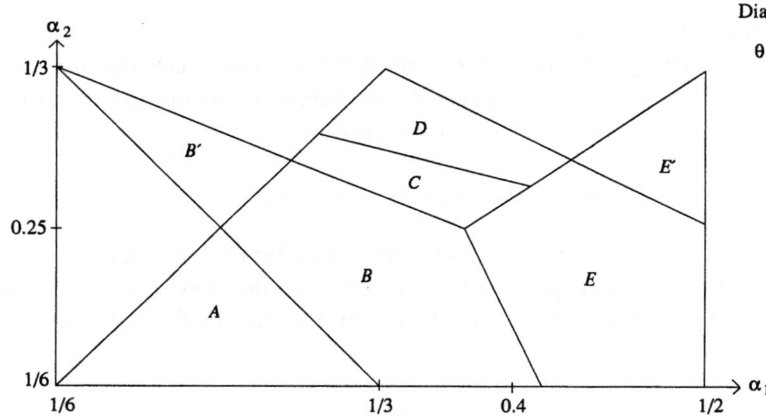

Diagram 2

θ = 0.52

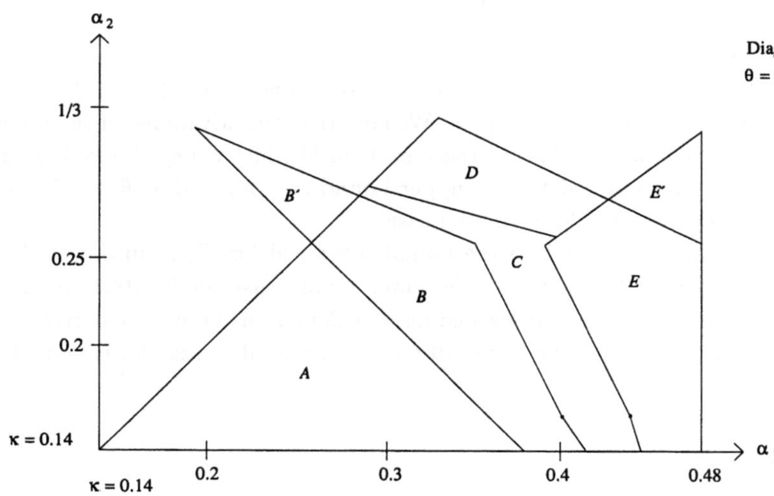

Diagram 3

$\theta = 0.54$

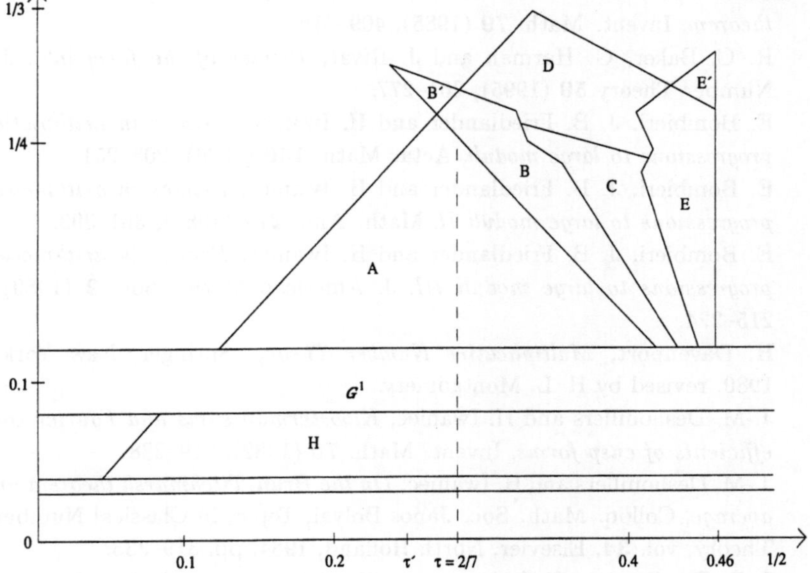

Diagram 4

$\theta = 0.51$

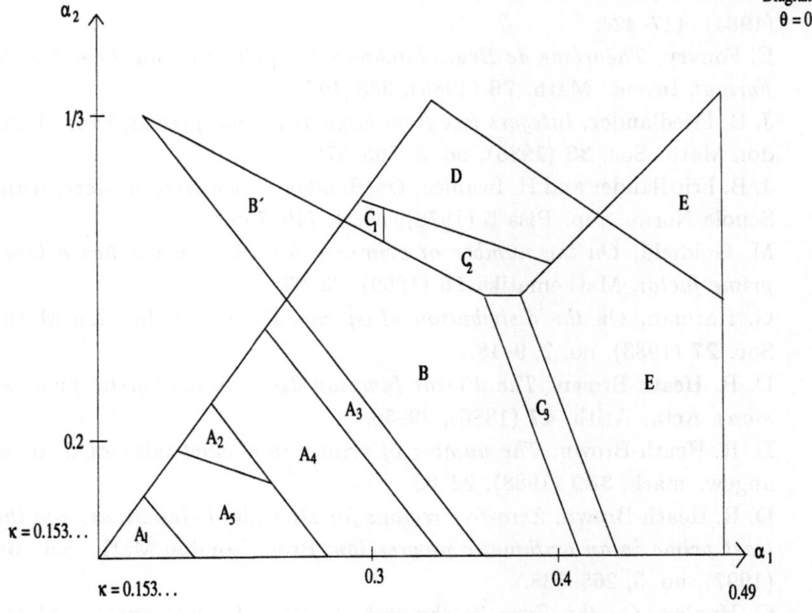

References

[1] L. M. Adleman and D. R. Heath-Brown, *The first case of Fermat's last theorem*, Invent. Math. **79** (1985), 409–416.

[2] R. C. Baker, G. Harman and J. Rivat, *Primes of the form* $[n^c]$, J. Number Theory **50** (1995), 261–277.

[3] E. Bombieri, J. B. Friedlander and H. Iwaniec, *Primes in arithmetic progressions to large moduli*, Acta. Math. **156** (1986), 203–251.

[4] E. Bombieri, J. B. Friedlander and H. Iwaniec, *Primes in arithmetic progressions to large moduli II*, Math. Ann. **277** (1987), 361–393.

[5] E. Bombieri, J. B. Friedlander and H. Iwaniec, *Primes in arithmetic progressions to large moduli III*, J. American Math. Soc. **2** (1989), 215–224.

[6] H. Davenport, *Multiplicative Number Theory*, Springer, New York, 1980, revised by H. L. Montgomery.

[7] J.-M. Deshouillers and H. Iwaniec, *Kloosterman sums and Fourier coefficients of cusp forms*, Invent. Math. **70** (1982), 219–288.

[8] J.-M. Deshouillers and H. Iwaniec, *On the Brun-Titchmarsh theorem on average*, Colloq. Math. Soc. Janos Bolyai, Topics in Classical Number Theory, vol. **34**, Elsevier, North Holland, 1984, pp. 319–333.

[9] J. B. Friedlander, *Shifted primes without large prime factors*, Number Theory and Applications (1989), Kluwer, 393–401.

[10] E. Fouvry, *Sur le théorème de Brun-Titchmarsh*, Acta. Arith. **43** (1984), 417–424.

[11] E. Fouvry, *Théorème de Brun-Titchmarsh; application au théorème de Fermat*, Invent. Math. **79** (1985), 383–407.

[12] J. B. Friedlander, *Integers free from large and small primes*, Proc. London Math. Soc. **33** (1986), no. 3, 565–576.

[13] J. B. Friedlander and H. Iwaniec, *On Bombieri's asymptotic sieve*, Ann. Scuola Norm. Sup. Pisa **5** (1978), no. 4, 719–756.

[14] M. Goldfeld, *On the number of primes* p *for which* $p + a$ *has a large prime factor*, Mathematika **16** (1969), 23–27.

[15] G. Harman, *On the distribution of* αp *modulo one*, J. London Math. Soc. **27** (1983), no. 2, 9–18.

[16] D. R. Heath-Brown, *The divisor function* $d_3(n)$ *in arithmetic progressions*, Acta. Arith. **47** (1986), 29–56.

[17] D. R. Heath-Brown, *The number of primes in a short interval*, J. reine angew. math. **389** (1988), 22–63.

[18] D. R. Heath-Brown, *Zero-free regions for Dirichlet L-functions, and the least prime in an arithmetic progression*, Proc. London Math. Soc. **64** (1992), no. 3, 265–338.

[19] C. Hooley, *On the Brun-Titchmarsh theorem*, J. reine angew. Math.

255 (1972), 60–79.

[20] C. Hooley, *On the largest prime factor of* $p+a$, Mathematika **20** (1973), 135–143.

[21] C. Hooley, *Applications of sieve methods to the theory of numbers* (1976), Cambridge University Press, Cambridge.

[22] H. Iwaniec, *A new form of the error term in the linear sieve*, Acta. Arith. **37** (1980), 307–320.

[23] H. Iwaniec, *On the Brun-Titchmarsh theorem*, J. Math. Soc. Japan **34** (1982), 95–123.

[24] H. L. Montgomery and R. C. Vaughan, *On the large sieve*, Mathematika **20** (1973), 119–134.

[25] Y. Motohashi, *A note on the least prime in an arithmetic progression with a prime difference*, Acta. Arith. **17** (1970), 283–285.

[26] B. Rousselet, *Inegalités de type Brun-Titchmarsh en moyenne*, Groupe de travaile en théorie analytique et elementaire des nombres 1986–1987 (1988), Université de Paris-Sud, 91–123.

R. C. Baker
Department of Mathematics
Brigham Young University
Provo, UT 84602

G. Harman
School of Mathematics
University of Wales
College of Cardiff,
Senghennydd Rd.
Cardiff CF2 4AG
UNITED KINGDOM

Congruences for Fourier coefficients of half-integral weight modular forms and special values of L-functions

Antal Balog, Henri Darmon and Ken Ono

Dedicated to Professor Heini Halberstam

Abstract. Congruences for Fourier coefficients of integer weight modular forms have been the focal point of a number of investigations. In this note we shall exhibit congruences for Fourier coefficients of a slightly different type. Let $f(z) = \sum_{n=0}^{\infty} a(n)q^n$ be a holomorphic half integer weight modular form with integer coefficients. If ℓ is prime, then we shall be interested in congruences of the form

$$a(\ell N) \equiv 0 \mod \ell$$

where N is any quadratic residue (resp. non-residue) modulo ℓ. For every prime $\ell > 3$ we exhibit a natural holomorphic weight $\frac{\ell}{2} + 1$ modular form whose coefficients satisfy the congruence $a(\ell N) \equiv 0 \mod \ell$ for every N satisfying $\left(\frac{-N}{\ell}\right) = 1$. This is proved by using the fact that the Fourier coefficients of these forms are *essentially* the special values of real Dirichlet $L-$series evaluated at $s = \frac{1-\ell}{2}$ which are expressed as generalized Bernoulli numbers whose numerators we show are multiples of ℓ. From the works of Carlitz and Leopoldt, one can deduce that the Fourier coefficients of these forms are almost always a multiple of the denominator of suitable Bernoulli numbers. Using these examples as a template, we establish sufficient conditions for which the Fourier coefficients of a half integer weight modular form are almost always divisible by a given positive integer M.

We also present two more examples of half-integer weight forms with such congruence properties, whose coefficients are determined by the special values at the center of the critical strip for the quadratic twists of the modular L-functions associated to the modular form Δ of weight 12 and level 1, and to the unique form $\eta^8(z)\eta^8(2z)$ of weight 8 and level 2. We suggest a conceptual

---

The second author wishes to acknowledge CICMA for providing a stimulating intellectual environment while this paper was written. His research was supported by a grant from NSERC and by a nouveaux chercheurs grant from FCAR. The third author is supported by grants DMS-9304580 and DMS-9508976.

explanation for these congruences by remarking that the twists of the mod p Galois representations ($p = 11$ and 7 respectively) associated to these two forms are isomorphic to the Galois representations associated to certain elliptic curves of odd analytic rank.

1. Congruences for Fourier coefficients

First we shall fix the following notation. If $D \equiv 0, 1 \mod 4$ is the fundamental discriminant of the quadratic field $\mathbb{Q}(\sqrt{D})$, then let χ_D denote the Kronecker character with conductor $|D|$. In particular, if p is a prime that splits (resp. is inert) in $\mathbb{Q}(\sqrt{D})$, then $\chi_D(p) = 1$ (resp. $\chi_D(p) = -1$). It turns out that χ_D, in terms of Jacobi symbols, is given by:

$$(0) \qquad \chi_D(n) := \begin{cases} \left(\frac{n}{|D|}\right) & \text{if } D \equiv 1 \mod 4 \\ \left(\frac{-1}{n}\right)\left(\frac{n}{|D|}\right) & \text{if } D = 4D_1,\ D_1 \equiv 3 \mod 4 \\ \left(\frac{2}{n}\right)\chi_{D_1}(n) & \text{if } D = 2D_1,\ D_1 \equiv 1 \mod 4 \\ \left(\frac{2}{n}\right)\chi_{4D_1}(n) & \text{if } D = 2D_1,\ D_1 \equiv 3 \mod 4. \end{cases}$$

Here $\left(\frac{-1}{n}\right) := (-1)^{\frac{n-1}{2}}$ if n is odd and is zero otherwise, and $\left(\frac{2}{n}\right) := (-1)^{\frac{n^2-1}{8}}$ if n is odd and is zero otherwise. By χ_0 we shall mean the identity character.

If χ is a Dirichlet character modulo N and k is a positive integer or $k \in \mathbb{Z}_{\geq 0} + \frac{1}{2}$, then let $M_k(N, \chi)$ (resp. $S_k(N, \chi)$) denote the finite dimensional complex vector space of holomorphic modular forms (resp. cusp forms) with respect to $\Gamma_0(N)$. Similarly let $M_k(N)$ (resp. $S_k(N)$) denote the space of holomorphic modular forms (resp. cusp forms) with respect to $\Gamma_1(N)$. If $f(z)$ is such a modular form, then it has a Fourier expansion in $q := e^{2\pi i z}$ of the form

$$f(z) = \sum_{n=0}^{\infty} a(n)q^n.$$

For more on the theory of modular forms see [12,13,23].

The congruence properties of Fourier coefficients have been investigated by a number of authors (see [21,22,25,26]). For example, if $\Delta(z) = \sum_{n=1}^{\infty} \tau(n)q^n$ is the unique normalized cusp form of weight 12 with respect to $SL_2(\mathbb{Z})$, then it is well known that

$$\tau(n) \equiv \sigma_{11}(n) \mod 691$$

and

(1) $$\tau(n) \equiv 0 \mod 23 \text{ if } \left(\frac{n}{23}\right) = -1$$

where $\sigma_{11}(n) := \sum_{0<d|n} d^{11}$. These congruences are explained by the theory of modular ℓ-adic Galois representations as developed by Deligne, Serre, and Swinnerton Dyer.

Still there are other examples of congruences for Fourier coefficients which have been the focus of some attention. For example, if $p(n)$ denotes the number of partitions of n, then it is well known that

$$p(5n+4) \equiv 0 \mod 5,$$

(2) $$p(7n+5) \equiv 0 \mod 7,$$

and

$$p(11n+6) \equiv 0 \mod 11$$

for every non-negative integer n. These congruences may be viewed as consequences of the action of certain Hecke operators. Here we illustrate this fact for (2). To see this we construct a holomorphic integer weight modular form whose coefficients inherit these congruence properties. Recall that Euler's generating function for $p(n)$ is given by the infinite product

$$\prod_{n=1}^{\infty} \frac{1}{1-q^n}.$$

Also recall that $\eta(z) := q^{\frac{1}{24}} \prod_{n=1}^{\infty}(1-q^n)$, the Dedekind eta-function, is a weight $\frac{1}{2}$ cusp form. If we define $F_7(z)$ and $a(n)$ by

$$F_7(z) = \frac{\eta^7(7z)}{\eta(z)} = q^2 \prod_{n=1}^{\infty} \frac{(1-q^{7n})^7}{(1-q^n)} = \sum_{n=2}^{\infty} a(n)q^n,$$

then (2) holds if and only if

$$a(7n) \equiv 0 \mod 7$$

for every positive integer n. However $F_7(z)$ is a holomorphic modular form in $M_3(7, \chi_{-7})$ and may be rewritten as

$$F_7(z) = \frac{1}{8}E_3(z) - \frac{1}{8}\eta^3(z)\eta^3(7z)$$

where $E_3(z) = \sum_{n=1}^{\infty} \sigma_{3,\chi_{-7}}(n)q^n$ and $\sigma_{3,\chi_{-7}}(n) = \sum_{0<d|n} \chi_{-7}(d)\frac{n^2}{d^2}$.

It is easy to verify that

$$F_7(z) \mid T_7 = \sum_{n=1}^{\infty} a(7n)q^n = \frac{49}{8}E_3(z) + \frac{7}{8}\eta^3(z)\eta^3(7z);$$

this implies that $a(7n) \equiv 0 \mod 7$ for all n which implies (2).

However there are other congruences that the partition function satisfies. For example it is known that

$$p(49n+19) \equiv p(49n+33) \equiv p(49n+40) \equiv 0 \mod 49$$

for every non-negative integer n (see [17]). In a more convenient form, $p(49n+7\delta-2) \equiv 0 \mod 49$ for all n if δ is a quadratic non-residue modulo 7. Following an argument similar to the one above, this means that the Fourier coefficients of the holomorphic integer weight modular form

$$\frac{\eta^{49}(49z)}{\eta(z)} = q^{100}\prod_{n=1}^{\infty}\frac{(1-q^{49n})^{49}}{1-q^n} = \sum_{n=0}^{\infty} a(n)q^n$$

satisfies the congruence

$$a(49n+7\delta) \equiv 0 \mod 49$$

for all n where δ is a quadratic non-residue modulo 7.

We shall be interested in the arithmetic implications of other congruences of this type.

Definition 1. *Let $F(n)$ be an integer valued arithmetic function, M a positive integer, and ℓ a prime. If*

$$F(\ell N) \equiv 0 \mod M$$

for every positive integer N that is a quadratic residue (resp. non-residue) modulo ℓ, then we say that F has a quadratic congruence modulo M of type $(\ell, +1)$ (resp. $(\ell, -1)$).

Therefore if $F(n) = p(n-2)$, then F has a quadratic congruence modulo 49 of type $(7, -1)$. Moreover by (1) we see that the Fourier coefficients of $\Delta(23z)$ satisfy the quadratic congruence modulo 23 of type $(23, -1)$.

In the following proposition we will give an explicit criterion for determining the truth of certain congruences of this type that will be used in the sequel.

Proposition 2. *Let $f(z) = \sum_{n=0}^{\infty} a(n)q^n \in M_k(N, \chi_0)$ be a holomorphic integer weight modular form with rational integer coefficients. If ℓ is a prime* *dividing N, then the coefficients possess a quadratic congruence of type $(\ell,+1)$ (resp. $(\ell,-1)$) if and only if*

$$a(\ell n)\equiv 0 \mod \ell$$

for every positive integer $n\leq C$ satisfying $\left(\frac{n}{\ell}\right)=1$ (resp. $\left(\frac{n}{\ell}\right)=-1$) where

$$C:=\frac{kN\ell^3}{12}\prod_{p|N}\left(1+\frac{1}{p}\right). \tag{3}$$

Proof of Proposition 2. If $f(z)=\sum_{n=0}^{\infty}a(n)q^n$, and $g(z)=\sum_{n=0}^{\infty}b(n)q^n\in M_k(N,\chi_0)$ have algebraic integer Fourier coefficients, then by a theorem of Sturm [25], $f(z)\equiv g(z) \mod M$ if $a(n)\equiv b(n)$ for every $n\leq \frac{kN}{12}\prod_{p|N}(1+\frac{1}{p})$.

Define $f_1(z)$ and $f_2(z)$ by

$$f_1(z):=f(z)|T_\ell\equiv\sum_{n=0}^{\infty}a(\ell n)q^n \mod \ell$$

and

$$f_2(z):=f(z)|T_{\ell^2}\equiv\sum_{n=0}^{\infty}a(\ell^2 n)q^n \mod \ell.$$

Since the Hecke operators preserve spaces of modular forms, we find that $f_1(z),f_2(z)\in M_k(N,\chi_0)$. If $f_3(z)=f_2(\ell z)\in M_k(N\ell,\chi_0)$, then $f_4(z):=f_1(z)-f_3(z)\in M_k(N\ell,\chi_0)$ and

$$f_4(z)\equiv\sum_{\substack{n=0\\(n,\ell)=1}}^{\infty}a(\ell n)q^n \mod \ell.$$

Now let $f_5(z)$ denote the modular form that is the *quadratic twist* of $f_4(z)$ by $\left(\frac{n}{\ell}\right)$; therefore we find that

$$f_5(z)\equiv\sum_{\substack{n=0\\(n,\ell)=1}}^{\infty}\left(\frac{n}{\ell}\right)a(\ell n)q^n \mod \ell.$$

It turns out that $f_5(z)\in M_k\left(N\ell^3,\chi_0\right)$. Therefore if we define the modular forms $f_+(z)$ and $f_-(z)$ by

$$f_+(z):=\frac{1}{2}\left(f_4(z)+f_5(z)\right)$$

and

$$f_{-}(z) := \frac{1}{2}\left(f_4(z) - f_5(z)\right),$$

then we will find that $f_{+}(z) \equiv 0 \mod \ell$ (resp. $f_{-}(z) \equiv 0 \mod \ell$) if and only if the Fourier coefficients of $f(z)$ satisfy a quadratic congruence modulo ℓ of type $(\ell, +1)$ (resp. $(\ell, -1)$).

The claim then follows from Sturm's theorem if we let $g(z) = 0$. □

2. Special values of $L(s, \chi_D)$

In this section we present a number of examples of such congruences which involve special values of real Dirichlet $L-$functions at negative integers.

If χ is a Dirichlet character with conductor f and n is any positive integer, then it is well known that $L(1-n, \chi) = -\frac{B_{n,\chi}}{n}$ where $B_{n,\chi}$ is the nth generalized Bernoulli number with character χ defined by

$$\sum_{a=1}^{f} \frac{\chi(a)te^{at}}{e^{ft}-1} = \sum_{n=0}^{\infty} B_{n,\chi}\frac{t^n}{n!}.$$

The properties of these numbers are important in the construction of $p-$adic L-functions (see [11,28]). These numbers also occur in congruences involving the special values of $L-$functions of elliptic curves with complex multiplication and also the class numbers of real quadratic fields (see [18]).

In [5] H. Cohen explicitly constructed holomorphic modular forms of half integer weight whose Fourier coefficients are explicit expressions involving the special values at negative integers of Dirichlet $L-$functions of quadratic characters.

Fix a positive integer r. If $N = 0$, then let $H(r, 0) := \zeta(1-2r)$. If N is a positive integer and $Dn^2 = (-1)^r N$ where D is the fundamental discriminant of a quadratic number field, then define $H(r, N)$ by

$$H(r,N) := L(1-r,\chi_D)\sum_{d|n}\mu(d)\chi_D(d)d^{r-1}\sigma_{2r-1}\left(\frac{n}{d}\right). \tag{4}$$

In particular, if $D = (-1)^r N$ is the discriminant of a quadratic field, then

$$H(r,N) = L(1-r,\chi_D) = -\frac{B_{r,\chi_D}}{r}. \tag{5}$$

In all other cases let $H(r, N) := 0$.

If $r \geq 2$ and $F_r(z) := \sum_{n=0}^{\infty} H(r,n)q^n$, then $F_r(z) \in M_{r+\frac{1}{2}}(4, \chi_0)$ (see [Th. 3.1,5]). We shall show that lots of these modular forms have the desired congruence properties. First we need the following lemma:

Lemma 3. *Let $p > 3$ be a prime and suppose that D is a fundamental discriminant of $\mathbb{Q}(\sqrt{D})$ of the form $D = (-1)^{\frac{p+1}{2}} pN$ where N is a positive integer satisfying $\left(\frac{-N}{p}\right) = 1$. Then*

$$\sum_{a=1}^{|D|} \chi_D(a) a^{\frac{p+1}{2}} \equiv 0 \mod p|D|.$$

Proof of Lemma 3. From (0) and the law of quadratic reciprocity, we may factor $\chi_D(n)$ as

$$\chi_D(n) = \left(\frac{n}{p}\right)\chi(n)$$

where χ is a non-trivial real character modulo N satisfying $\chi(p) = \left(\frac{-N}{p}\right)$. Therefore the sum, which we denote T, may be rewritten as

$$T := \sum_{a=1}^{|D|} \chi_D(a) a^{\frac{p+1}{2}} = \sum_{a=1}^{pN} \left(\frac{a}{p}\right)\chi(a) a^{\frac{p+1}{2}}.$$

We first show that $T \equiv 0 \mod N$, a fact which does not depend on the condition that $\left(\frac{-N}{p}\right) = 1$. We split T into residue classes modulo N, and by the Binomial Theorem deduce that

$$T = \sum_{b=0}^{p-1} \sum_{r=1}^{N} \left(\frac{bN+r}{p}\right) \chi(r)(bN+r)^{\frac{p+1}{2}} \equiv$$

$$\equiv \sum_{r=1}^{N} \chi(r) r^{\frac{p+1}{2}} \sum_{b=0}^{p-1} \left(\frac{bN+r}{p}\right) \mod N.$$

Since $\left(\frac{n}{p}\right)$ is a non-trivial Dirichlet character with conductor p and since $\gcd(N,p) = 1$, we find that

$$\sum_{b=0}^{p-1} \left(\frac{bN+r}{p}\right) = \sum_{d=0}^{p-1} \left(\frac{d}{p}\right) = 0.$$

Therefore it is easy to see that $T \equiv 0 \mod N$. Using the same argument, it is also easy to deduce that $T \equiv 0 \mod p$. However to complete the proof we need to establish that $T \equiv 0 \mod p^2$ since $p||D|$.

To establish this claim, we split the sum T into residue classes modulo p and from the Binomial Theorem we find that

$$T = \sum_{c=0}^{N-1} \sum_{s=1}^{p} \left(\frac{cp+s}{p}\right) \chi(cp+s)(cp+s)^{\frac{p+1}{2}} \equiv$$
$$\equiv \sum_{c=0}^{N-1} \sum_{s=1}^{p} \left(\frac{s}{p}\right) \chi(cp+s) \left(s^{\frac{p+1}{2}} + \frac{p+1}{2} \cdot cps^{\frac{p-1}{2}}\right) \mod p^2.$$

Since $\chi(n)$ is a non-trivial Dirichlet character with conductor N, and $\gcd(N,p) = 1$, we find that for every integer s

$$\sum_{c=0}^{N-1} \chi(cp+s) = \sum_{d=0}^{N-1} \chi(D) = 0. \tag{6}$$

Therefore we find that

$$2T \equiv 2\sum_{s=1}^{p} \left(\frac{s}{p}\right) s^{\frac{p+1}{2}} \sum_{c=0}^{N-1} \chi(cp+s)$$
$$+ \sum_{c=0}^{N-1} \sum_{s=1}^{p} \left(\frac{s}{p}\right) \chi(cp+s)(p+1)cps^{\frac{p-1}{2}} \mod p^2$$
$$\equiv (p+1) \sum_{c=0}^{N-1} \sum_{s=1}^{p} \left(\frac{s}{p}\right) \chi(cp+s)cps^{\frac{p-1}{2}} \mod p^2.$$

Since $\left(\frac{s}{p}\right) s^{\frac{p-1}{2}} \equiv 1 \mod p$, for every integer $s \not\equiv 0 \mod p$ then

$$2T \equiv (p+1) \sum_{c=0}^{N-1} \sum_{s=1}^{p-1} \chi(cp+s)cp \mod p^2.$$

Finally we show that

$$S := \sum_{c=0}^{N-1} \sum_{s=1}^{p-1} \chi(cp+s)c = (1 - \chi(p)) \sum_{r=1}^{N} \chi(r)r.$$

This implies that $2T \equiv pS \equiv 0 \mod p^2$ since $\chi(p) = \left(\frac{-N}{p}\right) = 1$ which would complete the proof.

Note that this identity holds for arbitrary Dirichlet characters. By (6) we have

$$S=\frac{1}{p}\sum_{c=0}^{N-1}\sum_{s=1}^{p-1}\chi(cp+s)cp+\frac{1}{p}\sum_{s=1}^{p-1}s\sum_{c=0}^{N-1}\chi(cp+s)=$$

$$=\frac{1}{p}\sum_{c=0}^{N-1}\sum_{s=1}^{p-1}\chi(cp+s)(cp+s)=$$

$$=\frac{1}{p}\sum_{c=0}^{N-1}\sum_{s=1}^{p}\chi(cp+s)(cp+s)-\frac{1}{p}\sum_{c=0}^{N-1}\chi(cp+p)(cp+p)$$

$$=\frac{1}{p}\sum_{a=1}^{pN}\chi(a)a-\chi(p)\sum_{r=1}^{N}\chi(r)r.$$

We split the first sum again into residue classes modulo N and obtain

$$\frac{1}{p}\sum_{a=1}^{pN}\chi(a)a=\frac{1}{p}\sum_{b=0}^{p-1}\sum_{r=1}^{N}\chi(r)\,(bN+r)=$$

$$=\frac{1}{p}\sum_{b=0}^{p-1}\sum_{r=1}^{N}\chi(r)r=\sum_{r=1}^{N}\chi(r)r.$$

Therefore we find that $2T\equiv pS=0 \mod p^2$; this completes the proof. □

Although the congruence properties of the denominator of ordinary and generalized Bernoulli numbers are well known by the Von Staudt-Clausen type theorems (see [4,10,28]), the nature of the prime divisors of the numerators seems to be quite elusive although some results in this direction are known. In the next theorem we present elementary circumstances for which a given prime $p>3$ must divide the numerator of the generalized Bernoulli number $B_{\frac{p+1}{2},\chi_D}$.

Theorem 4. *If $p>3$ is prime and D is a fundamental discriminant of $\mathbb{Q}(\sqrt{D})$ of the form $D=(-1)^{\frac{p+1}{2}}pN$ where N is a positive integer satisfying $\left(\frac{-N}{p}\right)=1$, then*

$$B_{\frac{p+1}{2},\chi_D}\equiv 0 \mod p.$$

Moreover the special value $L(\frac{1-p}{2},\chi_D)\equiv 0 \mod p$.

Proof Theorem 4. It is well known that

$$B_{\frac{p+1}{2},\chi_D}=|D|^{\frac{p-1}{2}}\sum_{a=1}^{|D|}\chi_D(a)B_{\frac{p+1}{2}}\left(\frac{a}{|D|}\right)$$

where $B_{\frac{p+1}{2}}(x)$ is the $\frac{p+1}{2}^{\text{st}}$ Bernoulli polynomial defined by

$$B_{\frac{p+1}{2}}(x) = \sum_{i=0}^{\frac{p+1}{2}} \binom{\frac{p+1}{2}}{i} B_i x^{\frac{p+1}{2}-i}.$$

Therefore one finds that

$$B_{\frac{p+1}{2},\chi_D} = \sum_{a=1}^{|D|} \chi_D(a) \sum_{i=0}^{\frac{p+1}{2}} \binom{\frac{p+1}{2}}{i} B_i |D|^{i-1} a^{\frac{p+1}{2}-i}.$$

Since $p > 3$ and $i \leq \frac{p+1}{2}$, by Von Staudt-Clausen it follows that all of the B_i in the above sum are $p-$integral (i.e. denominators are prime to p). Therefore since $|D| \equiv 0 \mod p$ we find that

$$B_{\frac{p+1}{2},\chi_D} \equiv \sum_{a=1}^{|D|} \chi_D(a) \left(B_0 |D|^{-1} a^{\frac{p+1}{2}} + \frac{p+1}{2} \cdot a^{\frac{p-1}{2}} B_1 \right) \mod p.$$

This reduces to

$$B_{\frac{p+1}{2},\chi_D} \equiv B_0 |D|^{-1} \sum_{a=1}^{|D|} \chi_D(a) a^{\frac{p+1}{2}} + \frac{p+1}{2} \cdot B_1 \sum_{a=1}^{|D|} \chi_D(a) \left(\frac{a}{p}\right) \mod p.$$

Since $\chi_D(n)\left(\frac{n}{p}\right)$ is a character modulo $|D|$, the second sum is identically zero and hence we find that

$$B_{\frac{p+1}{2},\chi_D} \equiv |D|^{-1} B_0 \sum_{a=1}^{|D|} \chi_D(a) a^{\frac{p+1}{2}} \mod p.$$

The result now follows as a consequence of (5) and Lemma 3. □

Corollary 5. *Let $\ell > 3$ be prime. Then the Fourier coefficients $H(\frac{\ell+1}{2}, n)$ of the weight $\frac{\ell}{2}+1$ modular form $F_{\frac{\ell+1}{2}}(z)$ satisfy a quadratic congruence modulo ℓ of type $(\ell, \left(\frac{-1}{\ell}\right))$.*

Proof of Corollary 5. By (5) and Theorem 4 we see that if D is a fundamental discriminant of the form $D = (-1)^{\frac{\ell+1}{2}} \ell N$ where N is a positive integer satisfying $\left(\frac{-N}{\ell}\right) = 1$, then

$$H\left(\frac{\ell+1}{2}, \ell N\right) = -\frac{2B_{\frac{\ell+1}{2},\chi_D}}{\ell+1} \equiv 0 \mod \ell.$$

Then by (4) it is easy to deduce that for every integer n that

$$H\left(\frac{\ell+1}{2},\ell N n^2\right) \equiv 0 \mod \ell.$$

Therefore it follows that

$$H\left(\frac{\ell+1}{2},M\right) \equiv 0 \mod \ell$$

for every positive integer $M = \ell m$ where $\left(\frac{-m}{\ell}\right) = 1$. However this is precisely the condition that $\left(\frac{m}{\ell}\right) = \left(\frac{-1}{\ell}\right)$. Therefore the modular form $F_{\frac{\ell+1}{2}}(z)$ satisfies a quadratic congruence modulo ℓ of type $(\ell, (\frac{-1}{\ell}))$. □

There are various other congruences for $H(r,N)$ which are not of this type which are also of interest. To illustrate this we now prove the following theorem.

Theorem 6. *For every positive integer $N \equiv 1 \mod 5$ the function $H(5,N)$ satisfies the congruence*

$$H(5,N) \equiv 0 \mod 5.$$

Proof of Theorem 6. Let $f_1(z) := F_5(z)\Theta(5z) \in M_6(20,\chi_5)$, its Fourier expansion is given by

$$f_1(z) = \sum_{n=0}^{\infty} b(n)q^n$$
$$-\frac{1}{132}\left(\Theta^{11}(z) - \frac{22\Theta^7(z)\eta^8(4z)}{\eta^4(2z)} + \frac{88\Theta^3(z)\eta^{16}(4z)}{\eta^8(2z)}\right)\Theta(5z).$$

Since $\Theta(5z) = 1 + 2\sum_{n=1}^{\infty} q^{5n^2}$, it is clear that it suffices to check that

$$b(n) \equiv 0 \mod 5$$

for every positive integer $n \equiv 1 \mod 5$. Now it is known that

$$f_3(z) = \sum_{n\equiv 1 \mod 5}^{\infty} b(n)q^n$$

is a weight 6 modular form with respect to $\Gamma_1(500)$. Therefore by Sturm's theorem it suffices to check that $b(n) \equiv 0 \mod 5$ for every $n \equiv 1 \mod 5$ up to 90000. The congruence has been verified with machine computation. □

In [Th. 4, 4] Carlitz proved that if χ is a primitive Dirichlet character with conductor f and p is a prime for which $p \nmid f$ and n is a positive integer for which $p^e \mid n$, then p^e divides the numerator of $B_{n,\chi}$. For example if $D = -N$ is the fundamental discriminant of the quadratic number field $\mathbb{Q}(\sqrt{D})$ where $N \equiv 1 \mod 5$, then this result implies that the numerator of B_{5,χ_D} is a multiple of 5. However by Theorem 6 we find that even more is true. We obtain:

Corollary 7. *Let $D = -N$ be the fundamental discriminant of $\mathbb{Q}(\sqrt{D})$ where $N \equiv 1 \mod 5$ is a positive integer. Then the numerator of B_{5,χ_D} is a multiple of 25.*

Proof of Corollary 7. By (5) and Theorem 6 it follows that

$$H(5, N) = L(-4, \chi_D) = -\frac{B_{5,\chi_D}}{5} \equiv 0 \mod 5.$$

This immediately implies that the numerator of B_{5,χ_D} is a multiple of 25. □

Now we make some observations regarding the divisibility of the Fourier coefficients of half-integer weight modular forms. In [22] Serre proved a remarkable theorem regarding the divisibility of the Fourier coefficients of holomorphic integer weight modular forms. Let $f(z) = \sum_{n=0}^{\infty} a(n)q^n$ be the Fourier expansion of a holomorphic integer weight modular form with respect to some congruence subgroup of $SL_2(\mathbb{Z})$ whose coefficients $a(n)$ are algebraic integers in a fixed number field. Then he proved that given a positive integer M, the set of non-negative integers n for which $a(n) \equiv 0 \mod M$ has arithmetic density one.

Unfortunately much less is known regarding the divisibility properties of the coefficients of holomorphic half-integer weight forms. Let r be a non-negative integer and let $f(z) = \sum_{n=0}^{\infty} a(n)q^n \in M_{r+\frac{1}{2}}(N, \chi)$ with rational integer coefficients. If $r = 0$, then by the Serre-Stark basis theorem (see [19]), it is known that $f(z)$ is a finite linear combination of theta functions. Moreover these functions are of the form $\Theta_{a,M}(dz)$ where d is a positive integer and

$$\Theta_{a,M}(z) := \sum_{n \equiv a \mod M} q^{n^2}.$$

In particular for all but a finite number of square-free positive integers t, it is the case that $a(tn^2) = 0$ for every integer n. Therefore the number of integers $n \leq x$ for which $a(n) \neq 0$ is $O(\sqrt{x})$. However if $r \geq 1$, then the situation is very different and is of significant interest. A thorough understanding of the divisibility properties of Fourier coefficients when $r = 1$ will shed some light on the divisors of class numbers of imaginary quadratic fields and the Shafarevich-Tate groups of twists of certain modular elliptic curves. Therefore it is worth examining any analogs of Serre's divisibility result which may hold for half-integer weight forms.

From the works of Carlitz and Leopoldt we find that the modular forms $F_r(z)$ provide us with an infinite family of interesting modular forms, which are not trivially zero modulo an integer M, for which we find obvious analogs of Serre's divisibility theorem. Both proved that if χ is a Dirichlet character with conductor f that is not a power of a prime p, then $L(1-r,\chi)$ is an algebraic integer. However if χ is a character with a conductor that is a power of a prime p, then the prime ideal divisors of the denominator of $L(1-r,\chi)$ are prime ideal divisors of p. However if $r \geq 2$ fixed, it is known that the denominators of $L(1-r,\chi_D)$ are bounded which implies that there are at most finitely many D for which $L(1-r,\chi_D)$ is not an integer. Therefore since the modular form $F_r(z)$ may be written as

$$F_r(z) := \zeta(1-2r)\sum_{n=0}^{\infty} a_r(n)q^n = -\frac{B_{2r}}{2r}\sum_{n=0}^{\infty} a_r(n)q^n$$

where the coefficients $a_r(n)$ are rational with denominators bounded by D_r, the least common multiple of all the denominators occurring in the $a_r(n)$, it follows that the numerator of almost every $a_r(n)$ is a multiple of the denominator of $\zeta(1-2r) = -\dfrac{B_{2r}}{2r}$. By Von Staudt-Clausen and the Voronoi congruences (see [15.2.4, 10]), $\prod_{(p-1)|2r} p$ divides the denominator of $\frac{B_{2r}}{2r}$.

We have proved:

Proposition 8. *Let $r \geq 2$ be a positive integer. Then for all but finitely many square-free integers t we find that*

$$a_r(tn^2) \equiv 0 \mod M$$

for every integer n where $M = \prod_{(p-1)|2r} p$. In particular, the number of non-negative integers $n \leq x$ for which $a_r(n) \not\equiv 0 \mod M$ is $O(\sqrt{x})$.

Using this proposition as a template we establish circumstances for which the Fourier coefficients of a half-integer weight modular form are almost always a multiple of a power of a prime p. From the discussion above it is clear that we only need to consider those half-integer weight forms with weight $\geq \frac{3}{2}$. First we note that the classical theta function $\Theta(z) = 1 + 2\sum_{n=1}^{\infty} q^{n^2} \equiv 1 \mod 2$. Therefore if $f(z) = \sum_{n=0}^{\infty} a(n)q^n$ is a holomorphic half-integer weight modular form with integer coefficients, then $f(z)\cdot\Theta(z) \equiv f(z) \mod 2$ is an integer weight holomorphic modular form for which the Fourier coefficients are almost always even by Serre's theorem. Therefore the coefficients $a(n)$ are almost always even. Therefore we may assume that p is an odd prime.

Let $f_p(z)$ be the weight $\frac{p-1}{2}$ modular form defined by

$$f_p(z) = \frac{\eta^p(z)}{\eta(pz)}.$$

It is easy to verify that $f_p(z) \in M_{\frac{p-1}{2}}(p, \chi_D)$ where $D := (-1)^{\frac{p-1}{2}}p$. More importantly if s is a positive integer, then since $1 - X^p \equiv (1-X)^p \mod p$, we find that

$$f_p^{p^s}(z) \equiv 1 \mod p^{s+1}. \tag{7}$$

Using this notation we observe:

Proposition 9. *Let $f(z) = \sum_{n=0}^{\infty} a(n)q^n \in M_{r+\frac{1}{2}}(N, \chi)$ with rational integer coefficients. Let p be an odd prime and let s and k be positive integers for which*

$$2r = kp^s(p-1).$$

If for every cusp $\frac{c}{d}$ of $\Gamma_0(Np)$

$$\mathrm{Ord}(f, \frac{c}{d}) \geq \frac{Nkp^{s+1}}{24 \cdot \gcd(d, Np/d)d}\left(p - \frac{\gcd(d,p)^2}{p}\right)$$

where $\mathrm{Ord}(f, \frac{c}{d})$ is the analytic order of $f(z)$ at the cusp $\frac{c}{d}$, then for all but finitely many square-free positive integers t

$$a(tn^2) \equiv 0 \mod p^{s+1}$$

for every integer n. In particular, the set of non-negative integers $n \leq x$ for which $a(n) \not\equiv 0 \mod p^{s+1}$ is $O(\sqrt{x})$.

Proof of Proposition 9. If $2r = kp^s(p-1)$, then by (7) the weight of $\left(f_p^{p^s}(z)\right)^k$ is exactly $\frac{1}{2}$ more than $r + \frac{1}{2}$, the weight of $f(z)$. The system of inequalities implies that the modular form

$$f(z) \cdot \left(\frac{\eta^{p^{s+1}}(z)}{\eta^{p^s}(pz)}\right)^{-k} \equiv f(z) \mod p^{s+1}$$

is holomorphic at all the cusps of $\Gamma_0(Np)$ since the order of $f_p^{p^s}(z)$ at a cusp $\frac{c}{d}$ is given by (see [2])

$$\frac{Np^{s+1}}{24 \cdot \gcd(d, Np/p)}\left(p - \frac{\gcd(d,p)^2}{p}\right).$$

Therefore this form is a holomorphic weight $\frac{1}{2}$ form, hence is a finite linear combination of theta functions by the Serre-Stark basis theorem. This completes the proof. □

By following an argument similar to the one above, it is easy to see that there will be half integer weight modular forms which are congruent modulo an integer m to a linear combination of weight $\frac{3}{2}$ theta series of the form

$$\Theta_{3,a,M}(z) = \sum_{n \equiv a \mod M} nq^{n^2}.$$

Clearly all such forms will have the property that almost all of their Fourier coefficients will be a multiple of m. Using these as templates of the more general case, it seems reasonable to make the following conjectures.

Conjecture A. *Let $f(z) = \sum_{n=0}^{\infty} a(n)q^n$ be a holomorphic modular form of weight $r + \frac{1}{2}$ with integer coefficients. If p is an odd prime that divides almost all (but not all) of the Fourier coefficients $a(n)$, then either*

- $p - 1 \mid 2r$

or

- $p - 1 \mid 2r - 2.$

A stronger version of this conjecture is:

Conjecture B. *Let $f(z) = \sum_{n=0}^{\infty} a(n)q^n$ be holomorphic modular form of weight $r + \frac{1}{2}$ with integer coefficients. If p is an odd prime and s is a non-negative integer for which*

$$a(n) \equiv 0 \mod p^{s+1}$$

for almost (but not all) every non-negative integer n, then either

- $p^s(p-1) \mid 2r$

or

- $p^s(p-1) \mid 2r - 2.$

Moreover if $a(N) \equiv 0 \mod p^{s+1}$ for almost all N, then for all but finitely many integers t the congruence

$$a(tn^2) \equiv 0 \mod p^{s+1}$$

holds for all n. In particular the set of non-negative integers $n \leq x$ for which $a(n) \not\equiv 0 \mod p^{s+1}$ is at most $O(\sqrt{x})$.

3. Special values of modular L-functions

In this section we investigate the the congruence properties of special values of quadratic twists of a modular L-function at the center of the critical strip on the real line. By the work of Kohnen, Shimura, Waldspurger, and Zagier, the Fourier coefficients of certain special half-integer weight forms are *essentially* (up to a transcendental factor) the square-root of these special values. We now present two simple examples for which congruences exist.

A congruence related to Δ: Let $L(\Delta, s)$ denote the modular L–function defined by

$$L(\Delta, s) := \sum_{n=1}^{\infty} \frac{\tau(n)}{n^s},$$

where $\tau(n)$ is Ramanujan's τ-function giving the Fourier expansion

$$\Delta(z) = \sum_{n=1}^{\infty} \tau(n) q^n$$

at infinity of the unique cusp form Δ of weight 12 and level 1.

If D is the fundamental discriminant of $\mathbb{Q}(\sqrt{D})$, we let $L(\Delta, D, s)$ denote the *twisted* L–function defined by

$$L(\Delta, D, s) := \sum_{n=1}^{\infty} \frac{\chi_D(n)\tau(n)}{n^s}.$$

Let $g(z) = \sum_{n=1}^{\infty} a(n)q^n$ be the weight $\frac{13}{2}$ eigenform defined by

$$g(z) := \frac{\Theta^9(z)\eta^8(4z)}{\eta^4(2z)} - \frac{18\Theta^5(z)\eta^{16}(4z)}{\eta^8(2z)} + \frac{32\Theta(z)\eta^{24}(4z)}{\eta^{12}(2z)}.$$

Using the Shimura correspondence, Kohnen and Zagier [14] proved a general theorem which in this case implies that if D is the fundamental discriminant of a real quadratic field, then

$$L(\Delta, D, 6) = \left(\frac{\pi}{D}\right)^6 \frac{\sqrt{D}}{5!} \frac{\langle \Delta(z), \Delta(z) \rangle}{\langle g(z), g(z) \rangle} \cdot (a(D))^2 \tag{8}$$

where $\langle \Delta(z), \Delta(z) \rangle$ and $\langle g(z), g(z) \rangle$ are the relevant Peterson scalar products. Therefore we shall refer to $(a(D))^2$ as the rational factor of $L(\Delta, D, 6)$. With this notation we prove the following congruences for the rational factors of $L(\Delta, D, 6)$:

Theorem 10. *If D_0 is a positive integer satisfying $\left(\frac{-D_0}{11}\right) = 1$, then the Fourier coefficient $a(11D_0)$ satisfies*

$$a(11D_0) \equiv 0 \mod 11.$$

Proof of Theorem 10. It suffices to check that the Fourier coefficients $a(n)$ satisfy a quadratic congruence modulo 11 of type $(11, -1)$. Let $T(z) \in$

$M_{12}(44, \chi_0)$ be defined by

$$T(z) := \sum_{n=1}^{\infty} c(n)q^n = g(z) \cdot \Theta(11z) \cdot \frac{\eta^{11}(z)}{\eta(11z)}.$$

Since the right hand factor is a modular form whose Fourier expansion is $\equiv 1 \mod 11$, it suffices to check that the Fourier expansion of $T(z)$ has a quadratic congruence modulo 11 of type $(11, -1)$. However by Proposition 2 it suffices to check that

$$c(11n) \equiv 0 \mod 11$$

for all $n \leq 95832$ that satisfy $\left(\frac{n}{11}\right) = -1$. This congruence has been verified by machine computation. □

Therefore by (8) we obtain:

Corollary 11. *Using the notation above, if $D = 11D_0$ is the fundamental discriminant of $\mathbb{Q}(\sqrt{D})$ where D_0 is a positive integer satisfying $\left(\frac{-D_0}{11}\right) = 1$, then the rational factor of $L(\Delta, D, 6)$ is a multiple of 121.*

Galois representations: We now propose a conceptual explanation for corollary 11 in terms of the Galois representations associated to cusp forms. Let $G_{\mathbb{Q}} = \mathrm{Gal}(\bar{\mathbb{Q}}/\mathbb{Q})$ be the absolute Galois group of $\mathbb{Q}$, and let

$$\rho_\Delta : G_{\mathbb{Q}} \longrightarrow \mathbf{GL}_2(\mathbb{Z}_{11})$$

be the 11-adic representation associated by Deligne to the cusp form Δ. Let $\chi : G_{\mathbb{Q}} \longrightarrow \mathbb{Z}_{11}^\times$ be the 11-adic cyclotomic character giving the action of $G_{\mathbb{Q}}$ on the 11-power roots of unity. Finally we identify χ_D with the character $G_{\mathbb{Q}} \longrightarrow \pm 1 \subset \mathbb{Z}_{11}^\times$ corresponding to the field $\mathbb{Q}(\sqrt{D})$.

Let E_{11} be the elliptic curve $X_0(11)$. It is the (unique, up to isogeny) elliptic curve of conductor 11, and its associated eigenform $f_{11} \in S_2(\Gamma_0(11))$ is

$$f_{11} = \eta(z)^2\eta(11z)^2 = q\prod(1-q^n)^2(1-q^{11n})^2 = \sum a_n(E_{11})q^n.$$

Denote by $\rho_{E_{11}}$ the Galois representation attached to this cusp form. (Equivalently, this is the Galois representation describing the action of $G_{\mathbb{Q}}$ on the 11-adic Tate module of E_{11}.) Then we have:

Lemma 12. *The 11-adic Galois representations $\rho_1 = \rho_\Delta \otimes \chi_D \otimes \chi^{-5}$ and $\rho_2 = \rho_{E_{11}} \otimes \chi_{-D_0}$ are residually absolutely irreducible, and their associated mod 11 representations are isomorphic.*

Proof. We have

$$\Delta = q\prod(1-q^n)^{24} \equiv q\prod(1-q^n)^2(1-q^{11n})^2 \equiv f_{11} \pmod{11},$$

so that $\tau(n) \equiv a_n(E_{11}) \pmod{11}$ for all n. Moreover, we have

$$\chi_D(p)p^{-5} \equiv \chi_{-D_0}(p) \pmod{11},$$

and the lemma follows directly from this.

Lemma 13. *If* 11 *is split in the field* $\mathbb{Q}(\sqrt{-D_0})$, *then*

$$L(E_{11}, \chi_{-D_0}, 1) = 0.$$

Proof. This follows from the calculation of the sign in the functional equation for $L(E_{11}, \chi_{-D_0}, s)$, which can be shown to be -1 when the conductor of E_{11} splits in the quadratic field $\mathbb{Q}(\sqrt{-D_0})$. (See for example [9].) Since 1 is the symmetry point for the functional equation, it follows that $L(E_{11}, \chi_{-D_0}, 1) = 0$, as was to be shown.

In some sense, lemmas 12 and 13 provide a conceptual explanation (although not a proof!) for the congruences of corollary 11. Indeed, lemma 12 leads us to expect that the special values of the L-functions $L(\Delta, D, 6+s)$ and $L(E_{11}, -D_0, 1+s)$, which are associated to the Galois representations $\rho_1 = \rho_\Delta \otimes \chi_D \otimes \chi^{-5}$ and $\rho_2 = \rho_{E_{11}} \otimes \chi_{-D_0}$, might be congruent modulo 11. This follows a philosophy suggested by Mazur. (Cf. [16], in particular the discussion on p. 208.) Lemma 13 states that $L(E_{11}, -D_0, 1) = 0$, so that one is led to expect that the rational factor of $L(\Delta, D, 6)$ is divisible by 11 (and hence, 121, since it is a square).

A congruence related to $\eta^8(z)\eta^8(2z)$: We present a second example of such congruences. In this example let $f(z) := \sum_{n=1}^{\infty} c(n)q^n = \eta^8(z)\eta^8(2z)$; hence $f(z)$ is the unique normalized weight 8 eigenform of level 2. Let $L(f,s)$ denote the modular $L-$function

$$L(f,s) := \sum_{n=1}^{\infty} \frac{c(n)}{n^s},$$

and let $L(f, D, s)$ be the twisted modular $L-$function defined as above.

Now let $h(z)$ denote the weight $\frac{9}{2}$ eigenform

$$h(z) := \sum_{n=1}^{\infty} d(n)q^n = \frac{\Theta^5(z)\eta^8(4z)}{\eta^4(2z)} - \frac{16\Theta(z)\eta^{16}(4z)}{\eta^8(2z)}.$$

The Shimura lift of $h(z)$ is $f(z)$, hence by [27] it turns out that if $N_1 \equiv N_2 \not\equiv 5 \mod 8$ are two positive square-free integers with corresponding quadratic characters χ_{D_1} and χ_{D_2} where $d(N_1) \neq 0$, then

$$L(f, D_2, 4) = \frac{d^2(N_2)L(f, D_1, 4)N_1^{\frac{7}{2}}}{d^2(N_1)N_2^{\frac{7}{2}}}. \tag{9}$$

With this notation we prove:

Theorem 14. *If D_0 is a positive integer satisfying $\left(\frac{-D_0}{7}\right) = 1$, then the Fourier coefficient $d(7D_0)$ satisfies the congruence*

$$d(7D_0) \equiv 0 \mod 7.$$

Proof of Theorem 14. It suffices to check that the Fourier coefficients $d(n)$ satisfy a quadratic congruence modulo 7 of type $(7, -1)$. Let $S(z) \in M_8(28, \chi_0)$ be the modular form defined by

$$S(z) := \sum_{n=1}^{\infty} b(n)q^n := h(z) \cdot \Theta(7z) \cdot \frac{\eta^7(z)}{\eta(7z)}.$$

Since the right hand factor is a modular form whose Fourier expansion is $\equiv 1 \mod 7$, by Proposition 1 it suffices to check that

$$b(7n) \equiv 0 \mod 7$$

for all $n \leq 10976$ where $\left(\frac{n}{7}\right) = -1$. This has been verified by machine computation. □

By (9) we obtain:

Corollary 15. *Let D_0 be a positive square-free integer for which $7D_0 \not\equiv 5 \mod 8$ and $\left(\frac{-D_0}{7}\right) = 1$, and let $D = 7D_0$. Then the rational factor of $L(f, D, 4)$ is a multiple of* 49.

Galois representations: We now propose as before an interpretation in terms of Galois representations. Suppose that $D = 7D_0$ with 7 not dividing D_0. Let E_{14} be the elliptic curve $X_0(14)$. It is the (unique) modular elliptic curve of conductor 14, and its associated eigenform $f_{14} \in S_2(\Gamma_0(14))$ is

$$f_{14} = \eta(z)\eta(2z)\eta(7z)\eta(14z) = q\prod(1 - q^n)(1 - q^{2n})(1 - q^{7n})(1 - q^{14n}).$$

Let ρ_f be the 7-adic Galois representation associated to f, and let $\rho_{E_{14}}$ be the 7-adic representation associated to f_{14}. Then we have:

Lemma 16. *The 7-adic representations $\rho_f \otimes \chi_D \otimes \chi^{-3}$ and $\rho_{E_{14}} \otimes \chi_{-D_0}$ are both residually irreducible, and their mod 7 reductions are isomorphic.*

Proof. We have

$$\begin{aligned} f &= q\prod(1-q^n)^8(1-q^{2n})^8 \\ &\equiv q\prod(1-q^n)(1-q^{2n})(1-q^{7n})(1-q^{14n}) = f_{14} \pmod{14}. \end{aligned}$$

The lemma follows directly from this, and from the fact that $\chi_D(p)p^{-3} \equiv \chi_{-D_0}(p) \pmod 7$.

Lemma 17. *If 2 and 7 are both split or both inert in the field $\mathbb{Q}(\sqrt{-D_0})$, then $L(E_{14}, -D_0, 1) = 0$.*

Proof. This follows from the calculation of the sign in the functional equation for $L(E_{14}, \chi_{-D_0}, s)$, which can be shown to be -1 when $\chi_{-D_0}(14) = 1$. (See for example [9].) Since 1 is the symmetry point for the functional equation, it follows that $L(E_{14}, -D_0, 1) = 0$.

As in the case of Δ, lemmas 16 and 17 suggest a conceptual explanation (although again not a proof!) for the congruences of corollary 15.

4. Some remarks on the Bloch-Kato conjectures

This section contains a few general remarks and questions, motivated by the examples of sec. 3, about the Bloch-Kato conjecture on special values of L-functions associated to arithmetic objects of a very general type (the "motives" in the sense of Grothendieck and Deligne, as defined for example in [6]).

Motives: It is beyond our scope to give a detailed account of motives. For the purpose of our discussion, a motive M (over $\mathbb{Q}$, with rational coefficients, of rank r) can be thought of as a piece of the cohomology of an algebraic variety over $\mathbb{Q}$, giving rise to:

• For each prime ℓ, an ℓ-adic representation M_ℓ of $G_\mathbb{Q} = \mathrm{Gal}(\bar{\mathbb{Q}}/\mathbb{Q})$, arising from ℓ-adic étale cohomology. The object M_ℓ is an r-dimensional $\mathbb{Q}_\ell$-vector space equipped with an action of $G_\mathbb{Q}$. Since $G_\mathbb{Q}$ acts through a compact quotient, it leaves stable a $\mathbb{Z}_\ell$- sublattice T_ℓ of M_ℓ, and one can define the mod ℓ representation associated to M_ℓ to be the $\mathbf{F}_\ell$-vector space $\bar{M}_\ell = T_\ell/\ell T_\ell$. This space depends on the choice of T_ℓ in general, but its semi-simplification does not, by the Brauer-Nesbitt theorem. To obtain a canonical object we simply define $\bar{M}_\ell$ to be the semi-simplification of $T_\ell/\ell T_\ell$.

• The system $\{M_\ell\}$ should form a compatible system of rational ℓ-adic representations in the sense of [20], §I.11. More precisely, the action of $G_\mathbb{Q}$ is unramified outside $S \cup \{\ell\}$, where S is a fixed finite set of primes not depending on ℓ. Let D_p be a decomposition group at p in $G_{\mathbb{Q}}$ and let I_p denote an inertia subgroup of D_p. Let Frob_p be the canonical ("frobenius") generator of D_p/I_p which gives the map $x \mapsto x^p$ on residue fields. If $p \neq \ell$ is a prime, then the characteristic polynomial $Z_p(M,T)$ of Frob_p acting on $M_\ell^{I_p}$

$$Z_p(M,T) = \det((1 - \text{Frob}_p T)|M_\ell^{I_p})$$

has integer coefficients and should not depend on the choice of $\ell \neq p$.
• A rational vector space M_B: the so called "Betti realization", arising from singular cohomology.
• A rational vector space M_{DR} coming from the algebraic DeRham cohomology, equipped with its natural Hodge structure.

The structures M_B and M_{DR} will not play an explicit role in our discussion, but are used in defining certain (possibly transcendental) periods associated to M as in [6].

The L-function: One defines the local L-function at p by $L_p(M,s) = Z_p(M,p^{-s})^{-1}$. By the compatibility axiom, $L_p(M,s)$ does not depend in the choice of the prime ℓ used to define it. One then defines the global L-function as a product over all primes p:

$$L(M,s) = \prod_p L_p(M,s).$$

This Euler product converges in a right half plane, by the Weil conjectures. It is conjectured that $L(M,s)$ has an analytic continuation and a functional equation (cf. [6], §1.2.) We will assume this. In the cases where M arises from a classical modular form, this is known to be true.

The Bloch-Kato conjectures: Under the assumption that the motive is "'critical" (in the sense of [6], Def. 1.3), Deligne has given a very general conjectural formula for the special value $L(M,0)$, modulo rational multiples, in terms of a certain period integral defined in terms of the structures M_B and M_{DR}. This conjecture has been refined by Bloch and Kato, and predicts that the "rational part" $L_{rat}(M,0)$ of $L(M,0)$ (i.e., the special value $L(M,0)$ divided by the Deligne period) can be interpreted as the order of a certain Selmer group $\text{Sel}(M) = \oplus_\ell \text{Sel}_\ell(M)$, where

$$\text{Sel}_\ell(M) := \ker\left(H^1(\mathbb{Q}, M_\ell \otimes \mathbb{Q}_\ell/\mathbb{Z}_\ell) \longrightarrow \oplus_v H^1_f(\mathbb{Q}_v, M_\ell \otimes \mathbb{Q}_\ell/\mathbb{Z}_\ell)\right).$$

The groups $H^1_f(\mathbb{Q}_v, M_\ell \otimes \mathbb{Q}_\ell/\mathbb{Z}_\ell)$ are certain subgroups of the local Galois cohomology groups $H^1(\mathbb{Q}_v, M_\ell \otimes \mathbb{Q}_\ell/\mathbb{Z}_\ell)$. When $v \notin S \cup \{\ell\}$ then these are exactly the unramified cohomology classes. When $v = \ell$, the definition of H^1_f

is more subtle and relies on the crystalline cohomology theory developed by Fontaine and Messing. See [3] for details.

The module Sel(M) is a torsion $\hat{\mathbb{Z}}$-module which is cofinitely generated over $\hat{\mathbb{Z}}$. In general it need not be finite. If it is infinite, one conjectures that $L_{rat}(M,0) = 0$.

Congruences between motives: We say that two motives M and N are congruent modulo ℓ if the local L-factors $Z_p(M,T)$ and $Z_p(N,T)$ are congruent mod ℓ for all primes $p \neq \ell$. In particular, it follows from the Chebotarev density theorem that the mod ℓ Galois representations $\bar{M}_\ell$ and $\bar{N}_\ell$ associated to M and N are isomorphic. For example, modular forms whose Fourier coefficients are congruent modulo ℓ give rise to congruent motives. (See, for example, the discussion in the introduction to [16].) Generally speaking, motivated by the philosophy expressed in [16], one might expect a mod ℓ congruence between two motives to translate into a congruence modulo ℓ between the special values of their associated L-functions, at least when one ignores the possible complications arising from the bad factors in the Euler product decomposition.

Question 18. *If two motives M and N are congruent modulo ℓ, when can one predict a mod ℓ congruence between the associated L-values $L_{rat}(M,0)$ and $L_{rat}(N,0)$?*

It would be interesting to formulate a precise, convincing conjecture along these lines, and to compare it with the conjectures of Bloch and Kato.

Acknowledgements

The authors thank Will Galway (University of Illinois) for his assistance regarding the machine computations which were required for this paper.

References

[1] G. Andrews, *The theory of partitions*, Addison-Wesley, 1976.

[2] A. Biagioli, *The construction of modular forms as products of transforms of the Dedekind Eta function*, Acta. Arith. **54** (1990), 273-300.

[3] S. Bloch, K. Kato, *L-functions and Tamagawa numbers of motives*, The Grothendieck festschrift, vol. 1, Birkhäuser, 1990.

[4] L. Carlitz, *Arithmetic properties of generalized Bernoulli numbers*, J. Reine und Angew. Math. **201-202** (1959), 173-182.

[5] H. Cohen, *Sums involving the values at negative integers of L−functions of quadratic characters*, Math. Ann. **217** (1975), 271-285.

[6] P. Deligne, *Valeurs speciales de fonctions L et periodes d'integrales*, Proc. Symp. Pure Math., Corvallis Proceedings, 33 **2** (1979), 313-346.

[7] D. Eichhorn and K. Ono, *Partition function congruences*, appearing in this volume.

[8] B. Gordon, *private communication.*

[9] B. Gross and D. Zagier, *Heegner points and derivatives of L-series*, Inv. Math. **84** (1986), 225-320.

[10] K. Ireland and M. Rosen, *A classical introduction to modern number theory*, Springer-Verlag, 1982.

[11] K. Iwasawa, *Lectures on p−adic L−functions*, Princeton Univ. Press, 1972.

[12] M. Knopp, *Modular functions in analytic number theory*, Markham, 1970.

[13] N. Koblitz, *Introduction to elliptic curves and modular forms*, Springer-Verlag, 1984.

[14] W. Kohnen and D. Zagier, *Values of L−series of modular forms at the center of the critical strip*, Invent. Math. **64** (1981), 173-198.

[15] H. Leopoldt, *Eine verallgemeinerung der Bernoullischen zahlen*, Abh. Math. Sem. Univ. Hamburg **22** (1958), 131-140.

[16] B. Mazur, *On the arithmetic of special values of L-functions*, Invent. Math. **55** (1979), 207-240.

[17] S. Ramanujan, *Congruence properties of partitions*, Proc. London Math. Soc. (2) **18** (1920), 19-20.

[18] K. Rubin, *Congruences for special values of L−functions of elliptic curves with complex multiplication*, Invent. Math. **71** (1983), 339-364.

[19] J.-P. Serre and H. Stark, *Modular forms of weight $\frac{1}{2}$*, Springer Lect. Notes Math. **627** (1977), 27-68.

[20] J-P. Serre, *Abelian ℓ-adic representations and elliptic curves*, Addison-Wesley, 1988.

[21] ______, *Congruences ét formes modulaires (d'aprés H.P.F. Swinnerton-Dyer)*, Seminaire Bourbaki **416** (1971).

[22] ______, *Divisibilité des coefficients des formes modulaires*, C.R. Acad. Sci. Paris (A) **279** (1974), 679-682.

[23] G. Shimura, *Introduction to the arithmetic theory of automorphic functions*, Princeton Univ. Press, 1971.

[24] ______On modular forms of half-integral weight, Ann. Math. **97** (1973), 440-481.

[25] J. Sturm, *On the congruence of modular forms*, Springer Lect. Notes Math. **1240** (1984).

[26] H.P.F. Swinnerton-Dyer, *On ℓ-adic representations and congruences for*

coefficients of modular forms, Springer Lect. Notes Math. **350** (1973).

[27] J.L. Waldspurger, *Sur les coefficients de Fourier des formes modulaires de poids demi-entier*, J. Math. Pures et Appl. **60** (1981), 375-484.

[28] L. Washington, *Introduction to cyclotomic fields*, Springer-Verlag, 1980.

Antal Balog
Mathematical Institute of the
Hungarian Academy of Sciences
P.O. Box 127
Budapest 1364
HUNGARY
h1165bal@ella.hu

Henri Darmon
Department of Mathematics
McGill University
Montréal, PQ H3A 2K6
CANADA
darmon@math.princeton.edu

Current address:
Department of Mathematics
Princeton University,
Princeton, NJ 08540

Ken Ono
School of Mathematics
Institute for Advanced Study,
Princeton, NJ 08540
ono@math.ias.edu

Department of Mathematics
Penn State University,
University Park, PN 16802
ono@math.psu.edu

The asymptotic formula for the number of representations of an integer as a sum of five squares

Paul T. Bateman

Dedicated to Heini and Doreen Halberstam

1. Introduction

In this paper we remark that the asymptotic formula for the number of representations of an integer as a sum of five squares can be derived by a simple elementary argument from Jacobi's formula for the number of representations of an integer as a sum of four squares. A by-product of our argument is an asymptotic formula for the sum

$$\sum_{|j|<n^{1/2}} \sigma(n-j^2),$$

where σ denotes the sum-of-divisors function.

If s is a positive integer, let $r_s(n)$ denote the number of solutions of $x_1^2 + x_2^2 + \cdots + x_s^2 = n$ in integers $x_1, \ldots, x_s$, so that

$$\left(\sum_{j=-\infty}^{\infty} z^{j^2}\right) = \sum_{n=0}^{\infty} r_s(n) z^n = 1 + \sum_{n=1}^{\infty} r_s(n) z^n \qquad (|z|<1)$$

and

$$r_{s+1}(n) = \sum_{|j|\le n^{1/2}} r_s(n-j^2). \tag{1}$$

By the use of the Hardy-Littlewood method (without minor arcs) and the classical transformation formula for the theta function given by the power series $\sum\limits_{j=-\infty}^{\infty} z^{j^2}$, it is easy to prove that, if $s \ge 5$, then

$$\begin{aligned} r_s(n) = {} & \frac{\pi^{s/2}}{\Gamma(s/2)} n^{s/2-1} \sum_{k=1}^{\infty} \sum_{\substack{h \bmod k \\ (h,k)=1}} \left(\frac{G(h,k)}{k}\right)^s e^{-2\pi i h n/k} \\ & + O(n^{s/4}) \end{aligned} \tag{2}$$

where

$$G(h,k) = \sum_{j=1}^{k} e^{2\pi i h j^2/k}.$$

(See [3], [7], or [8].)

The initial example of (2) is the formula

$$r_5(n) = (4/3)\pi^2 n^{3/2} \sum_{k=1}^{\infty} A_k(n) + O(n^{5/4}), \tag{3}$$

where

$$A_k(n) = \sum_{\substack{h \bmod k \\ (h,k)=1}} \{G(h,k)/k\}^5 e^{-2\pi i h n/k}.$$

(Since $|G(h,k)| \leq 2k^{1/2}$, the infinite series in (3) converges absolutely.) If for prime values of p we put

$$\chi_p(n) = \sum_{j=0}^{\infty} A_{p^j}(n),$$

then, since $A_k(n)$ is a multiplicative function of k for fixed n, (3) may be written

$$\begin{aligned} r_5(n) &= (4/3)\pi^2 n^{3/2} \chi_2(n) \sum_{k \text{ odd}>0} A_k(n) + O(n^{5/4}) \\ &= (4/3)\pi^2 n^{3/2} \prod_p \chi_p(n) + O(n^{5/4}). \end{aligned} \tag{4}$$

Using (1) and Jacobi's formula for $r_4(n)$, we shall prove (4) with the slightly improved error term $O(n^{1+\epsilon})$. The general idea of the elementary method used was kindly suggested to us by Harold Diamond.

Actually Hardy used modular forms to prove that the error term in (3) or (4) is identically zero and, more generally, that the error term in (2) is identically zero for $5 \leq s \leq 8$. (See [2], [4], or [5]; for the cases $s = 3, 4$ see [1].) However Hardy's exact formula seems beyond the scope of the simplistic arguments used here.

While $\sigma(n)$ denotes the sum of the positive integral divisors of n when n is a positive integer, it is also convenient and customary to put $\sigma(0) = -1/24$. For n a non-negative integer we put

$$\begin{aligned} \sigma^*(n) &= \sigma(n) & &\text{if } 4 \nmid n, \\ \sigma^*(n) &= \sigma(n) - 4\sigma(n/4) & &\text{if } 4 \mid n, \end{aligned}$$

so that in particular $\sigma^*(0) = 1/8$. Jacobi's formula for $r_4(n)$ is simply

$$r_4(n) = 8\sigma^*(n) \qquad (n = 0, 1, 2, \dots).$$

Using (1) we therefore have

$$\begin{aligned} r_5(n)/8 &= \sum_{|j| \le n^{1/2}} \sigma^*(n - j^2) \\ &= \sum_{|j| \le n^{1/2}} \sigma(n - j^2) - 4 \sum_{\substack{|j| \le n^{1/2} \\ j^2 \equiv n \pmod 4}} \sigma((n - j^2)/4). \end{aligned} \tag{5}$$

In the next two sections we shall prove the following two assertions:

$$\begin{aligned} \sum_{|j| \le n^{1/2}} \sigma(n - j^2) &= \frac{\pi^2}{6} n^{3/2} \left(\frac{5}{3} - \frac{1}{3}\chi_2(n)\right) \sum_{k \text{ odd} > 0} A_k(n) \\ &\quad + O(n(1 + \log n)^3) \end{aligned} \tag{A}$$

and

$$\begin{aligned} 4 \sum_{\substack{|j| \le n^{1/2} \\ j^2 \equiv n \pmod 4}} \sigma\left(\frac{n - j^2}{4}\right) &= \frac{\pi^2}{6} n^{3/2} \left(\frac{5}{3} - \frac{4}{3}\chi_2(n)\right) \sum_{k \text{ odd} > 0} A_k(n) \\ &\quad + O(n(1 + \log n)^3) \end{aligned} \tag{B}$$

(We remark that (B) is trivial when $n \equiv 2, 3 \pmod 4$, for in that case the sum on the left side of (B) is empty and the factor $5 - 4\chi_2(n)$ on the right side is zero.)

Combining (A) and (B) with (5), we obtain

$$\begin{aligned} r_5(n)/8 &= \sum_{|j| \le n^{1/2}} \sigma^*(n - j^2) \\ &= (\pi^2/6) n^{3/2} \chi_2(n) \sum_{k \text{ odd} > 0} A_k(n) + O(n(1 + \log n)^3), \end{aligned} \tag{C}$$

an improvement on (4) but a far cry from Hardy's exact formula.

2. Proof of (A)

Clearly

$$\begin{aligned}\sum_{|j|<n^{1/2}} \sigma(n-j^2) &= 2\sum_{0\le j<n^{1/2}}\ \sum_{d|(n-j^2)} (n-j^2)/d - \sigma(n)\\ &= 2\sum_{d=1}^{n}\ \sum_{\substack{0\le j<n^{1/2}\\ j^2\equiv n \pmod d}} (n-j^2)/d - \sigma(n)\\ &= 2\sum_{d=1}^{n}\ \sum_{\substack{0\le m\le d-1\\ m^2\equiv n \pmod d}}\ \sum_{\substack{0\le j<n^{1/2}\\ j\equiv m \pmod d}} (n-j^2)/d - \sigma(n).\end{aligned}$$

Now

$$-n < \sum_{\substack{0\le j<n^{1/2}\\ j\equiv m \pmod d}} (n-j^2) - d^{-1}\int_0^{n^{1/2}} (n-t^2)dt < n,$$

so that

$$\sum_{\substack{0\le j<n^{1/2}\\ j\equiv m \pmod d}} (n-j^2) = 2n^{3/2}/(3d) + O(n).$$

Let $f_d(n)$ denote the number of integers m such that

$$m^2 \equiv n \pmod d, 0 \le m \le d-1.$$

Then

$$\begin{aligned}\sum_{|j|<n^{1/2}} \sigma(n-j^2) &= 2\sum_{d=1}^{n} \frac{f_d(n)}{d}\left\{\frac{2n^{3/2}}{3d} + O(n)\right\} - \sigma(n)\\ &= \frac{4}{3}n^{3/2}\sum_{d=1}^{\infty}\frac{f_d(n)}{d^2} + O\left(n^{3/2}\sum_{d=n+1}^{\infty}\frac{g(d)}{d^2}\right)\\ &\quad + O\left(n\sum_{d=1}^{n}\frac{g(d)}{d}\right),\end{aligned}$$

where $g(d)$ is the maximum of $f_d(n)$ for all integer values of n. It is not difficult (although a bit tedious) to show that g is the multiplicative function given by

$$g(2) = 2,\ g(4) = 2,\ g(2^{2\alpha-1}) = g(2^{2\alpha}) = 2^{\alpha} \quad \text{for} \quad \alpha \ge 2,$$
$$g(p^{2\alpha-1}) = 2p^{\alpha-1} \quad \text{and} \quad g(p^{2\alpha}) = p^{\alpha} \quad \text{for} \quad \alpha \ge 1 \quad \text{and } p \text{ an odd prime.}$$

Thus

$$\sum_{d=1}^{n}\frac{g(d)}{d}<\prod_{p\le n}\left\{1+\sum_{j=1}^{\infty}\frac{g(p^j)}{p^j}\right\}=\frac{7}{2}\prod_{\substack{p\,\mathrm{odd}\\ p\le n}}\left(1+\frac{3}{p-1}\right)$$
$$\ll\prod_{p\le n}\left(1+\frac{1}{p-1}\right)^3\ll(1+\log n)^3,$$

and

$$\sum_{d=n+1}^{\infty}\frac{g(d)}{d^2}<\sum_{j=1}^{\infty}\frac{1}{2^{j-1}n}\sum_{d=2^{j-1}n+1}^{2^jn}\frac{g(d)}{d}$$
$$\ll\sum_{j=1}^{\infty}\frac{1}{2^{j-1}n}\{1+\log(2^jn)\}^3\ll\frac{(1+\log n)^3}{n}.$$

Hence

$$\sum_{|j|<n^{1/2}}\sigma(n-j^2)=\frac{4}{3}n^{3/2}\sum_{d=1}^{\infty}\frac{f_d(n)}{d^2}+O(n(1+\log n)^3). \tag{6}$$

Since $f_d(n)$ is a multiplicative function of d for fixed n, (6) may be written

$$\sum_{|j|<n^{1/2}}\sigma(n-j^2)=\frac{4}{3}n^{3/2}\psi(n)\sum_{d\,\mathrm{odd}>0}\frac{f_d(n)}{d^2}$$
$$+O(n(1+\log n)^3), \tag{7}$$

where we have put

$$\psi(n)=\sum_{\alpha=0}^{\infty}2^{-2\alpha}f_{2^\alpha}(n). \tag{8}$$

Now (cf. Lemma 2.12 of [6])

$$f_d(n)=\sum_{m=1}^{d}d^{-1}\sum_{j=1}^{d}e^{2\pi im(j^2-n)/d}$$
$$=\sum_{c|d}c^{-1}\sum_{\substack{h\bmod c\\(h,c)=1}}\sum_{j=1}^{c}e^{2\pi ih(j^2-n)/c}$$
$$=\sum_{c|d}\sum_{\substack{h\bmod c\\(h,c)=1}}\frac{G(h,c)}{c}e^{-2\pi ihn/c}. \tag{9}$$

If d is odd and $c \mid d$, then $G(h,c)^4 = c^2$ and so (9) gives (for d odd)

$$f_d(n) = \sum_{c|d} c^2 \sum_{\substack{h \bmod c \\ (h,c)=1}} \left(\frac{G(h,c)}{c}\right)^5 e^{-2\pi i h n/c} = \sum_{c|d} c^2 A_c(n), \tag{10}$$

so that

$$\sum_{d\,\mathrm{odd}>0} \frac{f_d(n)}{d^2} = \sum_{d\,\mathrm{odd}>0} \sum_{c|d} \left(\frac{d}{c}\right)^{-2} A_c(n) = \frac{\pi^2}{8} \sum_{c\,\mathrm{odd}>0} A_c(n).$$

Thus (7) may be written

$$\sum_{|j|<n^{1/2}} \sigma(n-j^2) = (\pi^2/6)n^{3/2}\psi(n) \sum_{k\,\mathrm{odd}>0} A_k(n) + O(n(1+\log n)^3). \tag{11}$$

If h is odd, $G(h,2) = 0$ and $G(h,2^j)^4 = -2^{2(j+1)}$ if $j > 1$. (Cf. [8].) Hence (9) gives

$$\begin{aligned} f_{2^\alpha}(n) &= \sum_{j=0}^{\alpha} \sum_{\substack{h \bmod 2^j \\ (h,2^j)=1}} \frac{G(h,2^j)}{2^j} e^{-2\pi i h n/2^j} \\ &= 1 - \sum_{j=1}^{\alpha} 2^{2(j-1)} \sum_{\substack{h \bmod 2^j \\ (h,2^j)=1}} \left(\frac{G(h,2^j)}{2^j}\right)^5 e^{-2\pi i h n/2^j} \\ &= 1 - \sum_{j=1}^{\alpha} 2^{2(j-1)} A_{2^j}(n). \end{aligned} \tag{12}$$

(When $\alpha = 1$, (12) reduces to $f_2(n) = 1 - A_2(n) = 1$.) Thus (8) and (12) give

$$\begin{aligned} \psi(n) &= \sum_{\alpha=0}^{\infty} 2^{-2\alpha} \{1 - \sum_{j=1}^{\alpha} 2^{2(j-1)} A_{2^j}(n)\} \\ &= \frac{4}{3} - \frac{1}{3} \sum_{j=1}^{\infty} A_{2^j}(n) = \frac{5}{3} - \frac{1}{3}\chi_2(n). \end{aligned} \tag{13}$$

In view of (11) and (13), (A) is proved.

We remark that it is often convenient to evaluate $\chi_2(n)$ by using the following consequence of (8) and (13):

$$\chi_2(n) = 5 - 3\psi(n) = 5 - 3\sum_{\alpha=0}^{\infty} 2^{-2\alpha} f_{2^\alpha}(n) = \frac{5}{4} - 3\sum_{\alpha=2}^{\infty} 2^{-2\alpha} f_{2^\alpha}(n), \tag{14}$$

and to evaluate $\chi_p(n)$ for odd primes p by using the following consequence of (10):

$$(15)\qquad \sum_{j=0}^{\infty} p^{-2j} f_{p^j}(n) = \sum_{j=0}^{\infty} p^{-2j} \sum_{i=0}^{j} p^{2i} A_{p^i}(n) = (1-p^{-2})^{-1}\chi_p(n).$$

3. Proof of (B)

As in the proof of (A) we have

$$\begin{aligned}
\sum_{\substack{|j|<n^{1/2} \\ j^2\equiv n \pmod 4}} 4\sigma\left(\frac{n-j^2}{4}\right) &= 2 \sum_{\substack{0\le j<n^{1/2} \\ j^2\equiv n \pmod 4}} 4 \sum_{d|(n-j^2)/4} \frac{n-j^2}{4d} + O(\sigma(n)) \\
&= 2\sum_{d=1}^{\lfloor n/4\rfloor} \sum_{\substack{0\le j<n^{1/2} \\ j^2\equiv n \pmod{4d}}} \frac{n-j^2}{d} + O(\sigma(n)) \\
&= 2\sum_{d=1}^{\lfloor n/4\rfloor} \sum_{\substack{0\le m\le 4d-1 \\ m^2\equiv n \pmod{4d}}} \sum_{\substack{0\le j<n^{1/2} \\ j\equiv m \pmod{4d}}} \frac{n-j^2}{d} + O(\sigma(n)) \\
&= 2\sum_{d=1}^{\lfloor n/4\rfloor} \frac{f_{4d}(n)}{d}\left(\frac{2}{3}\frac{n^{3/2}}{4d} + O(n)\right) + O(\sigma(n)) \\
&= \frac{1}{3}n^{3/2}\sum_{d=1}^{\infty}\frac{f_{4d}(n)}{d^2} + O\left(n^{3/2}\sum_{d=n+1}^{\infty}\frac{g(d)}{d}\right) + O\left(n\sum_{d=1}^{n}\frac{g(d)}{d}\right) \\
&= \frac{4}{3}n^{3/2}\rho(n)\sum_{d\,\mathrm{odd}>0}\frac{f_{4d}(n)}{d^2} + O(n(1+\log n)^3),
\end{aligned}$$

where

$$\rho(n) = \frac{1}{4}\sum_{\alpha=0}^{\infty}\frac{f_{2^{\alpha+2}}(n)}{2^{2\alpha}} = 4\sum_{\beta=2}^{\infty}\frac{f_{2^\beta}(n)}{2^{2\beta}} = 4\psi(n) - 5 = \frac{5}{3} - \frac{4}{3}\chi_2(n).$$

Thus (B) is proved.

4. Further remarks

Remark 1. If s is sufficiently large, it is almost trivial (and probably well-known) that the asymptotic formula for $r_s(n)$ implies the asymptotic formula for $r_{s+1}(n)$. For example, if $s \geq 6$ and if we know that

$$r_s(n) = \frac{\pi^{s/2}}{\Gamma(s/2)} n^{s/2-1} \sum_{k=1}^{\infty} \sum_{\substack{h \bmod k \\ (h,k)=1}} \left(\frac{G(h,k)}{k}\right)^s e^{-2\pi i h n/k}$$

$$+ O(n^{s/2-3/2}), \tag{16}$$

then it is easy to deduce that (16) also holds with s replaced by $s+1$. For if (16) holds, then by (1)

$$r_{s+1}(n) = \frac{\pi^{s/2}}{\Gamma(s/2)} \sum_{k=1}^{\infty} \sum_{\substack{h \bmod k \\ (h,k)=1}} \left(\frac{G(h,k)}{k}\right)^s e^{-2\pi i h n/k}.$$

$$\cdot \sum_{|j| \leq n^{1/2}} (n-j^2)^{s/2-1} e^{2\pi i h j^2/k} + O(n^{s/2-1}). \tag{17}$$

Now the arithmetic function $j \mapsto e^{2\pi i h j^2/k}$ has mean-value $G(h,k)/k$; more specifically we have the finite Fourier series

$$e^{2\pi i h j^2/k} = \sum_{\ell=1}^{k} e^{2\pi i h \ell^2/k} \quad k^{-1} \sum_{r=1}^{k} e^{2\pi i r(\ell - j)/k}$$

$$= k^{-1} G(h,k) + k^{-1} \sum_{r=1}^{k-1} e^{-2\pi i r j/k} \sum_{\ell=1}^{k} e^{2\pi i (h\ell^2 + r\ell)/k}.$$

Thus

$$\sum_{|j| \leq n^{1/2}} (n-j^2)^{s/2-1} e^{2\pi i h j^2/k} = k^{-1} G(h,k) \sum_{|j| \leq n^{1/2}} (n-j^2)^{s/2-1}$$

$$+ k^{-1} \sum_{r=1}^{k-1} \left(\sum_{\ell=1}^{k} e^{2\pi i (h\ell^2 + r\ell)/k} \right) \left(\sum_{|j| \leq n^{1/2}} (n-j^2)^{s/2-1} e^{-2\pi i r j/k} \right). \tag{18}$$

On the one hand

$$\sum_{|j| \leq n^{1/2}} (n-j^2)^{s/2-1} = \int_{-n^{1/2}}^{n^{1/2}} (n-t^2)^{s/2-1} dt + O(n^{s/2-1})$$

$$= \frac{\Gamma(s/2)\Gamma(1/2)}{\Gamma((s+1)/2)} n^{(s+1)/2-1} + O(n^{s/2-1}) \tag{19}$$

On the other hand, if $1 \leq r \leq k-1$, partial summation and the formula for the sum of a geometric progression give

$$\Big|\sum_{|j|\le n^{1/2}} (n-j^2)^{s/2-1}e^{-2\pi irj/k}\Big| \le n^{s/2-1} \max_{0\le t\le n^{1/2}} \Big|\sum_{|j|\le t} e^{-2\pi irj/k}\Big|$$

$$\le n^{s/2-1}/\sin(\pi r/k) \le \frac{kn^{s/2-1}}{\min(r,k-r)}. \tag{20}$$

Finally Hilfssatz 10 of §1 of [8] yields

$$\Big|\sum_{\ell=1}^{k} e^{2\pi i(h\ell^2+r\ell)/k}\Big| \le 2k^{1/2}. \tag{21}$$

Using (19), (20), and (21) in (18), we obtain

$$\sum_{|j|\le n^{1/2}} (n-j^2)^{s/2-1}e^{2\pi ihj^2/k} = \frac{G(h,k)}{k}\frac{\Gamma(s/2)\Gamma(1/2)}{\Gamma((s+1)/2)}n^{(s+1)/2-1}$$

$$+ O(k^{1/2}(1+\log k)n^{s/2-1}) \tag{22}$$

From (17) and (22) we readily get (16) with s replaced by $s+1$.

While (16) holds for all $s \ge 3$, the passage from s to $s+1$ is less immediate for the smaller values of s. The step from 4 and 5 was made possible by the simple nature of Jacobi's formula for $r_4(n)$ and by the simple formula for $G(h,k)^4$. The referee observes that it would probably be possible to push through the step from 5 to 6 by an elaboration of the above argument. The author thanks the referee for his helpful suggestions concerning this part of the paper.

Remark 2. From the calculations of $A_k(n)$ given in [2] or [4], or from the formulas (14) and (15), it is easy to deduce that

$$\chi_2(n) \le 5/4 \quad \text{with equality if and only if} \quad n \equiv 2,3 \pmod 4,$$
$$\chi_2(n) \ge 5/8 \quad \text{with equality if and only if} \quad n \equiv 1 \pmod 8,$$

and that for p an odd prime

$$\chi_p(n) \le 1+1/p^2 \quad \text{with equality if and only if} \quad p\nmid n \quad \text{and} \quad \left(\frac{n}{p}\right)=1$$

$$\chi_p(n) \ge 1-1/p^2 \quad \text{with equality if and only if} \quad p\nmid n \quad \text{and} \quad \left(\frac{n}{p}\right)=-1.$$

Since $\sum_{k\text{ odd}>0} A_k(n) = \prod_{p\text{ odd}} \chi_p(n)$, the last two inequalities give

$$8/\pi^2 = \prod_{p\text{ odd}}(1-1/p^2) < \sum_{k\text{ odd}>0} A_k(n) \le \prod_{p\text{ odd}}(1+1/p^2) = 12/\pi^2,$$

where equality is achieved in the upper bound if and only if n is an even power of 2. Thus we can determine the least upper bounds and the greatest lower bounds of $n^{-3/2}$ times the main terms on the right-hand sides of (A), (B), and (C).

Remark 3. We close with a conjecture. Hardy's exact formula for $r_5(n)/8 = \sum_{|j|\le n^{1/2}} \sigma^*(n-j^2)$, namely

$$\sum_{|j|\le n^{1/2}} \sigma^*(n-j^2) = (\pi^2/6)n^{3/2}\chi_2(n) \sum_{k \text{ odd}>0} A_k(n),$$

suggests that perhaps there is a similar exact formula for $\sum_{|j|\le n^{1/2}} \sigma(n-j^2)$. A plausible conjecture is that the main term on the right-hand side of (A), namely

$$\frac{\pi^2}{6} n^{3/2}\left(\frac{5}{3} - \frac{1}{3}\chi_2(n)\right) \sum_{k \text{ odd}>0} A_k(n)$$

is exactly equal to

$$\sum_{|j|\le n^{1/2}} \sigma(n-j^2) = \sum_{|j|< n^{1/2}} \sigma(n-j^2)$$

if n is not a perfect square, but is equal to

$$\sum_{|j|\le n^{1/2}} \sigma(n-j^2) + 2n = \sum_{|j|< n^{1/2}} \sigma(n-j^2) + 2(n-1/24)$$

if n is a perfect square. This conjecture is trivially true when $n \equiv 2, 3 \pmod 4$, for in that case $\sigma(n-j^2) = \sigma^*(n-j^2)$ and $\chi_2(n) = 5/3 - \chi_2(n)/3$. But it is easy to verify it in many other cases as well.

References

[1] P. T. Bateman, *On the representations of a number as the sum of three squares*, Trans. Amer. Math. Soc. **71** (1951), 70-101.

[2] L. E. Dickson, *Studies in the Theory of Numbers*, University of Chicago Press, 1930, Chapter 13.

[3] G. H. Hardy, *On the representation of a number as the sum of any number of squares, and in particular of five or seven*, Proc. Nat. Acad. Sci. U.S.A. **4** (1918), 189–193.

[4] G. H. Hardy, *On the representation of a number as the sum of any number of squares, and in particular of five*, Trans. Amer. Math. Soc. **21** (1920), 255–284; **29** (1927), 845–847.

[5] M. I. Knopp, *Modular Functions in Analytic Number Theory*, Markham, Chicago, 1970, Chapter 5.

[6] R. C. Vaughan, *The Hardy-Littlewood Method*, Cambridge University Press, 1981, Chapter 2.

[7] A. Walfisz, *Über Gitterpunkte in mehrdimensionalen Ellipsoiden*, Math. Zeit. **19** (1924), 300–307.

[8] A. Walfisz, *Gitterpunkte in mehrdimensionalen Kugeln* (1957), Monografie Matematyczne, Warszawa, Chapter 1.

Paul T. Bateman
Department of Mathematics
University of Illinois
Urbana, IL 61801
bateman@math.uiuc.edu

Distribution of the error term for the number of lattice points inside a shifted ball

P. Bleher and J. Bourgain

Dedicated to Heini Halberstam on the occasion of his retirement

1. Introduction

Define

$$N_x(R) = \#\left\{n \in \mathbf{Z}^d \;\middle|\; |x-n| \le R\right\} \qquad (x \in [0,1[^d) \tag{1.1}$$

and

$$F_x(R) = R^{-\frac{d-1}{2}}\ [N_x(R) - R^d\ \mathrm{Vol}_d\ B(1)] \tag{1.2}$$

where $B(1)$ is the unit ball in $\mathbf{R}^d$. Thus $N_x(R)$ is the number of lattice points in $\mathbf{Z}^d$ in the R-ball centered at x and $F_x(R)$ is the normalized difference with the volume of the ball. We are interested in the problem of the distribution of $F_x(R)$ as a function of R for given x.

In dimension $d = 2$ the problem was solved completely in [B-C-D-L]. The result here is independent of x.

Theorem 1.3 ($d = 2$) **[B-C-D-L].** *Let R be uniformly distributed on a segment $0 \le R \le T$. The random variable $R^{-1/2}[N_x(R) - \pi\ R^2]$ has a limit distribution as $T \to \infty$ which is absolutely continuous with respect to Lebesgue measure. The density $p_x(\lambda)$ is an entire function of λ which decays for real λ faster than* $\exp(-|\lambda|^{4-\varepsilon})$.

This result is essentially optimal. Recall in $d = 2$ the conjecture (the Gauss circle problem)

$$|N_0(R) - \pi\ R^2| \ll R^{\frac{1}{2}+\varepsilon}\,. \tag{1.4}$$

It is known that

$$\varliminf_{R\to\infty} \frac{N_0(R) - \pi\ R^2}{R^{1/2}\,(\log R)^{1/4}} < 0 \quad \text{and} \quad \varlimsup_{R\to\infty} \frac{N_0(R) - \pi\ R^2}{R^{1/2}} > 0\,. \tag{1.5}$$

In order to establish a similar result in $d \ge 3$, we need to assume certain diophantine properties for the center x. Let x satisfy a diophantine condition

$$\left(\prod_{j=1}^{d} |a_j|\right)^{1+\varepsilon} \left\|\sum_{j=1}^{d} a_j\ x_j\right\| \gg 1 \quad \text{for all} \quad \{a_j\}_{j=1}^{d} \in \mathbf{Z}^d \backslash \{0\} \tag{1.6}$$

(which is generically fulfilled).

Theorem 1.7.

(1) Let x satisfy (1.6). *Then $F_x(R)$ has for $R \to \infty$ a limit distribution ν_x satisfying a tail estimate $\nu_x[|\lambda'| > \lambda] < \exp(-\lambda^{4-\varepsilon})$.*

(2) For almost all x, ν_x is absolutely continuous with respect to Lebesgue measure with smooth density $p_x(\lambda) \ll \exp(-|\lambda|^{4-\varepsilon})$.

(3) For $d = 3$ and almost all x, $p_x(\lambda)$ extends to an entire function and one has moreover the lower estimate $p_x(\lambda) \gg \exp(-|\lambda|^{4+\varepsilon})$.

The proof of these facts is based on standard methods from analytic number theory. The main remaining open question is whether the statement (2) still holds under the diophantine condition (1.6). To show this would require certain improvements of some of the estimates presented below. The problems here are closely related to developing a "circle method" approach to the Linnik problem, which is of course a subject of independent interest.

2. Formal expression for the difference (1.2)

In expressing $F_x(R)$ we follow the standard approach based on the Poisson summation formula (as in [B-C-D-L])

$$\sum_{\xi \in \mathbf{Z}^d} f(\xi) = \sum_{\xi \in \mathbf{Z}^d} \widehat{f}(\xi) \tag{2.1}$$

which for

$$f = \chi_{B(x,R)} \tag{2.2}$$

yields essentially

$$\begin{aligned} F_x(R) &= R^{-\frac{d-1}{2}}\ [N_x(R) - \mathrm{Vol}_d(B(0,R))] \\ &= \sum_{\xi \neq 0} \widehat{f}(\xi) \sim \sum_{\xi \neq 0} |\xi|^{-\frac{d+1}{2}}\ \cos(2\pi\ |\xi|\ R - \phi)\ e^{2\pi i \langle x, \xi \rangle}\,. \end{aligned} \tag{2.3}$$

Of particular interest to us will be the behaviour of the expressions

$$r_\lambda(x) = \sum_{|\xi|^2 = \lambda} e^{2\pi i \langle x, \xi \rangle}\,. \tag{2.4}$$

Writing $|\xi| = k\ \sqrt{m}$, $k, m \in \mathbf{Z}_+$, m square free, one may rewrite

$$(2.5) \qquad (2.3) = \sum_{m \text{ square free}} f_m\left(R\ \sqrt{m}\right)$$

where thus

$$(2.6) \qquad f_m(t) = m^{-\frac{d+1}{4}} \sum_{k=1}^{\infty} k^{-\frac{d+1}{2}}\ r_{k^2m}(x)\ \cos(2\pi\ kt - \phi)$$

are periodic functions of t.

Recall the well-known fact that the set $\{\sqrt{m} \mid m \text{ square free}\}$ is linearly independent over $\mathbf{Q}$. Hence (2.5) is expected to behave like a sum of independent random variables.

In order to justify this, one needs to reduce (2.5) to a finite sum, thus establish an inequality

$$(2.7) \qquad \lim_{N\to\infty} \overline{\lim_{T\to\infty}}\ \frac{1}{T} \int_0^T \left| F_x(R) - \sum_{m\leq N} f_m\left(R\sqrt{m}\ ;\ x\right)\right|\ dR = 0\,.$$

We will also show that

(2.8) the f_m-series (2.6) converges uniformly.

Once the distribution of (2.5) reduced to that of the series

$$(2.9) \qquad \sum_{m \text{ square free}} f_m\left(t_m\right)$$

obtained by summing independent random-variables, the tail-estimate

$$(2.10) \qquad \gamma_x\ [|\lambda'| > \lambda] < \exp(-\lambda^{4-\varepsilon})$$

will result from an estimate

$$(2.11) \qquad \sum_{m\,sf} \|f_m\|_\infty^p < \infty \quad \text{for} \quad p > \frac{4}{3}$$

(using for instance interpolation between $p = 1$, $p = 2$).

A control of the density of γ_x is then achieved by estimating the Fourier transform

$$(2.12) \qquad \widehat{\gamma}_x\,(s) = \prod_m \int e^{2\pi i s f_m(t)}\,dt$$

as in [B-C-D-L].

3. Estimates on the $r_m(x)$

We will briefly indicate the proof of following facts. We always assume x satisfying the diophantine condition (1.6). For some of the statements, a weaker hypothesis is sufficient, however.

Lemma 3.1. $\sum_{m\leq M} |r_m(x)|^2 \ll M^{\frac{d}{2}+\varepsilon}$.

Lemma 3.2. $\sum_{m\leq M} |r_m(x)|^2 > c\, M^{\frac{d}{2}}$.

Lemma 3.3. $|r_m(x)| \ll M^{\frac{d-1}{4}+\varepsilon}$.

Remarks.

(1) The estimates (3.1), (3.2) are the correct ones. For (3.3), the correct estimate should be $|r_m(x)| \ll m^{\frac{d-2}{4}+\varepsilon}$, assuming x satisfies (1.6). In fact, for $d = 3$, the bound (3.3) is just the trivial estimate on the number of lattice points on a sphere of radius $\sqrt{m}$, independently of any further assumption on x.

(2) For $d \geq 2$ a diophantine assumption on x is necessary. For instance, let $d = 3$, $x = \left(\frac{p_1}{q_1}, \frac{p_2}{q_2}, \frac{p_3}{q_3}\right)$, q_1, q_2, q_3 odd. Then (cf. [B-D])

$$\lim_{M\to\infty} \frac{1}{M^2} \sum_{m\leq M} |r_m(x)|^2 = L(x) > 0 \tag{3.4}$$

where

$$L(x) = \frac{16\pi^2}{7} \frac{\zeta(2)}{\zeta(3)}\, q^{-2} \prod_{p|q} \frac{p^2}{p^2+p+1}. \tag{3.5}$$

4. Sketch of proof of (3.1) (upper estimate)

Let $N = [M^{1/2}]$ and let $S_N(x,t) = \sum_{n\leq N} e^{2\pi i(nx+n^2t)}$ be the Gauss sum. Then, if $m < M$,

$$r_m(x) = \int_0^1 \prod_{j=1}^d S_N(x_j,t)\, e^{2\pi i\, mt}\, dt \tag{4.1}$$

and from Parseval's identity

$$\sum_{m<M} |r_m(x)|^2 \leq \int_0^1 \prod_{j=1}^d |S_N(x_j,t)|^2\, dt\,. \tag{4.2}$$

Write

$$|S_N(x,t)|^2 = \sum_{n_1,n_2 \leq N} e^{i(n_1-n_2)[x+(n_1+n_2)t]} \tag{4.3}$$

$$\leq \sum_{k \leq N} \frac{1}{\|x+kt\| + \frac{1}{N}}\,. \tag{4.4}$$

Let $q = q(t)$ satisfy

$$t = \frac{a}{q} + \varepsilon \qquad |\varepsilon| < \frac{1}{Nq}\,. \tag{4.5}$$

Then one shows that

$$(4.4) < \frac{N \log N}{\|q(t)x\|}\,. \tag{4.6}$$

Hence

$$(4.2) < (N \log N)^d \, \frac{1}{N} \sum_{q=1}^{N} \left(\prod_{j=1}^{d} \frac{1}{\|q\,x_j\| + \frac{1}{N}} \right)\,. \tag{4.7}$$

Consider next the distribution of $\{\|q\,x_j\|\}_{j=1,\dots,d}$ for $q = 1,\dots,N$. One gets a bound

$$\frac{1}{N} \sum_{q=1}^{N} \left(\prod_{j=1}^{d} \frac{1}{\|q\,x_j\| + \frac{1}{N}} \right) \ll N^{\varepsilon} \tag{4.8}$$

provided

$$q^{1+\varepsilon'} \prod_{j=1}^{d} \|q\,x_j\| > 1 \quad \text{for all} \quad q \tag{4.9}$$

and hence (by Mahler's duality) if (1.6) holds. Inequality (3.1) follows.

5. Sketch of proof of (3.2) (lower estimate)

From the preceding, we need to show that

$$\int_0^1 \prod_{j=1}^{d} |S_N(x_j,t)|^2 \; dt \gg M^{\frac{d}{2}-\varepsilon}\,. \tag{5.1}$$

Choose small constant $\gamma > 0$. Define

$$E = \{q \sim \gamma\, N \mid \|q\, x_j\| < \gamma^2\ ,\ j = 1, \ldots, d\}\ . \tag{5.2}$$

By Weyl's equidistribution

$$\#\, E > \gamma^C\, N\,. \tag{5.3}$$

Write

$$\int \prod_{j=1}^{d} |S_N(x_j, t)|^2\ dt > \sum_{q \in E} \sum_{(a,q)=1} \int_{|t-\frac{a}{q}|<\frac{\gamma}{N^2}} \prod_{j=1}^{d} |S_N(x_j, t)|^2\ dt\,. \tag{5.4}$$

Let

$$S(a, b, q) = \frac{1}{q} \sum_{r=0}^{q-1} e^{2\pi i \frac{1}{q}(ar^2+br)} \tag{5.5}$$

be the Gauss sum. Assuming thus

$$\begin{cases} \left|x_j - \frac{b_j}{q}\right| < \frac{\gamma^2}{q} \\ \left|t - \frac{a}{q}\right| < \frac{\gamma}{N^2} \end{cases} \tag{5.6}$$

one gets

$$|S_N(x_j, t)| > N\ |S(a, b_j, q)| - 0(1)\ \frac{N}{\sqrt{q}} = N\,|S(1, 0, q)| - 0(1)\ \frac{N}{\sqrt{q}} \tag{5.7}$$

and substituting (5.7) in (5.4) clearly yields the lower bound (3.2)

$$(\gamma^C\ N)\,(\gamma\ N)\ \left(\frac{\gamma}{N^2}\right)\ N^d > \gamma^{C+2}\ N^d\,. \tag{5.8}$$

6. Sketch of proof of (3.3)

We use the circle method and Kloosterman's refinement. Recall (4.1), thus for $N \sim m^{1/2}$

$$r_m(x) = \int \prod_{j=1}^{d} S_N(x_j, t)\ e^{-2\pi imt}\ dt \tag{6.1}$$

redefining

$$S_N(x, t) = \sum \gamma\left(\frac{k}{N}\right)\ e^{2\pi i(kx+k^2t)} \tag{6.2}$$

where γ is a smooth bump function.

Define for $(a,q)=1$

$$V_{a,q} = \left\{ t \;\middle|\; \left| t - \frac{a}{q} \right| < \frac{1}{Nq} \right\} \tag{6.3}$$

and write the integral essentially as

$$\sum_q \sum_{(a,q)=1} \int_{V_{a,q}} \prod_{j=1}^{d} S_N(x_j,t)\, e^{-2\pi i m t}\, dt\,. \tag{6.4}$$

Let $t \in V_{a,q}$, $t = \frac{a}{q} + \varphi$, $|\varphi| < \frac{1}{Nq}$.

Write by Poisson summation

$$\begin{aligned} S_N(x,t) &= \sum_{r=0}^{q-1} e^{2\pi i r^2 \frac{a}{q}} \sum_{k \equiv r(q)} \gamma\left(\frac{k}{N}\right) e^{2\pi i (kx + k^2\varphi)} \\ &= \sum_{r=0}^{q-1} e^{2\pi i r^2 \frac{a}{q}} \sum_{\ell \in \mathbf{Z}} \gamma\left(\frac{r+\ell q}{N}\right) e^{2\pi i ((\ell q + r)x + (\ell q + r)^2 \varphi)} \\ &= \sum_{s \in \mathbf{Z}} \left\{ \frac{1}{q} \sum_{r=0}^{q-1} e^{\frac{2\pi i}{q}(ar^2 - sr)} \right\} \\ &\qquad \left\{ \int \gamma\left(\frac{y}{N}\right) e^{2\pi i ((x+\frac{s}{q})y + \varphi y^2)}\, dy \right\} \qquad (6.5) \\ &= \sum_{s \in \mathbf{Z}} S(a,s;q)\, J(x,\varphi,s;q) \qquad (6.6) \end{aligned}$$

$$\prod_{j=1}^{d} S_N(x_j,t)\, e^{imt} = \sum_{\substack{s=(s_1,\dots,s_d) \\ |qx+s| \ll m^{\varepsilon}}} \{\mathrm{I}\}\,\{\mathrm{II}\} \tag{6.7}$$

where

$$\{\mathrm{I}\} = \prod_j S(a,s_j;q)\, e^{-2\pi i m \frac{a}{q}} \tag{6.8}$$

$$\{\mathrm{II}\} = \prod_j J(x_j,\varphi,s_j;q)\, e^{-2\pi i m \varphi}\,. \tag{6.9}$$

From oscillatory integral estimates, the contribution of (6.9) is

$$\int_{|\varphi| < \frac{1}{Nq}} |\{\mathrm{II}\}|\, d\varphi < N^{\frac{d-3}{2}}\, q^{\frac{d-1}{2}}\, \|q\,x\|^{-\frac{d-1}{2}}\,. \tag{6.10}$$

For a given modulus q, estimate next the arithmetical factor

$$\sum_{(a,q)=1} \prod_j S(a,s_j;q)\, e^{-2\pi i m \frac{a}{q}}\,. \tag{6.11}$$

Assume q is a (power of) a prime. The general case follows from multiplicativity properties. Writing

$$S(a,s;q) = e^{-\frac{2\pi i}{q}\frac{a' s^2}{4}} \left(\frac{a}{q}\right) S(1,0,q) \tag{6.12}$$

where $\left(\frac{a}{q}\right)$ is the Jacobi symbol, (6.11) yields

$$S(1,0,q)^d \left[\sum_{(a,q)=1} \left(\frac{a}{q}\right)^d e^{-\frac{2\pi i}{q}\left(ma+\frac{|s|^2}{4}a'\right)}\right] \tag{6.13}$$

where the second factor in (6.13) is a Kloosterman or Salié sum according to whether d is even or odd. This yields, recollecting the preceding, essentially following bound on (6.4) and hence $|r_m(x)|$,

$$N^\varepsilon \sum_{q<N} q^{-\frac{d}{2}+\frac{1}{2}} \left\{N^{\frac{d-3}{2}}\, q^{\frac{d-1}{2}}\, \|q\,x\|^{-\frac{d-1}{2}}\right\} \ll N^{\frac{d-1}{2}+\varepsilon} = m^{\frac{d-1}{4}+\varepsilon} \tag{6.14}$$

again using the diophantine assumption on x.

Remark. The shortcoming in previous estimate is the fact that no cancelation in the q-summation is exploited. This problem appears also if one tries to use the circle method to study the ternary problem (Linnik problem) of representing n as

$$n = a_1\, n_1^2 + a_2\, n_2^2 + a_3\, n_3^2 \tag{6.15}$$

along the lines of what Kloosterman did with 4 squares. This would yield an alternative approach to the method of automorphic forms, which seems difficult to use in order to derive nontrivial bounds on $r_m(x) = \sum_{|\xi|=\sqrt{m}} e^{2\pi i\langle x,\xi\rangle}$ for diophantine x (unwinding the exponential $e^{2\pi i\langle x,\cdot\rangle}$ in spherical harmonies, of [S], does not seem to exploit such properties of x).

7. Convergence of the f_m-series

Recall (2.6), i.e.

$$f_m(t) = m^{-\frac{d+1}{4}} \sum_{k=1}^{\infty} k^{-\frac{d+1}{2}}\, r_{k^2 m}(x)\, \cos(2\pi k t - \phi)\,. \tag{7.1}$$

The estimate (3.3)

$$|r_{k^2m}(x)| \ll (mk^2)^{\frac{d-1}{4}+\varepsilon} \tag{7.2}$$

just falls short of establishing the absolute convergence of (7.1). In order to prove at least uniform convergence, we establish a further estimate on a partial sum

$$\sum \gamma\left(\frac{k}{K}\right) r_{k^2m}(x)\, e^{2\pi ikt} \tag{7.3}$$

assuming $K \gg m$. One has ($N \sim m^{1/2} K$)

$$(7.3) = \int \prod_{j=1}^{d} S_N(x_j, \tau) \left\{ \sum_k \gamma\left(\frac{k}{K}\right) e^{-2\pi i(k^2 m\tau - kt)} \right\} d\tau \tag{7.4}$$

where the second factor in (7.4) is again a quadratic exponential sum. Repeating the analysis from previous section (the case d odd is a bit harder), one may get an estimate

$$|(7.4)| < m^C K^{\frac{d+1}{2}-\delta} \tag{7.5}$$

for some $\delta > 0$. From (7.5), the uniform convergence of (7.1) clearly follows and in fact from (7.2), (7.5), we get

$$\|f_m\|_\infty \ll m^{-\frac{1}{2}+\varepsilon}\,. \tag{7.6}$$

Also, combining (3.1), (7.5) one easily deduces that

$$\sum_{m\le M} \|f_m\|_\infty^2 \ll M^{-\frac{1}{2}+\varepsilon} \tag{7.7}$$

and hence, by Hölder's inequality, (2.11)

$$\sum_{m=1}^{\infty} \|f_m\|_\infty^p < \infty \quad \text{for} \quad p > \frac{4}{3} \tag{7.8}$$

(assuming x satisfies (1.6)).

8. Proof of inequality (2.7)

It follows from the convergence of the f_m-series and (7.6) that

$$\lim_{N\to\infty} \overline{\lim_{T\to\infty}} \frac{1}{T} \int_0^T \left| F_x(R) - \sum_{m\le N} f_m\left(\sqrt{m}\, R; x\right) \right| dR = 0 \tag{8.1}$$

will follow from the estimate

$$\frac{1}{T}\int_0^T \left|\sum_{n>n_0} n^{-\frac{d+1}{4}}\, r_n(x)\, \cos\left(2\pi\,\sqrt{n}\,R-\phi\right)\right|^2 dR \tag{8.2}$$
$$< n_0^{-1/2+\varepsilon} \quad \text{for} \quad T > n_0^2.$$

Write $\sum\limits_{n>n_0}$ -summation in (8.2) as $\sum\limits_{n_0<n<T^2} + \sum\limits_{n\geq T^2} = \text{(I)} + \text{(II)}$.

From the separation of $\{\sqrt{n} \mid n < T^2\}$, orthogonality and (3.1), we get

$$\text{(I)} < \sum_{n_0<n<T^2} n^{-\frac{d+1}{2}}\, |r_n(x)|^2 \ll \max_{N>n_0}\, N^{-\frac{d+1}{2}}\, N^{\frac{d}{2}+\varepsilon} = n_0^{-\frac{1}{2}+\varepsilon}. \tag{8.3}$$

For the $\sum\limits_{n\geq T^2}$ -summation, fix a dyadic interval $n \sim N > T^2$ which we split in further subintervals of length $K \sim \frac{\sqrt{N}}{T}$. Then again by orthogonality

$$\frac{1}{T}\int_0^T \left|\sum_{n\sim N} n^{-\frac{d+1}{4}}\, r_n(x)\, \cos\left(2\pi\,\sqrt{n}\,R-\phi\right)\right|^2 dR \tag{8.4}$$
$$\sim N^{-\frac{d+1}{2}}\, K^{-1} \sum_{n\sim N} \left|\sum_{m=n}^{n+K} r_m(x)\right|^2.$$

Writing again

$$r_m(x) = \int \prod_{j=1}^{d} S_M(x_j,t)\, e^{-imt}\, dt \qquad (M \sim N^{1/2}) \tag{8.5}$$

(8.4) yields by Parseval

$$N^{-\frac{d+1}{2}}\, K^{-1} \int \prod_{j=1}^{d} |S_M(x_j,t)|^2 \left|\sum_{k=0}^{K} e^{-ikt}\right|^2 dt \tag{8.6}$$

which by (4.6) is bounded by

$$N^{-\frac{d+1}{2}}\, K^{-1} \int \prod_{j=1}^{d} \frac{M\log M}{\|q(t)x_j\| + \frac{1}{M}} \cdot \left|\sum_{k=0}^{K} e^{-ikt}\right|^2 dt \tag{8.7}$$

$$\leq N^{-1/2+\varepsilon} \sum_{q<M} \left(\prod_{j=1}^{d} \frac{1}{\|q\, x_j\| + \frac{1}{M}}\right) \int_{\Gamma_q} \frac{dt}{K\|t\|^2 + K^{-1}} \tag{8.8}$$

where

(8.9) $$\Gamma_q = \left\{ t \in \mathbf{T} \,\middle|\, \left| t - \frac{a}{q} \right| < \frac{1}{qM} \,,\ (a,q) = 1 \right\}.$$

Since thus $\int_{\Gamma_q} \frac{dt}{K\|t\|^2 + K^{-1}} \ll M^{-1+\varepsilon}$ and (4.8), (8.8) $\ll N^{-1/2+\varepsilon}\, M^{\varepsilon} = N^{-\frac{1}{2}+\varepsilon}$ and

(8.10) $$(\mathrm{II}) \ll T^{-1+\varepsilon} < n_0^{-2+\varepsilon}.$$

From (8.3), (8.10), the bound (8.2) follows.

9. Density function for typical x

Recall (3.2), for diophantine x

(9.1) $$\sum_{n \le M} |r_n(x)|^2 > c\, M^{\frac{d}{2}}.$$

From (9.1), one deduces for typical x that

(9.2) $$\sum_{\substack{M^{1-\varepsilon} < m < M \\ m\ sf,\, k < M^{\varepsilon}}} |r_{mk^2}(x)| > c\, M^{\frac{d}{2}}$$

(estimating from above the difference between the left members of (9.1), (9.2)).

Estimate the characteristic function of γ_x by

(9.3) $$\widehat{\gamma}_x(s) = \int \prod_{m\ sf} e^{-2\pi i s f_m(t_m)} \prod dt_m = \prod_{m\ sf} \int e^{-2\pi i s f_m(t)}\, dt.$$

For fixed $|s| > 1$, choose $M_0 \sim |s|^{2+\varepsilon}$ and estimate (9.3) by

(9.4) $$\prod_{m > M_0,\, m\ sf} \left| \int e^{-2\pi i s f_m} \right|.$$

Since $\int f_m = 0$ and $\|f_m\|_\infty \ll m^{-\frac{1}{2}+\varepsilon}$ by (7.6), one gets

(9.5) $$\left| \int e^{-2\pi i s f_m} \right| = 1 - (1 + 0(1))\, s^2 \int f_m^2$$

and

(9.6) $$(9.4) \le \exp c \left\{ - \sum_{m > M_0,\, m\ sf} \int f_m^2 \right\} s^2.$$

By (2.6), (9.2)

$$\sum_{m>M_0, m\ sf} \int f_m^2 \geq \sum_{\substack{m>M_0, m\ sf \\ k<M_0^\varepsilon}} m^{-\frac{d+1}{2}-\varepsilon} \, |r_{mk^2}(x)|^2 > M_0^{-\frac{1}{2}-\varepsilon} \tag{9.7}$$

so that

$$|\widehat{\gamma}_x(s)| < \exp - \left(|s|^{1-\varepsilon}\right) . \tag{9.8}$$

Hence p_x is smooth and Theorem 1.7 (2) follows.

10. Proof of $d = 3$ case

There is the following L^4-estimate on eigenfunctions of $-\Delta$ on $\mathbf{T}^3$:

$$\left\| \sum_{|n|^2=M} a_n \, e^{inx} \right\|_{L^4(\mathbf{T}^3)} \ll M^\varepsilon \, \left(\sum |a_n|^2 \right)^{1/2} . \tag{10.1}$$

This estimate follows from a uniform M^ε-bound on the number of lattice points on any ellipse in the plane of size M^ε.

Denote

$$M_0 \sim S^{\frac{4}{3}+\varepsilon} \tag{10.2}$$

and

$$\mathcal{M} = \Big\{ M_0^{1-\varepsilon} < n < M_0 \mid n = mk^2 \ , \ m \text{ square free}, \quad k \ll M_0^\varepsilon \quad \text{and} \quad |S| \ \|f_m\|_\infty = 0(1) \Big\} . \tag{10.3}$$

By (7.7), we have

$$\sum_{\substack{m\sim M \\ m\ sf}} \|f_m\|_\infty^2 \ll M^{-\frac{1}{2}+\varepsilon}; \tag{10.4}$$

hence

$$M_0 - (\# \mathcal{M}) < M_0^{1-\varepsilon} + s^2 \, M_0^{-\frac{1}{2}+\varepsilon} < M_0^{1-\varepsilon} . \tag{10.5}$$

Thus from Hölder's inequality and (10.1), for typical x we have

$$\sum_{\substack{n\leq M_0 \\ n\notin \mathcal{M}}} |r_n(x)|^2 \leq M_0^{\frac{1}{2}-\varepsilon} \left(\sum_{n\leq M_0} |r_n(x)|^4 \right)^{1/2} < M_0^{\frac{d}{2}-\varepsilon} \tag{10.6}$$

so that by (9.1)

$$\sum_{n\in\mathcal{M}} |r_n(x)|^2 > c\, M_0^{\frac{d}{2}}\,. \tag{10.7}$$

Hence by (10.7), (10.2)

$$\begin{aligned} |\widehat{\gamma}_x(s)| &\le \exp c \left\{ - \sum_{|S|\,\|f_m\|_\infty = 0(1)} \int f_m^2 \right\} s^2 \\ &< \exp \left\{ -M_0^{-\frac{d+1}{2}-\varepsilon} \cdot \sum_{n\in\mathcal{M}} |r_n(x)|^2 \right\} s^2 \\ &< \exp - \left(|S|^{\frac{4}{3}-\varepsilon} \right) \end{aligned} \tag{10.8}$$

implying that $p_x(\lambda)$ extends to an entire function of λ.

Complete proofs of the preceding will appear in a forthcoming paper.

References

[B-C-D-L] P. Bleher, Z. Cheng, F. Dyson, J. Lebowitz, *Distribution of the error term for the number of lattice points inside a shifted circle*, Commun. Math. Phys. **154** (1993), 433-469.

[H-B] Heath-Brown, *The distribution and moments of the error term in the Dirichlet divisor problem*, Acta Mathematica, Vol. **60**, 1992, N4, 389-415.

[L] Y. Linnik, *Ergodic properties of algebraic number fields*, Springer-Verlag, 1969.

[S] P. Sarnak, *Some applications of modular forms*, Cambridge Tracts in Mathematics, Cambridge University Press, 1990.

P. Bleher
Indiana University - Purdue
University Indianapolis
Indianapolis, IN 46202

J. Bourgain
School of Mathematics
Institute for Advanced Study
Princeton, NJ 08540

A probabilistic generalization of the Riemann zeta function

Nigel Boston[1]

To Heini Halberstam, for his generous welcome to Urbana

0. Introduction

There is an old problem that asks for the probability that two integers chosen at random are relatively prime. The informal solution goes as follows.

Let a and b be two integers. Assuming we have some method that picks integers with each congruence class modulo a prime p favored equally, the probability that p does not divide both a and b is $1 - 1/p^2$. If we next assume that divisibility by distinct primes is somehow independent, then the probability that no prime divides both a and b is

$$\prod_{\text{primes } p} (1 - 1/p^2) = 1/\zeta(2) = 6/\pi^2.$$

If we wish to put this argument on a more solid footing, we can. One approach is given in [2], theorem 332. Alternatively, let X be a compact Hausdorff topological group. Then X has a unique Haar measure normalized such that the measure of X, $\text{Meas}(X) = 1$.

Let U_i ($i \in I$) be a collection of open subsets of X and k be a positive integer. Denote $X \times ... \times X$ by X^k. Then the set $X^k - \cup U_i^k$ is closed, hence measurable, in X^k. We define

$$P(X, k) = \text{Meas}(X^k - \cup U_i^k).$$

Note that this definition depends on the choice of $\{U_i\}$, but there will usually be a clear choice in a given context. Since X^k has measure 1, we might think of it as a probability space. The function $P(X, k)$ was considered independently and earlier by Avinoam Mann [5] in the case of X a profinite group. In [4] something very similar was considered. Since writing this paper, Hildebrand has kindly pointed out earlier applications of compactifications of $\mathbf{Z}$ in number theory (see e.g. [7],[9]).

[1]Partially supported by the Sloan Foundation and NSF grant DMS 93-04277. I thank Andrea Lucchini, Avinoam Mann, Ram Murty, Jean-Pierre Serre, Aner Shalev, and John Shareshian for their comments on these ideas. I thank Steve Harding and Mike Slattery for their help with computational algebra packages. I thank God for leading me to these results.

Now let X be $\hat{\mathbf{Z}} := \{\underline{a} = (a_1, a_2, a_3, ...) : a_n \in \mathbf{Z}/n$ and $m \mid n \Rightarrow a_n \equiv a_m (\text{mod } m)\}$, a closed subset of $\prod_n \mathbf{Z}/n$, which by Tychonoff is compact if we give $\mathbf{Z}/n$ the discrete topology. So $X = \hat{\mathbf{Z}}$ is a compact, Hausdorff topological group. Note that $\mathbf{Z}$ embeds in $\hat{\mathbf{Z}}$ as a dense subset via

$$a \mapsto (a(\text{mod } 1), a(\text{mod } 2), a(\text{mod } 3), ...).$$

Let $U_n = \{\underline{a} \in \hat{\mathbf{Z}} : a_n = 0\}$ for each positive integer $n \geq 2$, an open subset of $\hat{\mathbf{Z}}$. The subset $U_n \cap \mathbf{Z}$ of $\mathbf{Z}$ picks out those integers divisible by n, so that $\mathbf{Z}^2 \cap (X^2 - \cup U_n^2) = \{(a,b) \in \mathbf{Z}^2 : \gcd(a,b) = 1\}$. This is the set we originally were after, and by enlarging this slightly to $X^2 - \cup U_n^2$ (remember $\mathbf{Z}$ is dense in X), we obtain a measurable set with measure $P(X, 2)$. By the Borel-Cantelli theorem, this is $1/\zeta(2)$ (this, to some extent, makes formal our original argument).

Similarly, $P(\hat{\mathbf{Z}}, k) = 1/\zeta(k)$ for all positive integers k (interpreting the righthand side as 0 when $k = 1$), so that the Riemann zeta function comes from suitably interpolating the values of $P(\hat{\mathbf{Z}}, k)$. This suggests considering the functions produced by other, natural choices of X as generalizations of the Riemann zeta function. In this paper, we attack the following three questions:

(1) Is there some formula for $P(X, k)$ (at least for certain kinds of X)?

(2) If so, can we interpolate this to obtain $P(X, s)$ for s a continuous real or even complex variable?

(3) What properties does the ensuing $P(X, s)$ have?

1. Finite X

Let us consider first the case of X being a finite set with the discrete topology. In this case $\text{Meas}(X^k - \cup U_i^k)$ is obtained by counting, using the inclusion-exclusion principle. We obtain that

$$P(X, k) = \sum_{V \in J} \mu(V)|V|^k/|X|^k (k \in \mathbf{Z}^+)$$

where the sum is over all subsets V of X of the form $U_{i_1} \cap ... \cap U_{i_r}$ (including the empty intersection X) and where μ is the Moebius function of the lattice J, defined inductively by $\mu(X) = 1$ and $\mu(V) = -\sum \mu(W)$, summing over all $W \in J$ properly containing V.

This answers question 1. As for question 2, we can clearly set

$$P(X, s) = \sum_{V \in J} \mu(V)|V|^s/|X|^s.$$

There are two desirable features about this case that we will want our spaces X (equipped with choice of open subsets $\{U_i\}$) to possess:

(A) that Meas(X)/Meas(V) is an integer for all $V \in J$ (J defined to be the set of finite intersections of the U_i's),

(B) that, given any m, there are only finitely many $V \in J$ with $m =$Meas(X)/Meas(V).

For the rest of this paper, we will restrict our attention to cases where (A) and (B) both hold. Note that in the case of X a finite space, $P(X, s)$ is a truncated Dirichlet series. The ring of truncated Dirichlet series with coefficients in **C** has the property of unique factorization, since it is a polynomial ring over **C** in $1/2^s, 1/3^s, 1/5^s, \ldots$ The factors of $P(X, s)$ will be called "generalized Euler factors" and our study will use them to give a partial answer to question (3).

2. X a group

One naturally arising situation in which (A) and (B) both hold is where X is a finitely generated profinite group (e.g. $\hat{\mathbf{Z}}$) and the U_i are all its proper open subgroups. In this case $P(X, k)$ measures the probability that a k-tuple of elements of X topologically generates X (cf. the introduction).

The definition of the Moebius function works fine for the lattice of open subgroups of X and, with an appropriate ordering of terms, we have (summing over all open subgroups)

$$P(X, k) = \sum_H \mu(H)/[X : H]^k \quad (k \in \mathbf{Z}^+).$$

It is then tempting to set $P(X, s) = \sum_H \mu(H)/[X : H]^s$. Thanks to (A) and (B) this is, at least formally, a Dirichlet series. As regards convergence, the appropriate ordering above does not coincide with the ordering in terms of increasing $[X : H]$, so unless we have e.g. absolute convergence, we cannot say that the Dirichlet series converges to $P(X, k)$ for $s = k \in \mathbf{Z}^+$.

Here are some examples of $P(X, s)$ factored into generalized Euler factors for X a finite group:

$$P(C_n, s) = \prod_{p|n, p \text{ prime}} (1 - 1/p^s)$$

$$P(C_p \times \ldots \times C_p, s) = (1 - 1/p^s)(1 - 1/p^{s-1})\ldots(1 - 1/p^{s-d+1})$$

$$P(A_4, s) = (1 - 1/2^{s-1})(1 + 1/2^{s-1})(1 - 1/3^s)$$

$$P(A_5, s) = 1 - 5/5^s - 6/6^s - 10/10^s + 20/20^s + 60/30^s - 60/60^s$$

$$P(S_5, s) = P(C_2, s)P(A_5, s)$$

$$P(A_5.2^{4^1}, s) = P(A_5, s)(1 - 1/16^{s-1})$$

This last group is a certain perfect group of order 960. In the second example, d equals the number of generators of the elementary abelian group. With the same use of d, we have (moving on to infinite X):

$$P(\hat{\mathbf{Z}} \times ... \times \hat{\mathbf{Z}}, s) = (\zeta(s)\zeta(s-1)...\zeta(s-d+1))^{-1}$$

$$P(C_2 * C_2, s) = (1 - 1/2^s)\zeta(s-1)^{-1}$$

$$\begin{aligned} P(C_2 * C_3, s) &= (1 - 1/2^s)(1 - 1/3^s)(1 - 1/3^{s-1})(1 - 1/2^{s-1})\cdot \\ &\quad \cdot (1 + 1/2^{s-1})P(A_5, s)(1 - 1/7^{s-1})... \\ &= 1 - 1/2^s - 4/3^s - 4/4^s - 5/5^s - 2/6^s + ... \end{aligned}$$

3. Properties of $P(X, s)$

There are several reasons to want to investigate what properties these functions have. On the one hand, in applications to group theory, one can attack questions such as that of Ribes and Wong [8] on profinite analogues of the Grushko-Neumann theorem, by looking at special values of $P(X, s)$ or, more promisingly, of its generalized Euler factors. One might investigate, more generally, what proportions of pairs of elements generate a group and so on (cf. [3],[6]). On the other hand, from the point of view of number theory, we have a new class of Dirichlet series to place on the map.

If N_i is a descending sequence of open normal subgroups of X with $\cap N_i = \{1\}$ and k is a positive integer, then $P(X, k) = \inf P(X/N_i, k)$. This reduces many of our considerations to the case of X being a finite group.

If X is a finite group and N is a minimal normal subgroup of X, then we can compare $P(X, k)$ and $P(X/N, k)$. For example, if $N \leq \Phi(X)$, the Frattini subgroup of X, then $P(X, k) = P(X/N, k)$, since the elements of $\Phi(X)$ are the non-generators of X. Otherwise, by Gaschütz [1],

$$P(X, k) = P(X/N, k)(1 + \sum_r (-1)^r \sum \epsilon_{i_1,...,i_r}/[X : M_{i_1} \cap ... \cap M_{i_r}]^k)$$

where the $M_1, ...$ are the maximal subgroups of X, the sum is over indices $i_1 < ... < i_r$, and $\epsilon_{i_1,...,i_r} = 1$ if $N(M_{i_1} \cap ... \cap M_{i_r}) = X$, 0 otherwise.

If X is infinite, the above leads to an expression for $P(X, k)$ (hence $P(X, s)$) as an infinite product of generalized Euler factors, with the same domain of convergence as the infinite sum, defining the same function in this domain.

Using the classification of simple groups (analogously to [6]), Shalev has shown (in private correspondence), using ideas of Pyber, that if A and B are finite groups with $|A|, |B| \geq 2, |A||B| \geq 5$, then $P(A * B, k) = 0$ for all positive integers k. On the other hand, if X is prosolvable (and topologically finitely generated), then [5] $P(X, k) \neq 0$ for some $k \in \mathbf{Z}^+$, the infinite product converges absolutely in some half plane, and all expressions for $P(X, s)$ obtained so far have the same domain of absolute convergence and define the same function in this domain. We are in good shape then in this case.

From a number theorist's point of view, the generalized Euler factors that arise in the factorization of $P(X, s)$ are of prime interest. By the observation at the start of this section, we may restrict our attention to the case X a finite group. In my talk at the Halberstam conference, I made the following a conjecture. It is easy to see that all the examples in section 2 above satisfy it.

Question. *Suppose X has composition factors $S_1, ..., S_m$. Then $P(X, s)$ factors as $\prod_{i=1}^{m} \tilde{P}(S_i, s)$ where the $\tilde{P}(S_i, s)$ are irreducible and in a certain list of possible "twists" of $P(S_i, s)$.*

This is true for X solvable. In this case, Gaschütz's formula above gives $P(X, k) = P(X/N, k)(1 - c(N)/|N|^k)$, where N is a minimal normal subgroup of X and $c(N)$ and $|N|$ are powers of the same prime. Factoring the term in brackets gives the following list of "twists" of $P(C_p, s)$:

$1 - \zeta/p^{s-d}$, where ζ is a root of unity and d a non-negative rational.

The conjectural space of twists is therefore parametrized by $\mathbf{Q}/\mathbf{Z} \times \mathbf{Q}^{\geq 0}$. One interesting thing to note here is that these generalized Euler factors are ones that do naturally occur in number theory (as opposed, say, to $P(A_5, s)$) - the ζ corresponds to twisting by finite characters and the $s - d$ to a Tate twist. Perhaps (a wild speculation) factors such as $P(A_5, s)$ have some, as yet unforeseen, number-theoretic interpretation.

The above question also has an affirmative action in the case $X = S \times ... \times S$ (d factors), where S is a finite nonabelian simple group. Namely,

$$P(X, s) = \prod_{r=0}^{d-1} (P(S, s) - r|\mathrm{Aut}(S)|/|S|^s).$$

This result was proven in [2] and also by Lucchini, Shalev (both private communication), and myself, all independently. The reason for so many names is not because the problem was so hard, but because it was so easy!

The question naturally leads to what might be thought of as an extension of the classification of finite simple groups, namely to classify all possible twists of $P(S, s)$ (S a nonabelian finite simple group). This assumes that what the question asserts holds, which we next show to be false. Thus we just ask for a classification of possible generalized Euler factors.

4. Counterexamples and a conjecture

Let X be $A_5 \wr C_2$, a group of order 7200, with a unique proper normal subgroup $N \cong A_5 \times A_5$. By Gaschütz, $P(X,s) = P(C_2,s)\tilde{P}(N,s)$, where $\tilde{P}(N,s)$ is some twisted version of $P(N,s)$. An affirmative answer to the question in section 3 would say that $\tilde{P}(N,s)$ breaks up into two irreducible factors, both of which should be twisted versions of $P(A_5,s)$. In fact this does not hold - $\tilde{P}(N,s)$ turns out to be itself irreducible. Using the computer software system CAYLEY, which efficiently calculates the lattice of subgroups of X and its Burnside matrix, it is straightforward to calculate the Moebius function of this lattice. We obtain that:

$$\begin{aligned}P(X,s) = P(C_2,s)(1 &- 25/25^s - 36/36^s - 120/60^s - 100/100^s \\ &+ 600/300^s + 720/360^s + 400/400^s + 1200/600^s \\ &+ 1800/900^s - 2400/1200^s - 7200/1800^s + 3600/3600^s).\end{aligned}$$

Using CASA, a Maple package from Linz, (though this can be, and was, checked by hand), one sees that the second factor is irreducible, completing the counterexample.

There are many more intriguing examples, for instance:

$$P(PSL(2,7),s) = 1 - 14/7^s - 8/8^s + 21/21^s + 28/28^s + 56/56^s - 84/84^s$$

whereas $P(PGL(2,7),s) = P(C_2,s)\tilde{P}(PSL(2,7),s)$ with

$$\tilde{P}(PSL(2,7),s) = 1 - 8/8^s - 21/21^s - 28/28^s + 56/56^s + 84/84^s$$

and

$$\begin{aligned}P(PSL(2,9),s) = 1 &- 12/6^s - 10/10^s - 30/15^s + 60/30^s + 36/36^s + 45/45^s \\ &+ 240/60^s + 90/90^s - 240/120^s - 900/180^s + 720/360^s\end{aligned}$$

whereas $P(PGL(2,9),s) = P(C_2,s)\tilde{P}(PSL(2,9),s)$ with

$$\tilde{P}(PSL(2,9),s) = 1 - 10/10^s - 36/36^s - 45/45^s + 90/90^s + 180/180^s.$$

The only primes p for which $P(PGL(2,p),s) = P(C_2,s)P(PSL(2,p),s)$ are apparently $p = 5$ and $p \equiv 3, 13, 27, 37 \pmod{40}$. Also, $P(S_n,s) =$

$P(C_2, s)P(A_n, s)$ for $n = 2, 5, 6$, but not for $n = 3, 4, 7, 8, 9$. These phenomena are being investigated by the author and Mann and reflect the existence (or non-existence) of maximal subgroups of the smaller group that stay maximal (if we ignore the smaller group) in the larger group.

Another part of the question in section 3 that fails is that regarding irreducibility of factors. One can factor $P(PSL(2,7), s)$ above as

$$(1 - 2/2^s)(1 + 2/2^s + 4/4^s - 14/7^s - 28/14^s - 28/28^s + 21/21^s + 42/42^s).$$

Note that both factors vanish at $s = 1$. The existence of a double zero at $s = 1$ is true for many (but not all) other $P(G, s)$ and leads to:

Conjecture. *If G is a finite, non-abelian simple group, then $P'(G, 1) = 0$.*

An equivalent statement of this is that $\prod [G : H]^{\mu(H)/[G:H]} = 1$, where the product is over all subgroups H of G. Since I wrote this paper, a more general version of this conjecture, namely that if a finite group G has no nontrivial normal p-subgroup, then $P'(G, 1) = 0$, has been proven by John Shareshian. This implies, in particular, that $P'(G, 1) = 0$ for every finite nonsolvable group G.

As a final curious observation, note that $P(A_6, s)$ $(= P(PSL(2,9), s)$ above) has a triple zero at $s = 1$. In the spirit of the conjecture of Birch and Swinnerton-Dyer on elliptic curves, perhaps there is some interesting internal characterization of $r(G)$, defined to be the order of vanishing of $P(G, s)$ at $s = 1$. Also, understanding the zeros of generalized Euler factors would have many group-theoretic applications.

References

[1] W. Gaschütz, *Die Eulersche Funktion endlicher auflösbarer Gruppen*, Ill. J. Math. **3** (1959), 469-476.

[2] G.H. Hardy and E.M. Wright, *An introduction to the theory of numbers*, 5th edition, Oxford Univ. Press, 1979.

[3] W.M. Kantor and A. Lubotzky, *The probability of generating a finite classical group*, Geo. Ded. **36** (1990), 67-87.

[4] J.P.S. Kung, M.R. Murty, and G.-C. Rota, *On the Rédei zeta function*, J. Number Theory **12** (1980), 421-436.

[5] A. Mann, *Positively finitely generated groups*, Forum Math. (to appear).

[6] A. Mann and A. Shalev, *Simple groups, maximal subgroups, and probabilistic aspects of profinite groups*, Israel J. Math (to appear).

[7] E.V. Novoselov, *A new method in probabilistic number theory*, Amer. Math. Soc. Transl. **52** (1966), 217-275.

[8] L. Ribes and K. Wong, *On the minimal number of generators of certain*

groups, Groups St. Andrews 1989, London Math. Soc. Lecture Note Series, vol. 160, pp. 408-421.

[9] W. Schwarz and J. Spilker, *Arithmetical functions: an introduction to elementary and analytic properties of arithmetic functions and to some of their almost-periodic properties*, Cambridge Univ. Press, LMS Lecture Notes 184.

Nigel Boston
Department of Mathematics
University of Illinois
Urbana, IL 61801
boston@math.uiuc.edu

A general Heine transformation and symmetric polynomials of Rogers

Douglas Bowman[1]

Dedicated to Heini Halberstam on the occasion of his retirement

1. Introduction

Our purpose here is first to give a multivariate generalization of the important transformation of Heine [6, pp. 305-306] (see also [1, ch.2]),

$$(1)\qquad {}_2\Phi_1\left(\begin{matrix} a,\ b \\ c \end{matrix}; z\right) = \frac{(az)_\infty (b)_\infty}{(c)_\infty (z)_\infty}\, {}_2\Phi_1\left(\begin{matrix} c/b,\ z \\ az \end{matrix}; b\right), \qquad |b|, |z| < 1,$$

and second to characterize the symmetric polynomials introduced by L. J. Rogers [8] in 1893.

Throughout we employ the standard notation:

$$(A)_n = (A;q)_n = (1-A)(1-Aq)\dots(1-Aq^{n-1}),$$

so that

$$(q)_n = (1-q)(1-q^2)\dots(1-q^n).$$

We always take $|q| < 1$. Let

$$(A_1, A_2, \dots, A_m)_n = (A_1)_n (A_2)_n \cdots (A_m)_n.$$

In (1) the ${}_2\Phi_1$ is the $m = 1$ case of the general basic hypergeometric function ${}_{m+1}\Phi_m$ defined by

$${}_{m+1}\Phi_m\left(\begin{matrix} a_1, a_2, \dots, a_{m+1} \\ b_1, b_2, \dots, b_m \end{matrix}; z\right) = \sum_{n\geq 0} \frac{(a_1)_n (a_2)_n \dots (a_{m+1})_n}{(b_1)_n (b_2)_n \dots (b_m)_n (q)_n} z^n.$$

For elementary facts about these functions we refer the reader to [5].

We introduce a convenient notation which we will use continually. Let

$$[x, y]_n = (x-y)(x-yq)\dots(x-yq^{n-1}), \qquad n \in \mathbb{Z}^+.$$

[1]Partially supported by National Science Foundation Grant DMS-9306089

If $x \neq 0$, then $[x,y]_n = x^n(y/x)_n$. For $x \neq 0$ we define $[x,y]_k$ for real k by the equation

$$[x,y]_k = \frac{(y/x)_\infty}{(yq^k/x)_\infty}. \tag{2}$$

Here and above $(A)_\infty$ is defined by

$$(A)_\infty = \lim_{n\to\infty}(A)_n.$$

We note that $(y)_n = [1,y]_n$ and define $[x]_n = [x,1]_n$. Clearly

$$[x,0]_n = x^n, \qquad [0,y]_n = (-1)^n q^{\binom{n}{2}} y^n. \tag{3}$$

Rogers's polynomials

$$h_j = h_j^{(m)}(\mathbf{a};\mathbf{b}) = h_j^{(m)}(a_1,\ldots,a_m;b_1,\ldots,b_m : q)$$

are defined explicitly by

$$h_j = \sum_{n_1+\cdots+n_m=j} \begin{bmatrix} j \\ n_1,\ldots,n_m \end{bmatrix} [b_1,a_1]_{n_1}\cdots[b_m,a_m]_{n_m},$$

where

$$\begin{bmatrix} j \\ n_1,\ldots,n_m \end{bmatrix} = \frac{(q)_j}{(q)_{n_1}\cdots(q)_{n_m}}.$$

Usually the argument q is implicit and is not written in the h_j. In applications we abbreviate $h_j(0,\ldots,0;b_1,\ldots,b_m : q)$ by $h_j(b_1,\ldots,b_m)$ or $h_j(\mathbf{b})$.

Rogers [7, 8] used his polynomials h_j for, among other things, his powerful symmetric expansion for Heine's ${}_2\Phi_1$:

If $f(\lambda_1,\lambda_2,\lambda_3,\lambda_4) = (\lambda_1\lambda_2)_\infty(\lambda_1^{-1}\lambda_4)_\infty\, {}_2\Phi_1\left(\begin{matrix}\lambda_1\lambda_2,\ \lambda_1\lambda_3 \\ \lambda_2\lambda_3\end{matrix}; \lambda_1^{-1}\lambda_4\right)$,

then

$$\begin{aligned} f(\lambda_1,\lambda_2,\lambda_3,\lambda_4) =& (\lambda_1^{-1}\lambda_2)_\infty(\lambda_1^{-1}\lambda_3)_\infty(\lambda_1^{-1}\lambda_4)_\infty(\lambda_1\lambda_2)_\infty(\lambda_1\lambda_3)_\infty(\lambda_1\lambda_4)_\infty \\ &\times \sum_{n\geq 0} \frac{h_n^{(2)}(\lambda_1^{-1},\lambda_1)h_n^{(3)}(\lambda_2,\lambda_3,\lambda_4)}{(q)_n}. \end{aligned} \tag{4}$$

Clearly $h_n^{(k)}(\lambda_1,\ldots,\lambda_k)$ is symmetric in its parameters. It is amazing how much information about special functions is contained in this one expansion; see Andrews [1, ch. 2].

In a recent book Fine [4, p.30] gave a q-multinomial theorem. However it is easily seen to be a special case of that given by Rogers [8]. We use the Rogers q-multinomial theorem to obtain an extension of the Heine transformation to an arbitrary number of variables. As applications we give some new expansions of the general ${}_{m+1}\Phi_m$ in terms of Rogers symmetric polynomials. (These expansions are not extensions of (4), but rather more generic.) Our characterization of Rogers's symmetric polynomials will reveal them to be discrete analogues of the general ${}_{m+1}\Phi_m$ function.

2. The Rogers q-multinomial theorem and a multivariate Heine transformation

Recall Cauchy's q-binomial theorem, [3], [5, pp. 7-8]:

$$\sum_{n\geq 0} \frac{(a)_n}{(q)_n} z^n = \frac{(az)_\infty}{(z)_\infty}, \qquad |z| < 1. \tag{5}$$

In the notation $[x, y]$, (5) may be written in the form

$$\sum_{n\geq 0} \frac{[b,a]_n}{(q)_n} z^n = \frac{(az)_\infty}{(bz)_\infty}, \qquad |bz| < 1. \tag{6}$$

We are now ready to discuss the q-multinomial theorem. Although it is implicit in [8], it does not seem to have received much attention. To obtain Rogers's theorem it is only necessary to multiply (6) together m times.

Theorem 2.1 q-Multinomial Theorem.

$$\frac{(a_1 z)_\infty \cdots (a_m z)_\infty}{(b_1 z)_\infty \cdots (b_m z)_\infty} = \sum_{j\geq 0} \frac{h_j}{(q)_j} z^j, \qquad |b_i z| < 1. \tag{7}$$

To obtain the theorem in Fine [4] put $a_1 = \cdots = a_m = 0$.

We now apply (7) to obtain an extension of the Heine transformation.

Theorem 2.2 Generalized Heine Transformation. *For $m \geq 1$, $|b_i| < 1$, $|d_i| < 1$ we have*

$$\sum_{k\geq 0} \frac{(d_1)_k \cdots (d_m)_k}{(c_1)_k \cdots (c_m)_k} \frac{h_k(\mathbf{a};\mathbf{b})}{(q)_k} =$$

$$\frac{(d_1)_\infty \cdots (d_m)_\infty (a_1)_\infty \cdots (a_m)_\infty}{(c_1)_\infty \cdots (c_m)_\infty (b_1)_\infty \cdots (b_m)_\infty} \sum_{k\geq 0} \frac{(b_1)_k \cdots (b_m)_k}{(a_1)_k \cdots (a_m)_k} \frac{h_k(\mathbf{c};\mathbf{d})}{(q)_k}.$$

The case $\mathbf{a} = \mathbf{b} = \mathbf{c} = \mathbf{0}$ of our theorem is equation (21.31) of [4]. Our theorem first appeared in the author's dissertation [2].

Proof. Observe that the sum

$$\sum_{k,l\geq 0} \frac{h_k(\mathbf{a};\mathbf{b})q^{kl}h_l(\mathbf{c};\mathbf{d})}{(q)_k(q)_l}$$

is invariant under the interchange of $\mathbf{a}$ with $\mathbf{c}$, $\mathbf{b}$ with $\mathbf{d}$ and converges absolutely for $|b_i|, |d_i| < 1$. Summing with respect to l and applying the q-multinomial theorem, Theorem 2.1, and (2), we find that

$$\frac{(c_1)_\infty \cdots (c_m)_\infty}{(d_1)_\infty \cdots (d_m)_\infty} \sum_{k\geq 0} \frac{(d_1)_k \cdots (d_m)_k}{(c_1)_k \cdots (c_m)_k} \frac{h_k(\mathbf{a};\mathbf{b})}{(q)_k}$$

is also invariant. This is equivalent to the statement of the theorem.

Letting $m = 1$ in the theorem gives

$$\sum_{k\geq 0} \frac{(d)_k}{(c)_k} \frac{h_k^{(1)}(a;b)}{(q)_k} = \frac{(d)_\infty (a)_\infty}{(c)_\infty (b)_\infty} \sum_{k\geq 0} \frac{(b)_k}{(a)_k} \frac{h_k^{(1)}(c;d)}{(q)_k}.$$

Using the fact that $h_k^{(1)}(x;y) = [y,x]_k = (x/y)_k y^k$ and replacing a by ab throughout, we obtain

$$\sum_{k\geq 0} \frac{(d)_k (a)_k}{(c)_k (q)_k} b^k = \frac{(d)_\infty (ab)_\infty}{(c)_\infty (b)_\infty} \sum_{k\geq 0} \frac{(b)_k (c/d)_k}{(ab)_k (q)_k} d^k,$$

which is the Heine transformation. Notice that in this case an additional symmetry was introduced due to the fact that h_k reduces to a q-factorial. This is the symmetry between a and d. For $m > 1$, h_k doesn't reduce to a factor of this type and the transformation group of substitutions of parameters induced by our theorem is isomorphic to $S_m^4 \times S_2$. Thus when $m = 1$ the group is $S_2 \times S_3$, Rogers [7], while for $m = 2$, the group is S_2^5. Before giving our characterization of $h_k^{(m)}$ we give some applications of our Heine extension. Note these results are valid when $|b|, |d_i| < 1$ for $1 \leq i \leq m$.

Corollary 2.1.

$$_{m+1}\Phi_m \left(\begin{matrix} a, d_1, d_2, \ldots, d_m \\ c_1, c_2, \ldots, c_m \end{matrix} ; b \right) = \frac{(d_1, \ldots, d_m, ab)_\infty}{(c_1, \ldots, c_m, b)_\infty} \sum_{k\geq 0} \frac{(b)_k}{(ab)_k (q)_k} h_k^{(m)}(\mathbf{c};\mathbf{d}).$$

Proof. In Theorem 2.2 put $\mathbf{a} = (ab, 0, \ldots, 0)$ and $\mathbf{b} = (b, 0, \ldots, 0)$ and the corollary follows.

Putting $c_m = a$ in the last result gives:

Corollary 2.2.

$$
{}_m\Phi_{m-1}\left(\begin{matrix} d_1, d_2, \ldots, d_m \\ c_1, c_2, \ldots, c_{m-1} \end{matrix}; b\right) = \frac{(d_1, \ldots, d_m, ab)_\infty}{(c_1, \ldots, c_{m-1}, a, b)_\infty} \sum_{k \geq 0} \frac{(b)_k}{(ab)_k (q)_k} h_k^{(m)}(a, \mathbf{c}; \mathbf{d}).
$$

Notice that the parameter a appears only on the right side. Putting $a = 0$ yields:

Corollary 2.3.

$$
{}_m\Phi_{m-1}\left(\begin{matrix} d_1, d_2, \ldots, d_m \\ c_1, c_2, \ldots, c_{m-1} \end{matrix}; b\right) = \frac{(d_1, \ldots, d_m)_\infty}{(c_1, \ldots, c_{m-1}, b)_\infty} \sum_{k \geq 0} \frac{(b)_k}{(q)_k} h_k^{(m)}(\mathbf{c}; \mathbf{d}).
$$

(In this last result notice that $\mathbf{c}$ has only $m-1$ elements; the first element of the first vector argument of h may be set equal to 0.)

We conclude this section stating the case $m = 2$ of these last three corollaries.

$$
(8) \qquad {}_3\Phi_2\left(\begin{matrix} a, d_1, d_2 \\ c_1, c_2 \end{matrix}; b\right) = \frac{(d_1, d_2, ab)_\infty}{(c_1, c_2, b)_\infty} \sum_{k \geq 0} \frac{(b)_k}{(ab)_k (q)_k} h_k^{(2)}(c_1, c_2; d_1, d_2).
$$

$$
(9) \qquad {}_2\Phi_1\left(\begin{matrix} d_1, d_2 \\ c \end{matrix}; b\right) = \frac{(d_1, d_2, ab)_\infty}{(c, a, b)_\infty} \sum_{k \geq 0} \frac{(b)_k}{(ab)_k (q)_k} h_k^{(2)}(a, c; d_1, d_2).
$$

$$
(10) \qquad {}_2\Phi_1\left(\begin{matrix} d_1, d_2 \\ c \end{matrix}; b\right) = \frac{(d_1, d_2)_\infty}{(c, b)_\infty} \sum_{k \geq 0} \frac{(b)_k}{(q)_k} h_k^{(2)}(c; d_1, d_2).
$$

3. The discrete linear n'th order q-difference equation and the polynomials $h_k^{(m)}$

In this section we prove a fundamental property of the polynomials $h_k^{(m)}$. We show that they are in fact discrete analogues of the basic hypergeometric functions ${}_m\Phi_{m-1}$. It is interesting that the symmetric expansions for basic series are in terms of these polynomials and certain operators. This has been explored somewhat in the author's dissertation [2] and will be the subject of a future long paper or memoir.

It is well known that the analytic solution at the origin of the q-difference equation

$$(1-z)f(z)+(\beta_1-\alpha_1 z)f(zq)+\cdots+(\beta_m-\alpha_m z)f(zq^m)=0.$$

with initial condition $f(0)=1$ is given by ${}_m\Phi_{m-1}$ (Thomae [9, pp. 108–112]). Here we study an analogous difference equation with a polynomial solution. This equation is

$$(1-q^n)H_n+(\beta_1-\alpha_1 q^{n-1})H_{n-1}+\cdots+(\beta_m-\alpha_m q^{n-m})H_{n-m}=0. \tag{11}$$

We specify the initial conditions $H_n=0$ for $n<0$ and $H_0=1$. It is easy to see that these conditions uniquely determine the polynomials H_n. We proceed to solve (11). We assume throughout that $\beta_0=\alpha_0=1$. We refer to (11) as the discrete q-difference equation because of the dependence on the discrete variable n leading to a solution which is a sequence.

Equation (11) can be written in the form

$$\sum_{n-m\leq j\leq n}(\beta_{n-j}-\alpha_{n-j}q^j)H_j=0,$$

or

$$\sum_{n-m\leq j\leq n}\beta_{n-j}H_j=\sum_{n-m\leq j\leq n}\alpha_{n-j}q^jH_j.$$

Multiplying both sides by z^n and summing over $n\geq 0$ gives

$$\sum_{n\geq 0}\sum_{n-m\leq j\leq n}\beta_{n-j}H_jz^n=\sum_{n\geq 0}\sum_{n-m\leq j\leq n}\alpha_{n-j}q^jH_jz^n.$$

Using the initial conditions, we get

$$\sum_{\substack{j\geq 0\\ 0\leq i\leq m}}\alpha_iH_jq^jz^{i+j}=\sum_{\substack{j\geq 0\\ 0\leq i\leq m}}\beta_iH_jz^{i+j},$$

or

$$P(z)\sum_{j\geq 0}H_jq^jz^j=Q(z)\sum_{j\geq 0}H_jz^j,$$

where

$$\alpha_0+\cdots+\alpha_mx^m=(1-a_1x)\cdots(1-a_mx),$$

and

$$\beta_0 + \cdots + \beta_m x^m = (1 - b_1 x) \cdots (1 - b_m x).$$

If $B(z) = \sum_{j \geq 0} H_j z^j$ is the generating function for the sequence H_j, then

$$B(z) = \frac{P(z)}{Q(z)} B(zq).$$

Iterating gives

$$B(z) = \frac{(a_1 z)_\infty \cdots (a_m z)_\infty}{(b_1 z)_\infty \cdots (b_m z)_\infty}.$$

Finally by Theorem 2.1,

$$\sum_{j \geq 0} H_j z^j = B(z) = \sum_{j \geq 0} \frac{h_j}{(q)_j} z^j.$$

Equating coefficients now gives

$$H_j = \frac{h_j}{(q)_j}. \tag{12}$$

This conclusion is summarized in the following theorem.

Theorem 3.1. *Let $\alpha_0 = \beta_0 = 1$ and let $m > 0$ be a fixed integer. Then the solution of the equation*

$$(\beta_0 - \alpha_0 q^n) H_n + (\beta_1 - \alpha_1 q^{n-1}) H_{n-1} + \cdots + (\beta_m - \alpha_m q^{n-m}) H_{n-m} = 0$$

with $H_n = 0$ for $-m < n < 0$ and $H_0 = 1$ is given by

$$H_j = \frac{h_j}{(q)_j} = \frac{1}{(q)_j} \sum_{n_1 + \cdots + n_m = j} \begin{bmatrix} j \\ n_1, \ldots, n_m \end{bmatrix} [b_1, a_1]_{n_1} \cdots [b_m, a_m]_{n_m},$$

where

$$\alpha_0 + \cdots + \alpha_m x^m = (1 - a_1 x) \cdots (1 - a_m x),$$

and

$$\beta_0 + \cdots + \beta_m x^m = (1 - b_1 x) \cdots (1 - b_m x).$$

Notice that when $q = 1$, $h_j = (b_1 + \cdots + b_m - a_1 - \cdots - a_m)^j$. For future reference we mention the easily proved relation

$$h_j(\mathbf{a};\mathbf{b}:q) = (-1)^j q^{\binom{j}{2}} h_j(\mathbf{b};\mathbf{a}:q^{-1}).$$

In the case $m = 1$ equation (11) reduces to $(1 - q^n)H_n = (b - aq^{n-1})H_{n-1}$ from which it follows immediately that $H_j = [b, a]_j/(q)_j$.

The case $m = 2$ of this theorem is important in that it gives a general class of polynomials which satisfy three-term recurrence relations. So for example, $h_n^{(2)}(\beta e^{i\theta}, \beta e^{-i\theta}; e^{i\theta}, e^{-i\theta})/(q)_j$ is the continuous q-ultraspherical polynomial $C_n(x;\beta|q)$, where $x = cos\theta$ (see [5, pp. 168-172]). The q-difference equation of our theorem is precisely the three-term recurrence for this orthogonal polynomial system.

References

[1] G.E. Andrews, *q-series: Their Development and Application in Analysis, Number Theory, Combinatorics, Physics, and Computer Algebra.* CBMS regional conference series 66, Amer. Math. Soc., Providence RI, 1986.

[2] D. Bowman, *q-Series, Partitions and Continued Fractions,* doctoral dissertation, University of California, Los Angeles, 1993.

[3] A.-L. Cauchy, *Mémoire sur les fonctions dont plusieurs valeurs sont liées entre elles par une équation linéaire, et sur diverses transformations de produits composés d'un nombre indéfini de facteurs,* C. R. Acad. Sci. Paris, T. (**XVII**), (1843), p. 523, *Oeuvres de Cauchy,* 1re série, T. (**VIII**), Gauthier-Villars, Paris, (1893), 42-50.

[4] N. Fine, *Basic Hypergeometric Series and Applications,* Amer. Math. Soc., Providence RI, 1991.

[5] G. Gasper and M. Rahman, *Basic Hypergeometric Series,* Cambridge, 1990.

[6] E. Heine, *Untersuchungen über die Reihe*

$$1 + \frac{(1-q^\alpha)(1-q^\beta)}{(1-q)(1-q^\gamma)}x + \frac{(1-q^\alpha)(1-q^{\alpha+1})(1-q^\beta)(1-q^{\beta+1})}{(1-q)(1-q^2)(1-q^\gamma)(1-q^{\gamma+1})}x^2 + \cdots,$$

J. reine angew. Math. (**34**), (1847), 285-328.

[7] L. J. Rogers, *On the three-fold symmetry in the elements of Heine's series,* Proc. Lond. Math. Soc. **24** (1893), 171-179.

[8] L. J. Rogers, *On the expansion of some infinite products,* Proc. Lond. Math. Soc. **24** (1893), 337-352.

[9] J. Thomae, *Les séries Heinéennes supérieures, ou les séries de la forme*

$$1+\sum_{n=1}^{\infty} x^n \frac{1-q^a}{1-q}\cdot\frac{1-q^{a+1}}{1-q^2}\cdots\frac{1-q^{a+n-1}}{1-q^n}\cdot\frac{1-q^{a'}}{1-q^{b'}}\cdot\frac{1-q^{a'+1}}{1-q^{b'+1}}\cdots$$
$$\frac{1-q^{a'+n-1}}{1-q^{b'+n-1}}\cdots\frac{1-q^{a^{(h)}}}{1-q^{b^{(h)}}}\cdot\frac{1-q^{a^{(h)}+1}}{1-q^{b^{(h)}+1}}\cdots\frac{1-q^{a^{(h)}+n-1}}{1-q^{b^{(h)}+n-1}},$$

Annali di Matematica Pura ed Applicato. **4** (1870), 105-138.

Douglas Bowman
Department of Mathematics
University of Illinois
Urbana, IL 61801
bowman@math.uiuc.edu

A sieve auxiliary function

David Bradley

Dedicated to Professor Heini Halberstam on the occasion of his retirement

Abstract. In the sieve theories of Rosser-Iwaniec and Diamond-Halberstam-Richert, the upper and lower bound sieve functions (F and f, respectively) satisfy a coupled system of differential-difference equations with retarded arguments. To aid in the study of these functions, Iwaniec introduced a conjugate difference-differential equation with an advanced argument, and gave a solution, q, which is analytic in the right half-plane. The analysis of the bounding sieve functions, F and f, is facilitated by an adjoint integral inner-product relation which links the local behaviour of $F - f$ with that of the sieve auxiliary function, q. In addition, q plays a fundamental role in determining the sieving limit of the combinatorial sieve, and hence in determining the boundary conditions of the sieve functions, F and f. The sieve auxiliary function, q, has been tabulated previously, but these data were not supported by numerical analysis, due to the prohibitive presence of high-order partial derivatives arising from the numerical quadrature methods used [15, 17]. In this paper, we develop additional representations of q. Certain of these representations are amenable to detailed error analysis. We provide this error analysis, and as a consequence, we indicate how q-values guaranteed to at least seven decimal places can be tabulated.

1. Introduction

In his seminal paper, Rosser's Sieve [11], Iwaniec introduced a pair of difference-differential equations which have been studied more recently by Diamond, Halberstam, and Richert [3–10], and by Wheeler [17, 18]. The equations appear as auxiliary equations in connection with the problem of estimating

$$S(\mathscr{A}, \mathscr{P}, x) := \#\{a \in \mathscr{A} : \gcd(a, \prod_{\substack{p<x \\ p\in\mathscr{P}}} p) = 1\},$$

where $\mathscr{P}$ is a set of primes and $\mathscr{A}$ is a finite set of integers. In the sieve theories of Rosser-Iwaniec and Diamond-Halberstam-Richert, the equations take the form

$$(uq_\kappa(u))' = \kappa q_\kappa(u) + \kappa q_\kappa(u+1) \tag{1.1}$$

and

$$(up_\kappa(u))' = \kappa p_\kappa(u) - \kappa p_\kappa(u+1), \tag{1.2}$$

where u, κ are real and positive. The parameter κ denotes the dimension of the sieve, or sifting density, and is a measure of the average number of residue classes per prime in the sequence being sifted. Iwaniec gave solutions to (1.1) and (1.2) involving the so-called complementary exponential integral [14, p.40] defined by

$$\mathrm{Ein}(z) := \int_0^z \frac{1-e^{-t}}{t}\,dt.$$

The solutions are

$$q_\kappa(u) = \frac{\Gamma(2\kappa)}{2\pi i} \int z^{-2\kappa} e^{uz} e^{\kappa\,\mathrm{Ein}(-z)}\,dz \tag{1.3}$$

and

$$p_\kappa(u) = \int_0^\infty e^{-xu-\kappa\,\mathrm{Ein}(x)}\,dx, \tag{1.4}$$

where in (1.3), the contour starts at $-\infty$, hugging the negative real axis, then circles the origin in the positive direction before returning to $-\infty$. In this paper, we focus on the problems presented by the function q_κ, since p_κ, being the Laplace transform of a positive function, is relatively simple to deal with. In [7], it is shown that the solutions (1.3), (1.4) are unique, subject to mild polynomial-like growth conditions at infinity. In Section 2 below, we prove that (1.3) is the unique solution in a class of functions representable as a Laplace/Mellin transform. In Section 3, an asymptotic expansion is derived, and a few properties of the coefficients are proved. In Section 4, we give a representation of q_κ in terms of an operator that arises in other contexts. Finally, it is of some interest to have values of q_κ tabulated. We take up this problem in Section 5. We remark that this paper is based in significant part on the author's Ph.D. thesis [2].

2. The function $q_\kappa(u)$

The difference-differential equation (1.1) can be rewritten in the form

$$\left(u^{1-\kappa}q_\kappa(u)\right)' = \kappa u^{-\kappa}q_\kappa(u+1),$$

so that the value of the function at u is given by an integral involving the function at larger values of the argument. Since integration is a smoothing operation, one expects repeated integrations to yield a C^∞ solution, given only mild assumptions on the behaviour of the function at infinity. In fact, it is easy to see that Iwaniec's solution (1.3) is analytic in the right half-plane and that $q_\kappa(u)$ as given by (1.3) is asymptotic to $u^{2\kappa-1}$ as u tends to infinity.

In [7], the solution (1.3) is shown to be unique in the class of normalized polynomial-like functions. In other words, (1.3) is the unique solution to (1.1) which satisfies $q_\kappa(u) \sim u^b$ as $u \to \infty$, for some constant b (and hence we must have $b = 2\kappa - 1$). In the sequel, we shall prove a uniqueness result of a somewhat different kind, which shows that (1.3) is unique in a class of functions representable as a Laplace/Mellin transform. For this task, it is profitable to view $q_\kappa(s)$ as a function of the complex variable κ, with s lying in the right half-plane, although for sieve applications, we are primarily concerned with positive real values of the parameters. But first, we need to recast (1.3) as an integral over the positive real axis.

Proposition 2.1. *Let n be a non-negative integer, and suppose $\Re(n+1-2\kappa) > 0$. Then*

$$\text{(2.1)}\quad q_\kappa(s) = \frac{(-1)^n}{\Gamma(n+1-2\kappa)}\int_0^\infty x^{n-2\kappa}\left(\frac{\partial}{\partial x}\right)^n e^{-sx}e^{\kappa\,\mathrm{Ein}(x)}\,dx, \quad \Re(s) > 0.$$

Remark. If κ is real and small enough so that $n = 0$ or $n = 1$ is permissible (i.e. $\kappa < 1/2$ in the former case, $\kappa < 1$ in the latter) then one can use the representation (2.1) to compute $q_\kappa(u)$ quite easily. However, as n increases, the higher order partial derivatives rapidly become cumbersome, and so for larger values of κ, the method of Section 5 is preferrable.

Proof Sketch. The case $n = 0$ can be found in Iwaniec [11, p.184]. Te Riele [15, p.6] and Wheeler [17, p.73] derive (2.1) from the $n = 0$ case by performing repeated integration by parts on the latter. One can also obtain (2.1) directly from (1.3), integrating by parts n times. The integrated terms all vanish due to the presence of e^{sz} as a factor in every derivative of $e^{sz}e^{\kappa\,\mathrm{Ein}(-z)}$. One can then collapse the contour onto the negative real axis, and after some minor simplifications, (2.1) results. See [2, p.12] for details.

We are now ready to prove that (2.1) is the unique solution to (1.1) in the class of functions representable as a Laplace/Mellin transform. For convenience, the following notation will be used. Let $\mathscr{G}$ denote the set of functions g such that $g^{(n)} \in L^1[0,\infty]$ for $n = 0,1,2,3$ and for which $g^{(n)}$ vanishes at both zero and infinity for $n = 0,1,2$. Also, for each real $B \geqslant 0$, put

$$\mathscr{F}_B := \{f : y \mapsto e^{-By} f(y) \in \mathscr{G}\}.$$

Theorem 1 (Uniqueness). *Let n be the least non-negative integer such that $\Re(2\kappa) < n+1$. Then up to multiplication by an arbitrary function of κ,*

$$q_\kappa(s) := \frac{(-1)^n}{\Gamma(n+1-2\kappa)} \int_0^\infty x^{n-2\kappa} \left(\frac{\partial}{\partial x}\right)^n e^{-sx} e^{\kappa \operatorname{Ein}(x)}\, dx, \quad \Re s > 0 \tag{2.2}$$

is the unique solution to the difference-differential equation

$$(sq_\kappa(s))' = \kappa q_\kappa(s) + \kappa q_\kappa(s+1)$$

which satisfies

(2.3) $\quad q_\kappa(s)$ *is an entire function of* κ,

(2.4)
$$\exists B \geqslant 0 \text{ such that } q_\kappa(s) = \int_0^\infty e^{-sy} f_\kappa(y)\, dy, \quad \Re(2\kappa) < -2, \quad \Re(s) > 0,$$

for some $f_\kappa \in \mathscr{F}_B$.

Remark. Note that f_κ can be very general. Of course, if $f_\kappa(x) \sim \dfrac{x^{-2\kappa}}{\Gamma(1-2\kappa)}$ as $x \to 0+$, then (2.4) would imply that $q_\kappa(u) \sim u^{2\kappa-1}$ as $u \to \infty$, but we do not assume this.

Proof. Suppose that (2.4) holds for some $B \geqslant 0$. We shall show that necessarily, $f_\kappa(x) = x^{-2\kappa} e^{\kappa \operatorname{Ein}(x)}$ up to an arbitrary constant multiple (depending on κ), and that (2.2) follows. The Laplace inversion theorem guarantees that for any $c > B$, we have

$$f_\kappa(x) = \frac{1}{2\pi i} \int_{c-i\infty}^{c+i\infty} e^{sx} q_\kappa(s)\, ds, \quad x > 0. \tag{2.5}$$

Our approach is to differentiate (2.5) with respect to x and then use the difference-differential equation satisfied by q_κ to obtain a differential equation for f_κ that can be solved explicitly. But first, we must justify differentiating (2.5) under the integral sign. We require the following lemmata.

Lemma 2.1. *Suppose (2.4) holds for some $B \geqslant 0$ and $f_\kappa \in \mathscr{F}_B$. Let $\Re(s) > B$. Then*

$$q_\kappa(s) = \frac{1}{s^p}\int_0^\infty e^{-sy} f_\kappa^{(p)}(y)\,dy, \quad p = 0, 1, 2, 3.$$

Proof. Integrate (2.4) by parts repeatedly. In each case, the integrated term vanishes by the hypotheses on f_κ and the definition of $\mathscr{G}$.

Lemma 2.2. *Suppose (2.4) holds for some $B \geqslant 0$. Let $c > B$. Then for all $x > 0$, the function*

$$F(x) := \int_{c-i\infty}^{c+i\infty} e^{sx} q_\kappa(s)\,ds \quad \textit{satisfies} \quad F'(x) = \int_{c-i\infty}^{c+i\infty} e^{sx}\, s\, q_\kappa(s)\,ds.$$

Proof. For positive integers n, define

$$F_n(x) := \int_{c-in}^{c+in} e^{sx} q_\kappa(s)\,ds, \quad x > 0.$$

It is clear that for each n, F_n' exists, is continous, and is given by

$$F_n'(x) = \int_{c-in}^{c+in} e^{sx}\, s\, q_\kappa(s)\,ds, \quad x > 0.$$

Let

$$G(x) := \int_{c-i\infty}^{c+i\infty} e^{sx}\, s\, q_\kappa(s)\,ds, \quad x > 0.$$

We shall show that the sequence F_n' converges uniformly to G on compact intervals. Let $0 < a < M$ and for now, make the restriction $x \in [a, M]$. By Lemma 2.1,

$$\begin{aligned}
|G(x) - F_n'(x)| &\leqslant \left|\int_{c+in}^{c+i\infty} e^{sx}\, s\, q_\kappa(s)\,ds\right| + \left|\int_{c-i\infty}^{c-in} e^{sx}\, s\, q_\kappa(s)\,ds\right| \\
&\ll e^{cM}\left|\int_{c+in}^{c+i\infty} s\, q_\kappa(s)\,ds\right| \\
&= e^{cM}\left|\int_{c+in}^{c+i\infty}\int_0^\infty e^{-sy} f_\kappa'''(y)\,dy\,\frac{ds}{s^2}\right| \\
&= e^{cM}\int_n^\infty\left|\int_0^\infty e^{-cy} f_\kappa'''(y) e^{-ity}\,dy\right|\frac{dt}{c^2+t^2} \\
&\leqslant e^{cM}\int_n^\infty\int_0^\infty |e^{-cy} f_\kappa'''(y)|\,dy\,\frac{dt}{c^2+t^2}. \qquad (2.6)
\end{aligned}$$

Now by the assumption $f_\kappa \in \mathscr{F}_B$, we have that $y \mapsto e^{-cy} f_\kappa'''(y) \in L^1[0,\infty]$. Thus the inner integral in (2.6) is finite, and as $n \to \infty$, it follows that $|G(x) - F_n'(x)| \to 0$ independently of $x \in [a, M]$. In other words, $F_n' \to G$ uniformly on $[a, M]$, and hence G is continuous there. Furthermore, by uniform convergence,

$$\int_a^x G(y)\,dy = \lim_{n\to\infty} \int_a^x F_n'(y)\,dy = \lim_{n\to\infty} (F_n(x) - F_n(a)) = F(x) - F(a).$$

By continuity of G and the fundamental theorem of calculus, $F'(x) = G(x)$ holds for all $x \in [a, M]$. Finally, since $0 < a < M$ were arbitrary, the lemma is proved.

Continuing with the proof of Theorem 1 we therefore have

$$\begin{aligned} f_\kappa'(x) &= \frac{1}{2\pi i} \int_{c-i\infty}^{c+i\infty} e^{sx} s\, q_\kappa(s)\,ds \\ &= \frac{e^{sx} s q_\kappa(s)}{2\pi i x} \bigg|_{c-i\infty}^{c+i\infty} - \frac{1}{2\pi i x} \int_{c-i\infty}^{c+i\infty} e^{sx} \left(s\, q_\kappa(s)\right)'\,ds. \end{aligned}$$

We claim the integrated term vanishes. Writing $s = c + iT$, we have

$$\lim_{T\to\infty} \left| e^{(c+iT)x} (c + iT) q_\kappa(c + iT) \right| = e^{cx} \lim_{T\to\infty} \left| \int_0^\infty e^{-(c+iT)y} f_\kappa'(y)\,dy \right| = 0,$$

by the Riemann-Lebesgue lemma. Therefore,

$$\begin{aligned} x f_\kappa'(x) &= -\frac{1}{2\pi i} \int_{c-i\infty}^{c+i\infty} e^{sx} \left(s q_\kappa(s)\right)'\,ds \\ &= -\frac{\kappa}{2\pi i} \int_{c-i\infty}^{c+i\infty} e^{sx} q_\kappa(s)\,ds - \frac{\kappa}{2\pi i} \int_{c-i\infty}^{c+i\infty} e^{sx} q_\kappa(s+1)\,ds \\ &= -\kappa f_\kappa(x) - \kappa e^{-x} \int_{c+1-i\infty}^{c+1+i\infty} e^{sx} q_\kappa(s)\,ds \\ &= -\kappa f_\kappa(x) - \kappa e^{-x} f_\kappa(x). \end{aligned}$$

Solving the separable first order differential equation for f_κ yields

$$f_\kappa(x) = A_\kappa\, x^{-2\kappa} e^{\kappa \operatorname{Ein}(x)}, \tag{2.7}$$

where A_κ is a constant depending only on κ. For convenience, put $C_\kappa = A_\kappa \Gamma(1 - 2\kappa)$. Since the difference-differential equation (1.1) satisfied by q_κ is homogeneous, i.e. insensitive to multiplication by an arbitrary function of κ, we may take $C_\kappa = 1$ without loss of generality. Also, note that f_κ as given by (2.7) does indeed lie in the set $\mathscr{F}$ for $\Re(2\kappa) < -2$, as stipulated. Thus we have shown that up to multiplication by an arbitrary function of κ,

$$(2.8)\quad q_\kappa(s) = \frac{1}{\Gamma(1-2\kappa)} \int_0^\infty x^{-2\kappa} e^{-sx} e^{\kappa\,\mathrm{Ein}(x)}\, dx, \quad \Re(2\kappa) < -2, \quad \Re s > 0.$$

However (2.8) is an analytic function of κ for $\Re(2\kappa) < 1$, so in fact, (2.8) provides an analytic continuation of $q_\kappa(s)$ to the open left half-plane $\Re(2\kappa) < 1$. To complete the proof of Theorem 1, we need to analytically continue q_κ to an entire function of κ. We now set

$$F_\kappa(x) = e^{-sx} e^{\kappa\,\mathrm{Ein}(x)}, \qquad G_\kappa(x) = x^{-2k}$$

and integrate (2.8) by parts n times, using the formula

$$\int_0^\infty F_\kappa(x) G_\kappa(x)\, dx = \sum_{j=0}^{n-1} (-1)^j F_\kappa^{(j)}(x)\, G_\kappa^{(-j-1)}(x) \Big|_0^\infty + (-1)^n \int_0^\infty F_\kappa^{(n)}(x)\, G_\kappa^{(-n)}(x)\, dx.$$

We claim the integrated terms all vanish. To see this, note that

$$G_\kappa^{(-j-1)}(x) = \frac{x^{j+1-2\kappa}}{(1-2\kappa)_{j+1}}.$$

Furthermore, since $F_\kappa(x)$ is an entire function of x,

$$\lim_{x\to 0+} F_\kappa^{(j)}(x) G_\kappa^{(-j-1)}(x) = F_\kappa^{(j)}(0) \lim_{x\to 0+} \frac{x^{j+1-2\kappa}}{(1-2\kappa)_{j+1}} = 0,$$

for $\Re(2\kappa) < 1$ and j a non-negative integer. On the other hand, e^{-sx} is a factor of every term in $F_\kappa^{(j)}(x)$, whereas for $x > 0$,

$$\left| \left(\frac{\partial}{\partial x} \right)^r e^{\kappa\,\mathrm{Ein}(x)} \right| \ll e^{\varepsilon x},$$

for every $r > 0$, $\varepsilon > 0$. Thus

$$\lim_{x\to\infty} G_\kappa^{(-j-1)}(x) F_\kappa^{(j)}(x) = 0.$$

It follows that for $\Re s > 0$,

$$
\begin{aligned}
q_\kappa(s) &= \frac{(-1)^n}{(1-2\kappa)_n\,\Gamma(1-2\kappa)}\int_0^\infty x^{n-2\kappa}\left(\frac{\partial}{\partial x}\right)^n e^{-sx}e^{\kappa\,\mathrm{Ein}(x)}\,dx \\
&= \frac{(-1)^n}{\Gamma(n+1-2\kappa)}\int_0^\infty x^{n-2\kappa}\left(\frac{\partial}{\partial x}\right)^n e^{-sx}e^{\kappa\,\mathrm{Ein}(x)}\,dx, \quad \Re(2\kappa) < 1.
\end{aligned} \tag{2.9}
$$

Now observe that (2.9) actually gives an analytic continuation of $q_\kappa(s)$ to the half-plane $\Re(2\kappa) < n+1$, which completes the proof.

3. The asymptotic expansion

Recalling (1.3) again, if 2κ is a positive integer, then the branch point in the integrand at $z = 0$ becomes a pole, and we can collapse the horizontal portions of the contour onto the negative real axis, so that

$$
q_\kappa(u) = \frac{\Gamma(2\kappa)}{2\pi i}\oint_{|z|=1} z^{-2\kappa}e^{uz}e^{\kappa\,\mathrm{Ein}(-z)}\,dz = \left(\frac{\partial}{\partial z}\right)^{2\kappa-1} e^{uz}e^{\kappa\,\mathrm{Ein}(-z)}\bigg|_{z=0}.
$$

This suggests that when 2κ is not a positive integer, we define the fractional derivative $(\partial/\partial z)^{2\kappa-1}$ by means of

$$
\left(\frac{\partial}{\partial z}\right)^{2\kappa-1} e^{uz}e^{\kappa\,\mathrm{Ein}(-z)}\bigg|_{z=0} := q_\kappa(u) = \frac{\Gamma(2\kappa)}{2\pi i}\int_{\Rightarrow\!\circ} z^{-2\kappa}e^{uz}e^{\kappa\,\mathrm{Ein}(-z)}\,dz.
$$

In other words, for general κ, it may be profitable to view $q_\kappa(u)$ as a fractional derivative. Thus, in some sense, soon to be made precise,

$$
q_\kappa(u) \sim \left(\frac{\partial}{\partial z}\right)^{2\kappa-1} e^{uz}e^{\kappa\,\mathrm{Ein}(-z)}\bigg|_{z=0}.
$$

For example, applying Leibniz's rule to the above yields the formal expansion

$$
\begin{aligned}
q_\kappa(u) &\sim \sum_{n=0}^\infty \binom{2\kappa-1}{n}\left(\frac{\partial}{\partial z}\right)^n e^{\kappa\,\mathrm{Ein}(-z)}\bigg|_{z=0}\cdot\left(\frac{\partial}{\partial z}\right)^{2\kappa-1-n} e^{uz}\bigg|_{z=0} \\
&= \sum_{n=0}^\infty \binom{2\kappa-1}{n} b_n(\kappa)\,u^{2\kappa-1-n},
\end{aligned} \tag{3.1}
$$

which, in view of our previous remarks, gives a true equality when 2κ is a positive integer. Here, $b_n(\kappa)$ is the nth degree polynomial in κ defined by

$$
b_n(\kappa) := \left(\frac{\partial}{\partial z}\right)^n e^{\kappa\,\mathrm{Ein}(-z)}\bigg|_{z=0} \tag{3.2}
$$

and the fractional derivative of e^{uz} is given, of course, by

$$\left(\frac{\partial}{\partial z}\right)^{2\kappa-1-n} e^{uz}\bigg|_{z=0} = \frac{\Gamma(2\kappa-n)}{2\pi i}\int z^{n-2\kappa}e^{uz}\,dz = u^{2\kappa-1-n}.$$

It turns out that (3.1) is a valid asymptotic expansion for u-values tending to positive infinity.

Theorem 2. *Let $b_n(\kappa)$ be defined as in (3.2), κ real. Then the asymptotic expansion*

$$q_\kappa(u) \sim \sum_{n=0}^{\infty}\binom{2\kappa-1}{n} b_n(\kappa)\,u^{2\kappa-1-n}, \quad u\to\infty$$

is valid.

Proof Sketch. The asymptotic expansion with remainder after n terms appears in essentially the above guise in both [11, p.183] and [17, p.37]. The main idea is to expand $e^{\kappa\,\mathrm{Ein}(-z)}$ into its Taylor series, and then use Hankel's contour formula for the reciprocal of the gamma function. For an analysis of the size of the remainder term in (3.1) see [2, Chapter 8, Theorem 6].

It is interesting to deduce some properties of the coefficients $b_n(\kappa)$ of the asymptotic expansion provided by Theorem 2. Rewriting (3.2) in the form

$$e^{\kappa\,\mathrm{Ein}(-z)} = \sum_{n=0}^{\infty}\frac{z^n}{n!}b_n(\kappa), \tag{3.3}$$

it is immediate that $b_0(\kappa) = 1$. For positive integers n, we have the following recurrence formula which gives $b_n(\kappa)$ in terms of $b_0(\kappa), b_1(\kappa), \ldots, b_{n-1}(\kappa)$.

Proposition 3.1. *If n is a positive integer and $b_n(\kappa)$ is given by (3.3), then*

$$b_n(\kappa) = -\frac{\kappa}{n}\sum_{j=0}^{n-1}\binom{n}{j}b_j(\kappa).$$

Proof. Applying the operator $z\cdot d/dz$ to (3.3), one obtains

$$\sum_{n=0}^{\infty} nb_n(\kappa)\frac{z^n}{n!} = -\kappa\,(e^z-1)\,e^{\kappa\,\mathrm{Ein}(-z)}.$$

The result now follows on comparing coefficients of $z^n/n!$.

For concreteness, the first few b polynomials are listed below.

$$b_0 = 1, \quad b_1 = -\kappa, \quad b_2 = \kappa^2 - \kappa/2, \quad b_3 = -\kappa^3 + 3\kappa^2/2 - \kappa/3.$$

The following result shows a connection with the Bernoulli numbers, defined by

$$\frac{z}{e^z - 1} = \sum_{n=0}^{\infty} \frac{z^n}{n!} B_n, \quad |z| < 2\pi.$$

Theorem 3. *Let α be an arbitrary constant. Then for all non-negative integers n,*

$$\sum_{j=0}^{n-2} \binom{n}{j} b_{j+1}(\alpha) B_{n-j} = (n/2 - \alpha)\, b_n(\alpha) - b_{n+1}(\alpha).$$

Remark. The sum on the left vanishes when $n = 0, 1$. For general n, $b_n(\alpha)$ is a polynomial in α of degree n. Thus the coefficients of α^n and of α^{n+1} on the right hand side both must vanish.

Proof. Let $[z^n/n!]F(z)$ denote the coefficient of $z^n/n!$ in the analytic function $F(z)$. We proceed by developing

$$Q_n(u, \alpha) := \left(\frac{\partial}{\partial z}\right)^n e^{uz} e^{\alpha \operatorname{Ein}(-z)}\bigg|_{z=0} = [z^n/n!]\, e^{uz} e^{\alpha \operatorname{Ein}(-z)}$$

as a Fourier series in u on $]0, 1[$. Note that $b_n(\alpha) = Q_n(0, \alpha)$. We require the following

Lemma 3.1. *Let α be an arbitrary constant. Then for all non-negative integers n,*

$$Q_n(1, \alpha) = (1 - n/\alpha)\, Q_n(0, \alpha) = (1 - n/\alpha)\, b_n(\alpha).$$

Proof. By definition,

$$\begin{aligned} Q_n(1, \alpha) &= [z^n/n!]\, e^z e^{\alpha \operatorname{Ein}(-z)} \\ &= [z^n/n!]\,(e^z - 1)\, e^{\alpha \operatorname{Ein}(-z)} + [z^n/n!]\, e^{\alpha \operatorname{Ein}(-z)} \\ &= n \left[\frac{z^{n-1}}{(n-1)!}\right] \left(\frac{e^z - 1}{z}\right) e^{\alpha \operatorname{Ein}(-z)} + b_n(\alpha) \\ &= -\frac{n}{\alpha} \left[\frac{z^{n-1}}{(n-1)!}\right] \frac{d}{dz} e^{\alpha \operatorname{Ein}(-z)} + b_n(\alpha) \\ &= (1 - n/\alpha)\, b_n(\alpha). \end{aligned}$$

Returning to the proof of Theorem 3, we will compute the Fourier coefficients of $Q_n(u,\alpha)$. We note that $Q_n(u,\alpha)$ is sufficiently smooth to be representable as the sum of its Fourier series for $0 < u < 1$. Since

$$\alpha \int_0^1 e^{uz} e^{\alpha\,\mathrm{Ein}(-z)}\,du = \alpha\left(\frac{e^z-1}{z}\right) e^{\alpha\,\mathrm{Ein}(-z)} = -\frac{d}{dz}\,e^{\alpha\,\mathrm{Ein}(-z)}$$

and

$$\begin{aligned}\alpha \int_0^1 e^{uz} e^{\alpha\,\mathrm{Ein}(-z)} e^{-2\pi i m u}\,du &= \frac{\alpha\, e^{\alpha\,\mathrm{Ein}(-z)}}{z-2\pi i m}\,(e^z-1)\\ &= -\frac{z}{z-2\pi i m}\cdot\frac{d}{dz}\,e^{\alpha\,\mathrm{Ein}(-z)},\end{aligned}$$

it follows that for $0 < u < 1$,

$$\begin{aligned}\alpha Q_n(u,\alpha) &= -\left[\frac{z^n}{n!}\right]\frac{d}{dz}\,e^{\alpha\,\mathrm{Ein}(-z)} - \sum_{m\neq 0} e^{2\pi i m u}\left[\frac{z^n}{n!}\right]\frac{z}{z-2\pi i m}\cdot\frac{d}{dz}\,e^{\alpha\,\mathrm{Ein}(-z)}\\ &= -b_{n+1}(\alpha) + \sum_{m\neq 0} e^{2\pi i m u}\sum_{j=0}^{n-1}\binom{n}{j} b_{j+1}(\alpha)\frac{(n-j)!}{(2\pi i m)^{n-j}}.\end{aligned}$$

When $u = 0$, the Fourier series converges to the arithmetic mean of the function evaluated at the two end-points $u = 0, 1$. Thus applying Lemma 3.1, we have

$$\begin{aligned}-\sum_{m\neq 0}\sum_{j=0}^{n-1}\binom{n}{j} b_{j+1}(\alpha)\frac{(n-j)!}{(2\pi i m)^{n-j}} &= -\frac{\alpha}{2}Q_n(0,\alpha) - \frac{\alpha}{2}Q_n(1,\alpha) - b_{n+1}(\alpha)\\ &= -\frac{\alpha}{2}b_n(\alpha) - \frac{\alpha}{2}\left(1-\frac{n}{\alpha}\right)b_n(\alpha) - b_{n+1}(\alpha)\\ &= (n/2-\alpha)\,b_n(\alpha) - b_{n+1}(\alpha).\end{aligned}$$

The result now follows on applying Euler's famous evaluation of the Riemann zeta function in the form

$$-\sum_{m\neq 0}\frac{(n-j)!}{(2\pi i m)^{n-j}} = \begin{cases} B_{n-j}, & n-j \geqslant 2,\\ 0, & n-j = 1.\end{cases}$$

Next, we analyze the behaviour of $b_n(\kappa)$ for large n. Since the generating function $e^{\kappa\,\mathrm{Ein}(-z)}$ is entire, it follows by Hadamard's root test that $\limsup_{n\to\infty}\sqrt[n]{|\,b_n(\kappa)/n!\,|} < 1/R$ for every $R > 0$, i.e. $\limsup_{n\to\infty}\sqrt[n]{|\,b_n(\kappa)/n!\,|} = 0$. However, since this information does not reveal how quickly or how slowly $|\,b_n(\kappa)/n!\,|$ tends to zero, we seek a concrete upper bound.

Theorem 4. *Let $\kappa > 0$. Then, for all non-negative integers n, we have*

$$\left|\frac{b_n(\kappa)}{n!}\right| \leqslant \left(\frac{e}{\log(1+n/\kappa)}\right)^n. \tag{3.4}$$

A somewhat more precise inequality is given by

$$\left|\frac{b_n(\kappa)}{n!}\right| \leqslant \exp\left\{\kappa \int_0^L t^{-1}\left(e^t-1\right)\,dt\right\} \Big/ \left\{\log(1+n/\kappa)\right\}^n, \tag{3.5}$$

where $L := \log(1+n/\kappa)$.

Proof. For any $r > 0$, Cauchy's inequality gives

$$\left|\frac{b_n(\kappa)}{n!}\right| \leqslant r^{-n} \max_{|z|=r} \left|e^{\kappa \operatorname{Ein}(-z)}\right|.$$

Now on the circle $|z| = r$,

$$\left|e^{\kappa \operatorname{Ein}(-z)}\right| = \left|\exp\left\{-\kappa \sum_{n=1}^{\infty} \frac{z^n}{n!\,n}\right\}\right| \leqslant \exp\left\{\kappa \sum_{n=1}^{\infty} \frac{r^n}{n!\,n}\right\} = \exp\left\{\kappa \int_0^r \frac{e^t-1}{t}\,dt\right\}.$$

Thus

$$\left|\frac{b_n(\kappa)}{n!}\right| \leqslant \inf_{r>0} r^{-n} \exp\left\{\kappa \int_0^r t^{-1}\left(e^t-1\right)\,dt\right\} = \inf_{r>0} r^{-n} e^{-\kappa \operatorname{Ein}(-r)}. \tag{3.6}$$

Minimizing with respect to $r > 0$, we find on taking the logarithmic derivative that $\kappa r^{-1}\left(e^r-1\right) - r^{-1}n = 0$ so that $r = \log(1+n/\kappa) = L$. The inequality (3.5) now follows on substituting the optimal value of r into (3.6). If we use the fact that

$$\int_0^r \frac{e^t-1}{t}\,dt = \sum_{n=1}^{\infty} \frac{r^n}{n!\,n} \leqslant e^r - 1,$$

then (3.5) with $r = L$ yields

$$\left|\frac{b_n(\kappa)}{n!}\right| \leqslant \frac{\exp\{\kappa(1+n/\kappa-1)\}}{\log^n(1+n/\kappa)} = \left(\frac{e}{\log(1+n/\kappa)}\right)^n,$$

which is (3.4).

Improvements. Put $M(r) := \max_{|z|=r} \left|e^{\kappa \operatorname{Ein}(-z)}\right|$. Our proof of (3.4) used the inequality

$$M(r) = \max_{|z|=r} \left| \exp\left\{ -\kappa \sum_{n=1}^{\infty} \frac{z^n}{n!\, n} \right\} \right| \leqslant \exp\left\{ \kappa \sum_{n=1}^{\infty} \frac{r^n}{n!\, n} \right\}, \tag{3.7}$$

which at first glance may appear wasteful, since at no point on the circle $|z| = r$ do we have $\arg(z^n) = \pi$ for every positive integer n. However, an attractive argument of Andrew Odlyzko [13] shows that the estimate (3.7) cannot be substantially improved. Consider $z = x + \pi i$, where $x > 0$ is large. The main contribution to the sum comes from terms with n close to x, and for such terms,

$$z^n = x^n \left(1 + \frac{\pi i}{x}\right)^n \sim x^n e^{\pi i} = -x^n, \quad x \to \infty.$$

Thus it would seem that any significant improvement on Theorem 4 cannot be based on Cauchy's inequality. However, if one considers the Cauchy integral formula

$$\frac{b_n(\kappa)}{n!} = \frac{1}{2\pi i} \oint_{|z|=r} z^{-n-1} e^{\kappa \operatorname{Ein}(-z)}\, dz,$$

one observes that $\left|e^{\kappa \operatorname{Ein}(-z)}\right|$ is close to its maximum $M(r)$ for only a small portion of z-values on the circle. This suggests that Theorem 4 can be improved using saddle-point asymptotics. Using this approach, the author was able to improve on Theorem 4 by the factor $\left(2\pi(\kappa + n)\log\left(1 + n/\kappa\right)\right)^{-1/2}$. See [2, Chapter 9, Theorem 8] for details.

4. An operator representation

We begin this section with an informal argument which should help motivate what follows. Let $f(z)$ be a formal power series in z and let $D = d/du$. Since $D^n e^{uz} = z^n e^{uz}$ for all non-negative integers n, it follows by linearity that the equation $f(z)\, e^{uz} = f(D)\, e^{uz}$ holds, at least in the formal sense. If we now apply this observation to the contour representation (1.3) with $f(z) :=$

$e^{\kappa\,\mathrm{Ein}(-z)}$, we obtain

$$q_\kappa(u) = \frac{\Gamma(2\kappa)}{2\pi i}\int z^{-2\kappa} e^{\kappa\,\mathrm{Ein}(-D)} e^{uz}\,dz$$

$$= e^{\kappa\,\mathrm{Ein}(-D)}\,\frac{\Gamma(2\kappa)}{2\pi i}\int z^{-2\kappa} e^{uz}\,dz$$

$$= e^{\kappa\,\mathrm{Ein}(-D)}\,u^{2\kappa-1}. \tag{4.1}$$

Pulling the differential operator outside the integral requires justification, and we shall do this shortly. But for the moment, a few remarks about (4.1) are in order. Recalling the power series representation for Ein, we may write

$$\mathrm{Ein}(-D) = -\sum_{n=1}^{\infty}\frac{D^n}{n!\,n}$$

and hence expanding the operator into powers of D, we formally obtain the expression

$$q_\kappa(u) \sim \sum_{n=0}^{\infty} b_n(\kappa)\frac{D^n}{n!}u^{2\kappa-1} = \sum_{n=0}^{\infty}\binom{2\kappa-1}{n} b_n(\kappa)\,u^{2\kappa-1-n} \tag{4.2}$$

in agreement with Theorem 2. In particular, the $n = 0$ term gives the known asymptotic formula $q_\kappa(u) \sim u^{2\kappa-1}$ as $u \to \infty$.

Next, we point out that the differential operator $\mathrm{Ein}(-D)$ can be recast in the form of an integral operator. For the sake of brevity, we put $T := \mathrm{Ein}(-D)$. Recalling the integral representation for Ein, we have

$$T := \mathrm{Ein}(-D) = \int_0^1 \frac{1-e^{tD}}{t}\,dt.$$

Now if f is analytic in a disk centred at u with radius $r > 0$, and $|t| < r$, then Taylor's theorem gives

$$e^{tD}f(u) = \sum_{n=0}^{\infty}\frac{t^n}{n!}D^n f(u) = \sum_{n=0}^{\infty}\frac{t^n}{n!}f^{(n)}(u) = f(u+t).$$

Thus for those functions f which are analytic in a disk centred at u with radius $r \geqslant 1$,

$$Tf(u) = \int_0^1 \left(1-e^{tD}\right) f(u)\frac{dt}{t} = \int_0^1 \frac{f(u)-f(u+t)}{t}\,dt. \tag{4.3}$$

Now the integral on the far right of (4.3) makes sense if f is integrable on $[u, u+1]$ and for some $\varepsilon > 0$, we have $|f(u) - f(u+t)| \ll t^{\varepsilon}$ as $t \to 0+$. For such f, we can define $Tf(u)$ by (4.3), and if T is defined this way, then $Tf(u)$ makes sense for a larger class of functions than merely those functions which are analytic in a suitably large disk centred at u. Thus there is no need to view T as a power series in D in order to determine $Tf(u)$. In the case of interest, $f(u) = u^{2\kappa-1}$ and (4.1) becomes

$$q_{\kappa}(u) = e^{\kappa T} u^{2\kappa-1} = \sum_{n=0}^{\infty} \frac{\kappa^n}{n!} T^n u^{2\kappa-1} \tag{4.4}$$

and so it makes sense to study the iterated integral operator T^n. We shall take this up after first proving the representation (4.4) rigorously.

Theorem 5. *Let $D := d/du$ and $T := \mathrm{Ein}(-D)$. Then for any complex number κ, and real $u > 0$,*

$$q_{\kappa}(u) = e^{\kappa T} u^{2\kappa-1} = \sum_{n=0}^{\infty} \frac{\kappa^n}{n!} T^n u^{2\kappa-1}.$$

Aside. In sieve applications, we are concerned primarily with positive real values of κ, and $\kappa > 1$ in particular.

Proof. Fix κ and consider the function of the complex variable w defined by

$$g_w(u) := \int_{\rightleftarrows\!\circlearrowleft} z^{-2\kappa} e^{uz} e^{w\,\mathrm{Ein}(-z)}\, dz, \quad u > 0.$$

By Taylor's theorem,

$$g_w(u) = \sum_{n=0}^{\infty} \frac{w^n}{n!} \left(\frac{\partial}{\partial w}\right)^n g_w(u)\bigg|_{w=0}.$$

On the other hand, from (1.3) and the definition of $g_w(u)$,

$$q_{\kappa}(u) = \frac{\Gamma(2\kappa)}{2\pi i} g_w(u)\bigg|_{w=\kappa}.$$

Thus,

$$q_{\kappa}(u) = \frac{\Gamma(2\kappa)}{2\pi i} \sum_{n=0}^{\infty} \frac{\kappa^n}{n!} \left(\frac{\partial}{\partial w}\right)^n g_w(u)\bigg|_{w=0}$$

and it remains only to show that for all non-negative integers n,

$$T^n u^{2\kappa-1} = \frac{\Gamma(2\kappa)}{2\pi i}\left(\frac{\partial}{\partial w}\right)^n g_w(u)\bigg|_{w=0}. \tag{4.5}$$

For $n = 0$, we have

$$\frac{\Gamma(2\kappa)}{2\pi i} g_0(u) = \frac{\Gamma(2\kappa)}{2\pi i}\int_{\multimap} z^{-2\kappa} e^{uz}\,dz = u^{2\kappa-1} = T^0 u^{2\kappa-1}.$$

Suppose now that (4.5) holds up to $n-1$, where n is a positive integer. Then

$$\begin{aligned}
&\frac{\Gamma(2\kappa)}{2\pi i}\left(\frac{\partial}{\partial w}\right)^n g_w(u)\bigg|_{w=0}\\
&= \frac{\Gamma(2\kappa)}{2\pi i}\int_{\multimap} z^{-2\kappa} e^{uz}\left(\mathrm{Ein}(-z)\right)^n dz\\
&= \frac{\Gamma(2\kappa)}{2\pi i}\int_{\multimap} z^{-2\kappa} e^{uz}\left(\mathrm{Ein}(-z)\right)^{n-1}\int_0^1 \frac{1-e^{tz}}{t}\,dt\,dz\\
&= \int_0^1 \frac{1}{t}\frac{\Gamma(2\kappa)}{2\pi i}\int_{\multimap} z^{-2\kappa} e^{uz}\left(1-e^{tz}\right)\left(\mathrm{Ein}(-z)\right)^{n-1} dz\,dt\\
&= \int_0^1 t^{-1}\{T^{n-1}u^{2\kappa-1} - T^{n-1}(u+t)^{2\kappa-1}\}\,dt\\
&= T^n u^{2\kappa-1},
\end{aligned}$$

by induction.

Having proved the representation (4.4), it is natural to ask how rapidly the series of T-iterates converges to the function q_κ. To this end, we prove the following

Theorem 6. *Let $\kappa > 0$, $u > 0$, and let n be a non-negative integer satisfying $n > 2\kappa - 1$. Define*

$$S_n := q_\kappa(u) - \sum_{j=0}^{n-1}\frac{\kappa^j}{j!}T^j\,u^{2\kappa-1}.$$

Then as $n \to \infty$, we have, with $c := \mathrm{Ein}(1) = 0.796599\ldots$,

$$S_n \ll_\kappa \frac{\kappa^n}{n!} + \kappa^n\left(\frac{e^c}{u}\right)^{n/\log n}\left(\frac{\log n}{n}\right)^n \Gamma\left(\frac{n}{\log n} + 1 - 2\kappa\right) u^{2\kappa-1}.$$

Remark. If $2\kappa \geqslant 1$, then Stirling's formula provides the simplification

$$S_n \ll_\kappa \frac{\kappa^n}{n!} + \kappa^n \left(\frac{e^c}{ue}\right)^{n/\log n} \left(\frac{\log n}{n}\right)^{n(1-1/\log n+1/2n)} u^{2\kappa-1}.$$

Proof. We have

$$\begin{aligned} S_n &= \sum_{j=n}^{\infty} \frac{\kappa^j}{j!} T^j\, u^{2\kappa-1} = \sum_{j=n}^{\infty} \frac{\Gamma(2\kappa)}{2\pi i} \left(\frac{\partial}{\partial w}\right)^j g_w(u)\bigg|_{w=0} \\ &= \sum_{j=n}^{\infty} \frac{\kappa^j}{j!} \frac{\Gamma(2\kappa)}{2\pi i} \int z^{-2\kappa} e^{uz}\, \mathrm{Ein}^j(-z)\, dz \\ &= \frac{\Gamma(2\kappa)}{2\pi i} \int z^{-2\kappa} e^{uz} \sum_{j=n}^{\infty} \frac{\kappa^j}{j!}\, \mathrm{Ein}^j(-z)\, dz. \end{aligned}$$

Since $n > 2\kappa - 1$ and $\sum_{j=n}^{\infty} \frac{\kappa^j}{j!} \mathrm{Ein}^j(-z) \ll |z|^n$ as $z \to 0$, we may collapse the contour onto the real half-line, obtaining

$$S_n = \pi^{-1}\Gamma(2\kappa)\sin(2\pi\kappa) \int_0^\infty x^{-2\kappa} e^{-ux} \sum_{j=n}^{\infty} \frac{\kappa^j}{j!}\, \mathrm{Ein}^j(x)\, dx.$$

Let

$$\begin{aligned} I_n &:= \int_0^1 x^{-2\kappa} e^{-ux} \sum_{j=n}^{\infty} \frac{\kappa^j}{j!}\, \mathrm{Ein}^j(x)\, dx, \\ J_n &:= \int_1^\infty x^{-2\kappa} e^{-ux} \sum_{j=n}^{\infty} \frac{\kappa^j}{j!}\, \mathrm{Ein}^j(x)\, dx. \end{aligned}$$

Note that for any $y \geqslant 0$,

$$\sum_{j=n}^{\infty} \frac{y^j}{j!} = \frac{y^n}{n!}\left(1 + \frac{y}{n+1} + \frac{y^2}{(n+1)(n+2)} + \cdots\right) \leqslant \frac{y^n}{n!} e^y.$$

Thus, as $0 \leqslant \mathrm{Ein}(x) \leqslant x$, we have

$$I_n \leqslant \frac{\kappa^n}{n!} \int_0^1 x^{n-2\kappa} e^{-ux} e^{\kappa x}\, dx \leqslant \frac{\kappa^n}{n!} \cdot \frac{e^\kappa}{n+1-2\kappa}. \tag{4.6}$$

With I_n now satisfactorily estimated, we turn to J_n. Recall the formula [14, p.40]

$$\mathrm{Ein}(x) = \log x + \gamma + \mathrm{E}_1(x), \quad x > 0,$$

where E_1 denotes the exponential integral defined by

$$\mathrm{E}_1(x) = \int_x^\infty e^{-t}\frac{dt}{t}, \quad x > 0,$$

and γ denotes Euler's constant. To estimate J_n, we use $\mathrm{Ein}(x) \leqslant \log x + \gamma + \mathrm{E}_1(1) \leqslant \log x + c$, where $c := \mathrm{Ein}(1) = 0.7965995\ldots \leqslant 1$. Thus,

$$J_n \leqslant \int_1^\infty x^{-2\kappa} e^{-ux} \sum_{j=n}^\infty \frac{(\kappa\alpha)^j}{j!}\, dx, \tag{4.7}$$

where $\alpha := \alpha(x) = c + \log x$. We now estimate the sum $\sum_{j=n}^\infty \frac{(\kappa\alpha)^j}{j!}$. Let $r > \kappa$. Then

$$\sum_{j=n}^\infty \frac{(\kappa\alpha)^j}{j!} = \sum_{j=n}^\infty \left(\frac{\kappa}{r}\right)^j \frac{(\alpha r)^j}{j!} \leqslant \left(\frac{\kappa}{r}\right)^n \sum_{j=0}^\infty \frac{(\alpha r)^j}{j!} = \left(\frac{\kappa}{r}\right)^n e^{\alpha r}. \tag{4.8}$$

Substituting (4.8) into the estimate (4.7) for J_n yields

$$\begin{aligned} J_n &\leqslant \left(\frac{\kappa}{r}\right)^n \int_1^\infty x^{-2\kappa} e^{-ux} e^{\alpha r}\, dx \\ &= \left(\frac{\kappa}{r}\right)^n e^{cr} \int_1^\infty x^{r-2\kappa} e^{-ux}\, dx \\ &\leqslant \left(\frac{\kappa}{r}\right)^n e^{cr} \frac{\Gamma(r+1-2\kappa)}{u^{r+1-2\kappa}}. \end{aligned} \tag{4.9}$$

For fixed $r > \max(\kappa, 2\kappa - 1)$, it is clear from (4.9) that $\lim_{n\to\infty} J_n = 0$. For example, taking $r = 2\kappa$ and using $c = \mathrm{Ein}(1) < 1$, we see that $J_n \leqslant e^{2\kappa} u^{-1} 2^{-n}$. But one can do better by choosing r more carefully, say by minimizing the right hand side of (4.9) with respect to r. By taking the logarithmic derivative, the optimal r is seen to satisfy

$$c - n/r + \psi(r + 1 - 2\kappa) - \log u = 0,$$

or

$$n = r\{\psi(r + 1 - 2\kappa) + c - \log u\}, \tag{4.10}$$

where as customary, $\psi = \Gamma'/\Gamma$. Inverting (4.10) and using the fact that $\psi(r+1-2\kappa) \sim \log r$ as $r \to \infty$, we find that r is approximately $n/\log n$. With this choice, (4.9) yields

$$J_n \leqslant \kappa^n \left(\frac{e^c}{u}\right)^{n/\log n} \left(\frac{\log n}{n}\right)^n \Gamma\left(\frac{n}{\log n} + 1 - 2\kappa\right) u^{2\kappa-1}. \tag{4.11}$$

The theorem now follows on combining (4.6) with (4.11) and absorbing the factors which are independent of n into the $\ll_\kappa$ constant.

We now return to study the integral operator T and its iterates in more detail. Recall that for $f \in L^1[u, u+1]$ satisfying some Lipschitz condition at u, we may define

$$Tf(u) = \int_0^1 \frac{f(u) - f(u+t)}{t}\, dt, \tag{4.12}$$

which is consistent with the definition of T as a differential operator for f analytic at u. We also note that the representation (4.12) has occurred in important contexts outside sieve theory. For example,

$$\begin{aligned} T(\log u) &= -\int_0^1 \frac{\log(u+t) - \log u}{t}\, dt = -\int_0^1 \log(1 + t/u)\, \frac{dt}{t}, \qquad w = -t/u \\ &= -\int_0^{-1/u} \log(1-w)\, \frac{dw}{w} \\ &= \mathrm{Li}_2\left(-1/u\right), \end{aligned}$$

where

$$\mathrm{Li}_2(z) := -\int_0^z \log(1-w)\, \frac{dw}{w} = \sum_{n=1}^{\infty} \frac{z^n}{n^2}, \qquad |z| < 1,$$

is the ubiquitous dilogarithm. For $f \in C^1[u, u+1]$, we can integrate (4.12) by parts. Thus

$$Tf(u) = \{f(u) - f(u+t)\} \log t \Big|_0^1 + \int_0^1 f'(u+t) \log t\, dt. \tag{4.13}$$

But

$$\begin{aligned} \lim_{t\to 0+} \{f(u) - f(u+t)\} \log t &= \lim_{t\to 0+} \frac{f(u) - f(u+t)}{t} \cdot t \log t = f'(u) \lim_{t\to 0+} t \log t \\ &= 0. \end{aligned}$$

Thus (4.13) becomes

$$Tf(u) = \int_0^1 f'(u+t)\log t\,dt. \tag{4.14}$$

Since $\int_0^1 \log t\,dt = -1$, it follows that Tf can be viewed as a weighted average of $-f'$. More generally, for n a positive integer, one can show that

$$T^n f(u) = \int_{[0,1]^n} f^{(n)}(u+t_1+\cdots+t_n)\log t_1\cdots\log t_n\,dt_1\cdots dt_n$$

if $f \in C^n[u,u+1]$.

The integrands in the representations (4.12) and (4.14) both contain apparent singularities at the origin. We develop an additional representation in which the corresponding singularity is hidden. Again, assume f possesses the requisite derivatives.

Proposition 4.1. *Let $f \in C^1[u,u+1]$. Then $Tf(u) = -\int_0^1\int_0^1 f'(u+st)\,ds\,dt$. More generally, if n is any positive integer and $f \in C^n[u,u+1]$, then*

$$T^n f(u) = (-1)^n \int_{[0,1]^n}\int_{[0,1]^n} f^{(n)}(u+s_1t_1+\cdots+s_nt_n)\,ds_1\cdots ds_n\,dt_1\cdots dt_n.$$

Proof. By definition

$$\begin{aligned} Tf(u) &= \int_0^1 \frac{f(u)-f(u+t)}{t}\,dt = -\int_0^1\int_u^{u+t} f'(x)\,dx\,\frac{dt}{t} \\ &= -\int_0^1\int_0^t f'(u+r)\,dr\,\frac{dt}{t} \\ &= -\int_0^1\int_0^1 f'(u+st)\,ds\,dt. \end{aligned} \tag{4.15}$$

The general case is handled inductively.

Note that the apparent singularity at the origin has been hidden. Also, (4.15) shows that Tf is a weighted average of $-f'$ over the unit square.

5. Numerical analysis

Here, we give a detailed numerical analysis of a Simpson's rule-based scheme for computing $q_\kappa(u)$ for the range $1 \leqslant \kappa < 3$, $1 \leqslant u \leqslant 6$. Following Iwaniec [11], we set

$$e^{\kappa\,\mathrm{Ein}(-z)} = \sum_{j=0}^{n} b_j(\kappa)\,\frac{z^j}{j!} + S_n(-z)$$

and rewrite (1.3) (via Hankel's formula for $1/\Gamma$) in the form

$$q_\kappa(u) = \sum_{j=0}^{n} \binom{2\kappa - 1}{j} b_j(\kappa) u^{2\kappa-1-j} + \frac{\Gamma(2\kappa)}{2\pi i} \int z^{-2\kappa} e^{uz}\, S_n(-z)\, dz.$$

Next, since $|S_n(-z)| \ll |z|^{n+1}$ as $z \to 0$, the integral around the circular portion tends to 0 with the radius, as long as $n+1 > 2\kappa - 1$. Here $1 \leqslant \kappa < 3$, so we require $n \geqslant 4$. Then the contour may be collapsed onto the negative real axis, and after routine calculations, one obtains

$$(5.1)\quad q_\kappa(u) = \sum_{j=0}^{4} \binom{2\kappa - 1}{j} b_j(\kappa)\, u^{2\kappa-j-1} + \frac{\sin(2\pi\kappa)}{\pi} \Gamma(2\kappa) \int_0^\infty x^{-2\kappa} e^{-ux} \{e^{\kappa \operatorname{Ein}(x)} - P(x)\}\, dx,$$

where $P(x) := \sum_{j=0}^{4} (-1)^j\, b_j(\kappa)\, \frac{x^j}{j!}$ is the fourth-degree Maclaurin polynomial for $e^{\kappa \operatorname{Ein}(x)}$. We use the representation (5.1) to compute $q_\kappa(u)$ for various values of κ, u. We have written C code to perform these computations, but to guarantee their accuracy, an accompanying error analysis is necessary. Since $\binom{2\kappa-1}{j} b_j(\kappa)$ is a polynomial in κ for non-negative integers j, we regard computations of the summation term in (5.1) as exact and henceforth focus exclusively on the computational problems that the integral in (5.1) presents. To avoid additional complications in the corresponding error analysis, we do not make use of any numerical integration packages. For the same reason, we avoid using built-in special functions such as the incomplete gamma function. The only non-elementary function we do use is the complete gamma function, since we are satisfied that its implementation gives the correct answers to within the number of significant digits specified.

Let $I = I_1 + I_2 + I_3 - I_4$ denote the integral in (5.1), where

$$I_1 := \int_0^1 x^{-2\kappa} e^{-ux} \{e^{\kappa \operatorname{Ein}(x)} - P(x)\}\, dx, \quad I_2 := \int_1^{20} x^{-2\kappa} e^{-ux} e^{\kappa \operatorname{Ein}(x)}\, dx$$

$$I_3 := \int_{20}^\infty x^{-2\kappa} e^{-ux} e^{\kappa \operatorname{Ein}(x)}\, dx, \quad I_4 := \int_1^\infty x^{-2\kappa} e^{-ux} P(x)\, dx.$$

In I_3, we have chosen 20 as the lower limit of integration so as to make the integral negligible in the rectangle of interest, namely $1 \leqslant \kappa < 3$, $1 \leqslant u \leqslant 6$. See Theorem 7 below for details. In I_1, we have chosen 1 as the upper limit of integration in order to make computation via Taylor series expansion feasible. It turns out (Theorem 9) that at least 35 terms are needed to make the error negligible. Accordingly, we approximate the braced expression in I_1 by its Taylor polynomial of degree 35. By definition of $P(x)$, the lowest degree term in this polynomial has x to the fifth power. We then approximate I_1 by incomplete gamma functions. Complementary incomplete gamma functions are used to approximate I_4. This leaves I_2, which we treat by splitting up the interval $[1, 20]$ into appropriate subintervals and then applying Simpson's rule.

Theorem 7. *Again, let*

$$\mathrm{E}_1(x) := \int_x^\infty \frac{e^{-t}}{t}\, dt, \quad x > 0$$

denote the exponential integral [1, p.228], [14, p.40]. *Then,*

$$I_3 \leqslant u^{-1}\, 20^{-\kappa} e^{\kappa\gamma} e^{\kappa\, \mathrm{E}_1(20)} e^{-20u}. \tag{5.3}$$

Furthermore, on any rectangle $1 \leqslant u_1 \leqslant u \leqslant u_2 \leqslant 6$, $1 \leqslant \kappa_1 \leqslant \kappa \leqslant \kappa_2 \leqslant 3$, *we have*

$$I_3 \leqslant u_1^{-1}\, 20^{-\kappa_1} e^{\kappa_1\gamma} e^{\kappa_1\, \mathrm{E}_1(20)} e^{-20u_1} \leqslant 0.184 \times 10^{-9}, \qquad \textit{with } u_1 = \kappa_1 = 1. \tag{5.4}$$

Proof. We apply once more the relationship [14, p.40]

$$\mathrm{Ein}(x) = \log x + \gamma + \mathrm{E}_1(x), \quad x > 0,$$

where γ denotes Euler's constant. Thus,

$$I_3 = \int_{20}^\infty x^{-2\kappa} e^{-ux} e^{\kappa\, \mathrm{Ein}(x)}\, dx = \int_{20}^\infty x^{-\kappa} e^{-ux} e^{\kappa(\gamma + \mathrm{E}_1(x))}\, dx.$$

Note that the factors $x^{-\kappa}$ and $e^{\kappa(\gamma+\mathrm{E}_1(x))}$ in the integrand are decreasing functions of x. Therefore,

$$I_3 \leqslant 20^{-\kappa} e^{\kappa(\gamma + \mathrm{E}_1(20))} \int_{20}^\infty e^{-ux}\, dx,$$

which gives (5.3). Since $\gamma + \mathrm{E}_1(20) - \log 20 < 0$, the expression on the right in (5.3) is decreasing in κ and u, from which (5.4) follows.

We can compute values of $\mathrm{Ein}(x)$ for $x > 1$ using $\mathrm{Ein}(x) = \log x + \gamma + \mathrm{E}_1(x)$, and a continued fraction which can approximate E_1 to within any specified amount. Our approximation is based on a continued fraction expansion for the complementary incomplete gamma function [1, p.263], [16, p.356]

$$\Gamma(a,u) := \int_u^\infty t^{a-1}e^{-t}\,dt = \cfrac{e^{-u}u^a}{u + \cfrac{1-a}{1 + \cfrac{1}{u + \cfrac{2-a}{1 + \cfrac{2}{u + \cfrac{3-a}{1 + \cfrac{3}{u + \ddots}}}}}}} \tag{5.5}$$

defined for all a and $u > 0$. Note that $\mathrm{E}_1(x) = \Gamma(0,x)$ for $x > 0$.

We now wish to determine the error committed in using an approximation E_1^* in place of E_1 in the evaluation of I_2.

Theorem 8. *Let* $0 < \varepsilon < 0.048$ *be given and suppose that* $|\,\mathrm{E}_1^*(x) - \mathrm{E}_1(x)| < \varepsilon$ *for* $1 \leqslant x \leqslant 20$. *Let*

$$\Delta I_2 := \left| I_2 - \int_1^{20} x^{-\kappa} e^{-ux} e^{\kappa(\gamma + \mathrm{E}_1^*(x))}\,dx \right|.$$

Then for all $u \geqslant 1$, $\kappa \geqslant 1$,

$$\Delta I_2 \leqslant (e^{\kappa\varepsilon} - 1)e^{\kappa\varepsilon}\{e^{\kappa(\gamma+\mathrm{E}_1^*(1))}e^{-u} + e^{\kappa(\gamma+\mathrm{E}_1^*(2)-\log 2)}e^{-2u}/(1-e^{-u})\}. \tag{5.6}$$

Furthermore, on any rectangle $1 \leqslant u_1 \leqslant u \leqslant u_2 \leqslant 6$, $1 \leqslant \kappa_1 \leqslant \kappa \leqslant \kappa_2 \leqslant 3$, *we have*

$$\begin{aligned}\Delta I_2 &\leqslant (e^{\kappa_2\varepsilon} - 1)e^{\kappa_2\varepsilon}\{e^{\kappa_2(\gamma+\mathrm{E}_1^*(1))}e^{-u_1} + e^{\kappa_1(\gamma+\mathrm{E}_1^*(2)-\log 2)}e^{-2u_1}/(1-e^{-u_1})\}\\ &\leqslant 0.127 \times 10^{-12}, \qquad \text{with } \varepsilon = 10^{-14},\ u_1 = \kappa_1 = 1,\ \kappa_3 = 3.\end{aligned} \tag{5.7}$$

Proof. For (5.6) we have

$$\begin{aligned}\Delta I_2 &= \left| \int_1^{20} x^{-\kappa} e^{-ux} e^{\kappa\gamma}(e^{\kappa\,\mathrm{E}_1^*(x)} - e^{\kappa\,\mathrm{E}_1(x)})\,dx \right|\\ &\leqslant \int_1^{20} x^{-\kappa} e^{-ux} e^{\kappa\gamma} e^{\kappa\,\mathrm{E}_1(x)} \left| e^{\kappa(\mathrm{E}_1^*(x)-\mathrm{E}_1(x))} - 1 \right| dx\\ &\leqslant (e^{\kappa\varepsilon} - 1)\sum_{j=1}^{19} j^{-\kappa} e^{-ju} e^{\kappa\gamma} e^{\kappa\,\mathrm{E}_1(j)},\end{aligned}$$

since the rest of the integrand is decreasing in x. Thus

$$\begin{aligned}\Delta I_2 &\leqslant (e^{\kappa\varepsilon}-1)\sum_{j=1}^{19} j^{-\kappa}e^{-ju}e^{\kappa\gamma}e^{\kappa\,\mathrm{E}_1^*(j)}e^{\kappa(\mathrm{E}_1(j)-\mathrm{E}_1^*(j))}\\ &\leqslant (e^{\kappa\varepsilon}-1)e^{\kappa\varepsilon}\sum_{j=1}^{19} j^{-\kappa}e^{-ju}e^{\kappa\gamma}e^{\kappa\,\mathrm{E}_1^*(j)}.\end{aligned}$$

It is easy to show that $\gamma+\mathrm{E}_1(x)-\log x < -0.048$ if $x \geqslant 2$. Thus if $x \geqslant 2$, then $\gamma+\mathrm{E}_1^*(x)-\log x < 0$. It follows that

$$\begin{aligned}\Delta I_2 &\leqslant (e^{\kappa\varepsilon}-1)e^{\kappa\varepsilon}\{e^{\kappa(\gamma+\mathrm{E}_1^*(1))}e^{-u}+e^{\kappa(\gamma+\mathrm{E}_1^*(2)-\log 2)}(e^{-2u}+e^{-3u}+\ldots)\}\\ &= (e^{\kappa\varepsilon}-1)e^{\kappa\varepsilon}\{e^{\kappa(\gamma+\mathrm{E}_1^*(1))}e^{-u}+e^{\kappa(\gamma+\mathrm{E}_1^*(2)-\log 2)}e^{-2u}/(1-e^{-u})\},\end{aligned}$$

which completes the proof of (5.6). For (5.7), we note that the expression above is majorized by the expression in the statement of (5.7) over the given rectangle. The reason for this is that since $\varepsilon < \gamma$, $\gamma+\mathrm{E}_1^*(x)-\log x > 0$ if $x=1$, and $\gamma+\mathrm{E}_1^*(x)-\log x < 0$ if $x \geqslant 2$ as we have seen.

Theorem 9. *Let $n \geqslant 6$, and let $R_n(x)$ be the Taylor polynomial of degree n for $e^{\kappa\,\mathrm{Ein}(x)}-P(x)$. Put $I_1^* = \int_0^1 x^{-2\kappa}e^{-ux}R_n(x)\,dx$. Then*

$$|I_1-I_1^*| \leqslant \frac{e^{-\kappa\,\mathrm{Ein}(-r)}}{r^n(r-1)}\int_0^1 x^{n-5}e^{-ux}\,dx \tag{5.8}$$

holds for every $r>1$. Furthermore, on any rectangle $1 \leqslant u_1 \leqslant u \leqslant u_2 \leqslant 6$, $1 \leqslant \kappa_1 \leqslant \kappa \leqslant \kappa_2 \leqslant 3$, we have

$$|I_1-I_1^*| \leqslant \frac{e^{-\kappa_2\,\mathrm{Ein}(-r)}}{r^n(r-1)}\int_0^1 x^{n-5}e^{-u_1x}\,dx \tag{5.9}$$

for every $r>1$.

Note that the integrals can be evaluated exactly in terms of elementary functions. For the full rectangle (i.e. $u_1=1, u_2=6, \kappa_1=1, \kappa_2=3$), it is possible to take $r=2.6, n=35$ and thereby obtain $|I_1-I_1^*| \leqslant 0.173... \times 10^{-8}$.

Proof. Let $r>1$. By definition of I_1^* and $R_n(x)$, we have

$$|I_1-I_1^*| = \left|\int_0^1 x^{-2\kappa}e^{-ux}\sum_{j=n+1}^{\infty}(-1)^j\,b_j(\kappa)\,\frac{x^j}{j!}\,dx\right|.$$

We now invoke a bound on $|b_j(\kappa)/j!|$ which was developed in the course of proving Theorem 4. Thus, by inequality (3.6) with $r > 1$, we have

$$\begin{aligned}|I_1 - I_1^*| &\leqslant \int_0^1 x^{-2\kappa} e^{-ux} \sum_{j=n+1}^{\infty} e^{-\kappa \operatorname{Ein}(-r)} \left(\frac{x}{r}\right)^j dx \\ &\leqslant e^{-\kappa \operatorname{Ein}(-r)} \int_0^1 x^{-6} e^{-ux} x^{n+1} \sum_{j=n+1}^{\infty} r^{-j}\, dx \\ &= \frac{e^{-\kappa \operatorname{Ein}(-r)}}{r^n (r-1)} \int_0^1 x^{n-5} e^{-ux}\, dx\end{aligned}$$

which completes the proof of (5.8). Since $-\kappa \operatorname{Ein}(-r) = \kappa \sum_{n=1}^{\infty} \frac{r^n}{n!\, n} > 0$, (5.9) follows from (5.8).

In practice, we take $n = 35$ in Theorem 9 and we compute

$$\begin{aligned}I_1^* &= \sum_{j=5}^{35} (-1)^j \frac{b_j(\kappa)}{j!} \int_0^1 x^{n-2\kappa} e^{-ux}\, dx \\ &= \sum_{j=5}^{35} (-1)^j \frac{b_j(\kappa)}{j!} u^{2\kappa-n-1} \int_0^u t^{n-2\kappa} e^{-t}\, dt\end{aligned}$$

using the incomplete gamma function

$$\gamma(a,u) := \int_0^u t^{a-1} e^{-t}\, dt = u^a \sum_{n=0}^{\infty} \frac{(-u)^n}{n!\,(n+a)},$$

which is defined for all $u > 0$ and $\Re(a) > 0$. We calculate values of $\gamma(a,u)$ using a truncation of the series.

Theorem 10. *Let $6 \leqslant L$ be an integer and let $r > 1$. Then the error committed in computing I_1^* using L terms of the series representation of $\gamma(a,u)$ is less than*

$$\frac{6^{L+1} e^{-\kappa \operatorname{Ein}(-r)}}{(L+1)!\,(L+1)\, r^4\, (r-1)}.$$

When $L = 41, r = \frac{3}{2}$, the expression above is $0.3381... \times 10^{-17}$.

Proof. Recall $I_1^* = \sum_{j=5}^{35} (-1)^j \frac{b_j(\kappa)}{j!} u^{2\kappa-j-1} \gamma(j - 2\kappa + 1, u)$. Put

$$\widetilde{I}_1 = \sum_{j=5}^{35} (-1)^j \frac{b_j(\kappa)}{j!} \sum_{n=0}^{L} \frac{(-u)^n}{n!\,(n+j-2\kappa+1)}.$$

Then

$$|I_1^* - \widetilde{I}_1| \leqslant \sum_{j=5}^{35} \left| \frac{b_j(\kappa)}{j!} \sum_{n=L+1}^{\infty} \frac{(-u)^n}{n!\,(n+j-2\kappa+1)} \right|.$$

Since $1 \leqslant u \leqslant 6$ and $L \geqslant 6$, the terms $\dfrac{u^n}{n!\,(n+j-2\kappa+1)}$ are decreasing to zero, and so by the alternating series test

$$|I_1^* - \widetilde{I}_1| \leqslant \sum_{j=5}^{35} \left| \frac{b_j(\kappa)}{j!} \right| \frac{u^{L+1}}{(L+1)!\,(L+1+j-2\kappa+1)}.$$

Once again invoking inequality (3.6) with $r > 1$, we have

$$\begin{aligned} |I_1^* - \widetilde{I}_1| &\leqslant \frac{6^{L+1}}{(L+1)\,(L+1)!} \sum_{j=5}^{\infty} e^{-\kappa\,\mathrm{Ein}(-r)} r^{-j} \\ &= \frac{6^{L+1} e^{-\kappa\,\mathrm{Ein}(-r)}}{(L+1)!\,(L+1)\,r^4\,(r-1)}. \end{aligned}$$

Recall that $P(x) = \sum_{j=0}^{4} (-1)^j\, b_j(\kappa)\, \dfrac{x^j}{j!}$ and

$$\begin{aligned} I_4 = \int_1^\infty x^{-2\kappa} e^{-ux} P(x)\,dx &= \sum_{j=0}^{4} (-1)^j \frac{b_j(\kappa)}{j!} \int_1^\infty x^{j-2\kappa} e^{-ux}\,dx \\ &= \sum_{j=0}^{4} (-1)^j \frac{b_j(\kappa)}{j!} u^{2\kappa-j-1} \int_u^\infty t^{j-2\kappa} e^{-t}\,dt. \end{aligned}$$

Therefore, we can compute I_4 using the complementary incomplete gamma function (5.5). We have implemented a continued fraction algorithm which computes $\Pi^*(a,u)$, an approximation to $\Pi(a,u) := e^u u^{-a} \Gamma(a,u)$, to within any specified tolerance.

Theorem 11. *Let $\Pi(a,u) := e^u u^{-a}\Gamma(a,u)$ and let Π^* be a function which satisfies $|\Pi(a,u) - \Pi^*(a,u)| < 10^{-11}$ for all values of $a = j - 2\kappa + 1$, $1 \leqslant \kappa \leqslant 3$, and $1 \leqslant u \leqslant 6$. Put $I_4^* := \sum_{j=0}^{4} (-1)^j \dfrac{b_j(\kappa)}{j!} e^{-u} \Pi^*(j-2\kappa+1, u)$. Then $|I_4 - I_4^*| < 4.016... \times 10^{-11}$ on the full rectangle $1 \leqslant u \leqslant 6, 1 \leqslant \kappa \leqslant 3$.*

Proof. We have

$$I_4 = \sum_{j=0}^{4} (-1)^j \frac{b_j(\kappa)}{j!} u^{2\kappa-j-1}\Gamma(j-2\kappa+1,u)$$
$$= \sum_{j=0}^{4} (-1)^j \frac{b_j(\kappa)}{j!} e^{-u}\Pi(j-2\kappa+1,u).$$

Thus, $|I_4 - I_4^*| \leqslant e^{-u} \sum_{j=0}^{4} \left| \frac{b_j(\kappa)}{j!} \right| 10^{-11}$. The remainder of the proof consists in improving our bounds on $|\, b_j(\kappa)/j!\,|$ for $j = 0, 1, 2, 3, 4$.

From Proposition 3.1, we infer that

$$b_0 = 1, \quad b_1 = -\kappa, \quad \frac{b_2}{2!} = \frac{\kappa^2}{2} - \frac{\kappa}{4}, \quad \frac{b_3}{3!} = -\frac{\kappa^3}{6} + \frac{\kappa^2}{4} - \frac{\kappa}{18},$$

$$\frac{b_4}{4!} = \frac{\kappa^4}{24} - \frac{\kappa^3}{8} + \frac{25\kappa^2}{288} - \frac{\kappa}{96}.$$

Thus $|b_0| = 1, |b_1| = |\kappa| \leqslant 3$. Since $b_2'(\kappa) = 2\kappa - 1/2 > 0$ on $[1,3]$, it follows that $b_2(1) \leqslant b_2 \leqslant b_2(3)$ whence $|b_2/2!| \leqslant 15/4$. A similar argument gives $|b_3/3!| \leqslant 29/12$. Unfortunately, b_4 is not monotone on $[1,3]$. However, it is not difficult to show that b_4 has precisely one extremum in $]1,3[$. The extremum is a local minimum and lies in $[1.5, 2.0]$. The graph of $b_4(\kappa)/4!$ is easily shown to be convex on this subinterval and so we construct tangents at $\kappa = 1.5$ and $\kappa = 2.0$ and take the height of their intersection as a possible lower bound on $b_4(\kappa)/4!$. In this way it is easily shown that $|b_4/4!| \leqslant b_4(3)/4! = 3/4$ on $[1,3]$. From this the theorem follows immediately.

It now remains to show how

$$I_2 = \int_1^{20} x^{-2\kappa} e^{-ux} e^{\kappa\,\mathrm{Ein}(x)}\, dx$$

can be effectively computed. For this, we use Simpson's rule. Recall that the error committed in using Simpson's rule to estimate $\int_a^b r(x)\,dx$ is no more than $M(b-a)^5/2880n^4$, where $M := \max\{|r^{(4)}(x)| : x \in [a,b]\}$, and $a = x_0 < x_1 < \ldots < x_{2n} = b$ are evenly spaced. For our application, we partition the interval $[1,20]$ into 15 subintervals chosen so that the number of function evaluations needed to commit an error $\leqslant 10^{-10}$ for each subinterval is approximately the same. Thus approximately 60 ($n = 30$ above) function evaluations are needed for each subinterval. If for given κ, u, we put $r(x) = x^{-2\kappa} e^{-ux} e^{\kappa\,\mathrm{Ein}(x)}$, then each pair of parameter values for κ, u requires a separate computation of I_2.

For each κ, u, and subinterval [a, b], we estimate M and hence determine n so that $M(b-a)^5/2880n^4 < 10^{-10}$. The estimation of the various M values is not as difficult as it may appear, since it so happens that $r^{(4)}$ is decreasing in x for $x \geqslant 1, \kappa > 0, u > 0$. In fact, $r^{(5)}(x) = -x^{-2\kappa}e^{-ux}e^{\kappa\,\mathrm{Ein}(x)}S(x)/x^5$, where $S(x)$ is a polynomial in x, κ, u, e^{-x} consisting entirely of non-negative terms, a result that can be most easily verified using MAPLE's symbolic capabilities. For a more elegant approach, see [2, p.64]. Thus, it is easy to estimate M and hence to determine the number of function evaluations needed to keep the quadrature error $\leqslant 10^{-10}$ for each subinterval. With 15 subintervals, the total quadrature error is then $\leqslant 15 \times 10^{-10}$. If we combine this error with the error bounds of Theorems 7, 8, 9, and 11, we see that $q_\kappa(u)$ can be computed with an absolute error $\leqslant 0.5 \times 10^{-8}$ if $1 \leqslant u \leqslant 6$, and $1 \leqslant \kappa < 3$.

6. Table of q_κ Values

Based on the methods of Section 5, we have written C code to evaluate the function $q_\kappa(u)$ in the rectangle $1 \leqslant u \leqslant 6$, $1 \leqslant \kappa < 3$. The following table of q_κ values, rounded to seven decimal places, was created using this code. In view of the analysis we have given, the accuracy of these data is guaranteed to within 10^{-7}.

We note that various sieve functions, including q_κ, have been tabulated previously [12, 15, 17]. Although these previous calculations were not supported by numerical analysis, they appear to be in good agreement with our tables.

$u\backslash\kappa$	1.0	1.1	1.2	1.3	1.4
1.0	0.0000000	−0.2463298	−0.4517757	−0.5827670	−0.6070362
1.1	0.1000000	−0.1554682	−0.3861985	−0.5584687	−0.6368735
1.2	0.2000000	−0.0599213	−0.3103561	−0.5184864	−0.6472184
1.3	0.3000000	0.0397676	−0.2252552	−0.4640109	−0.6389713
1.4	0.4000000	0.1431639	−0.1317106	−0.3960188	−0.6128812
1.5	0.5000000	0.2499139	−0.0303940	−0.3153254	−0.5695819
1.6	0.6000000	0.3597240	0.0781317	−0.2226219	−0.5096179
1.7	0.7000000	0.4723479	0.1933882	−0.1185025	−0.4334625
1.8	0.8000000	0.5875756	0.3149645	−0.0034833	−0.3415321
1.9	0.9000000	0.7052267	0.4425035	0.1219827	−0.2341962
2.0	1.0000000	0.8251444	0.5756920	0.2574941	−0.1117856
2.1	1.1000000	0.9471912	0.7142531	0.4026928	0.0254016
2.2	1.2000000	1.0712460	0.8579405	0.5572571	0.1770947
2.3	1.3000000	1.1972009	1.0065331	0.7208959	0.3430465
2.4	1.4000000	1.3249594	1.1598317	0.8933446	0.5230304
2.5	1.5000000	1.4544349	1.3176559	1.0743612	0.7168372
2.6	1.6000000	1.5855488	1.4798409	1.2637237	0.9242733
2.7	1.7000000	1.7182301	1.6462364	1.4612274	1.1451587
2.8	1.8000000	1.8524137	1.8167041	1.6666826	1.3793255
2.9	1.9000000	1.9880400	1.9911164	1.8799134	1.6266165
3.0	2.0000000	2.1250544	2.1693554	2.1007557	1.8868843
3.1	2.1000000	2.2634064	2.3513115	2.3290561	2.1599897
3.2	2.2000000	2.4030493	2.5368827	2.5646710	2.4458019
3.3	2.3000000	2.5439398	2.7259739	2.8074655	2.7441967
3.4	2.4000000	2.6860376	2.9184959	3.0573122	3.0550567

Table 1. Values of $q_\kappa(u)$

$u\backslash\kappa$	1.0	1.1	1.2	1.3	1.4
3.5	2.5000000	2.8293050	3.1143652	3.3140913	3.3782701
3.6	2.6000000	2.9737070	3.3135030	3.5776892	3.7137307
3.7	2.7000000	3.1192105	3.5158354	3.8479984	4.0613372
3.8	2.8000000	3.2657846	3.7212924	4.1249166	4.4209926
3.9	2.9000000	3.4134002	3.9298079	4.4083469	4.7926046
4.0	3.0000000	3.5620300	4.1413192	4.6981965	5.1760844
4.1	3.1000000	3.7116478	4.3557668	4.9943774	5.5713470
4.2	3.2000000	3.8622293	4.5730942	5.2968052	5.9783107
4.3	3.3000000	4.0137512	4.7932477	5.6053993	6.3968969
4.4	3.4000000	4.1661916	5.0161759	5.9200823	6.8270301
4.5	3.5000000	4.3195293	5.2418298	6.2407803	7.2686374
4.6	3.6000000	4.4737447	5.4701628	6.5674220	7.7216485
4.7	3.7000000	4.6288186	5.7011300	6.8999392	8.1859955
4.8	3.8000000	4.7847330	5.9346886	7.2382662	8.6616129
4.9	3.9000000	4.9414706	6.1707975	7.5823395	9.1484370
5.0	4.0000000	5.0990149	6.4094173	7.9320981	9.6464066
5.1	4.1000000	5.2573501	6.6505100	8.2874832	10.1554620
5.2	4.2000000	5.4164612	6.8940393	8.6484378	10.6755454
5.3	4.3000000	5.5763335	7.1399700	9.0149071	11.2066010
5.4	4.4000000	5.7369531	7.3882684	9.3868379	11.7485742
5.5	4.5000000	5.8983068	7.6389019	9.7641788	12.3014121
5.6	4.6000000	6.0603815	7.8918391	10.1468799	12.8650634
5.7	4.7000000	6.2231651	8.1470496	10.5348929	13.4394780
5.8	4.8000000	6.3866455	8.4045041	10.9281711	14.0246072
5.9	4.9000000	6.5508113	8.6641742	11.3266691	14.6204036

Table 1. Values of $q_\kappa(u)$ (continued)

$u\backslash\kappa$	1.5	1.6	1.7	1.8	1.9
1.0	−0.5000000	−0.2527189	0.1194642	0.5692400	1.0130470
1.1	−0.5900000	−0.3981341	−0.0624276	0.3857664	0.8775667
1.2	−0.6600000	−0.5279768	−0.2392354	0.1907910	0.7095828
1.3	−0.7100000	−0.6407309	−0.4075077	−0.0103330	0.5155997
1.4	−0.7400000	−0.7350924	−0.5642137	−0.2127803	0.3017026
1.5	−0.7500000	−0.8099226	−0.7066537	−0.4121459	0.0736383
1.6	−0.7400000	−0.8642136	−0.8323939	−0.6043663	−0.1631256
1.7	−0.7100000	−0.8970635	−0.9392178	−0.7856610	−0.4033546
1.8	−0.6600000	−0.9076573	−1.0250893	−0.9524867	−0.6420132
1.9	−0.5900000	−0.8952530	−1.0881243	−1.1015022	−0.8742366
2.0	−0.5000000	−0.8591695	−1.1265684	−1.2295403	−1.0953091
2.1	−0.3900000	−0.7987782	−1.1387787	−1.3335854	−1.3006458
2.2	−0.2600000	−0.7134953	−1.1232092	−1.4107544	−1.4857782
2.3	−0.1100000	−0.6027764	−1.0783990	−1.4582820	−1.6463414
2.4	0.0600000	−0.4661112	−1.0029623	−1.4735077	−1.7780642
2.5	0.2500000	−0.3030198	−0.8955804	−1.4538653	−1.8767600
2.6	0.4600000	−0.1130491	−0.7549941	−1.3968734	−1.9383194
2.7	0.6900000	0.1042299	−0.5799982	−1.3001275	−1.9587037
2.8	0.9400000	0.3492247	−0.3694364	−1.1612938	−1.9339393
2.9	1.2100000	0.6223233	−0.1221966	−0.9781029	−1.8601125
3.0	1.5000000	0.9238962	0.1627929	−0.7483444	−1.7333655
3.1	1.8100000	1.2542974	0.4865664	−0.4698631	−1.5498923
3.2	2.1400000	1.6138667	0.8501242	−0.1405542	−1.3059353
3.3	2.4900000	2.0029301	1.2544352	0.2416395	−0.9977823
3.4	2.8600000	2.4218014	1.7004388	0.6787316	−0.6217639

Table 1. Values of $q_\kappa(u)$ (continued)

$u\backslash\kappa$	1.5	1.6	1.7	1.8	1.9
3.5	3.2500000	2.8707826	2.1890474	1.1726943	−0.1742506
3.6	3.6600000	3.3501654	2.7211480	1.7254611	0.3483491
3.7	4.0900000	3.8602314	3.2976040	2.3389294	0.9495909
3.8	4.5400000	4.4012534	3.9192567	3.0149622	1.6329963
3.9	5.0100000	4.9734951	4.5869266	3.7553904	2.4020547
4.0	5.5000000	5.5772126	5.3014147	4.5620143	3.2602248
4.1	6.0100000	6.2126543	6.0635033	5.4366053	4.2109358
4.2	6.5400000	6.8800616	6.8739578	6.3809072	5.2575890
4.3	7.0900000	7.5796692	7.7335268	7.3966378	6.4035591
4.4	7.6600000	8.3117055	8.6429433	8.4854900	7.6521950
4.5	8.2500000	9.0763932	9.6029258	9.6491329	9.0068210
4.6	8.8600000	9.8739492	10.6141785	10.8892130	10.4707379
4.7	9.4900000	10.7045851	11.6773922	12.2073550	12.0472238
4.8	10.1400000	11.5685076	12.7932452	13.6051631	13.7395349
4.9	10.8100000	12.4659185	13.9624034	15.0842214	15.5509066
5.0	11.5000000	13.3970153	15.1855212	16.6460949	17.4845536
5.1	12.2100000	14.3619907	16.4632418	18.2923302	19.5436715
5.2	12.9400000	15.3610338	17.7961978	20.0244564	21.7314366
5.3	13.6900000	16.3943295	19.1850117	21.8439855	24.0510070
5.4	14.4600000	17.4620588	20.6302960	23.7524132	26.5055234
5.5	15.2500000	18.5643994	22.1326540	25.7512194	29.0981088
5.6	16.0600000	19.7015255	23.6926798	27.8418686	31.8318700
5.7	16.8900000	20.8736077	25.3109589	30.0258108	34.7098975
5.8	17.7400000	22.0808138	26.9880684	32.3044815	37.7352662
5.9	18.6100000	23.3233083	28.7245773	34.6793027	40.9110356

Table 1. Values of $q_\kappa(u)$ (continued)

$u\backslash\kappa$	2.0	2.1	2.2	2.3	2.4
1.0	1.3333333	1.3908464	1.0499836	0.2185501	−1.1001099
1.1	1.3043333	1.5243262	1.3821575	0.7439733	−0.4513199
1.2	1.2213333	1.5881626	1.6432278	1.2199458	0.1997599
1.3	1.0903333	1.5852899	1.8299174	1.6340082	0.8299406
1.4	0.9173333	1.5193991	1.9406981	1.9763980	1.4191551
1.5	0.7083333	1.3948118	1.9755306	2.2397103	1.9501799
1.6	0.4693333	1.2163857	1.9356654	2.4186357	2.4084172
1.7	0.2063333	0.9894412	1.8234878	2.5097531	2.7817205
1.8	−0.0746667	0.7197033	1.6423938	2.5113625	3.0602523
1.9	−0.3676667	0.4132552	1.3966896	2.4233480	3.2363655
2.0	−0.6666667	0.0765007	1.0915091	2.2470638	3.3045045
2.1	−0.9656667	−0.2838678	0.7327443	1.9852380	3.2611200
2.2	−1.2586667	−0.6608950	0.3269873	1.6418900	3.1045958
2.3	−1.5396667	−1.0473851	−0.1185195	1.2222605	2.8351853
2.4	−1.8026667	−1.4359209	−0.5959269	0.7327492	2.4549552
2.5	−2.0416667	−1.8188800	−1.0968160	0.1808610	1.9677358
2.6	−2.2506667	−2.1884493	−1.6122308	−0.4248418	1.3790769
2.7	−2.4236667	−2.5366372	−2.1327077	−1.0747824	0.6962076
2.8	−2.5546667	−2.8552849	−2.6483007	−1.7584074	−0.0719997
2.9	−2.6376667	−3.1360758	−3.1486047	−2.4642212	−0.9150621
3.0	−2.6666667	−3.3705447	−3.6227762	−3.1798173	−1.8209158
3.1	−2.6356667	−3.5500851	−4.0595517	−3.8919068	−2.7759442
3.2	−2.5386667	−3.6659566	−4.4472651	−4.5863441	−3.7650031
3.3	−2.3696667	−3.7092911	−4.7738629	−5.2481510	−4.7714437
3.4	−2.1226667	−3.6710987	−5.0269181	−5.8615382	−5.7771352

Table 1. Values of $q_\kappa(u)$ (continued)

$u\backslash\kappa$	2.0	2.1	2.2	2.3	2.4
3.5	−1.7916667	−3.5422730	−5.1936439	−6.4099264	−6.7624842
3.6	−1.3706667	−3.3135958	−5.2609050	−6.8759644	−7.7064546
3.7	−0.8536667	−2.9757418	−5.2152288	−7.2415469	−8.5865848
3.8	−0.2346667	−2.5192823	−5.0428162	−7.4878312	−9.3790045
3.9	0.4923333	−1.9346896	−4.7295503	−7.5952520	−10.0584507
4.0	1.3333333	−1.2123398	−4.2610060	−7.5435362	−10.5982823
4.1	2.2943333	−0.3425167	−3.6224581	−7.3117162	−10.9704944
4.2	3.3813333	0.6845856	−2.7988889	−6.8781425	−11.1457315
4.3	4.6003333	1.8788594	−1.7749958	−6.2204963	−11.0933004
4.4	5.9573333	3.2502815	−0.5351981	−5.3158000	−10.7811818
4.5	7.4583333	4.8089097	0.9363567	−4.1404290	−10.1760423
4.6	9.1093333	6.5648813	2.6557865	−2.6701210	−9.2432453
4.7	10.9163333	8.5284105	4.6394688	−0.8799864	−7.9468612
4.8	12.8853333	10.7097863	6.9040356	1.2554829	−6.2496778
4.9	15.0223333	13.1193709	9.4663676	3.7624043	−4.1132096
5.0	17.3333333	15.7675971	12.3435899	6.6674961	−1.4977075
5.1	19.8243333	18.6649676	15.5530666	9.9980699	1.6378328
5.2	22.5013333	21.8220522	19.1123970	13.7820221	5.3356619
5.3	25.3703333	25.2494871	23.0394107	18.0478272	9.6392685
5.4	28.4373333	28.9579729	27.3521639	22.8245302	14.5933716
5.5	31.7083333	32.9582735	32.0689354	28.1417400	20.2439128
5.6	35.1893333	37.2612142	37.2082227	34.0296230	26.6380487
5.7	38.8863333	41.8776811	42.7887388	40.5188963	33.8241440
5.8	42.8053333	46.8186194	48.8294083	47.6408222	41.8517646
5.9	46.9523333	52.0950320	55.3493642	55.4272017	50.7716704

Table 1. Values of $q_\kappa(u)$ (continued)

$u\backslash\kappa$	2.5	2.6	2.7	2.8	2.9
1.0	−2.7500000	−4.3749277	−5.4031157	−5.0933651	−2.6742323
1.1	−2.1292333	−4.0168146	−5.6002049	−6.1323890	−4.7273899
1.2	−1.4230667	−3.4677596	−5.5018414	−6.8127828	−6.4550111
1.3	−0.6639000	−2.7630670	−5.1339481	−7.1330788	−7.8073186
1.4	0.1182667	−1.9381278	−4.5271187	−7.1028117	−8.7521505
1.5	0.8958333	−1.0278948	−3.7153951	−6.7406573	−9.2729269
1.6	1.6436000	−0.0664595	−2.7352620	−6.0728617	−9.3668596
1.7	2.3387667	0.9132980	−1.6248025	−5.1318880	−9.0433475
1.8	2.9609333	1.8800020	−0.4229747	−3.9552308	−8.3225166
1.9	3.4921000	2.8040314	0.8310157	−2.5843615	−7.2338741
2.0	3.9166667	3.6577369	2.0982680	−1.0637805	−5.8150554
2.1	4.2214333	4.4156297	3.3406780	0.5598468	−4.1106475
2.2	4.3956000	5.0545501	4.5213805	2.2384820	−2.1710760
2.3	4.4307667	5.5538166	5.6051508	3.9234466	−0.0515459
2.4	4.3209333	5.8953616	6.5587711	5.5661059	2.1889718
2.5	4.0625000	6.0638529	7.3513685	7.1185086	4.4887136
2.6	3.6542667	6.0468063	7.9547276	8.5339910	6.7840623
2.7	3.0974333	5.8346880	8.3435819	9.7677480	9.0104561
2.8	2.3956000	5.4210094	8.4958864	10.7773759	11.1032698
2.9	1.5547667	4.8024151	8.3930734	11.5233914	12.9986727
3.0	0.5833333	3.9787654	8.0202946	11.9697269	14.6344649
3.1	−0.5079000	2.9532125	7.3666497	12.0842073	15.9508937
3.2	−1.7057333	1.7322733	6.4254038	11.8390077	16.8914526
3.3	−2.9945667	0.3258968	5.1941944	11.2110963	17.4036647
3.4	−4.3564000	−1.2524712	3.6752293	10.1826614	17.4398505

Table 1. Values of $q_\kappa(u)$ (continued)

$u\backslash\kappa$	2.5	2.6	2.7	2.8	2.9
3.5	−5.7708333	−2.9858275	1.8754758	8.7415261	16.9578827
3.6	−7.2150667	−4.8535534	−0.1931574	6.8815496	15.9219285
3.7	−8.6639000	−6.8313594	−2.5136458	4.6030180	14.3031799
3.8	−10.0897333	−8.8912332	−5.0636803	1.9130250	12.0805736
3.9	−11.4625667	−11.0013886	−7.8155053	−1.1741585	9.2415006
4.0	−12.7500000	−13.1262176	−10.7357614	−4.6367217	5.7825066
4.1	−13.9172333	−15.2262444	−13.7853338	−8.4449618	1.7099835
4.2	−14.9270667	−17.2580801	−16.9192046	−12.5609376	−2.9591476
4.3	−15.7399000	−19.1743805	−20.0863101	−16.9381317	−8.1967609
4.4	−16.3137333	−20.9238044	−23.2294023	−21.5211192	−13.9628580
4.5	−16.6041667	−22.4509739	−26.2849130	−26.2452429	−20.2049068
4.6	−16.5644000	−23.6964356	−29.1828233	−31.0362942	−26.8571873
4.7	−16.1452333	−24.5966230	−31.8465346	−35.8102002	−33.8401435
4.8	−15.2950667	−25.0838211	−34.1927441	−40.4727159	−41.0597410
4.9	−13.9599000	−25.0861302	−36.1313225	−44.9191218	−48.4068313
5.0	−12.0833333	−24.5274328	−37.5651946	−49.0339264	−55.7565201
5.1	−9.6065667	−23.3273598	−38.3902232	−52.6905724	−62.9675421
5.2	−6.4684000	−21.4012585	−38.4950944	−55.7511493	−69.8816401
5.3	−2.6052333	−18.6601612	−37.7612061	−58.0661078	−76.3229490
5.4	2.0489333	−15.0107546	−36.0625587	−59.4739802	−82.0973846
5.5	7.5625000	−10.3553503	−33.2656472	−59.8011036	−86.9920362
5.6	14.0062667	−4.5918551	−29.2293566	−58.8613468	−90.7745645
5.7	21.4534333	2.3862568	−23.8048582	−56.4558415	−93.1926025
5.8	29.9796000	10.6899718	−16.8355081	−52.3727157	−93.9731609
5.9	39.6627667	20.4347639	−8.1567475	−46.3868313	−92.8220372

Table 1. Values of $q_\kappa(u)$ (continued)

References

[1] M. Abramowitz and I. A. Stegun, *Handbook of Mathematical Functions*, Dover, New York, 1972.

[2] D. Bradley, *A Sieve Auxiliary Function*, Ph.D. Thesis, University of Illinois, Urbana, 1995.

[3] H. Diamond, H. Halberstam, and H.-E. Richert, *Combinatorial sieves of dimension exceeding one*, J. Number Theory **28** (1988), 306–346.

[4] H. Diamond, H. Halberstam, and H.-E. Richert, *Sieve auxiliary functions I*, in "Number Theory, Proceedings of the First Conference of the Canadian Number Theory Association" (R. Mollin, ed.), W. de Gruyter, Berlin, 1990, pp. 99–113.

[5] H. Diamond, H. Halberstam, and H.-E. Richert, *A boundary value problem for a pair of differential delay equations related to sieve theory, I*, in "Analytic Number Theory: Proceedings of a Conference in honour of P. T. Bateman" (B. Berndt, et al. eds.), Birkhäuser, Boston, 1990, pp. 133–157.

[6] H. Diamond, H. Halberstam, and H.-E. Richert, *A boundary value problem for a pair of differential delay equations related to sieve theory, II*, J. Number Theory **45** (1993), 129–185.

[7] H. Diamond, H. Halberstam, and H.-E. Richert, *Sieve auxiliary functions II*, in "A tribute to Emil Grosswald: Number theory and related analysis" (M. Knopp and M. Sheingorn, eds.), Contemporary Math., Vol. 143, Amer. Math. Soc., Providence, RI, 1993, pp. 247–253.

[8] H. Diamond, H. Halberstam, and H.-E. Richert, *A boundary value problem for a pair of differential delay equations related to sieve theory, III*, J. Number Theory **47** (1994), 300–328.

[9] H. Diamond, H. Halberstam, and H.-E. Richert, *Estimation of the sieve auxiliary functions q_κ in the range* $1 < \kappa < 2$, Analysis **14** (1994), 75–102.

[10] H. Diamond, H. Halberstam, and H.-E. Richert, *Combinatorial sieves of dimension exceeding one II*, These Proceedings, 265–308.

[11] H. Iwaniec, *Rosser's sieve*, Acta Arith. **36** (1980), 171–202.

[12] H. Iwaniec, J. van de Lune, and H. J. J. te Riele, *The limits of Buchstab's iteration sieve*, Nederl.-Akad.-Wetensch.-Indag.-Math. **42** (1980), no. 4, 409-417.

[13] A. Odlyzko, *e-mail communication*, July 9, 1993.

[14] F. W. J. Olver, *Asymptotics and Special Functions*, Academic Press, New York, 1974.

[15] H. J. J. te Riele, *Numerical Solution of two coupled nonlinear equations related to the limits of Buchstab's iteration sieve*, Afdeling Numerieke Wiskunde [Department of Numerical Mathematics], 86, Mathematisch Centrum, Amsterdam, 1980.

[16] H. S. Wall, *Analytic Theory of Continued Fractions*, Chelsea, New York, 1967.

[17] F. Wheeler, *On Two Differential-Difference Equations Arising in Analytic Number Theory*, Ph.D. Thesis, University of Illinois, Urbana, 1988.

[18] F. Wheeler, *Two differential-difference equations*, Trans. Amer. Math. Soc. **318** (1990), 491–523.

David Bradley
Centre for Experimental
and Constructive Mathematics
Simon Fraser University
Burnaby B.C. V5A 1S6
CANADA
dbradley@cecm.sfu.ca

Bounds on complete exponential sums

Todd Cochrane

Dedicated to Heini Halberstam

Abstract. We give a brief survey on complete exponential sums of the type $S(g) = \sum_{\mathbf{x}} e_p(g(\mathbf{x}))$, with g a polynomial in n variables over the finite field $\mathbb{F}_p$, and prove several new results including the following. If g is non-composite, that is, not of the type $g(\mathbf{x}) = f(h(\mathbf{x}))$ with $\deg f \geq 2$, and $p \geq c_1(d,n)$ then the weights of the characteristic values of $S(g)$ are all $\leq 2n-2$. For homogeneous g we can take $c_1(d,n) = 2$. A partial converse is also given. Next, if g is homogeneous, absolutely irreducible of degree d, and has a singular locus of dimension ℓ in $\mathbb{P}^{n-1}$, then the characteristic values of $S(g)$ have weight $\leq n+\ell+1$, and so $|S(g)| \leq (4d+5)^n p^{\frac{n+\ell+1}{2}}$.

1. Introduction

Let p be a prime, $\mathbb{F}_p$ the finite field in p elements, and $g(\mathbf{x}) \in \mathbb{F}_p[x_1, .., x_n]$, be a polynomial of degree $d = d_g$. Complete exponential sums of the type,

$$S(g) = \sum_{\mathbf{x}} e_p(g(\mathbf{x})),$$

where the sum is over all points in $\mathbb{F}_p^n$ and, as usual $e_p(*) = e^{(2\pi i */p)}$, have a rich history dating back to Gauss. Gauss established the identity

$$(1) \qquad G_p := \sum_x e_p(x^2) = \begin{cases} \sqrt{p} & \text{if } p \equiv 1 \pmod 4 \\ i\sqrt{p} & \text{if } p \equiv -1 \pmod 4), \end{cases}$$

and the upper bound

$$|\sum_x e_p(x^d)| \leq ((d, p-1) - 1)p^{1/2}$$

for any positive integer d. Weil (1948) proved the more general bound

$$(2) \qquad |\sum_x e_p(g(x))| \leq (d-1)p^{1/2},$$

for any polynomial $g(x)$ in one variable over $\mathbb{F}_p$, of degree d not divisible by p.

For any positive integer j set

$$S_j(g) = \sum_{\mathbf{x}\in\mathbb{F}_{p^j}^n} \psi_j(g(\mathbf{x}))$$

where $\psi_j() = e_p(Tr_j())$ and Tr_j is the trace mapping from $\mathbb{F}_{p^j}$ to $\mathbb{F}_p$. Then, by the rationality of the L-function, there is a finite set of complex numbers $\alpha_1, \ldots, \alpha_\ell, \beta_1, \ldots, \beta_m$, with $\alpha_i \neq \beta_k$ for all i and k, such that

$$S_j(g) = \alpha_1^j + \cdots + \alpha_\ell^j - \beta_1^j - \cdots - \beta_m^j, \tag{3}$$

for any positive integer j. The values α_i and β_j are characteristic values of the L-function associated with the S_j. For simplicity we shall simply call them the characteristic values of $S(g)$. Deligne (1980, Weil II) proved a very general version of the Riemann Hypothesis, which in this setting states that the modulus of each of the characteristic values is a power of $\sqrt{p}$, the power being called the weight of the characteristic value. Bombieri (1978), proved that the total number of characteristic values is bounded above by $(4d+5)^n$, and so we have the upper bound

$$|S(g)| \leq (4d+5)^n p^{M/2} \tag{4}$$

where M is the maximum weight of the characteristic values. Thus to bound the exponential sum $S(g)$, it suffices to determine the maximum weight of the characteristic values.

The trivial Bound for $S(g)$ is just $|S(g)| \leq p^n$. If g is not of the form $h^p - h + c$, with h a polynomial over $\mathbb{F}_p$, then one saves a factor of $p^{1/2}$ and obtains

$$|S(g)| \leq (d-1)p^{n-\frac{1}{2}}. \tag{5}$$

For polynomials of the form $h^p - h + c$ one has trivially $|S(g)| = p^n$. The upper bound in (5) follows easily from the one variable bound of Weil (2) for polynomials of degree d not divisible by p. Deligne (1977, Proposition 3.8) proved the general case. Deligne also established a sharp upper bound for polynomials nonsingular at infinity.

Theorem. *Deligne (1973, Theorem 8.4). Let g be a polynomial of degree d not divisible by p, whose maximal homogeneous part is nonsingular. Then $S(g)$ has $(d-1)^n$ characteristic values each of modulus $p^{n/2}$. In particular,*

$$|S(g)| \leq (d-1)^n p^{n/2}. \tag{6}$$

When $n = 1$ the inequality in (6) is just the result of Weil (2). The theorem of Deligne was generalized by Katz (1980, Theorem 5.1.1), Adolphson and

Sperber (1989) and Denef and Loeser (1991, Theorem 9.2). The latter two papers deal with exponential sums over Tori, and give more general conditions under which one can reach the conclusion of Deligne's theorem.

Example 1. (Adapted from an example in Adolphson and Sperber (1989)). Let

$$g(\mathbf{x}) = a_1 x_1^{d_1} + \cdots + a_n x_n^{d_n} + h(\mathbf{x}),$$

and suppose that none of the exponents d_i are divisible by p, and that every nonzero monomial $c_i x_1^{i_1} \cdots x_n^{i_n}$ in h satisfies $\sum_{j=1}^{n} \frac{i_j}{d_j} < 1$. Then $S(g)$ has $\Pi_{i=1}^{n}(d_i - 1)$ characteristic values, each of weight n. Since the exponents d_i may be distinct the polynomial g may have bad singularities at infinity.

There is still much middle ground to be covered between the upper bounds of (5) and (6). Can one characterize those polynomials g for which one saves an extra factor of $p^{1/2}$ in (5)? It would appear that this question has not been addressed in the literature, although the result may be known. An obvious case where one cannot save the factor of $p^{1/2}$ is when g degenerates to a polynomial in a single variable, that is, $g(\mathbf{x}) = f(L(\mathbf{x}))$ for some polynomial f in a single variable of degree greater than one, and linear polynomial $L(\mathbf{x})$. Davenport and Lewis (1962) showed that for a nondegenerate cubic polynomial g, $S(g) \ll p^{5/4}$ when $n = 2$, whence by the Riemann Hypothesis (and the analogous upper bound for any extension of $\mathbb{F}_p$) we now obtain $S(g) \ll p$, and $S(g) \ll p^2$ when $n = 3$. Does one always obtain the extra savings of $p^{1/2}$ for nondegenerate polynomials?

Example 2.

$$\sum_{x,y} e_p(x^2y^2) = p + \sum_{y \neq 0} \Big(\sum_{x} e_p(x^2y^2) \Big) = p + (p-1)G_p,$$

where G_p is the Gaussian sum in (1). This sum has three characteristic values, of weights 1, 2 and 3 respectively.

The polynomial in Example 2 is nondegenerate and yet does not admit an extra savings of $p^{1/2}$. However, x^2y^2 is a composite polynomial. As we shall see in Theorem 1, this is the characteristic that identifies when one cannot obtain the extra savings. We call the polynomial g composite if it can be expressed in the manner $g(\mathbf{x}) = f(h(\mathbf{x}))$ with f a polynomial in one variable over $\overline{\mathbb{F}}_p$ of degree at least 2 and h any polynomial in n variables over $\overline{\mathbb{F}}_p$.

Theorem 1. *If g is non-composite and $p \geq c_1(d,n)$ then the weights of the characteristic values of $S(g)$ are all $\leq 2n-2$, and so*

$$|S(g)| \leq (4d_g + 5)^n p^{n-1}. \tag{7}$$

If g is homogeneous then we may take $c_1(d,n) = 2$. Conversely, if g is a composite polynomial of the form $g(\mathbf{x}) = f(h(\mathbf{x}))$, with h homogeneous and $p \nmid d_f := deg(f)$, then there are at least $d_f - 1$ characteristic values of weight $2n-1$.

It is quite possible that the hypothesis $p \geq c_1(d,n)$ is only due to our method of proof. We do not know at present if the converse statement in Theorem 1 holds when h is not homogeneous.

Theorem 1 is of particular interest when $n = 2$. In this case we obtain $|S(g)| \ll p$, for non-composite g, which is best possible in general. Another interesting special case is when g is of prime degree. In this case g is composite if and only if g degenerates to a polynomial in a single variable of degree at least 2. Thus, the result of Davenport and Lewis mentioned above is just a special case of Theorem 1.

It is worth pointing out that one occasionally saves a factor of $p^{1/2}$ (or more) at the ground field level (due to cancellation in the sum (3)), even though all of the characteristic values have weight $2n-1$. Indeed, if $(d, p-1) = 1$ then $\sum_x e_p(x^d) = 0$! But if $p \nmid d$ then this sum has $d-1$ characteristic values each of weight 1. Theorem 2 provides further examples where this phenomenon occurs.

Is it possible to characterize those polynomials for which one saves yet another factor of $p^{1/2}$? We start by considering a special class of homogeneous polynomials.

Theorem 2. *Suppose that g is a homogeneous polynomial over $\mathbb{F}_p$ of degree d with $(d, p-1) = 1$, and having the factorization*

$$g = P_1^{e_1} P_2^{e_2} \dots P_s^{e_s},$$

with the P_i distinct (nonassociate) irreducible polynomials over $\mathbb{F}_p$. If exactly r of the factors P_i are absolutely irreducible then

$$S(g) = (r-1)p^{n-1} + O^{\dagger}(p^{n-\frac{3}{2}}). \tag{8}$$

In particular, if exactly one of the factors is absolutely irreducible then,

$$|S(g)| \ll^{\dagger} p^{n-\frac{3}{2}}. \tag{9}$$

It is important to point out that one cannot conclude from (9) that all of the weights of the characteristic values are $\leq 2n-3$. Indeed, g could be a

†The constants in the big "Oh's" and $\ll$ depend only on n and the degrees of the polynomials.

composite polynomial of the form $g = P^e$ with P absolutely irreducible, in which case some of the weights are $2n-1$.

It is reasonable to ask whether (9) holds unconditionally for any absolutely irreducible homogeneous polynomial. The answer is no. One need only take a sum of two composite polynomials in distinct variables to obtain a counterexample.

Example 3. Let $g(x,y,u,v) = (x^2+y^2)^3 + (u^3+v^3)^2$ and assume $p \neq 2,3$. Since X^3+Y^2 is absolutely irreducible it follows from Schinzel's decomposability theorem (see Schinzel (1980), section 8, Theorem 11), that g is absolutely irreducible over any field. Now, since the number of solutions of the equation $x^2+y^2=\lambda$ is $p - \left(\frac{-1}{p}\right)$ for any nonzero λ we obtain,

$$\sum_{x,y} e_p((x^2+y^2)^3) = p\left[\left(\frac{-1}{p}\right)\sum_{\lambda} e_p(\lambda^3)\right].$$

Also, by (21) we have

$$\sum_{u,v} e_p((u^3+v^3)^2) = p\sum_{\lambda} e_p(\lambda^2) + O(p).$$

Thus

$$S(g) = p^2 G_p \sum_{\lambda} e_p(\lambda^3) + O(p^{\frac{5}{2}}).$$

The polynomial in the previous example is absolutely irreducible but it has a singular locus of dimension 1 in projective 3-space. In the next Theorem we give an upper bound for $S(g)$ in terms of the dimension of the singular locus. Let g be any polynomial over $\mathbb{F}_p$, not necessarily homogeneous, with maximal homogeneous part g_d. The singular locus of g at infinity is the algebraic subset of projective (n-1)-space $\mathbb{P}^{n-1}$ defined by,

$$\frac{\partial g_d}{\partial x_i} = 0, \qquad 1 \le i \le n \qquad g_d = 0.$$

Let ℓ denote the dimension of the singular locus, putting $\ell = -1$ if g is nonsingular at infinity. Birch (1962) proved that

$$|S(g)| \ll p^{n-2^{1-d}(n-\ell)}. \tag{10}$$

We sharpen the upper bound in (10) and at the same time generalize the result of Deligne (6).

Theorem 3. *Suppose that $p > c_2(d,n)$. Let g be a polynomial of degree d over $\mathbb{F}_p$ such that its maximal homogeneous part is absolutely irreducible and has a singular locus of dimension ℓ. Then all of the characteristic values of $S(g)$ have weight $\le n+\ell+1$ and we have*

$$|S(g)| \le (4d+5)^n p^{\frac{n+\ell+1}{2}}. \tag{11}$$

If g is homogeneous, then we may take $c_2(d,n) = 2$.

If $n \ge 3$ and the maximal homogeneous part of g is nonsingular, then g_d is necessarily absolutely irreducible and we obtain the best possible exponent $\frac{n}{2}$ in (11). In this case the upper bound in (11) coincides with the upper bound of Deligne in (6), up to the constant factor. Skorobogatov (1992, Theorem 3.2) obtained (11) for the case of homogeneous g with $(d, p-1) = 1$.

The upper bound in (11) is best possible for polynomials g that degenerate to polynomials nonsingular at ∞ in fewer variables. A nondegenerate polynomial is given in Example 3 for which the upper bound in (11) is best possible. The results of Adolphson and Sperber, and Denef and Loeser provide special classes of polynomials for which one can improve on the upper bound in (11) when singularities are present at infinity; see Example 1.

An upper bound very similar to (10) can be given in terms of a factor called the h-invariant of g, the smallest positive integer h such that g_d may be written as

$$g_d = A_1B_1 + A_2B_2 + \cdots + A_hB_h \tag{12}$$

for some forms $A_i, B_i \in \mathbb{F}_p[\mathbf{x}]$ of positive degree. Schmidt (1984, Theorem 1) established

$$|S(g)| \ll p^{n-2^{1-d}\lceil h/\Phi(d)\rceil}, \tag{13}$$

where $\lceil\alpha\rceil$ denotes the smallest integer $> \alpha$, and one can take $\Phi(2) = \Phi(3) = 1$, $\Phi(4) = 3$, $\Phi(5) = 13$ and $\Phi(d) < (\log 2)^{-d} d!$ in general. Davenport and Lewis (1962) much earlier obtained (13) for the case of cubic polynomials. In general (13) is weaker than (11). Are there examples where the upper bound in (13) is sharper than (11)? When $d = 3$ this would require $2\ell + h + 3 > 2n$. We note that one always has $h > (n-\ell-1)/2$ since any solution of $A_1 = B_1 = \cdots = A_h = B_h = 0$ is a singular point of g_d.

The author wishes to express his thanks to Shreeram Abhyankar for fruitful discussions on algebraic geometry and to the many others who helped steer him to the relevant sources.

2. Lemmas

Lemma 1. *If h is a non-composite polynomial defined over $\overline{\mathbb{F}}_p$ of degree d_h then for all but $c_3(d_h, n)$ values of $\tau \in \overline{\mathbb{F}}_p$, the polynomial $h(\mathbf{x}) - \tau$ is absolutely irreducible. If moreover h is homogeneous (and non-composite) then $h(\mathbf{x}) - \tau$ is absolutely irreducible for all nonzero values of τ.*

Proof. Since h is non-composite polynomial, it follows from a theorem of Bertini (see Shafarevic (1994), Chapter II, section 6.1, Theorem 1) that for all but finitely many values of τ, the fibres $h^{-1}(\tau)$, in affine n-space, are irreducible. That the number depends only on d_h and n follows from elimination theory. By a theorem of Noether (see Schmidt (1976), Chapter V, Theorem 2A), we can take $c_3(d_h, n) = m^{2^m}$ where $m = \binom{n+d_h-1}{n}$. For homogeneous h it is easy to see that for any nonzero τ, $h - \tau$ is reducible over $\overline{\mathbb{F}}_p$ if and only if $h - 1$ is reducible over $\overline{\mathbb{F}}_p$. Since $h - \tau$ is reducible for at most finitely many values of τ it follows that it is absolutely irreducible for all nonzero τ.

A more direct proof can be given for homogeneous h. We may assume $\tau = 1$. Let $k = \overline{\mathbb{F}}_p$, and $H(z, \mathbf{x}) = z^d - h(\mathbf{x})$, with $d = d_h$. Certainly $h - 1$ is reducible over k if and only if H is reducible over k. View H as a polynomial in z with coefficients in $k(\mathbf{x})$. By the theorem of Capelli (1898) and Rédei (1959), (see also Schinzel (1982), section 13, Theorem 21), H is reducible over $k(\mathbf{x})$ if and only if for some prime divisor p of d, $h = b^p$, with $b \in k(\mathbf{x})$. But this implies that h is composite. ■

It would be of interest to find the smallest possible value for $c_3(d_h, n)$. When $n = 2$ the best one can hope for is $c_3(d_h, 2) = d_h - 1$ as the following example illustrates.

Example 4. The polynomial $h(x, y) = y(x - r_1)(x - r_2) \ldots (x - r_d) + x - \tau$ is reducible if and only if $\tau = r_i$ for some i.

Y. Stein (1989) was able to establish this best possible value for the case $n = 2$ over fields of characteristic 0. One should also see the work of Ruppert (1986) and section 11 of Schinzel's book (1982) on polynomials for related material.

Lemma 2. *Let $g(\mathbf{x})$ be an absolutely irreducible homogeneous polynomial over $\mathbb{F}_p$ with singular locus of dimension ℓ in $\mathbb{P}^{n-1}$. Then for any $\tau \in \mathbb{F}_p$ the number $N(\tau)$ of solutions of the equation $g(\mathbf{x}) = \tau$ with coordinates in $\mathbb{F}_p$ satisfies*

$$|N(\tau) - p^{n-1}| \ll \begin{cases} p^{\frac{n+\ell}{2}}, & \tau \neq 0 \\ p^{\frac{n+\ell+1}{2}}, & \tau = 0. \end{cases} \tag{14}$$

Proof. Let $N_{\mathbb{P}^n}(\tau)$ denote the number of projective solutions of the equation $g(\mathbf{x}) = \tau x_0^d$ with coordinates in $\mathbb{F}_p$, and $N_{\mathbb{P}^{n-1}}(0)$ denote the number of projective solutions of the equation $g(\mathbf{x}) = 0$, with coordinates in $\mathbb{F}_p$. Then it is easy to see that for any nonzero τ,

$$\begin{aligned} N(\tau) &= N_{\mathbb{P}^n}(\tau) - N_{\mathbb{P}^{n-1}}(0) \\ &= (p^{n-1} + \cdots + p + 1) + O(p^{\frac{n+\ell}{2}}) - (p^{n-2} + \cdots + p + 1) + O(p^{\frac{n-1+\ell}{2}}) \\ &= p^{n-1} + O(p^{\frac{n+\ell}{2}}), \end{aligned}$$

by Theorem 1 of Hooley (1991). If $\tau = 0$ then we use $N(\tau) = (p-1)N_{\mathbb{P}^{n-1}}(0)+1$ together with Hooley's theorem to obtain (14). ■

Let χ be any multiplicative character on $\mathbb{F}_p^*$ of exponent d; $\chi^d = \chi_o$, where χ_o is the principal character. Then for any homogeneous polynomial $h(\mathbf{x})$ of degree d the character sum

$$\sum_{\mathbf{x} \in \mathbb{P}^{n-1}} \chi(h(\mathbf{x})), \tag{15}$$

over all points of $\mathbb{P}^{n-1}$ with coordinates in $\mathbb{F}_p$, is well defined and has an associated set of characteristic values of integral weight.

Lemma 3. *(i) Let h be any non-composite homogeneous polynomial in $n \geq 2$ variables over $\mathbb{F}_p$ of degree d. Then for any nonprincipal character χ of exponent d, the weights of the characteristic values of the sum in (15) are all $\leq 2n-3$.*

(ii) If moreover h is absolutely irreducible with singular locus of dimension ℓ then the weights are all $\leq n+\ell$. In particular

$$\sum_{\mathbf{x} \in \mathbb{P}^{n-1}} \chi(h(\mathbf{x})) \ll p^{\frac{n+\ell}{2}}. \tag{16}$$

Proof. For any value τ in $\mathbb{F}_p$ let $N(\tau)$ denote the number of solutions of the equation $h(\mathbf{x}) = \tau$ with coordinates in $\mathbb{F}_p$, and write

$$N(\tau) = p^{n-1} + E(\tau). \tag{17}$$

Since h is homogeneous it is easy to see that $N(\tau)$ and $E(\tau)$ depend only on the residue class $\overline{\tau}$ of τ in the quotient group $\mathbb{F}_p^*/(\mathbb{F}_p^*)^d$, where here $(\mathbb{F}_p^*)^d$ denotes the set of dth powers. Let $\{\tau_1, \ldots, \tau_{(p-1,d)}\}$ be a set of coset representatives for these residue classes. Then for any nonprincipal character χ of exponent d,

$$\sum_{\mathbf{x} \in \mathbb{P}^{n-1}} \chi(h(\mathbf{x})) = \frac{1}{p-1} \sum_{\tau \neq 0} \chi(\tau) N(\tau) = \frac{1}{p-1} \sum_{\tau \neq 0} \chi(\tau) E(\tau),$$

and so

$$\sum_{\mathbf{x}\in\mathbb{P}^{n-1}} \chi(h(\mathbf{x})) = \frac{1}{(p-1,d)} \sum_{i=1}^{(p-1,d)} \chi(\tau_i)E(\tau_i). \tag{18}$$

Now, if h is non-composite, then by Lemma 1 $h(\mathbf{x})-\tau$ is absolutely irreducible for any nonzero τ whence by the theorem of Lang and Weil (1954), $|E(\tau)| \ll p^{n-3/2}$. It follows from (18) that

$$|\sum_{\mathbf{x}\in\mathbb{P}^{n-1}} \chi(h(\mathbf{x}))| \ll p^{n-\frac{3}{2}}.$$

Since the analogous upper bound holds when $\mathbb{F}_p$ is replaced by $\mathbb{F}_{p^j}$ and χ by $\chi_j = \chi(\mathcal{N}_{\mathbb{F}_{p^j}/\mathbb{F}_p})$, we conclude by the theorem of Bombieri (1978, Theorem 3) that all of the characteristic values have weight $\leq 2n-3$. The second part of the lemma follows from (18) and Lemma 2 in exactly the same manner. ■

3. Formulae for $S(g)$

Let g be a polynomial of the form $g(\mathbf{x}) = f(h(\mathbf{x}))$ with f and h polynomials over $\mathbb{F}_p$ and h homogeneous of degree $d = d_h$. Then using the notation of section 2 and (18) we have,

$$S(g) = \sum_{\tau} e_p(f(\tau))N(\tau) = e_p(f(0))N(0) + \sum_{i=1}^{(p-1,d)} N(\tau_i) \sum_{\overline{\tau}=\overline{\tau}_i} e_p(f(\tau))$$

$$= e_p(f(0))N(0) + \frac{1}{(p-1,d)} \sum_{i=1}^{(p-1,d)} N(\tau_i) \sum_{\tau\neq 0} (\sum_{\chi^d=1} \chi(\tau/\tau_i))e_p(f(\tau))$$

$$= e_p(f(0))N(0) + \sum_{\chi^d=1} [\sum_{\mathbf{x}\in\mathbb{P}^{n-1}} \overline{\chi}(h(\mathbf{x}))] \sum_{\tau\neq 0} \chi(\tau)e_p(f(\tau)).$$

Pulling off the $\chi = \chi_o$ term and writing

$$N_{\mathbb{P}}(h) = \frac{p^{n-1}-1}{p-1} + E_{\mathbb{P}}(h), \tag{19}$$

for the number of projective zeros of h with coordinates in $\mathbb{F}_p$, we obtain

$$S(g) = pe_p(f(0))E_{\mathbb{P}}(h) + (p^{n-1} + E_{\mathbb{P}}(h)) \sum_{\tau} e_p(f(\tau))$$

$$+\sum_{\substack{\chi^d=\chi_o \\ \chi\neq\chi_o}} [\sum_{\mathbf{x}\in\mathbb{P}^{n-1}} \overline{\chi}(h(\mathbf{x}))] \sum_{\tau\neq 0} \chi(\tau) e_p(f(\tau)). \tag{20}$$

The identity in (20) provides a way of expressing the characteristic values of $S(g)$ in terms of the characteristic values associated with the terms on the righthand side. Of course, one must first pass to an extension $\mathbb{F}_q$ of $\mathbb{F}_p$ with $d|q-1$ in order to pick up a full set of d multiplicative characters of exponent d. An alternate way of writing (20) is

$$S(g) = p^{n-1} \sum_{\tau} e_p(f(\tau)) + \frac{p}{p-1} e_p(f(0)) E(0)$$

$$+(p-1,d)^{-1} \sum_{i=1}^{(p-1,d)} E(\tau_i) \sum_{\lambda} e_p(f(\tau_i \lambda^d)). \tag{21}$$

If g itself is homogeneous then we may take $f(x) = x$ and $h = g$. The identities in (20) and (21) simplify to

$$S(g) = pE_{\mathbb{P}}(g) + \sum_{\substack{\chi^d=\chi_o \\ \chi\neq\chi_o}} [\sum_{\mathbf{x}\in\mathbb{P}^{n-1}} \chi(g(\mathbf{x}))] \sum_{\tau\neq 0} \chi(\tau) e_p(\tau), \tag{22}$$

and

$$S(g) = \frac{p}{p-1} E(0) + (p-1,d)^{-1} \sum_{i=1}^{(p-1,d)} E(\tau_i) \sum_{\lambda} e_p(\tau_i \lambda^d), \tag{23}$$

respectively.

4. Proof of Theorem 1

Suppose first that g is any non-composite homogeneous polynomial. Then by (22), Lemma 3(i) and the the fact that all of the characteristic values associated with $E_{\mathbb{P}}$ have weight $\leq 2n-4$, it follows that the characteristic values of $S(g)$ all have weight $\leq 2n-2$.

Suppose next that g is a composite polynomial of the form $g(\mathbf{x}) = f(h(\mathbf{x}))$ with f and h polynomials over some extension $\mathbb{F}_q$ of $\mathbb{F}_p$, with h homogeneous and $d_f \geq 2$. By modifying f and h if need be we may assume that h is non-composite. Furthermore, we claim that by multiplying f and h by constants if need be we may assume that f and h have coefficients in $\mathbb{F}_p$. Indeed, the maximal homogeneous part of g, $ah(\mathbf{x})^{d_f}$, for some constant $a \in \mathbb{F}_q$, has coefficients in $\mathbb{F}_p$. Thus, if $P(\mathbf{x})$ is any irreducible factor of $h(\mathbf{x})$ over $\mathbb{F}_q$ then all of its conjugates must also be factors of $h(\mathbf{x})$. But the product of the distinct conjugates is a polynomial over $\mathbb{F}_p$, and so h can be expressed as a product of polynomials with coefficients in $\mathbb{F}_p$ up to a constant factor.

If $p \nmid d_f$ then the sum $S(f) = \sum_\tau e_p(f(\tau)))$ has $d_f - 1$ characteristic values, each of weight 1. Thus the term $p^{n-1} \sum_\tau e_p(f(\tau))$ in (20) gives rise to $d_f - 1$ characteristic values of $S(g)$, each of weight $2n-1$, while the weights associated with the remaining terms in (20) are all $\leq 2n - 2$.

Suppose now that g is any non-composite polynomial. We shall apply Hooley's method of bounding weights by looking at second moments. For $\lambda \in \mathbb{F}_{p^j}$ let

$$S_j(\lambda) = \sum_{\mathbf{x}} \psi_j(\lambda g(\mathbf{x})), \tag{24}$$

so that $S_j(1) = S_j$, and let M_j denote the second moment,

$$M_j = \sum_{\lambda \in \mathbb{F}_{p^j}, \lambda \neq 0} |S_j(\lambda)|^2.$$

The following proposition may be gleaned from Hooley's (1982) work; the proof is given in Cochrane (1994, section 4.4).

Proposition . *Suppose that N and C are positive real numbers such that $M_j \leq Cp^{Nj}$ for all j sufficiently large , and that $p > \frac{C}{B} + 1$, where B is the number of characteristic values of maximal weight. Then all of the characteristic values associated with the sums S_j have weights strictly less than N.*

Now, for any fixed constant c,

$$S_j(\lambda) = \sum_\tau \psi_j(\lambda\tau)(N_j(\tau) - c), \tag{25}$$

where $N_j(\tau)$ is the number of solutions of the equation $g(\mathbf{x}) = \tau$ with coordinates in $\mathbb{F}_{p^j}$. Defining $S_j(0)$ formally by the expression in (25), it follows from Parsevals identity that

$$M_j \leq \sum_{\lambda \in \mathbb{F}_{p^j}} |S_j(\lambda)|^2 = p^j \sum_{\tau \in \mathbb{F}_{p^j}} |N_j(\tau) - c|^2. \tag{26}$$

By Lemma 1, the polynomial $g(\mathbf{x}) - \tau$ is absolutely irreducible for all but $c_3(d,n)$ values of τ. Putting $c = p^{j(n-1)}$ in (26) it follows from the estimate of Lang and Weil (1954) and the elementary upper bound $N_j(\tau) \ll p^{j(n-1)}$ for any τ, that

$$M_j \ll p^j \sum_{\substack{\tau, g(\mathbf{x})=\tau \\ abs.irred.}} p^{2j(n-\frac{3}{2})} + O(p^{j(2n-1)}) \ll p^{j(2n-1)}.$$

Thus, by the Proposition, all of the characteristic values have weight strictly less than $2n-1$, for p sufficiently large.

5. Proof of Theorem 2

If $(p-1,d)=1$ then (22) is just the well known identity,

$$S(g) = pE_{\mathbb{P}}(g) = (p-1)^{-1}(pN(g) - p^n), \tag{27}$$

where $N(g)$ is the number of zeros of g with coordinates in $\mathbb{F}_p$, and $E_{\mathbb{P}}(g)$ is as defined in (19). The earliest reference that I know of to (27) is the work of Chalk and Williams (1965, Lemma 15). Skorobogatov (1992) more recently used a version of (27) in his work on exponential sums.

Suppose that g admits the factorization

$$g = P_1^{e_1} \cdots P_s^{e_s},$$

with the P_i irreducible polynomials over $\mathbb{F}_p$, and say that exactly r of the polynomials P_i are absolutely irreducible. We first remark that by (27) the value of $S(g)$ depends only on the the irreducible factors P_i and not at all on the exponents e_i, (of course the e_i must be such that the degree of g is relatively prime to $p-1$.) Let $N(P_i)$ denote the number of zeros of P_i with coordinates in $\mathbb{F}_p$. If P_i is absolutely irreducible then by the theorem of Lang and Weil (1954),

$$N(P_i) = p^{n-1} + O(p^{n-\frac{3}{2}}).$$

If P_i is not absolutely irreducible, then the set of zeros of P_i with coordinates in $\mathbb{F}_p$ is contained in an algebraic subset of codimension 2 and so $N(P_i) \ll p^{n-2}$. Also, the number of common zeros of two or more of the factors P_i is $O(p^{n-2})$. Thus

$$N(g) = r(p^{n-1} + O(p^{n-\frac{3}{2}})) + O(p^{n-2}) = rp^{n-1} + O(p^{n-\frac{3}{2}}),$$

and so by (27),

$$S(g) = (r-1)p^{n-1} + O(p^{n-\frac{3}{2}}). \tag{28}$$

6. Proof of Theorem 3

Suppose that g is homogeneous and absolutely irreducible, with singular locus of dimension ℓ in $\mathbb{P}^{n-1}$. Then by Lemma 2, all of the characteristic values associated with $E_{\mathbb{P}}(g)$ have weight $\leq n + \ell - 1$. It follows from Lemma 3(ii) and (22) that all of the characteristic values of $S(g)$ have weight $\leq n + \ell + 1$, proving Theorem 3 for the case of homogeneous polynomials.

For nonhomogeneous g we again turn to the method of Hooley and the Proposition in section 4. The details have been worked out in section 4.6 of Cochrane (1994).

References

A. Adolphson and S. Sperber, *Exponential sums and Newton polyhedra, cohomology and estimates*, Ann. Math. **130** (1989), 367–406.

B.J. Birch, *Forms in many variables*, Proc. Roy. Soc. (A) **265** (1962), 245–263.

E. Bombieri, *On exponential sums in finite fields II*, Inventiones Math. **47** (1978), 29–39.

A. Capelli, *Sulla riduttibilità delle equazioni algebriche*, Nota secunda, ibidem (3) **4** (1898), 84–90.

J.H.H. Chalk and K.S. Williams, *The distribution of solutions of congruences*, Mathematika **12** (1965), 176–192.

T. Cochrane, *Exponential Sums and the Distribution of Solutions of Congruences*, Inst. of Math., Academia Sinica, Taipei, 1994.

H. Davenport and D.J. Lewis, *Exponential sums in many variables*, Amer. J. Math. **84** (1962), 649–665.

P. Deligne, *Cohomologie Etale (SGA* $4\frac{1}{2}$ *)*, Springer Lecture Notes in Math. 569, Springer-Verlag, 1977.

———, *La conjecture de Weil II*, Publ. Math. I.H.E.S. **52** (1980), 137–252.

J. Denef and F. Loeser, *Weights of exponential sums, intersection cohomology and Newton polyhedra*, Invent. math. **106** (1991), 275–294.

C. Hooley, *On exponential sums and certain of their applications*, Journées Arith. 1980, London Math. Soc. Lec. Note Ser. 56, J.V. Armitage (ed.), Cambridge Univ. Press, 1982, pp. 92–122.

———, *On the number of points on a complete intersection over a finite field*, J. of Number Theory **38** (1991), 338–358.

N.M. Katz, *Sommes exponentielles*, Astérisque **79** (1980).

S. Lang and A. Weil, *Number of points on varieties in finite fields*, Amer. J. Math. **76** (1954), 819–827.

A. Rédei, *Algebra, I Teil*, Akademische Verlaggesellschaft, Leipzig, 1959.

W. Ruppert, *Reduzibilität ebener kurven*, J. Reine Angew. Math. **369** (1986), 167–191.

A. Schinzel, *Selected Topics on Polynomials*, The University of Michigan Press, Ann Arbor, 1982.

W.M. Schmidt, *Bounds for exponential sums*, Acta Arithmetica **44** (1984), 282–297.

I.R. Shafarevich, *Basic Algebraic Geometry, 1, Second, Revised and Expanded Edition*, Springer-Verlag, New York, 1994.

A.N. Skorobogatov, *Exponential sums, the geometry of hyperplane sections, and some diophantine problems*, Israel J. of Math. **80** (1992), 359–379.

Y. Stein, *The total reducibility order of a polynomial in two variables*, Israel J. Math. **68** (1989), no. 1, 109–122.

A. Weil, *On some exponential sums*, Proc. Nat. Acad. Sci. U.S.A. **34** (1948), 204–207.

Todd Cochrane
Department of Mathematics
Kansas State University
Manhattan KS 66506
cochrane@math.ksu.edu

A note on the fourth power moment of the Riemann zeta-function

J. B. Conrey[1]

Dedicated to Professor Heini Halberstam

Abstract. We give explicit formulae for all of the terms in the asymptotic expansion of the mean fourth power of the Riemann zeta-function on the critical line.

Heath-Brown has evaluated the 4th power moment of the Riemann zeta-function on the critical line via

$$\int_0^T |\zeta(1/2+it)|^4 \, dt = TP_4\left(\log \frac{T}{2\pi}\right) + O(T^{7/8+\epsilon})$$

where P_4 is a 4th degree polynomial. Ingham had previously discovered the leading coefficient of P_4; Heath-Brown gave an explicit formula for the second coefficient of P_4 and expressed the other coefficients as infinite sums of infinite integrals. In this note we find an explicit expression for P_4 in terms of special values of $\zeta(s)$. In particular, we show that the coefficients of P_4 are in the field

$$Q(\gamma_0, \gamma_1, \gamma_2, \gamma_3, \zeta(2), \zeta'(2), \zeta''(2))$$

where the γ's are the coefficients in the Laurent expansion

$$\zeta(s) = \frac{1}{s-1} + \gamma_0 + \gamma_1(s-1) + \dots$$

of $\zeta(s)$ at $s=1$.

Theorem. *With P_4 defined implicitly above, we have*

$$P_4(x) = g_1(x) + g_0(x)$$

where

$$g_1(x) = 2 \operatorname*{Res}_{s=0} \frac{x^s \zeta(s+1)^4}{s(s+1)\zeta(2s+2)}$$

[1]Research supported in part by a grant from the NSF

and

$$g_0(x) = \left(\frac{d}{ds}\right)^2 \frac{(xe^{2\gamma_0})^s(\frac{1}{2}\zeta(s+1)^2 - \frac{1}{s}\zeta(2s+1))}{(s+1)\zeta(s+2)}\Bigg|_{s=0}.$$

Proof. We refer to Heath-Brown's paper. There he shows that P_4 is naturally expressed as a sum of $g_1 + g_0$ where g_1 arises from the diagonal terms and is exactly as above. The other term, g_0, arises from the off diagonal terms and it is the explicit calculation of g_0 that is the purpose of this paper. Following Heath-Brown [H-B] (see pages 404, 406, and 407), we write

$$\sum_{n\le x} d(n)d(n+r) = m(x,r) + E(x,r)$$

where $m(x,r)/x$ is a polynomial in $\log x$ of degree 2 in x with coefficients depending on r. In fact,

$$m(x,r) = \operatorname*{Res}_{s=1} D(s,r)\frac{x^s}{s}$$

with

$$D(s,r) = \sum_{n=1}^{\infty} \frac{d(n)d(n+r)}{n^s}.$$

Then, as Heath-Brown shows,

$$g_0\left(\frac{T}{2\pi}\right) = \frac{2}{T}\sum_{r=1}^{\infty}\frac{1}{r}\int_0^{T/(2\pi)} m'(x,r)\sin(Tr/x)\,dx.$$

Lemma. *With the above notation,*

$$m(x,r) = x\sum_{s=1}^{\infty}\frac{\mu(s)}{s^2}\sum_{d|r}\frac{1}{d}\left(\left(\log\frac{x}{d^2s^2} + 2\gamma_0 - 1\right)^2 + 1\right)$$

and

$$m'(x,r) = \sum_{s=1}^{\infty}\frac{\mu(s)}{s^2}\sum_{d|r}\frac{1}{d}\left(\log\frac{x}{d^2s^2} + 2\gamma_0\right)^2$$

Proof. According to Heath-Brown, $m(x,r)$ may be calculated by considering the main terms in

$$2\sum_{q\le x^{1/2}} R(x,q,r) - \sum_{q\le x^{1/2}} R(qx^{1/2},q,r)$$

where

$$R(x,q,r)=\frac{x}{q^2}\sum_{d|(q,r)}\sum_{\delta|q/d} d\delta\mu(q/d\delta)\left(\log\frac{x\delta^2}{q^2}+2\gamma_0-1\right).$$

Writing $q=sd\delta$ we have

$$\sum_{q\le x^{1/2}} R(x,q,r)=x\sum_{d|r}\frac{1}{d}\sum_{s\le x^{1/2}/d}\frac{\mu(s)}{s^2}\left(\log\frac{x}{d^2s^2}+2\gamma_0-1\right)\sum_{\delta\le x^{1/2}/sd}\frac{1}{\delta}.$$

Using

$$\sum_{\delta\le y}\frac{1}{\delta}=\log y+\gamma_0+O(1/y)$$

(and we note

$$\sum_{\delta\le y}\frac{\log\delta}{\delta}=\frac{1}{2}\log^2 y-\gamma_1+O(\log y/y),$$

for future reference) we find that the above sum is

$$=\frac{x}{2}\sum_{d|r}\frac{1}{d}\sum_{s\le x^{1/2}/d}\frac{\mu(s)}{s^2}\left(\log\frac{x}{d^2s^2}+2\gamma_0-1\right)\left(\log\frac{x}{d^2s^2}+2\gamma_0\right)$$

$$+O\left(x^{1/2}\sum_{d|r}\sum_{s\le x^{1/2}}\frac{1}{s}\right).$$

Extending the first sum over s to ∞, we find that the above is

$$=\frac{x}{2}\sum_{s=1}^{\infty}\frac{\mu(s)}{s^2}\sum_{d|r}\frac{1}{d}\left(\log\frac{x}{d^2s^2}+2\gamma_0-1\right)\left(\log\frac{x}{d^2s^2}+2\gamma_0\right)+O(x^{1/2}\log x d(r)).$$

Similarly, using

$$\sum_{\delta\le y}1=y+O(1)$$

and

$$\sum_{\delta\le y}\log\delta=y\log y-y+O(\log y)$$

we find that the main part of

$$\sum_{q\le x^{1/2}} R(x^{1/2}q,q,r)$$

is given by

$$x\sum_{s=1}^{\infty}\frac{\mu(s)}{s^2}\sum_{d|r}\frac{1}{d}\left(\log\frac{x}{d^2s^2}+2\gamma_0-2\right).$$

Putting these two calculations together, we obtain the formula for $m(x,r)$. Differentiating with respect to x gives the rest of the lemma.

Now we compute

$$g_0 = g_0(T/(2\pi)) = \frac{2}{T}\sum_{r=1}^{\infty}\frac{1}{r}\int_0^{T/(2\pi)} m'(x,r)\sin(Tr/x)\,dx.$$

To begin with, we change variables, letting $y = Tr/x$. Then, substituting the expression for $m'(x,r)$ from the lemma gives

$$g_0 = 2\sum_{r=1}^{\infty}\int_{2\pi r}^{\infty}\sum_{s=1}^{\infty}\frac{\mu(s)}{s^2}\sum_{d|r}\frac{1}{d}\left(\log\frac{Tr}{s^2d^2y}+2\gamma_0\right)^2\frac{\sin y}{y^2}\,dy.$$

We bring the sum over s to the front, and interchange the summation over r and the integration, and replace d by n and write $r = mn$. Then we have

$$g_0 = 2\sum_{s=1}^{\infty}\frac{\mu(s)}{s^2}\int_{2\pi}^{\infty}\sum_{mn\le y/2\pi}\frac{1}{n}\left(\log\frac{Tm}{s^2ny}+2\gamma_0\right)^2\frac{\sin y}{y^2}\,dy.$$

Next, we interchange the sum over m and the integration and find that

$$g_0 = 2\sum_{s=1}^{\infty}\frac{\mu(s)}{s^2}\sum_{m=1}^{\infty}\int_{2\pi m}^{\infty}\sum_{n\le y/2\pi m}\frac{1}{n}\left(\log\frac{Tm}{s^2ny}+2\gamma_0\right)^2\frac{\sin y}{y^2}\,dy.$$

A change of variables in the integration, $u = y/2\pi m$, gives

$$g_0 = 2\sum_{s=1}^{\infty}\frac{\mu(s)}{s^2}\sum_{m=1}^{\infty}\int_1^{\infty}\sum_{n\le u}\frac{1}{n}\left(\log\frac{T/2\pi}{s^2un}+2\gamma_0\right)^2\frac{\sin 2\pi um}{2\pi um}\,\frac{du}{u}.$$

Now,

$$-\sum_{m=1}^{\infty}\frac{\sin 2\pi um}{\pi m} = u-[u]-1/2 = ((u)).$$

Thus, bringing the sum over m to the inside, we obtain

$$g_0 = -\sum_{s=1}^{\infty} \frac{\mu(s)}{s^2} \int_1^{\infty} \sum_{n \le u} \frac{1}{n} \left(\log \frac{T/2\pi}{s^2 un} + 2\gamma_0 \right)^2 ((u)) \frac{du}{u^2}.$$

Interchanging the sum over n and the integration, and changing s into m leads to

$$g_0 = -\sum_{m=1}^{\infty} \frac{\mu(m)}{m^2} \sum_{n=1}^{\infty} \int_n^{\infty} \frac{1}{n} \left(\log \frac{T/2\pi}{m^2 un} + 2\gamma_0 \right)^2 ((u)) \frac{du}{u^2}.$$

Now we express the logs via differentiation:

$$g_0 = -\left(\frac{d}{ds}\right)^2 \left(\sum_{m=1}^{\infty} \frac{\mu(m)}{m^{2+s}} \left(\frac{Te^{2\gamma_0}}{2\pi} \right)^s \sum_{n=1}^{\infty} \frac{1}{n^{1+s}} \int_n^{\infty} ((u)) \frac{du}{u^{2+s}} \right|_{s=0}.$$

It is well known that

$$-\int_n^{\infty} \frac{((u))}{u^{s+2}} du = \frac{1}{s+1} \left(\zeta(s+1) - \frac{1}{sn^s} - \sum_{m=1}^{n} \frac{1}{m^{s+1}} + \frac{1}{2n^{s+1}} \right).$$

Thus,

$$g_0 = \lim_{N \to \infty} -\left(\frac{d}{ds}\right)^2 \left(\left(\frac{Te^{2\gamma_0}}{2\pi} \right)^s \frac{1}{(s+1)\zeta(s+2)} \right.$$
$$\left. \sum_{n=1}^{N} \frac{1}{n^{1+s}} \left(\zeta(s+1) - \frac{1}{sn^s} - \sum_{m=1}^{n} \frac{1}{m^{s+1}} + \frac{1}{2n^{s+1}} \right) \right|_{s=0}.$$

Now

$$\sum_{n=1}^{N} \frac{1}{n^{s+1}} \sum_{m=1}^{n} \frac{1}{m^{s+1}} = \sum_{m=1}^{N} \frac{1}{m^{s+1}} \sum_{n=m}^{N} \frac{1}{n^{s+1}}$$

so that

$$2\sum_{n=1}^{N} \frac{1}{n^{s+1}} \sum_{m=1}^{n} \frac{1}{m^{s+1}} = \left(\sum_{n=1}^{N} \frac{1}{n^{s+1}} \right)^2 + \sum_{n=1}^{N} \frac{1}{n^{2s+2}}.$$

Thus, the sum over n in the above expression is

$$= \zeta(s+1) \sum_{n=1}^{N} \frac{1}{n^{s+1}} - \frac{1}{s} \sum_{n=1}^{N} \frac{1}{n^{2s+1}} - \frac{1}{2} \left(\sum_{n=1}^{N} \frac{1}{n^{s+1}} \right)^2.$$

If we assume that $s > 0$, we can let $N \to \infty$ here, getting

$$\frac{1}{2}\zeta(s+1)^2 - \frac{1}{s}\zeta(2s+1).$$

Inserting this into our prior expression for g_0, we obtain the Theorem.

References

[H-B] D.R. Heath-Brown, *The fourth power moment of the Riemann zeta function*, J. London Math. Soc (3) **38** (1979), 385-422.

J. B. Conrey
Department of Mathematics
Oklahoma State University,
Stillwater, OK 74078
conrey@math.okstate.edu

Effective estimates of exponential sums over primes

Hédi Daboussi

Dedicated to Professor Heini Halberstam

1. Introduction

The asymptotic behavior of the sum

$$S(x, \alpha) = \sum_{n \le x} \Lambda(n) e(n\alpha),$$

where α is real, $e(\alpha) = e^{2\pi i \alpha}$, and Λ is the von Mangoldt function, has been extensively studied by many authors. It plays a central role in Vinogradov's solution of the 3-primes conjecture [10]. It is also a main tool in the study of the equidistribution of the sequence $\{p\alpha, p \text{ prime}\}$ modulo 1.

Let h, q be integers; we set

$$\Psi(x, q, h) = \sum_{n \le x, n \equiv h(q)} \Lambda(n).$$

Clearly, if a and q are coprime integers, then

$$S\left(x, \frac{a}{q}\right) = \sum_{h=1}^{q} e\left(h\frac{a}{q}\right) \Psi(x, q, h).$$

By the prime number theorem for arithmetic progressions this yields an asymptotic formula for $S(x, a/q)$ when q is fixed.

When α is irrational, one can approximate α (via Dirichlet's theorem) by rational numbers a/q, but the denominators q depend on x.

Our knowledge of $\Psi(x, q, h)$ when q depends on x is rather poor.

- Siegel's class-number theorem [7] (see also [2] or [6]) yields the following: *Given abitrary constants $c_1 > 0$ and $c_2 > 0$, we have uniformly for $q < (\log x)^{c_1}$*

$$\Psi(x, q, h) = \frac{x}{\varphi(q)} + O\left(\frac{x}{(\log x)^{c_2}}\right).$$

However, the O-constant is ineffective (due to the possible existence of a Siegel zero).

- The only known effective estimate follows from Page's theorem [5]:
Let $\epsilon > 0, c_3 > 0, c_4 > 0$. Then

$$\Psi(x,q,h) = \frac{x}{\varphi(q)}\left(1 + O\left(\frac{1}{(\log x)^{c_3}}\right)\right) \tag{1}$$

holds for all $q < (\log x)^{2-\epsilon}$. Moreover, the asymptotic formula (1) *is valid for all $q < (\log x)^{c_4}$ except possibly those values of q which are multiples of some number q_0 satisfying $q_0 > (\log x)^{2-\epsilon}$.*
(Here and in the following, the numbers c_k denote positive constants and ϵ is an arbitrarily small number).
- A non-trivial upper bound for S was obtained by I. M. Vinogradov. He proved in [10] that if

$$\alpha = \frac{a}{q} + \frac{\beta}{q^2}, \quad |\beta| \le 1, \quad q \le x, \tag{2}$$

then

$$S(x,\alpha) \ll x(\log x)^{11/2}\left(\sqrt{\frac{1}{q} + \frac{q}{x}} + \exp\left(-\frac{1}{2}\sqrt{\log x}\right)\right).$$

This estimate is non-trivial for $q > (\log x)^{11}$.
- Recently, Vaughan [9] obtained better estimates by using a remarkable identity. He proved that if α satisfies (2) then

$$S(x,\alpha) \ll (\log x)^4\left(\frac{x}{\sqrt{\varphi(q)}} + x^{\frac{4}{5}} + \sqrt{xq}\right),$$

which is non-trivial for $q > (\log x)^8$.

In both Vinogradov's and Vaughan's methods the main idea is a connection between exponential sums over primes and bilinear forms of two types:

$$(\textit{Type I}) : \quad \sum_m \sum_n a(m)e(mn\alpha) \tag{3}$$

and

$$(\textit{Type II}) : \quad \sum_m \sum_n a(m)b(n)e(mn\alpha). \tag{4}$$

We quote the following from Vaughan [9] :

Suppose that the range for n in (3) is $[1,x]$. Then as a fairly general principle the estimates for (3) are good provided that $m < M$ with

M small compared with x, and those for (4) are good provided that $m < M, n < N$ with both M and N small compared with x.

Vinogradov's proof in [10] used the function

$$\sum_{d|n, d|P(\sqrt{x})} \mu(d), \text{ where } P(z) = \prod_{p \leq z} p \text{ and } n \leq x.$$

In the same monograph, he also considered the more general function

$$u_z(n) = \sum_{d|n, d|P(z)} \mu(d), \text{ where } z \text{ is a parameter,}$$

and applied some results from Brun's sieve.

Vaughan's method is much more elementary than Vinogradov's method in that it requires only a classical lemma (see Lemma 3 below). The coefficients of Vaughan's bilinear forms are estimated almost trivially (although in [3] a slight improvement is obtained by a more careful estimation of one of the coefficients).

The purpose of this paper is to improve Vaughan's estimate in essentially the range $q < \exp(c_5\sqrt{\log x})$. Our result is less elementary (again sieve estimates are required), and for larger values of q it is much weaker than Vaughan's result.

Our main tool is a very simple *multiplicative identity*, built (as in Vinogradov [10]) upon the function u_z and producing sums of type I and II, but with coefficients $a(m)$ and $b(n)$, having nice arithmetic properties.

Theorem 1. *Let $c \in \,]0, 1/2[$ and suppose that*

$$\alpha = \frac{a}{q} + \frac{\beta}{q^2}, \quad |\beta| \leq 1, \quad q \leq x.$$

Then

$$S(x, \alpha) \ll x(\log x)^{3/4}\sqrt{\log\log x}\sqrt{\frac{1}{q} + \frac{q\log q}{x}} + x\exp\left(-c\sqrt{\log x}\right).$$

This is trivial unless q is in the range

$$(\log x)^{(3/2)+\epsilon} \leq q \leq x/(\log x)^{(5/2)+\epsilon}. \tag{5}$$

This slight improvement in the power of the logarithm, when combined with Page's theorem, leads to an effective upper bound for $S(x, \alpha)$:

Corollary 2. *Let $\alpha \in \mathbf{R}$, $x \in \mathbf{R}$. There exists an effective constant $C(x,\alpha)$, such that*

$$\frac{1}{x}|S(x,\alpha)| \le C(x,\alpha),$$

and for irrational α

$$C(x,\alpha) \to 0 \text{ when } x \to \infty.$$

Proof. Let $\alpha \in \mathbf{R}$, $x \ge 1, Q = x/(\log x)^3$. By Dirichlet's theorem there exists a rational approximation a/q such that

$$\alpha = \frac{a}{q} + \frac{\gamma}{qQ}, (a,q) = 1, \quad |\gamma| \le 1, \quad q \le Q.$$

- If $q > (\log x)^{2-\epsilon}$, we apply Theorem 1 to obtain

$$\frac{1}{x}|S(x,\alpha)| \ll \frac{1}{(\log x)^{1/4-\epsilon}}.$$

- If $q \le (\log x)^{2-\epsilon}$, we apply Page's theorem (1) to obtain, with an arbitrary constant c_3,

$$S(x,\frac{a}{q}) = \frac{x}{\varphi(q)} \sum_{h=1,(h,q)=1}^{q} e\left(\frac{ha}{q}\right) + O\left(\frac{x}{(\log x)^{c_3}}\right),$$

noting that, when $(a,q)=1$,

$$\sum_{h=1,(h,q)=1}^{q} e\left(\frac{ha}{q}\right) = c_q(a) = \mu(q).$$

So for $q \le (\log x)^{2-\epsilon}$,

$$S\left(x,\frac{a}{q}\right) = x\frac{\mu(q)}{\varphi(q)} + O\left(\frac{x}{(\log x)^{c_3}}\right). \tag{6}$$

Set $\beta = \alpha - a/q$, so that $|\beta| \le 1/qQ$. Clearly

$$S(x,\alpha) = e(x\beta)S\left(x,\frac{a}{q}\right) - 2\pi i\beta \int_1^x e(u\beta)S\left(u,\frac{a}{q}\right)du.$$

Applying the relation (6) on the interval $[\sqrt{x}, x]$ and integrating by parts again, we see that

$$S(x,\alpha) = \frac{\mu(q)}{\varphi(q)} \sum_{n\le x} e(n\beta) + O\left(\frac{x}{(\log x)^{c_3}}\right) + O(\sqrt{x}) + O\left(|\beta| \frac{x^2}{(\log x)^{c_3}}\right).$$

The last error term is at most

$$\ll \frac{1}{q} \frac{x(\log x)^3}{(\log x)^{c_3}}.$$

We choose c_3 sufficiently large in order to get

$$|S(x,\alpha)| \le \frac{x}{\varphi(q)} + O\left(\frac{x}{(\log x)^2}\right)$$

and finally

$$\frac{1}{x}|S(x,\alpha)| \ll \frac{1}{\varphi(q)}.$$

Therefore we can choose the constant $C(x,\alpha)$ such that

$$C(x,\alpha) = O\left(\frac{1}{\varphi(q)} + \frac{1}{(\log x)^{1/4-\epsilon}}\right).$$

When α is irrational, the denominators q tend to infinity as $x \to \infty$, so we obtain $C(x,\alpha) \to 0$.

2. Notations

Given $z \ge 2$, we define the two completely multiplicative function v_z and u_z by

$$v_z(p) = \begin{cases} 1 & \text{if } p \le z \\ 0 & \text{if } p > z \end{cases}, \quad u_z(p) = \begin{cases} 0 & \text{if } p \le z \\ 1 & \text{if } p > z \end{cases}.$$

We let

$$\kappa(n) = \begin{cases} 0 & \text{if } n = 1 \\ 1 & \text{if } n \ge 2 \end{cases}$$

and set

$$||t|| = \min_{n\in\mathbf{Z}} |n-t|.$$

Finally, we denote by **1** the arithmetic function that is equal to one on every positive integer.

3. Some convolution identities

Clearly

$$v_z * u_z = \mathbf{1},$$

and by the Möbius inversion formula

$$u_z = v_z \mu * \mathbf{1}.$$

The von Mangoldt function Λ satisfies the identities

$$v_z \Lambda + u_z \Lambda = \Lambda, \quad \Lambda * \mathbf{1} = \log . \tag{7}$$

Multiplying the last identity by the completely multiplicative function u_z gives

$$u_z \Lambda * u_z = u_z \log$$

and finally

$$u_z \Lambda + u_z \Lambda * u_z \kappa = u_z \log, \tag{8}$$

which is our multiplicative identity.

A similar identity appears already in our paper [1], where we multiply the identity $\Lambda = \mu * \log$ by u_z, getting the less efficient identity

$$u_z \Lambda = u_z \mu \kappa * u_z \log + u_z \log .$$

4. Sums of type I and II

Applying (7), we see that

$$S(x, \alpha) = \sum_{n \leq x} \Lambda(n) e(n\alpha) = S_1 + S_2$$

where

$$S_1 = \sum_{n \leq x} v_z(n) \Lambda(n) e(n\alpha), \quad S_2 = \sum_{n \leq x} u_z(n) \Lambda(n) e(n\alpha).$$

Using the identity (8) we split S_2 into $B_1 - B_2$, where

$$B_1 = \sum_{n\le x} u_z(n)\log(n)e(n\alpha), \quad B_2 = \sum_{n\le x}\sum_{d|n} u_z\left(\frac{n}{d}\right)\Lambda\left(\frac{n}{d}\right)u_z(d)\kappa(d)e(n\alpha).$$

S_1 is estimated trivially by

$$S_1 \ll z\log x.$$

B_2 is a bilinear sum of type II. In fact, $u_z(\frac{n}{d})\Lambda(\frac{n}{d})$ is non-zero only if $\frac{n}{d}\ge z$ and $u_z(d)\kappa(d)$ is non-zero only if $d\ge z$, so the range of summation for d can be restricted to to $z\le d\le n/z$. Thus

$$B_2 = \sum_{z\le d\le \frac{x}{z}} u_z(d) \sum_{z\le m\le \frac{x}{d}} u_z(m)\Lambda(m)e(md\alpha).$$

5. Some lemmas

Lemma 3. *Let $U\ge 3, \alpha = a/q+\beta/q^2, |\beta|\le 1$. Then*

(9) $$\sum_{n\le U}\min(M, |\sin\pi n\alpha|^{-1}) \ll \frac{U}{q}M + M + U\log q + q\log q$$

and

(10) $$\sum_{n\le U}\min\left(\frac{t}{n}, |\sin\pi n\alpha|^{-1}\right) \ll \frac{t}{q}\log U + (U+q)\log q.$$

This lemma corresponds to Vinogradov's Lemmas 8a and 8b in [10].

The following lemmas are classical applications of Brun's sieve or Selberg's sieve (see [4] or [6]):

Lemma 4. *If $U\ge \log z$, then*

$$\sum_{n\le U} u_z(n) \ll \frac{U}{\log z}.$$

Lemma 5. *When $U\to\infty$, then uniformly in $|h|\le U, h\ne 0$ we have*

$$\sum_{m\le U}\Lambda(m)\Lambda(m+h) \ll U\frac{|h|}{\varphi(|h|)}.$$

6. The type I sum

We write

$$B_1(x,\alpha) = \sum_{n\le x} u_z(n)e(n\alpha)\int_1^n \frac{dt}{t} = \int_1^x \frac{1}{t}\sum_{t\le n\le x} u_z(n)e(n\alpha)dt.$$

Introducing

$$T_1(t,\alpha) = \sum_{n\le t} u_z(n)e(n\alpha),$$

we see that

$$B_1(x,\alpha) \ll \log x \sup_{t\le x}|T_1(t,\alpha)|. \tag{11}$$

Lemma 6. *Let* $z<y<x$, $\alpha = a/q+\beta/q^2$, $|\beta|\le 1$. *Then*

$$B_1(x,\alpha) \ll x\log x\log z\exp\left(-\frac{\log y}{\log z}\right) + (y+q)\log x\log q + \frac{x}{q}\log x\log y.$$

Proof. We suppose that $t>y$ since we have trivially

$$\sup_{t\le y}|T_1(t,\alpha)| \le \sup_{t\le y} t = y.$$

By the Möbius inversion formula

$$T_1(t,\alpha) = \sum_{n\le t}\sum_{nd\le t} v_z(n)\mu(n)e(nd\alpha).$$

Let

$$T_{1,1}(t,\alpha) = \sum_{n\le y}\sum_{nd\le t} v_z(n)\mu(n)e(nd\alpha),$$
$$T_{1,2}(t,\alpha) = \sum_{n>y}\sum_{nd\le t} v_z(n)\mu(n)e(nd\alpha).$$

Clearly

$$T_{1,1}(t,\alpha) \le \sum_{n\le y}\min\left(\frac{t}{n}, |\sin\pi n\alpha|^{-1}\right).$$

Applying the inequality (10) we get

$$T_{1,1}(t,\alpha) \ll \frac{t}{q}\log y + (y+q)\log q. \tag{12}$$

Next, for $t \leq x$ we have

$$T_{1,2}(t,\alpha) \leq \sum_{n>y} v_z(n)\mu^2(n) \sum_{d\leq \frac{t}{n}} 1 \leq t\sum_{n>y} \frac{v_z(n)}{n}\mu^2(n) \leq x\sum_{n>y}\frac{v_z(n)}{n}\mu^2(n).$$

By Rankin's trick (see, for instance [8, Ch. III.5]), we have for any $\eta > 0$

$$\sum_{n>y}\frac{v_z(n)}{n}\mu^2(n) \leq \sum_{n}\frac{v_z(n)}{n}\mu^2(n)\frac{n^\eta}{y^\eta}.$$

Let $\eta = 1/\log z$. Then

$$\sum_{n}\frac{v_z(n)n^\eta}{n}\mu^2(n) = \prod_{p\leq z}\left(1+\frac{p^\eta}{p}\right) \ll \exp\left(\sum_{p\leq z}\frac{1}{p} + O\left(\eta\sum_{p\leq z}\frac{\log p}{p}\right)\right),$$

so

$$\sum_{n}\frac{v_z(n)n^\eta}{n}\mu^2(n) \ll \log z.$$

Finally,

$$T_{1,2}(t,\alpha) \ll x\log z \exp\left(-\frac{\log y}{\log z}\right). \tag{13}$$

The lemma follows from inequalities (11), (12) and (13).

7. The type II sum

Lemma 7. *Suppose that $\alpha = a/q + \beta/q^2, |\beta| \leq 1$. Then*

$$B_2 \ll \frac{x\log x}{\sqrt{z}}\sqrt{\frac{\log x}{\log z}} + \frac{x\log x\sqrt{\log\log x}}{\sqrt{\log z}}\left(\sqrt{\frac{1}{q}+\frac{q\log q}{x}} + \sqrt{\frac{\log q}{z}}\right).$$

Proof. Before applying the Cauchy-Schwarz inequality, it is convenient to split up the sum B_2 into $O(\log x)$ sums over intervals of the type $[M, M']$, $M = 2^j z$, $j = 0, 1, 2\ldots$ and $M < M' \leq 2M$.

We define

$$T_2(M) = \sum_{M \le d < M'} u_z(d) \sum_{z \le m \le \frac{x}{d}} u_z(m)\Lambda(m)e(md\alpha),$$

so that

$$|B_2(x,\alpha)| \ll \log x \sup_{z \le M \le x/z} |T_2(M)|. \tag{14}$$

By the Cauchy-Schwarz inequality,

$$|T_2(M)|^2 \le \sum_{d \le 2M} u_z(d) \left(\sum_{M \le d \le 2M} \left| \sum_{z \le m \le \frac{x}{d}} u_z(m)\Lambda(m)e(dm\alpha) \right|^2 \right).$$

Applying Lemma 4, we see that

$$|T_2(M)|^2 \ll \frac{M}{\log z} \sum_{M \le d \le 2M} \sum_{z \le m \le \frac{x}{d}} \sum_{z \le m' \le \frac{x}{d}} u_z(m)\Lambda(m) \times$$
$$\times u_z(m')\Lambda(m')e(d(m-m')\alpha).$$

We sum first over d:

$$|T_2(M)|^2 \ll \frac{M}{\log z} \sum_{z \le m \le \frac{x}{M}} u_z(m)\Lambda(m) \sum_{z \le m' \le \frac{x}{M},} u_z(m')\Lambda(m') \times$$
$$\times \sum_{d \in I(m,m')} e(d(m-m')\alpha),$$

where $I(m,m')$ is an interval (depending on m, m') contained in $[M, 2M]$, thus satisfying $|I(m,m')| \le M$.

We write $m' = m + h$. Then

$$|T_2(M)|^2 \ll \frac{M}{\log z} \sum_{|h| \le \frac{x}{M}} \sum_{z \le m \le \frac{x}{M}} u_z(m)\Lambda(m) \times$$
$$\times u_z(m+h)\Lambda(m+h) \left| \sum_{d \in I(m,m+h)} e(dh\alpha) \right|.$$

Now $I(m, m+h)$ is an interval, so

$$\left| \sum_{d \in I(m,m+h)} e(dh\alpha) \right| \ll \min(|I(m,m+h)|, |\sin(\pi h\alpha)|^{-1}),$$
$$\ll \min(M, |\sin(\pi h\alpha)|^{-1}).$$

Therefore

$$|T_2(M)|^2 \ll \frac{M}{\log z} \sum_{|h|\le \frac{x}{M}} \sum_{z\le m\le \frac{x}{M}} u_z(m)\Lambda(m)\times$$
$$\times\, u_z(m+h)\Lambda(m+h)\min(M, |\sin(\pi h\alpha)|^{-1}).$$

We sum now first over m to obtain

$$|T_2(M)|^2 \ll \frac{M}{\log z} \sum_{|h|\le \frac{x}{M}} \min(M, |\sin(\pi h\alpha)|^{-1}) \sum_{z\le m\le \frac{x}{M}} \Lambda(m)\Lambda(m+h).$$

We distinguish between the values $h=0$ and $h\ne 0$, introducing

$$|T_{2,1}(M)|^2 = \frac{M^2}{\log z} \sum_{z\le m\le \frac{x}{M}} \Lambda(m)^2$$

and

$$|T_{2,2}(M)|^2 = \frac{M}{\log z} \sum_{|h|\le \frac{x}{M}, h\ne 0} \min(M, |\sin(\pi h\alpha)|^{-1}) \sum_{z\le m\le \frac{x}{M}} \Lambda(m)\Lambda(m+h),$$

so that

$$|T_2(M)|^2 \ll |T_{2,1}(M)|^2 + |T_{2,2}(M)|^2.$$

Clearly

$$|T_{2,1}(M)|^2 \ll xM\frac{\log x}{\log z},$$

and by Lemma 5

$$|T_{2,2}(M)|^2 \ll \frac{M}{\log z}\frac{x}{M} \sum_{|h|\le \frac{x}{M}, h\ne 0} \frac{|h|}{\varphi(|h|)} \min(M, |\sin(\pi h\alpha)|^{-1}).$$

It is well known that if $0<h\le x$,

$$\frac{h}{\varphi(h)} \ll \log\log x,$$

so

$$|T_{2,2}(M)|^2 \ll \frac{x}{\log z}\log\log x \sum_{0<h\le \frac{x}{M}} \min(M, |\sin(\pi h\alpha)|^{-1}).$$

By (9) it follows that

$$|T_{2,2}(M)|^2 \ll \frac{x}{\log z} \log\log x \left(\frac{x}{q} + M + \frac{x \log q}{M} + q \log q\right)$$

or

$$|T_{2,2}(M)|^2 \ll \frac{x^2}{\log z} \log\log x \left(\frac{1}{q} + \frac{M}{x} + \frac{\log q}{M} + \frac{q \log q}{x}\right).$$

Since $z < M < x/z$, we obtain

$$|T_{2,1}(M)|^2 \ll \frac{x^2}{z} \frac{\log x}{\log z}$$

and

$$|T_{2,2}(M)|^2 \ll \frac{x^2}{\log z} \log\log x \left(\frac{1}{q} + \frac{\log q}{z} + \frac{q \log q}{x}\right).$$

Thus

$$|T_2(M)| \ll \frac{x}{\sqrt{z}} \sqrt{\frac{\log x}{\log z}} + \frac{x\sqrt{\log\log x}}{\sqrt{\log z}} \left(\sqrt{\frac{1}{q} + \frac{q \log q}{x}} + \sqrt{\frac{\log q}{z}}\right).$$

This combined with (14) gives

$$B_2 \ll \frac{x \log x}{\sqrt{z}} \sqrt{\frac{\log x}{\log z}} + \frac{x \log x}{\sqrt{\log z}} \sqrt{\log\log x} \left(\sqrt{\frac{1}{q} + \frac{q \log q}{x}} + \sqrt{\frac{\log q}{z}}\right)$$

and completes the proof of the lemma.

8. Proof of Theorem 1

We can suppose that q satisfies (5) for otherwise the result is trivial.

We choose $z = \exp(\sqrt{\log x})$. Then, for any $c \in]0, 1/2[$,

$$\frac{x}{\sqrt{z}} \log x \sqrt{\frac{\log x}{\log z}} + \frac{x \log x}{\sqrt{\log z}} \sqrt{\log\log x} \sqrt{\frac{\log q}{z}} \ll x \exp\left(-c\sqrt{\log x}\right).$$

Thus

$$B_2 \ll x(\log x)^{3/4} \sqrt{\log\log x} \sqrt{\frac{1}{q} + \frac{q \log q}{x}} + x \exp -c\sqrt{\log x},$$

which is admissible.

• If $q \leq \exp(\log x)^\epsilon$, we choose $\log y = \sqrt{\log x}\log q = \log z \log q$. Then we have, using (5),

$$B_1(x,\alpha) \ll \frac{x}{q}(\log x)^{3/2} + \left(\exp\left(\sqrt{\log x}\log q\right)\right)\log^2 x + \frac{x}{q}(\log x)^{3/2}\log q$$
$$\ll \frac{x}{q}(\log x)^{3/2}\log q \ll x(\log x)^{3/4}\sqrt{\frac{1}{q}}.$$

• For larger values of q, we choose $y = x\exp\left(-\sqrt{\log x}\right)$. Then

$$B_1(x,\alpha) \ll x(\log x)^2\exp\left(-\sqrt{\log x}\right) + q\log q\log x + \frac{x}{q}\log^2 x,$$

so for any c satisfying $0 < c < 1/2$ we obtain

$$B_1(x,\alpha) \ll x\exp\left(-c\sqrt{\log x}\right) + q\log q\log x + \frac{x}{q}\log^2 x.$$

The fact that q is large gives

$$\frac{x}{q}\log^2 x \ll x(\log x)^{3/4}\sqrt{\frac{1}{q}},$$

and our assumption(5) gives

$$q\log q\log x \ll x(\log x)^{3/4}\sqrt{\frac{q\log q}{x}}.$$

Hence $B_1(x,\alpha)$ is admissible, and the proof of the theorem is complete.

Acknowledgement. I warmly thank P. Elliott for many discussions on this subject and, in particular, for drawing my attention to the role of Page's theorem in the work of Vinogradov.

References

[1] H. Daboussi, *On a convolution method*, Congreso de teoria de los numeros, Universidad del Pais Vasco (1984), 110–138.

[2] H. Davenport, *Multiplicative Number Theory,* Springer Verlag, New York 1980.

[3] F. Dress, H. Iwaniec, G. Tenenbaum *Sur une somme liée à la fonction de Möbius*, J. Reine Angew. Math **340** (1983), 53–58.

[4] H. Halberstam and H. -E. Richert, *Sieve Methods*, Academic Press, London 1974.

[5] A. Page, *On the number of primes in an arithmetic progression,* Proc. London Math. Soc. **39** (2) (1935), 116–141.

[6] K. Prachar, *Primzahlverteilung*, Springer-Verlag, Berlin 1957.

[7] C. L. Siegel, *Über die Klassenzahl quadratischer Körper*, Acta Arithmetica **1** (1936), 83–86.

[8] G. Tenenbaum, *Introduction à la théorie analytique et probabiliste des nombres,* Revue de l'Institut Elie Cartan **13** (1990), Université de Nancy I.

[9] R. Vaughan, *Sommes trigonométriques sur les nombres premiers,* C. R. Acad. Sci. Paris, Sér A **285** (1977), 981–983.

[10] I. M. Vinogradov, *The method of trigonometrical sums in the theory of numbers*, Interscience Publ., London (1954).

Hédi Daboussi
Département de Mathématiques
Bâtiment 425
Université Paris Sud
91405 Orsay
FRANCE
daboussi@matups.math.u-psud.fr

On products of multiplicative functions of absolute value at most 1 which are composed with linear functions

Hubert Delange

Dedicated to my friend Heini Halberstam

1. Introduction

Let $\Phi(n) = \prod_{j=1}^{q} f_j(a_j n + b_j)$, where $a_1, \dots, a_q$, $b_1, \dots, b_q$ are integers with $a_j \geq 1$ and $a_j + b_j \geq 1$ (so that $a_j n + b_j$ is a positive integer for every $n \geq 1$), and $f_1, \dots, f_q$ are multiplicative functions of absolute value ≤ 1. We are interested here in the behavior of $\frac{1}{x} \sum_{n \leq x} \Phi(n)$ when x tends to infinity.

To avoid complications we suppose that $b_j/a_j \neq b_k/a_k$ for $j \neq k$. We will give an asymptotic formula with an error term $o(1)$ which holds under a simple hypothesis.

We will first state the result without proof in the general case and derive some interesting consequences. Afterwards, for simplicity, we will give the proof in the case when $q = 2$; there is no essential difference in the general case.

The letters m, n, q always denote positive integers. The letter p denotes primes. $V_p(m)$ is the exponent of the prime p in the factorization of m.

2.

2.1. It is convenient to recall some facts concerning complex–valued functions of several positive integers, for which we refer to our paper [2].

In the set of complex–valued functions of q positive integers, the convolution is defined by

$$(f * g)(n_1, \dots, n_q) = \sum_{\substack{d_1 | n_1 \\ \cdots \\ d_q | n_q}} f(d_1, \dots, d_q) g\left(\frac{n_1}{d_1}, \dots, \frac{n_q}{d_q}\right).$$

It is commutative, associative and distributive with respect to addition.

The function f is said to be multiplicative if

(a) $$f(1,\dots,1)=1,$$

(b) $$f(n_1'n_1'',\dots,n_q'n_q'')=f(n_1',\dots,n_q')f(n_1'',\dots,n_q'')$$
$$\text{when } (n_1'\dots n_q',\, n_1''\dots n_q'')=1.$$

Such a function is completely determined by the values of $f(p^{r_1},\dots,p^{r_q})$ for all primes p and all systems $(r_1,\dots,r_q)$ of non-negative integers not all zero, and we have

$$f(n_1,\dots,n_q)=\prod_{p|n_1\dots n_q} f(p^{V_p(n_1)},\dots,p^{V_p(n_q)}).$$

If f and g are multiplicative, then $f*g$ is multiplicative.

2.2. We also recall the following simple lemma, the proof of which may be found in [3, p. 108–109].

Lemma 1. *Let $u_1, u_2,\dots,u_n,\dots$ and $v_1,v_2,\dots,v_n,\dots$ be complex-valued functions whose domain contains a set E. Suppose that, for every $t\in E$,*

$$|u_n(t)|\le U_n \quad \textit{and} \quad |u_n(t)-v_n(t)|\le V_n$$

where $\Sigma U_n^2<\infty$ and $\Sigma V_n<\infty$. Then the infinite product $\prod_{n=1}^{\infty}(1+u_n(t))e^{-v_n(t)}$ converges uniformly on E.

3.

We now turn to the statement of our results.

3.1. We fix integers $a_1,\dots,a_q,\ b_1,\dots,b_q$ with $a_j\ge 1$, $a_j+b_j\ge 1$ and $b_j/a_j\ne b_k/a_k$ if $j\ne k$ and set

$$N=a_1\dots a_q\prod_{1\le j<k\le q}(a_jb_k-a_kb_j),$$

so that $N\ne 0$.

Given positive integers $m_1,\dots,m_q$, the set of positive integers n such that $a_jn+b_j\equiv 0 \pmod{m_j}$ for each j has a density $\delta(m_1,\dots,m_q)$. We have

(1) $$\delta(m_1,\dots,m_q)\le \min_j \frac{a_j}{m_j},$$

because the number of positive integers $n\le x$ for which $m_j|a_jn+b_j$ is obviously $\le (a_jx+b_j)/m_j$.

Obviously $\delta(1,\dots,1)=1$. Although we do not need this now, we mention that δ is a multiplicative function of $m_1,\dots,m_q$.

Note that, for any given non-negative integers $r_1,\dots,r_q$, the set of positive integers n for which $V_p(a_jn+b_j)=r_j$ for each j has a density $D_p(r_1,\dots,r_q)$ because "$V_p(m)=r$" is equivalent to "$p^r|m$ and $p^{r+1}\nmid m$".

3.2. Now consider q multiplicative functions $f_1,\dots,f_q$ of absolute value ≤ 1. We will denote by f_j^* the function $f_j*\mu$, where μ is the Möbius function. The function f_j^* is multiplicative and for each prime p and each $r\ge 1$ we have $f_j^*(p^r)=f_j(p^r)-f_j(p^{r-1})$. Set

$$S_p(f_1,\dots,f_q)=\sum_{r_1,\dots,r_q\ge 0}\delta(p^{r_1},\dots,p^{r_q})f_1^*(p^{r_1})\cdots f_q^*(p^{r_q}).$$

This multiple series is absolutely convergent because (1) gives $\delta(p^{r_1},\dots,p^{r_q})\le \min_j a_j/p^{r_j}$ and we have $|f_j^*(p^r)|\le 2$. One can prove that $|S_p(f_1,\dots,f_q)|\le 1$.

Let $\Phi(n)=\prod_{j=1}^{q} f_j(a_jn+b_j)$. Our main result is the following

Theorem. *If*

$$\sum\frac{q-Re\,(f_1(p)+\cdots f_q(p))}{p}<\infty,$$

then, as x tends to infinity,

$$\frac{1}{x}\sum_{n\le x}\Phi(n)=\prod_{p\le x}S_p(f_1,\dots,f_q)+o(1). \tag{2}$$

We see that, when $p\nmid N$, we have

$$S_p(f_1,\dots,f_q)=1+\sum_{r=1}^{\infty}\frac{f_1^*(p^r)+\cdots+f_q^*(p^r)}{p^r}. \tag{3}$$

In fact, p cannot divide more than one of the numbers a_jn+b_j because if p divides a_jn+b_j and a_kn+b_k with $j\ne k$, then it divides $a_jb_k-a_kb_j$ and therefore it divides N. Therefore $\delta(p^{r_1},\dots,p^{r_q})=0$ if $r_j>0$ for more than one j. If $r_j>0$ and $r_k=0$ for $k\ne j$, then $\delta(p^{r_1},\dots,p^{r_q})=1/p^{r_j}$.

Note that by formula (3) it is obvious that $|S_p(f_1,\dots,f_q)|\le 1$ in the case when $p\nmid N$, for

$$1+\sum_{r=1}^{\infty}\frac{f_1^*(p^r)+\cdots+f_q^*(p^r)}{p^r}=1-\frac{q}{p}+\left(1-\frac{1}{p}\right)\sum_{r=1}^{\infty}\frac{f_1(p^r)+\cdots+f_q(p^r)}{p^r}.$$

The important point is that (2) may be written as

$$(4)\qquad \frac{1}{x}\sum_{n\le x}\Phi(n)=\Big(\prod_{p|N}S_p(f_1,\dots,f_q)\Big)\prod_{\substack{p\le x\\ p\nmid N}}(1+u_p)+o(1),$$

where

$$u_p=\sum_{r=1}^{\infty}\frac{f_1^*(p^r)+\cdots+f_q^*(p^r)}{p^r}.$$

We have

$$|u_p|\le 2q\sum_{r=1}^{\infty}\frac{1}{p^r}=\frac{2q}{p-1}.$$

Moreover, setting

$$v_p=\frac{f_1^*(p)+\cdots+f_q^*(p)}{p}=\frac{f_1(p)+\cdots+f_q(p)-q}{p},$$

we have

$$|u_p-v_p|=\left|\sum_{r=2}^{\infty}\frac{f_1^*(p^r)+\cdots+f_q^*(p^r)}{p^r}\right|\le\frac{2q}{p(p-1)}.$$

Since $\sum\left(\frac{2q}{p-1}\right)^2<\infty$ and $\sum\frac{2q}{p(p-1)}<\infty$, Lemma 1 shows that the infinite product $\prod_{p\nmid N}(1+u_p)e^{-v_p}$ converges uniformly in $f_1,\dots,f_q$. The same is true for the product $\prod S_p(f_1,\dots,f_q)e^{-v_p}$ because the factors corresponding to the primes p which divide N are bounded.

Let P be the value of the latter product. As x tends to infinity, we have

$$\prod_{p\le x}S_p(f_1,\dots,f_q)e^{-v_p}=P+o(1)$$

and therefore

$$(5)\quad \prod_{p\le x}S_p(f_1,\dots,f_q)=(P+o(1))\exp\left(\sum_{p\le x}v_p\right)=P\exp\left(\sum_{p\le x}v_p\right)+o(1),$$

because $\operatorname{Re}\left(\sum_{p\le x}v_p\right)\le 0$.

3.3. A first consequence is that, if the series $\sum \frac{q-(f_1(p)+\cdots+f_q(p))}{p}$ converges, then so does the infinite product $\prod S_p(f_1, \dots, f_q)$. By (2) this implies that $\frac{1}{x}\sum_{n\le x}\Phi(n)$ tends to the value of this product when x tends to infinity. This holds in particular if each of the series $\sum \frac{1-f_j(p)}{p}$ converges (and thus, in particular, if each f_j has a non zero mean value).

Note that, if the functions f_j depend upon a parameter t and if the series $\sum \frac{q-(f_1(p)+\cdots+f_q(p))}{p}$ converges uniformly on a set E of values of t, then the infinite product $\prod S_p(f_1, \dots, f_q)$ converges uniformly on E. This observation leads to the following result:

If $F_1, \dots, F_q$ are real-valued additive functions having limit distributions, then the function Ψ defined by

$$\Psi(n) = \sum_{j=1}^{q} F_j(a_j n + b_j)$$

has a limit distribution.

To obtain this result, we take $f_j(m) = \exp(itF_j(m))$. Then each of the series $\sum \frac{1-f_j(p)}{p}$ converges uniformly in t on every bounded interval (see the proof of the theorem of Erdös–Wintner that we gave in [1] (§ 6.3.1 and § 6.3.2)). It follows that, when x tends to infinity, $\frac{1}{x}\sum_{n\le x}\exp(it\Psi(n))$ converges to $\prod S_p(f_1, \dots, f_q)$, which is a continuous function of t. Now, $\frac{1}{x}\sum_{n\le x}\exp(it\Psi(n))$ is the characteristic function of the distribution of the numbers $\Psi(n)$, $n \le x$. Hence this distribution converges to a limit distribution whose characteristic function is $\prod S_p(f_1, \dots, f_q)$. We note that it is the limit distribution of the sum of a sequence of discrete random variables. A well known theorem of Paul Lévy shows that *it is continuous if and only if at least one of the limit distributions of the functions F_j is continuous.*

3.4. Suppose only that we have

$$\sum \frac{q - \operatorname{Re}(f_1(p) + \cdots + f_q(p))}{p} < \infty.$$

As x tends to infinity,

$$\sum_{p\le x} v_p = -\sum \frac{q - \operatorname{Re}(f_1(p) + \cdots + f_q(p))}{p} + o(1)$$
$$+ i\sum_{p\le x} \frac{\operatorname{Im}(f_1(p) + \cdots + f_q(p))}{p}.$$

Thus, (2) and (5) give

$$\frac{1}{x}\sum_{n\le x}\Phi(n)=C\exp(iA(x))+o(1), \tag{6}$$

where

$$C=P\exp\left(-\sum\frac{q-\operatorname{Re}(f_1(p)+\cdots+f_q(p))}{p}\right),$$
$$A(x)=\sum_{p\le x}\frac{\operatorname{Im}(f_1(p)+\cdots+f_q(p))}{p}.$$

We now show that

$$\sup_{x<t\le x^2}|A(t)-A(x)|=o(1) \tag{7}$$

as x tends to infinity.

Proof. For $t>x$

$$A(t)-A(x)=\sum_{x<p\le t}\frac{\operatorname{Im}(f_1(p)+\cdots+f_q(p))}{p}.$$

By Cauchy's inequality

$$|A(t)-A(x)|^2\le\left(\sum_{x<p\le t}\frac{1}{p}\right)\left(\sum_{x<p\le t}\frac{(\operatorname{Im}(f_1(p)+\cdots+f_q(p)))^2}{p}\right)$$
$$\le\left(\sum_{x<p\le x^2}\frac{1}{p}\right)\left(\sum_{x<p\le x^2}\frac{(\operatorname{Im}(f_1(p)+\cdots+f_q(p)))^2}{p}\right)$$

if $t\le x^2$. We know that

$$\sum_{x<p\le x^2}\frac{1}{p}=O(1).$$

Now

$$(\operatorname{Im}(f_1(p)+\cdots+f_q(p)))^2\le q((\operatorname{Im}f_1(p))^2+\cdots+(\operatorname{Im}f_q(p))^2)$$

and since

$$(\operatorname{Im}f_j(p))^2\le 1-(\operatorname{Re}(f_j(p)))^2\le 2(1-\operatorname{Re}(f_j(p))),$$

we have

$$(\operatorname{Im}(f_1(p)+\cdots+f_q(p)))^2 \leq 2q(q-\operatorname{Re}(f_1(p)+\cdots+f_q(p))).$$

Therefore

$$\sum_{x<p\leq x^2} \frac{(\operatorname{Im}(f_1(p)+\cdots+f_q(p)))^2}{p} \leq 2q \sum_{x<p\leq x^2} \frac{q-\operatorname{Re}(f_1(p)+\cdots+f_q(p))}{p},$$

which is $o(1)$ when x tends to infinity.

Note that (7) shows that $\exp(iA(x))$ is a slowly varying function of $\log x$, of absolute value 1.

From this we deduce the following result:

Suppose that, for each j, there exists a real number u_j such that

$$\sum \frac{1-Re(f_j(p)p^{-iu_j})}{p} < \infty .$$

Then, as x tends to infinity,

$$\frac{1}{x}\sum_{n\leq x} \Phi(n) = C\,x^{i\alpha}L(\log x) + o(1)\,, \tag{8}$$

where C is a constant, $\alpha = u_1 + \cdots + u_q$, and L is a slowly varying function of absolute value 1. More precisely, $L(\log x) = \exp(iA(x))$, where

$$A(x) = \sum_{p\leq x} \frac{1}{p} \operatorname{Im}(f_1(p)p^{-iu_1} + \cdots + f_q(p)p^{-iu_q}).$$

Proof. Let $g_j(m) = f_j(m)m^{-iu_j}$. Applying the preceding result with g_j in place of f_j we obtain

$$\frac{1}{x}\sum_{n\leq x} \frac{\Phi(n)}{(a_1n+b_1)^{iu_1}\cdots(a_qn+b_q)^{iu_q}} = C_0\,\exp(iA(x)) + o(1) \tag{9}$$

where C_0 is a constant and

$$A(x) = \sum_{p\leq x} \frac{1}{p} \operatorname{Im}(f_1(p)p^{-iu_1} + \cdots + f_q(p)p^{-iu_q}),$$

which satisfies (7). As n tends to infinity, we have

$$\frac{\Phi(n)}{(a_1 n + b_1)^{iu_1} \cdots (a_q n + b_q)^{iu_q}} = \frac{1}{a_1^{iu_1} \cdots a_q^{iu_q}} \cdot \frac{\Phi(n)}{n^{i\alpha}}(1 + o(1))$$

$$= \frac{1}{a_1^{iu_1} \cdots a_q^{iu_q}} \cdot \frac{\Phi(n)}{n^{i\alpha}} + o(1),$$

where $\alpha = u_1 + \cdots + u_q$. Hence, the left–hand side of (9) is

$$\frac{1}{a_1^{iu_1} \cdots a_q^{iu_q}} \cdot \frac{1}{x} \sum_{n \le x} \frac{\Phi(n)}{n^{i\alpha}} + o(1)$$

as x tends to infinity. Thus,

$$\frac{1}{x} \sum_{n \le x} \frac{\Phi(n)}{n^{i\alpha}} = C' \exp(iA(x)) + o(1) \tag{10}$$

where $C' = C_0 a_1^{iu_1} \cdots a_q^{iu_q}$.

Starting from the relation

$$\sum_{n \le x} \Phi(n) = \int_{1/2}^{x} t^{i\alpha} d \left(\sum_{n \le t} \frac{\Phi(n)}{n^{i\alpha}} \right)$$

$$= x^{i\alpha} \sum_{n \le x} \frac{\Phi(n)}{n^{i\alpha}} - i\alpha \int_{1}^{x} t^{i\alpha - 1} \left(\sum_{n \le t} \frac{\Phi(n)}{n^{i\alpha}} \right) dt$$

and taking account of (10) and (7), we obtain

$$\sum_{n \le x} \Phi(n) = C' \frac{i\alpha}{1 + i\alpha} x^{1 + i\alpha} \exp(iA(x)) + o(x),$$

which is equivalent to (8) with $C = \frac{i\alpha}{1+i\alpha} C'$.

Remark. If we add the hypothesis that, for each j, $(a_j, b_j) = 1$, then we have the following result :

Suppose that, for each j, there exists a character $\chi_j \mod a_j$ and a real number u_j such that

$$\sum \frac{1 - Re\,(f_j(p)\chi_j(p)p^{-iu_j})}{p} < \infty.$$

Then formula (8) *holds with*

$$L(\log x) = \exp(iA(x)),$$
$$A(x) = \sum_{p\le x} \frac{1}{p} \operatorname{Im}\left(f_1(p)\chi_1(p)p^{-iu_1} + \cdots + f_q(p)\chi_q(p)p^{-iu_q}\right).$$

This is obtained by applying the previous result with $f_j\chi_j$ in place of f_j.

4.

For the proof of the theorem we require the following very simple lemmas.

Lemma 2. *Let* $u_1, \ldots, u_q$, $v_1, \ldots, v_q$ *be complex numbers of absolute value* ≤ 1. *Then* $|u_1 \cdots u_q - v_1 \cdots v_q| \le \sum_{j=1}^{q} |u_j - v_j|$.

Proof. For $q = 2$ we have

$$|u_1u_2 - v_1v_2| = |u_2(u_1 - v_1) + v_1(u_2 - v_2)| \le |u_1 - v_1| + |u_2 - v_2|.$$

The general case follows by induction on q.

Lemma 3. *If* $|z| \le 1$, *then* $|z-1|^2 \le 2(1 - \operatorname{Re} z)$.

Proof. We have

$$|z-1|^2 = (z-1)(\overline{z}-1)) = z\overline{z} - (z + \overline{z}) + 1 = |z|^2 - 2\operatorname{Re} z + 1$$

Lemma 4. *If* $\operatorname{Re} z \le M$, *where* $M > 0$, *then* $|e^z - 1| \le \frac{e^M - 1}{M}|z|$.

Proof. We have $e^z - 1 = \int_0^1 ze^{zt}dt$ and $|e^{zt}| \le e^{Mt}$ for $0 \le t \le 1$.

Lemma 5. *If* $|z| \le 1$, *then* $|e^{z-1} - z| \le 1 - \operatorname{Re} z$.

Proof. Let $\zeta = z - 1$. By a form of Taylor's formula,

$$e^\zeta = 1 + \zeta + \zeta^2 \int_0^1 (1-t)e^{t\zeta}dt.$$

As $\operatorname{Re}\zeta \le 0$, we have $|(1-t)e^{t\zeta}| \le 1 - t$ for $0 \le t \le 1$. Thus,

$$\left|\int_0^1 (1-t)e^{t\zeta}dt\right| \le \int_0^1 (1-t)dt = \frac{1}{2}$$

and $|e^\zeta - 1 - \zeta| \le \frac{1}{2}|\zeta|^2$, that is $|e^{z-1} - z| \le \frac{1}{2}|z-1|^2$, which is $\le 1 - \operatorname{Re} z$ by Lemma 3.

5. Proof of the theorem in the case $q = 2$

5.1. Denote by $\mathcal{S}(m_1, m_2)$ the system of congruences

$$\begin{cases} a_1 n + b_1 \equiv 0 \pmod{m_1}, \\ a_2 n + b_2 \equiv 0 \pmod{m_2}). \end{cases}$$

It is an elementary fact that either there is no solution to this system or the system is equivalent to a single congruence $n \equiv n_0(\mod m)$ where m is a divisor of the product $m_1 m_2$. In any case, the set of positive solutions has a density, $\delta(m_1, m_2)$ say (namely, 0 or $\frac{1}{m}$). Moreover we have

$$\#\{n \leq x : n \text{ satisfies } \mathcal{S}(m_1, m_2)\} = \delta(m_1, m_2)x + R_x(m_1, m_2) \tag{11}$$

with $|R_x(m_1, m_2)| \leq 1$. As we have seen in general in § 3.1, we have $\delta(m_1, m_2) \leq \min(a_1/m_1, a_2/m_2)$.

Next, we show that δ is a multiplicative function of m_1 and m_2.

Obviously $\delta(1, 1) = 1$. Moreover we will see that, if $(m_1' m_2', m_1'' m_2'') = 1$, then $\delta(m_1' m_1'', m_2' m_2'') = \delta(m_1', m_2') \cdot \delta(m_1'', m_2'')$. Clearly, $\mathcal{S}(m_1' m_1'', m_2' m_2'')$ is equivalent to "$\mathcal{S}(m_1', m_2')$ and $\mathcal{S}(m_1'', m_2'')$" because $(m_1', m_1'') = (m_2', m_2'') = 1$.

If $\mathcal{S}(m_1, m_2')$ or $\mathcal{S}(m_1'', m_2'')$ has no solution, then $\delta(m_1', m_2') = 0$ or $\delta(m_1'', m_2'') = 0$, and $\delta(m_1' m_1'', m_2' m_2'') = 0$ because $\mathcal{S}(m_1' m_1'', m_2' m_2'')$ has no solution. Otherwise, $\mathcal{S}(m_1', m_2')$ is equivalent to a single congruence $n \equiv n_0'$ (mod m') where $m' | m_1' m_2'$, and $\mathcal{S}(m_1'', m_2'')$ is equivalent to a single congruence $n \equiv n_0''$ (mod m'') where $m'' | m_1'' m_2''$. Since $(m', m'') = 1$, $\mathcal{S}(m_1' m_1'', m_2' m_2'')$ is equivalent to a congruence modulo $m' m''$. Thus, $\delta(m_1', m_2') = \frac{1}{m'}$, $\delta(m_1'', m_2'') = \frac{1}{m''}$ and $\delta(m_1' m_1'', m_2' m_2'') = \frac{1}{m' m''}$.

5.2. Since $V_p(m) = r$ is equivalent to "$p^r | m$ and $p^{r+1} \nmid m$", it follows that the set of positive integers n for which $V_p(a_1 n + b_1) = r$ and $V_p(a_2 n + b_2) = s$ where $r, s \geq 0$, has density

$$D_p(r, s) = \delta(p^r, p^s) + \delta(p^{r+1}, p^{s+1}) - \delta(p^{r+1}, p^s) - \delta(p^r, p^{s+1}).$$

Obviously $\sum_{r,s \geq 0} D_p(r, s) \leq 1$. Actually, we have $\sum_{r,s \geq 0} D_p(r, s) = 1$. To see this, we note that $1 - \sum_{\substack{0 \leq r \leq k \\ 0 \leq s \leq k}} D_p(r, s)$ is the density of the set of positive integers n such that p^{k+1} divides $a_1 n + b_1$ or $a_2 n + b_2$. The number of elements $\leq x$ of this set is obviously $\leq \frac{(a_1 + a_2)x + b_1 + b_2}{p^{k+1}}$, so that its density is $\leq \frac{a_1 + a_2}{p^{k+1}}$, which tends to zero as k tends to infinity.

5.3. We now show that

$$S_p(f_1, f_2) = \sum_{\substack{r \geq 0 \\ s \geq 0}} D_p(r, s) f_1(p^r) f_2(p^s),$$

which implies $|S_p(f_1, f_2)| \le 1$.

Proof. Setting $f_1(p^{-1}) = f_2(p^{-1}) = 0$, we have for all $r \ge 0$

$$f_j^*(p^r) = f_j(p^r) - f_j(p^{r-1}) \qquad (j = 1 \text{ or } 2).$$

Thus, for $r, s \ge 0$,

$$\begin{aligned} f_1^*(p^r) f_2^*(p^s) &= f_1(p^r) f_2(p^s) + f_1(p^{r-1}) f_2(p^{s-1}) \\ &\qquad - f_1(p^{r-1}) f_2(p^s) - f_1(p^r) f_2(p^{s-1}) \\ &= \sum_{\varepsilon_1, \varepsilon_2 = 0 \text{ or } 1} (-1)^{\varepsilon_1 + \varepsilon_2} f_1(p^{r-\varepsilon_1}) f_2(p^{s-\varepsilon_2}). \end{aligned}$$

It follows that

$$S_p(f_1, f_2) = \sum_{r,s \ge 0} \left(\sum_{\varepsilon_1, \varepsilon_2 = 0 \text{ or } 1} (-1)^{\varepsilon_1 + \varepsilon_2} \delta(p^r, p^s) f_1(p^{r-\varepsilon_1}) f_2(p^{s-\varepsilon_2}) \right).$$

Now, (1) of § 3.1 gives $\delta(p^r, p^s) \le \min(a_1/p^r, a_2/p^s)$. It follows that for each pair $(\varepsilon_1, \varepsilon_2)$ the double series

$$\sum_{r,s \ge 0} (-1)^{\varepsilon_1 + \varepsilon_2} \delta(p^r, p^s) f_1(p^{r-\varepsilon_1}) f_2(p^{s-\varepsilon_2})$$

is absolutely convergent. Therefore we have

$$S_p(f_1, f_2) = \sum_{\varepsilon_1, \varepsilon_2 = 0 \text{ or } 1} \left(\sum_{r,s \ge 0} (-1)^{\varepsilon_1 + \varepsilon_2} \delta(p^r, p^s) f_1(p^{r-\varepsilon_1}) f_2(p^{s-\varepsilon_2}) \right).$$

Now

$$\begin{aligned} &\sum_{r,s \ge 0} (-1)^{\varepsilon_1 + \varepsilon_2} \delta(p^r, p^s) f_1(p^{r-\varepsilon_1}) f_2(p^{s-\varepsilon_2}) \\ &\qquad = \sum_{\substack{r \ge -\varepsilon_1 \\ s \ge -\varepsilon_2}} (-1)^{\varepsilon_1 + \varepsilon_2} \delta(p^{r+\varepsilon_1}, p^{s+\varepsilon_2}) f_1(p^r) f_2(p^s) \\ &\qquad = \sum_{r,s \ge 0} (-1)^{\varepsilon_1 + \varepsilon_2} \delta(p^{r+\varepsilon_1}, p^{s+\varepsilon_2}) f_1(p^r) f_2(p^s) \end{aligned}$$

since $f_1(p^{-1}) = f_2(p^{-1}) = 0$. Thus

$$\begin{aligned} S_p(f_1, f_2) &= \sum_{r,s \ge 0} f_1(p^r) f_2(p^s) \left(\sum_{\varepsilon_1, \varepsilon_2 = 0 \text{ or } 1} (-1)^{\varepsilon_1 + \varepsilon_2} \delta(p^{r+\varepsilon_1}, p^{s+\varepsilon_2}) \right) \\ &= \sum_{r,s \ge 0} D_p(r, s) f_1(p^r) f_2(p^s). \end{aligned}$$

5.4. We will now see that, to prove the theorem, it is sufficient to prove that the conclusion holds if there exists a positive integer k such that $f_1(p^r) = f_2(p^r) = 0$ when $r > k$.

Suppose that this has been proved. Given f_1 and f_2 satisfying the hypotheses of the theorem, define for each positive integer k the functions $f_1^{(k)}$ and $f_2^{(k)}$ by

$$f_j^{(k)}(n) = \prod_{\substack{p^r \| n \\ r \le k}} f_j(p^r).$$

$f_1^{(k)}$ and $f_2^{(k)}$ are multiplicative functions of absolute value ≤ 1 and we have

$$\sum \frac{2 - \operatorname{Re}(f_1^{(k)}(p) + f_2^{(k)}(p))}{p} < \infty.$$

Set $\Phi^{(k)}(n) = f_1^{(k)}(a_1 n + b_1) f_2^{(k)}(a_2 n + b_2)$.

Since $f_1^{(k)}(p^r) = f_2^{(k)}(p^r) = 0$ when $r > k$, the difference

$$\frac{1}{x}\sum_{n\le x}\Phi^{(k)}(n) - \prod_{p\le x} S_p(f_1^{(k)}, f_2^{(k)})$$

tends to zero as x tends to infinity for each fixed k. This implies that

$$\frac{1}{x}\sum_{n\le x}\Phi(n) - \prod_{p\le x} S_p(f_1, f_2)$$

also tends to zero, since the difference

$$\frac{1}{x}\sum_{n\le x}\Phi^{(k)}(n) - \prod_{p\le x} S_p(f_1^{(k)}, f_2^{(k)}) - \left(\frac{1}{x}\sum_{n\le x}\Phi(n) - \prod_{p\le x} S_p(f_1, f_2)\right)$$

converges to zero uniformly in x when k tends to infinity.

Proof. The difference $\Phi^{(k)}(n) - \Phi(n)$ has absolute value ≤ 2 and can be nonzero only if there is a prime p such that p^{k+1} divides $a_1 n + b_1$ or $a_2 n + b_2$. For $x \ge 1$ the number of $n \le x$ for which p^{k+1} divides $a_j n + b_j$ is obviously $\le \frac{(a_j+b_j)x}{p^{k+1}}$. It follows that, for $x \ge 1$,

$$\left|\frac{1}{x}\sum_{n\le x}\Phi^{(k)}(n) - \frac{1}{x}\sum_{n\le x}\Phi(n)\right| \le 2(a_1 + a_2 + b_1 + b_2)\sum \frac{1}{p^{k+1}}.$$

Moreover, it follows from Lemma 2 that

$$\left| \prod_{p \le x} S_p(f_1^{(k)}, f_2^{(k)}) - \prod_{p \le x} S_p(f_1, f_2) \right| \le (a_1 + a_2) \sum \frac{1}{p^{k+1}}$$

since

$$\left| S_p(f_1, f_2) - S_p(f_1^{(k)}, f_2^{(k)}) \right| = \left| \sum_{r \text{ or } s > k} D_p(r,s) f_1(p^r) f_2(p^s) \right|$$

$$\le \sum_{r \text{ or } s > k} D_p(r,s) = 1 - \sum_{\substack{0 \le r \le k \\ 0 \le s \le k}} D_p(r,s),$$

and we have seen in § 5.2 that this is $\le \frac{a_1+a_2}{p^{k+1}}$.

The asserted result follows from the fact that $\sum \frac{1}{p^{k+1}}$ tends to zero as k tends to infinity.

5.5. Now, suppose that the functions f_1 and f_2 satisfy the hypotheses of the theorem and that, for a positive integer k, $f_1(p^r) = f_2(p^r) = 0$ when $r > k$.

Given $y \ge 2$ we define the "truncated" functions $f_{1,y}$ and $f_{2,y}$ by

$$f_{j,y}(m) = \prod_{\substack{p^r \| m \\ p \le y}} f_j(p^r).$$

$f_{1,y}$ and $f_{2,y}$ are multiplicative functions of absolute value ≤ 1. We have

$$(f_{j,y} * \mu)(p^r) = \begin{cases} 0 & \text{if } p > y \text{ or } r > k+1, \\ f_j^*(p^r) & \text{if } p \le y \text{ and } r \le k+1. \end{cases}$$

Therefore

$$(f_{j,y} * \mu)(m) = \begin{cases} 0 & \text{if } m \nmid P_k = \prod_{p \le y} p^{k+1}, \\ f_j^*(m) & \text{if } m | P_k, \end{cases}$$

and it follows that

$$f_{j,y}(m) = \sum_{\substack{d | P_k \\ d | m}} f_j^*(d).$$

This gives

$$\sum_{n\le x} f_{1,y}(a_1n+b_1)f_{2,y}(a_2n+b_2)=\sum_{n\le x}\left(\sum_{\substack{d_1,d_2|P_k\\ d_1|a_1n+b_1\\ d_2|a_2n+b_2}} f_1^*(d_1)f_2^*(d_2)\right)$$

$$=\sum_{d_1,d_2|P_k} f_1^*(d_1)f_2^*(d_2)\ \#\{n\le x : n \text{ satisfies } \mathcal{S}(d_1,d_2)\}.$$

Using (11) it follows that

$$\sum_{n\le x} f_{1,y}(a_1n+b_1)f_{2,y}(a_2n+b_2)=x\sum\nolimits_1+\sum\nolimits_2, \tag{12}$$

where

$$\sum\nolimits_1=\sum_{d_1,d_2|P_k} f_1^*(d_1)f_2^*(d_2)\delta(d_1,d_2)$$

and

$$\sum\nolimits_2=\sum_{d_1,d_2|P_k} f_1^*(d_1)f_2^*(d_2)R_x(d_1,d_2).$$

Let K be the function of two positive integers defined by

$$K(n_1,n_2)=f_1^*(n_1)f_2^*(n_2)\delta(n_1,n_2).$$

Then $\sum_1 = H(P_k,P_k)$, where H is the convolution of K and the function equal to 1 for all pairs (n_1,n_2).

The function K is multiplicative, and therefore H is multiplicative as well. Thus,

$$H(P_k,P_k)=\prod_{p|P_k} H\left(p^{V_p(P_k)},p^{V_p(P_k)}\right)=\prod_{p\le y} H(p^{k+1},p^{k+1}).$$

Now

$$H(p^{k+1},p^{k+1})=\sum_{\substack{0\le r\le k+1\\ 0\le s\le k+1}} K(p^r,p^s)$$

$$=\sum_{\substack{0\le r\le k+1\\ 0\le s\le k+1}} f_1^*(p^r)f_2^*(p^s)\delta(p^r,p^s)=S_p(f_1,f_2).$$

Finally $\sum_1=\prod_{p\le y} S_p(f_1,f_2)$.

To estimate $\sum_2$, we use

$$\left|\sum\nolimits_2\right| \le \sum_{d_1,d_2|P_k} |f_1^*(d_1)f_2^*(d_2)|.$$

Since $|f_1^*(d_1)|$ and $|f_2^*(d_2)|$ are ≤ 2, this gives

$$\left|\sum\nolimits_2\right| \le 4\left(\sum_{d|P_k} 1\right)^2 = 4(k+2)^{2\pi(y)}.$$

which is $o(x)$ if $y = \log x$.

Thus (12) shows that, if $y = \log x$, then, as x tends to infinity,

$$\text{(13)} \qquad \frac{1}{x}\sum_{n\le x} f_{1,y}(a_1n+b_1)f_{2,y}(a_2n+b_2) = \prod_{p\le y} S_p(f_1,f_2) + o(1).$$

In what follows we will suppose that $y = \log x$, so that (13) holds.

From (13) we deduce that, as x tends to infinity,

$$\text{(14)} \qquad \frac{1}{x}\sum_{n\le x} f_{1,y}(a_1n+b_1)f_{2,y}(a_2n+b_2)e^{-A(y)} = P + o(1),$$

where P is the value of the infinite product

$$\prod S_p(f_1,f_2)\exp\left(\frac{2-f_1(p)-f_2(p)}{p}\right)$$

(which is convergent by § 3.2) and

$$A(u) = \sum_{p\le u} \frac{f_1(p)+f_2(p)-2}{p}.$$

5.6. To complete the proof of the theorem we will show that, as x tends to infinity,

$$\text{(15)} \quad \left(\sum_{n\le x}\Phi(n)\right)e^{-A(x)} - \left(\sum_{n\le x} f_{1,y}(a_1n+b_1)f_{2,y}(a_2n+b_2)\right)e^{-A(y)} = o(x).$$

Combined with (14), this will give

$$\left(\frac{1}{x}\sum_{n\le x}\Phi(n)\right)e^{-A(x)} = P + o(1)$$

$$= \prod_{p\le x} S_p(f_1,f_2)\exp\left(\frac{2-f_1(p)-f_2(p)}{p}\right) + o(1)$$

$$= \left(\prod_{p\le x} S_p(f_1,f_2)\right)e^{-A(x)} + o(1)\,,$$

which implies (2) since $\operatorname{Re} A(x) \le 0$.

Since

$$-\operatorname{Re} A(y) \le \sum \frac{2-\operatorname{Re}(f_1(p)+f_2(p))}{p},$$

to prove (15) it suffices to prove that

$$\text{(16)} \qquad \sum_{n\le x}\left|\Phi(n)e^{A(y)-A(x)} - f_{1,y}(a_1n+b_1)f_{2,y}(a_2n+b_2)\right| = o(x)\,.$$

We transform this relation by introducing the multiplicative functions $f_{1,y}^+$ and $f_{2,y}^+$ determined by

$$f_{j,y}^+(p^r) = \begin{cases} 1 & \text{if } p \le y\,, \\ f_j(p^r) & \text{if } p > y\,. \end{cases}$$

These functions have absolute value ≤ 1 and we have

$$f_j(m) = f_{j,y}(m)\cdot f_{j,y}^+(m)\,.$$

Thus $\Phi(n)e^{A(y)-A(x)} - f_{1,y}(a_1n+b_1)f_{2,y}(a_2n+b_2)$ is equal to

$$f_{1,y}(a_1n+b_1)f_{2,y}(a_2n+b_2)\left(f_{1,y}^+(a_1n+b_1)f_{2,y}^+(a_2n+b_2)e^{A(y)-A(x)} - 1\right).$$

Since $|f_{1,y}(a_1n+b_1)f_{2,y}(a_2n+b_2)| \le 1$, to prove (16) it suffices to prove that

$$\text{(17)} \qquad \sum_{n\le x}\left|f_{1,y}^+(a_1n+b_1)f_{2,y}^+(a_2n+b_2)e^{A(y)-A(x)} - 1\right| = o(x)\,.$$

The left–hand side of (17) is of the form $\sum_{n\le x}|z_1z_2-1|$, where $z_j = z_j(n) = f^+_{j,y}(a_jn+b_j)e^{A_j(y)-A_j(x)}$, with $A_j(u)=\sum_{p\le u}\frac{f_j(p)-1}{p}$. We have

$$|z_1z_2-1| = |z_2(z_1-1)+z_2-1| \le |z_2|\cdot|z_1-1|+|z_2-1|,$$

and $|z_2|\le \exp\left(\sum\frac{1-\operatorname{Re} f_2(p)}{p}\right)$. We thus see that, to prove (17) it suffices to prove that, for $j=1$ and for $j=2$,

$$\sum_{n\le x}\left|f^+_{j,y}(a_jn+b_j)e^{A_j(y)-A_j(x)}-1\right| = o(x)\,.$$

Finally, since the numbers a_jn+b_j, $n\le x$, are distinct integers $\le a_jx+b_j$, to prove the latter relation if suffices to show that

$$\sum_{m\le x}\left|f^+_{j,y}(m)e^{A_j(y)-A_j(x)}-1\right| = o(x)\,. \tag{18}$$

5.7. Now let F_j be the stongly additive function determined by

$$F_j(p)=f^+_{j,y}(p)-1\,.$$

$e^{F_j(m)}$ is a multiplicative function of m, of absolute value ≤ 1. Then it follows from Lemma 2 that, for each m,

$$\left|e^{F_j(m)}-f^+_{j,y}(m)\right| \le \sum_{p^r\|m}\left|e^{F_j(p^r)}-f^+_{j,y}(p^r)\right|\,.$$

For $p\le y$, $e^{F_j(p^r)}=f^+_{j,y}(p^r)=1$. For $p>y$, Lemma 5 gives

$$|e^{F_j(p)}-f^+_{j,y}(p)| \le 1-\operatorname{Re} f^+_{j,y}(p) = 1-\operatorname{Re} f_j(p)$$

and for $r>1$ we have $|e^{F_j(p^r)}-f^+_{j,y}(p^r)|\le 2$. We thus see that, for each m,

$$\left|e^{F_j(m)}-f^+_{j,y}(m)\right| \le \sum_{\substack{p|m\\p>y}}(1-\operatorname{Re} f_j(p))+2\sum_{\substack{p^2|m\\p>y}}1\,.$$

Therefore

$$\sum_{m\le x}\left|e^{F_j(m)}-f^+_{j,y}(m)\right| \le x\left(\sum_{p>y}\frac{1-\operatorname{Re} f_j(p)}{p}+2\sum_{p>y}\frac{1}{p^2}\right).$$

This is $o(x)$ as x tends to infinity.

Since

$$\operatorname{Re}(A_j(y) - A_j(x)) \leq \sum \frac{1 - \operatorname{Re} f_j(p)}{p},$$

it follows that, as x tends to infinity,

$$\sum_{m \leq x} \left| f^+_{j,y}(m) e^{A_j(y) - A_j(x)} - 1 \right| = \sum_{m \leq x} |\exp(F_j(m) + A_j(y) - A_j(x)) - 1| + o(x).$$

Thus, (18) is equivalent to

$$\sum_{m \leq x} |\exp(F_j(m) + A_j(y) - A_j(x)) - 1| = o(x).$$

Since $\operatorname{Re}(F_j(m) + A_j(y) - A_j(x)) \leq M_j$ where M_j is any positive number $\geq \sum \frac{1 - \operatorname{Re} f_j(p)}{p}$, it follows from Lemma 4 that

$$\left|\exp(F_j(m) + A_j(y) - A_j(x)) - 1\right| \leq \frac{e^{M_{j-1}}}{M_j} \left|F_j(m) + A_j(y) - A_j(x)\right|.$$

Therefore, to prove (18) it suffices to prove that, as x tends to infinity,

$$\sum_{m \leq x} \left|F_j(m) + A_j(y) - A_j(x)\right| = o(x). \tag{19}$$

By Cauchy's inequality

$$\sum_{m \leq x} \left|F_j(m) + A_j(y) - A_j(x)\right| \leq x^{1/2} \Big(\sum_{m \leq x} \left|F_j(m) + A_j(y) - A_j(x)\right|^2\Big)^{1/2}.$$

Now we remark that

$$F_j(m) + A_j(y) - A_j(x) = F_j(m) - \sum_{p \leq x} \frac{F_j(p)}{p}.$$

Thus the Turan–Kubilius inequality (see [4, p. 153]) gives

$$\sum_{m \leq x} \left|F_j(m) + A_j(y) - A_j(x)\right|^2 \leq 16x \sum_{p \leq x} \frac{|F_j(p)|^2}{p}.$$

This is $o(x)$ because

$$\sum_{p \leq x} \frac{|F_j(p)|^2}{p} = \sum_{y < p \leq x} \frac{|f_j(p) - 1|^2}{p} \leq 2 \sum_{y < p \leq x} \frac{1 - \operatorname{Re} f_j(p)}{p}$$

by Lemma 3. We thus obtain (19).

References

[1] H. Delange, *Sur les fonctions arithmétiques multiplicatives*, Ann. Sci. Ecole Norm. Sup. (3) **78** (1961), 273–304.

[2] H. Delange, *Sur les fonctions de plusieurs entiers strictement positifs*, Enseign. Math. **15** (1969), 77–88.

[3] H. Delange, *Sur des formules de Atle Selberg*, Acta Arith. **19** (1971), 105–146.

[4] P.D.T.A. Elliott, *Probabilistic Number Theory I*, Springer Verlag 1979.

Hubert Delange
Université de Paris-Sud
Département de Mathématiques
Bât. 425
91405 Orsay Cedex
FRANCE

Combinatorial sieves of dimension exceeding one II

H.G. Diamond, H. Halberstam, and H.–E. Richert[1]

Abstract. In an earlier paper bearing the same title, [DHR 1], we showed how to construct a class of sieves of dimension $\kappa > 1$ on the assumption of a certain analytic result, referred to there as Theorem 0, for a boundary value problem involving a pair of simultaneous linear differential delay equations, with κ as a parameter. In subsequent papers we proved Theorem 0 for various ranges of values of κ, and here we complete the proof, thereby validating unconditionally the main result of [DHR 1].

1. Introduction

We showed in [DHR 1] how to construct sieves of dimension $\kappa > 1$ on the assumption of a certain analytic result, Theorem 0 below, and in subsequent papers we have proved Theorem 0, by stages, for various ranges of values of κ: in [DHR 5] for all $\kappa \geq 2$ and for $\kappa = 1.5$; in [DHR 6] for all κ satisfying $1 < \kappa < 1.5$; and in [DHR 8] for all κ satisfying $1.5 < \kappa \leq 1.82$ and $1.85 \leq \kappa < 2$. In this paper we complete the proof of Theorem 0 by showing that it holds for the residual range

$$1.82 < \kappa < 1.85. \tag{1.1}$$

Here, to remind us, is a statement of

Theorem 0. *Let $\kappa > 1$ and let σ be the function defined in (2.15) and (2.16) below. Then there exist numbers $\alpha = \alpha_\kappa > 2$ and $\beta = \beta_\kappa > 2$ such that the simultaneous differential-delay system*

$$\begin{cases} \text{(i)} & F(u) = 1/\sigma(u), \quad 0 < u \leq \alpha, \\ \text{(ii)} & f(u) = 0, \quad 0 < u \leq \beta, \\ \text{(iii)} & (u^\kappa F(u))' = \kappa u^{\kappa-1} f(u-1), \quad \alpha < u, \\ \text{(iv)} & (u^\kappa f(u))' = \kappa u^{\kappa-1} F(u-1), \quad \beta < u, \end{cases}$$

[1]Professor Richert died on November 25, 1993.

has continuous solutions $F = F_\kappa$, $f = f_\kappa$ with the properties

$$F(u) = 1 + O(e^{-u}), \quad f(u) = 1 + O(e^{-u}),$$

$$F(u) \text{ decreases monotonically towards } 1 \text{ as } u \to \infty$$

and

$$f(u) \text{ increases monotonically towards } 1 \text{ as } u \to \infty.$$

The truth of Theorem 0 was, in effect, conjectured in [ILR] and, independently, in [Raw], and its proof has rested on the pioneering researches of Iwaniec (e.g. [Iw]) and on a great deal of technical preparation (e.g. [DHR 2, 3, 4] and the Appendices in [DHR 2, 5, 6], [Grp] and [GR]). If we were to write out a proof now, we could give a shorter and more elegant account; but it would still be rather long and complicated. To us at least, Theorem 0 still seems a difficult result. We hope that some of the techniques we have developed in the course of proving it will be of help in establishing superior κ-dimensional sieves such as one foreshadowed in [Sel] (or [Brü-F] when $\kappa = 2$).

The strategy for proving Theorem 0 is described in [DHR 5], sections 2 and 3; but before we review it we remind the reader of the application towards which Theorem 0 was directed.

Let $\mathcal{A}$ be a finite integer sequence whose elements are not necessarily distinct or of one sign. Let $\mathcal{P}$ be a set of primes, $z \geq 2$ a real number, and write

$$P(z) = \prod_{\substack{p<z \\ p\in\mathcal{P}}} p, \quad S(\mathcal{A}, \mathcal{P}, z) = |\{a \in \mathcal{A} : (a, P(z)) = 1\}|,$$

$$\mathcal{A}_d = \{a \in \mathcal{A} : a \equiv 0 \mod d\}.$$

For us to say anything worthwhile about $S(\mathcal{A}, \mathcal{P}, z)$ we have to know that $\mathcal{A}$ is fairly well distributed among the residue classes $0 \mod d$, $d \mid P(z)$ and d not too large. The distribution requirement is not demanding: we postulate only that there exist an approximation X to $|\mathcal{A}|$ and a multiplicative function $\omega(d)$ satisfying (i) $\omega(p) = 0$ if $p \notin \mathcal{P}$ and otherwise $0 < \omega(p) < p$, and (ii) the Iwaniec condition

$$(\Omega(\kappa)) \qquad \prod_{w_1 \leq p < w} \left(1 - \frac{\omega(p)}{p}\right)^{-1} \leq \left(\frac{\log w}{\log w_1}\right)^{\kappa} \left(1 + \frac{A}{\log w_1}\right), \quad 2 \leq w_1 \leq w$$

for some constants $\kappa > 1$, $A > 2$; such that the remainders

$$R_d := |\mathcal{A}_d| - \frac{\omega(d)}{d} X$$

are small on average over $d \mid P(z)$, d not exceeding a parameter y (that is chosen, in practice, to be as large as possible and certainly larger than z). In these circumstances

$$S(\mathcal{A},\mathcal{P},z) \le XV(z)\left\{F_\kappa\left(\frac{\log y}{\log z}\right) + O\left(\frac{\log\log y}{(\log y)^{1/(2\kappa+2)}}\right)\right\} + \sum_{\substack{d|P(z)\\ d<y}} c^+(d)R_d$$

and

$$S(\mathcal{A},\mathcal{P},z) \ge XV(z)\left\{f_\kappa\left(\frac{\log y}{\log z}\right) + O\left(\frac{\log\log y}{(\log y)^{1/(2\kappa+2)}}\right)\right\} - \sum_{\substack{d|P(z)\\ d<y}} c^-(d)R_d\,,$$

where $V(z) := \prod_{p<z}\left(1 - \frac{\omega(p)}{p}\right)$ and the constants implied by the O-notation depend at most on κ and A, and the coefficients $c^\pm(d)$ satisfy $|c^\pm(d)| \le 1+4^{\nu(d)}$ ($\nu(d)$ being the number of distinct prime factors of d); *provided we can show that the functions F_κ, f_k satisfy Theorem 0.* The significance of the numbers α_κ and β_κ is that the upper estimate for $S(\mathcal{A},\mathcal{P},z)$ improves on [AO] only if $z < y^{1/\alpha_\kappa}$, and the lower bound is non-trivial only if $z < y^{1/\beta_\kappa}$ (for this reason β_κ is called the *sieving limit*). It is usual to refer to κ as the *sieve dimension.* (For a more detailed historical discussion see [DHR 1].)

Its relevance to the construction of sieves apart, we hope the reader will find Theorem 0 of some independent interest as a problem in linear control theory.

Now for a sketch of the strategy we used to attack Theorem 0. It is necessary to introduce some notation, but this will be needed later in any case. The auxiliary functions $p = p_\kappa$ and $q = q_\kappa$ satisfy respectively the differential difference equations

$$(1.2)\qquad (up(u))' = \kappa p(u) - \kappa p(u+1), \quad u > 0,$$

$$(1.3)\qquad (uq(u))' = \kappa q(u) + \kappa q(u+1), \quad u > 0,$$

and the normalizing conditions

$$(1.4)\qquad p_\kappa(u) \sim u^{-1}, \quad q_\kappa(u) \sim u^{2\kappa-1}, \quad u \to \infty.$$

We have accumulated a great deal of information about these functions, and we shall need to find out even more about them both later. For the moment we remark that $p_\kappa(u)$ is positive and decreasing in u, and that $q_\kappa(u)$ has fewer than 2κ real zeros and is actually a polynomial in u of degree $2\kappa - 1$ when 2κ is an integer. The largest real zero $\rho = \rho_\kappa$ of $q_\kappa(u)$ has played a major role throughout this investigation. We know that $\alpha_\kappa > \rho_\kappa$ and that $\rho_\kappa \sim c_0\kappa$ $(\kappa \to \infty)$ where $c_0 = 3.5911\ldots$ satisfies $c_0 \log(c_0/e) = 1$. For $1 < \kappa < 2$ we have established rather sharp bounds for ρ_κ (see [DHR 7], Thm 0.2).

With p, q and σ (from the statement of Theorem 0) we form the functions

$$\Pi_\kappa(u,v) = \frac{up(u)}{\sigma(u)} + \kappa \int_{v-1}^{u} \frac{p(t+1)}{\sigma(t)}\,dt, \quad u > 0, \quad v > 1, \tag{1.5}$$

$$\chi_\kappa(u,v) = \frac{uq(u)}{\sigma(u)} - \kappa \int_{v-1}^{u} \frac{q(t+1)}{\sigma(t)}\,dt, \quad u > 0, \quad v > 1, \tag{1.6}$$

$$\tilde{\Pi}_\kappa(u) = \Pi_\kappa(u, u-1), \tag{1.7}$$

and

$$\tilde{\chi}_\kappa(u) = \chi_\kappa(u, u-1). \tag{1.8}$$

For Theorem 0 to be true, the numbers α_κ, β_κ necessarily must satisfy, depending on their relative size, one of two pairs of simultaneous transcendental equations:

$$\Pi(\alpha,\beta) = 2 \quad \textit{and} \quad \chi(\alpha,\beta) = 0 \quad \textit{if} \quad \beta < \alpha \le \beta + 1, \tag{1.9}$$

or

$$\tilde{\Pi}(\alpha) + (\alpha-1)p(\alpha-1)f(\alpha-1) = 2 \quad \textit{and} \quad \tilde{\chi}(\alpha) = (\alpha-1)q(\alpha-1)f(\alpha-1)$$
(1.10)
$\textit{if} \quad \beta + 1 \le \alpha,$

where $f(\alpha-1)$ *is given by*

$$(\alpha-1)^\kappa f(\alpha-1) = \kappa \int_\beta^{\alpha-1} \frac{t^{\kappa-1}}{\sigma(t-1)}\,dt \tag{1.11}$$

(see [ILR] or [DHR 5], Lem. 2.1).
The main difficulty in proving Theorem 0 has been to show conversely that each pair of equations possesses a unique solution pair α_κ, β_κ of the appropriate relative size, and to determine for each $\kappa > 1$ which of the two pairs is the right one.

It took us a long time to discover an effective criterion for classifying the numbers $\kappa > 1$. To describe it we need to report from [DHR 5] that

(1.12) $\tilde{\Pi}(u) - 2$ *is strictly decreasing in* $u > 2$ *and possesses a unique zero* $z_{\tilde{\Pi}}(\kappa) > 2$, and that

(1.13) $\tilde{\chi}(u)$ *is strictly increasing in* $u > \rho + 1$ *and possesses a unique zero* $z_{\tilde{\chi}}(\kappa) > \rho_\kappa + 1$.

Let

$$\mathcal{K}_1 := \{\kappa > 1 : z_{\tilde{\chi}(\kappa)} \le z_{\tilde{\Pi}}(\kappa)\}, \quad \mathcal{K}_2 := \{\kappa > 1 : z_{\tilde{\chi}(\kappa)} > z_{\tilde{\Pi}}(\kappa)\}.$$

We proved in [DHR 5] that *for each* $\kappa \in \mathcal{K}_1$ *the equations in* (1.9) *do possess a unique solution* α, β *with* $2 < \beta < \alpha \le z_{\tilde{\chi}}(\kappa) \le \beta + 1 \le z_{\tilde{\Pi}}(\kappa)$; *and that for each* $\kappa \in \mathcal{K}_2$ *the equations in* (1.10) *possess a unique solution* α, β *with* $\alpha > \max(\beta + 1, z_{\tilde{\chi}}(\kappa))$.

Our task came down, therefore, to studying the relative positions of $z_{\tilde{\Pi}}(\kappa)$ and $z_{\tilde{\chi}}(\kappa)$, a problem in hard analysis that we were able to deal with so long as the two zeros were well enough separated to interpose between them some known function of κ (for example, if $\kappa \ge 3.5$ we showed that $z_{\tilde{\Pi}}(\kappa) < \rho_\kappa + 1 < z_{\tilde{\chi}}(\kappa)$). We had, thanks to numerical work of H.-E. Richert, Wheeler and Bradley much evidence to guide us. In this way we showed in [DHR 5] that all $\kappa \ge 2$ belong to $\mathcal{K}_2$ and that $\kappa = 1.5$ lies in $\mathcal{K}_1$, and in [DHR 6] that all κ between 1 and 1.5 lie in $\mathcal{K}_1$. This left the numbers κ between 1.5 and 2 to be classified. Here numerical evidence shows $z_{\tilde{\Pi}}(\kappa)$ and $z_{\tilde{\chi}}(\kappa)$ to be close and, indeed, to change relative position at the point $\kappa = \kappa_0 = 1.8344\ldots$ where the two coincide; and we had no option but to study the two zeros as functions of κ, with all that this involved. We had proved already in [DHR 5], Lemma Π2, that $z_{\tilde{\Pi}}(\kappa)$ is strictly increasing; after preparing the ground in [DHR 7] (a study of $q_\kappa(u)$ as a function of κ for $1 < \kappa < 2$), we proved in [DHR 8] the much harder Proposition that $z_{\tilde{\chi}}(\kappa)$ also is strictly increasing. We went on, in [DHR 8], to combine the monotonicities of the two zeros with a finite number of evaluations of $z_{\tilde{\Pi}}(\kappa)$ and $z_{\tilde{\chi}}(\kappa)$ to show by a step procedure that the numbers κ in the interval $(1.5, 1.82]$ lie in $\mathcal{K}_1$ and the κ in $[1.85, 2)$ in $\mathcal{K}_2$. So, finally, we came to address the remaining interval $[1.82, 1.85]$ (containing the number κ_0), and this we accomplish in the present paper. For a long time this task struck us as very daunting, and the argument below shows that it is, indeed, formidable; but by this time we had prepared the ground enough to see that, with an effort, we could show the growth rates of the two zeros to be, in a very precise way, so significantly different that their graphs intersect at precisely one point, $\kappa = \kappa_0$.

We do not fully understand the significance of κ_0, where the sets $\mathcal{K}_1$ and $\mathcal{K}_2$ abut; there does not appear to be any abrupt change in the behavior of β_κ or α_κ as κ passes from one set to the other. All that we can say is that these sets correspond respectively to the cases $\beta_\kappa < \alpha \le \beta_\kappa + 1$ and $\beta_\kappa + 1 < \alpha_\kappa$; and that this distinction appears both in the analytic formulation of the κ-dimensional sieves (see (1.9) and (1.10) above) and in our combinatorial construction via inequalities of Rosser-Iwaniec type (see [DHR1, (2.20) et. seq]).

Our final argument is simple in outline—although some of the details are complicated. In section 2 we prove that $\dot{z}_{\tilde{\Pi}}(\kappa) < 2.19$ throughout the interval $[1.82, 1.85]$. (Throughout this article, a dot denotes differentiation w.r.t. κ.) Actually, the graph of $z_{\tilde{\Pi}}(\kappa)$ is practically a straight line with slope about 2.17 on the entire interval $(1.5, 2)$, and we can show that the cited inequality for $\dot{z}_{\tilde{\Pi}}$ holds even on this larger interval.

In section 3 we use this information in conjunction with a more delicate step-by-step procedure to show that $[1.82, 1.834] \subset \mathcal{K}_1$ and $[1.835, 1.85] \subset \mathcal{K}_2$, so that it remains only to compare $z_{\tilde{\Pi}}(\kappa)$ and $z_{\tilde{\chi}}(\kappa)$ on the tiny interval $[1.834, 1.835]$. It came as an agreeable and welcome surprise that we could focus in the end on such a small neighborhood of κ_0.

We have remarked already that in this neighborhood $\dot{z}_{\tilde{\Pi}}(\kappa) < 2.19$, and in section 4 we begin to deal with $\dot{z}_{\tilde{\chi}}(\kappa)$ on the interval $I_0 := [1.834, 1.835]$. From differentiating the equation

$$\tilde{\chi}_\kappa(z_{\tilde{\chi}}(\kappa)) = 0$$

with respect to κ we see that

$$\dot{z}_{\tilde{\chi}}(\kappa) = -\dot{\tilde{\chi}}_\kappa(u)/\tilde{\chi}'_\kappa(u)\big|_{u=z_{\tilde{\chi}}(\kappa)}. \tag{1.14}$$

With κ restricted to I_0, $z_{\tilde{\chi}}(\kappa)$ too is confined to a short interval, and we show in section 4 that $-\dot{\tilde{\chi}}_\kappa(u)$ is large on this interval. This is the most complicated part of our argument, but the groundwork for it was laid in [DHR 8], (4.3) et seq.

Finally, in a short section 5 we derive a lower bound for the positive function $\tilde{X}'_\kappa(u)$ on the short u-interval, with κ in I_0 of course, and in this way arrive at the inequality $\dot{z}_{\tilde{\chi}} > 2.41$, $\kappa \in I_0$. Since $z_{\tilde{\chi}}(\kappa) - z_{\tilde{\Pi}}(\kappa)$ is negative to the left of I_0 and positive to the right, the graphs of the continuous functions $z_{\tilde{\Pi}}(\kappa)$ and $z_{\tilde{\chi}}(\kappa)$ cross inside I_0; and since $\dot{z}_{\tilde{\chi}}(\kappa) > \dot{z}_{\tilde{\Pi}}(\kappa)$ throughout I_0 the graphs cannot intersect more than once. Thus the uniqueness of the threshold κ_0 is established: we know now that $(1, \kappa_0] = \mathcal{K}_1$ and $(\kappa_0, \infty) = \mathcal{K}_2$, and with this the proof of Theorem 0 is complete.

It remains only to remark that a rather lengthy Appendix provides accurate information about the behavior of $\dot{q}_\kappa(u)$, $\dot{q}'_\kappa(u)$ and $\dot{q}''_\kappa(u)$ for $\kappa \in I_0$ and for various ranges of values of u, that is crucial for the success of arguments in sections 4 and 5.

We are much indebted to Dr. David Bradley for his supporting computations. We remarked several times in earlier papers of this series that the accuracy of numerical evidence about $q_\kappa(u)$ and $\tilde{\chi}_\kappa(u)$ was then unsecured by numerical analysis. On the limited (κ, u)-ranges relevant to this paper,

Dr. Bradley was able to fill this gap, so that any such data cited here are correct to the degree specified (and, incidentally, are in good agreement with the earlier tables).

The authors will provide on request data or details of calculations.

2. An upper bound for $\dot{z}_{\tilde{\Pi}}(\kappa)$

Writing

$$\text{Ein } z := \int_0^z (1 - e^{-t})t^{-1}\,dt,$$

the function $p(u) = p_\kappa(u)$ defined by

$$p_\kappa(u) := \int_0^\infty e^{-ux-\kappa \text{ Ein } x}\,dx, \quad \kappa > 1, \quad u > 0, \tag{2.1}$$

is easily seen to satisfy the differential difference equation (1.2) and is the only function to do so and to have behavior at infinity as given in (1.4) (see [DHR 4]). From (2.1) we see at once that

$$(-1)^{\mu+\nu}\left(\frac{\partial}{\partial u}\right)^\mu \left(\frac{\partial}{\partial u}\right)^\nu p_\kappa(u) > 0$$

for any pair of non-negative integers μ and ν, and, in particular, we record the following useful monotonicities:

Lemma 2.1. *We have*

(2.2)

$p_\kappa(u) > 0$ *and* $p''_\kappa(u) > 0$, *and each of these functions is strictly decreasing both in* u *and in* κ;

(2.3) $p'_\kappa(u) < 0$, *and* $p'_\kappa(u)$ *is strictly increasing both in* u *and in* κ;

(2.4)

$\dot{p}'_\kappa(u) > 0$, *and* $\ddot{p}_\kappa(u) > 0$ *and each is strictly decreasing both in* u *and in* κ;

and

(2.5) $\dot{p}_\kappa(u) < 0$, *and* $\dot{p}_\kappa(u)$ *is strictly increasing both in* u *and* κ.

From (2.1) we have also, using the Cauchy-Schwarz inequality,

$$\begin{aligned}(p'(u))^2 &= \left(-\int_0^\infty e^{-ux-\kappa \operatorname{Ein} x} x\, dx\right)^2 \\ &< \left(\int_0^\infty e^{-ux-\kappa \operatorname{Ein} x}\, dx\right)\left(\int_0^\infty e^{-ux-\kappa \operatorname{Ein} x} x^2\, dx\right),\end{aligned}$$

so that

$$(p_\kappa'(u))^2 < p_\kappa(u) p_\kappa''(u); \tag{2.6}$$

and in the same way it follows also that

$$(p_\kappa''(u))^2 < p_\kappa'(u) p_\kappa'''(u). \tag{2.7}$$

Inequalities (2.6) and (2.7) may be restated as $(p'(u)/p(u))' > 0$ and $(p''(u)/p'(u))' > 0$ respectively, whence, by (2.2) and (2.3),

$$\frac{p'}{p}(u) < 0 \quad \textit{and is strictly increasing in} \quad u, \tag{2.8}$$

and

$$\frac{p''}{p'}(u) < 0 \quad \textit{and is strictly increasing in} \quad u. \tag{2.9}$$

Lemma 2.2. *We have*

$$\dot{p}_\kappa(u) = -\int_0^1 (p_\kappa(u) - p_\kappa(u+t))\frac{dt}{t},$$

and the following monotonicities hold:

$$\frac{\dot{p}_\kappa}{p_\kappa}(u) < 0 \quad \textit{and is strictly increasing in both} \quad u \quad \textit{and} \quad \kappa; \tag{2.10}$$

and

$$\frac{\dot{p}_\kappa'}{p_\kappa'}(u) < 0 \quad \textit{and is increasing in} \quad u \quad \textit{and strictly increasing in} \quad \kappa. \tag{2.11}$$

Proof. The integral representation of $\dot{p}_\kappa(u)$ first occurs in [G-R], (3.17), but follows directly from (2.1) (differentiate (2.1) with respect to κ, then substitute for Ein x and interchange order of integration).

The first assertions in each of (2.10), (2.11) follow from (2.5), (2.4) and (2.3) above.

From the integral representation of $\dot{p}_\kappa(u)$ we have

$$\begin{aligned}\left(\frac{\dot{p}_\kappa}{p}(u)\right)' &= \int_0^1 \frac{p'(u+t)p(u) - p(u+t)p'(u)}{p^2(u)}\frac{dt}{t}\\ &= \int_0^1 \left(\frac{p'}{p}(u+t) - \frac{p'}{p}(u)\right)\frac{p(u+t)}{p(u)}\frac{dt}{t}\\ &> 0\end{aligned}$$

by (2.8). This proves the second assertion in (2.10), and the third is derived via an application of Cauchy-Schwarz: by (2.1), $(\dot{p}_\kappa(u))^2 < p_\kappa(u)\ddot{p}_\kappa(u)$.

Again,

$$\frac{\dot{p}'_\kappa}{p'}(u) = \int_0^1 \left(\frac{p'(u+t)}{p'(u)} - 1\right)\frac{dt}{t},$$

whence

$$\left(\frac{\dot{p}'_\kappa}{p'}(u)\right)' = \int_0^1 \left(\frac{p''}{p'}(u+t) - \frac{p''}{p'}(u)\right)\frac{p'(u+t)}{p'(u)}\cdot\frac{dt}{t} > 0$$

by (2.9). This proves the second assertion in (2.11) and the third follows from $(\dot{p}'_\kappa(u))^2 \le (-p'(u))(-\ddot{p}'(u))$, by yet another application of Cauchy-Schwarz. ∎

Lemma 2.3. *When* $1.5 \le \kappa \le 2$ *and* $u \ge 3$, *the function*

$$\frac{1}{\kappa}p_\kappa(u) + \dot{p}_\kappa(u)$$

is strictly decreasing in u.

Proof. From differentiating (2.1) and then using that Ein $x \le x$ we see readily that

$$\left(\frac{1}{\kappa}p(u) + \dot{p}_\kappa(u)\right)' \le \frac{1}{\kappa}p'(u) + p''(u).$$

It suffices therefore to prove that

$$\frac{1}{\kappa}p'(u) + p''(u) < 0 \tag{2.12}$$

in the specified ranges. Differentiating (1.2) we get

$$\begin{aligned}up''(u) &= (\kappa-2)p'(u) - \kappa p'(u+1)\\ &< (\kappa-2)p'(u) - \kappa\{p'(u) + p''(u) + \frac{1}{2}p'''(u)\}\end{aligned}$$

since $p^{(4)}(u) > 0$; and therefore

$$(u+\kappa)p''(u) < -2p'(u) - \frac{\kappa}{2}p'''(u). \tag{2.13}$$

Differentiating (1.2) twice we have, by the same reasoning,

$$up'''(u) = (\kappa-3)p''(u) - \kappa p''(u+1) > (\kappa-3)p''(u) - \kappa p''(u) = -3p''(u),$$

and when we substitute in (2.13) we obtain

$$-2p'(u) > \left(u+\kappa-\frac{3\kappa}{2u}\right)p''(u). \tag{2.14}$$

But if $u \geq 3$ then $u+\kappa-3\kappa/(2u) \geq 3+\kappa-\kappa/2 = 3+\kappa/2$; and if $\kappa \leq 2$, $3+\kappa/2 \geq 2\kappa$. Thus

$$-2p'(u) > 2\kappa p''(u), \quad \kappa \leq 2, u \geq 3,$$

which is a restatement of (2.12). ■

Lemma 2.4. *When* $1.5 \leq \kappa \leq 2$ *and* $u \geq 3$,

$$\left(\frac{\dot{p}}{p}\right)_\kappa(u) \geq -\left(1-\frac{\frac{1}{2}-\frac{1}{3u}}{u+\kappa}\right)\left(u+\kappa-\kappa\left(u+\kappa-\frac{3\kappa}{2u}\right)^{-1}\right)^{-1}.$$

Proof. Since $p^{(4)}(t) > 0$ we deduce by Taylor's Theorem that

$$\begin{aligned}\left(\frac{\dot{p}}{p}\right)_\kappa(u) &= \int_0^1 \frac{p(u+t)-p(u)}{p(u)} \cdot \frac{dt}{t} \\ &\geq \frac{1}{p(u)}\int_0^1 \left\{p'(u)t + p''(u)\frac{t^2}{2} + p'''(u)\frac{t^3}{6}\right\}\frac{dt}{t} \\ &= \frac{1}{p(u)}\left\{p'(u) + \frac{1}{4}p''(u) + \frac{1}{18}p'''(u)\right\}.\end{aligned}$$

Differentiating (1.2) we get, since $p'''(t) < 0$,

$$up''(u) = (\kappa-2)p'(u) - \kappa p'(u+1) \geq (\kappa-2)p'(u) - \kappa\{p'(u)+p''(u)\}$$

so that

$$(u+\kappa)p''(u) \geq -2p'(u);$$

and from the proof of Lemma 2.3 we have

$$up'''(u) \geq -3p''(u).$$

Hence

$$\left(\frac{\dot{p}}{p}\right)_\kappa (u) \geq \left\{1 - \left(\frac{1}{2} - \frac{1}{3u}\right)(u+\kappa)^{-1}\right\}\frac{p'}{p}(u).$$

Now

$$up'(u) = (\kappa - 1)p(u) - \kappa p(u+1) \geq (\kappa-1)p(u) - \kappa\left\{p(u) + p'(u) + \frac{1}{2}p''(u)\right\}$$

by (1.2) using $p'''(t) < 0$ again, so that

$$(u+\kappa)p'(u) \geq -p(u) - \frac{\kappa}{2}p''(u) \geq -p(u) + \frac{\kappa}{u + \kappa - \frac{3\kappa}{2u}}p'(u)$$

by (2.14). ■

Corollary. $\left(\frac{\dot{p}}{p}\right)_{1.82}(3.84) > -0.1752$ *and* $\left(\frac{\dot{p}}{p}\right)_{1.82}(4.84) > -0.1471.$

We require also some information about the function $\sigma(u) = \sigma_\kappa(u)$, the continuous solution of the differential-delay problem

(2.15)
$$\sigma_\kappa(u) = 0, \quad u \leq 0; \quad \sigma_\kappa(u) = A_\kappa^{-1}u^\kappa \quad (0 < u \leq 2; A_\kappa = (2e^\gamma)^\kappa\Gamma(\kappa+1))$$

(2.16) $$u\sigma_\kappa'(u) = \kappa\sigma_\kappa(u) - \kappa\sigma_\kappa(u-2), \quad u > 2.$$

For the most part this information is already on record. When $u > 0$, $\sigma_\kappa(u)$ is positive and strictly increasing in u. From [DHR 2], Lemma $\sigma 1$,

(2.17) $$\frac{\sigma_\kappa'}{\sigma_\kappa}(u) > 0 \textit{ and is strictly decreasing in } u;$$

from [G-R], Theorem 1,

(2.18) $$\dot{\sigma}_\kappa(u) = -\int_0^2 (\sigma_\kappa(u) - \sigma_\kappa(u-t))\frac{dt}{t} < 0;$$

from [DHR 8], Proposition 4.3, or directly from (2.18),

(2.19) $$\frac{\dot{\sigma}_\kappa}{\sigma_\kappa}(u) < 0 \textit{ and is strictly increasing in } u;$$

further, by (2.17),

$$\left(\frac{\dot\sigma_\kappa}{\sigma_\kappa}(u)\right)^{\cdot} = \int_0^2 \left(\frac{\dot\sigma_\kappa}{\sigma_\kappa}(u-t) - \frac{\dot\sigma_\kappa}{\sigma_\kappa}(u)\right)\frac{\sigma(u-t)}{\sigma(u)} \cdot \frac{dt}{t} < 0,$$

so that
(2.20)

$$\frac{\dot\sigma_\kappa}{\sigma_\kappa}(u) \textit{ is strictly decreasing in } \kappa \textit{ or, equivalently, } \left(\frac{\dot\sigma_\kappa}{\sigma_\kappa}(u)\right)^2 > \frac{\ddot\sigma_\kappa}{\sigma_\kappa}(u).$$

Finally, we recall from [G-R] (§2 and Thm. 2) that $\sigma_\kappa''(t) \le 0$ if $t \ge U_2(\kappa)$, where $U_2(\kappa)$—the unique simple zero of $\sigma_\kappa''(t)$—is a strictly increasing function of κ; so that, by (2.18) and Taylor's formula,

$$-\left(\frac{\dot\sigma}{\sigma}\right)_\kappa (u) \ge 2\left(\frac{\sigma'}{\sigma}\right)_\kappa (u), \quad u \ge U_2(\kappa) + 2. \tag{2.21}$$

Lemma 2.5. $\left(-\frac{\dot\sigma}{\sigma}\right)_{1.82}(4.93) > 0.3565.$

Proof. By (2.21) and (2.16) the left side exceeds

$$\frac{3.64}{4.93}\left(1 - \frac{\sigma_{1.82}(2.93)}{\sigma_{1.82}(4.93)}\right),$$

and we read off the two σ-values from our tables, accurate to 9 decimal places.

From numerical data, $U_2(1.82) = 2.89965\ldots$; but actually it is sufficient to check that $\sigma_{1.82}''(2.93) < 0$. By (2.16) and (2.15) this reduces to checking that

$$u\sigma_\kappa''(u)\Big|_{\substack{\kappa=1.82\\u=2.93}} = \left\{\frac{(\kappa-1)\kappa}{u}\sigma_\kappa(u) - \left(\frac{(\kappa-1)\kappa}{u} + \frac{\kappa^2}{u-2}\right)\sigma_\kappa(u-2)\right\}_{\substack{\kappa=1.82\\u=2.93}}$$
$$= -0.004887\ldots < 0,$$

where we have used $\sigma_{1.82}(0.93) = 0.0509078\ldots$, $\sigma_{1.82}(2.93) = 0.3972938\ldots$ and, in the main calculation, also $\sigma_{1.82}(4.93) = 0.7683852\ldots$. ■

We have programs to compute the numerical values of the functions $p_\kappa(u)$, $\sigma_\kappa(u)$ and $\tilde\Pi_\kappa(u)$ for κ between 1 and 3 and u between 2 and 6 (in the case of p and σ for $0 < u < 6$), all guaranteed to eight decimal place accuracy. Later we shall need to estimate $\dot{\tilde\Pi}_\kappa(u)$ from above, numerically, at a small number of points (u, κ). In Lemma 2.6 below we indicate how, when eventually we quote them, such estimates have been arrived at.

Lemma 2.6. *An upper bound for* $\dot{\tilde{\Pi}}_\kappa(u)$ *is given by*

$$\dot{\tilde{\Pi}}_\kappa(u) < \epsilon^{-1}\left(\tilde{\Pi}_{\kappa+\epsilon}(u) - \tilde{\Pi}_\kappa(u)\right), \quad \epsilon = 10^{-3}.$$

Proof. By the mean value theorem (in κ), the quotient on the right is equal to $\dot{\tilde{\Pi}}_{\kappa+\theta\epsilon}(u)$ for some $\theta \in (0,1)$, and since $\ddot{\tilde{\Pi}}_\kappa(u) > 0$ (see Lemma 2.7 below), the result follows. ∎

We now formulate the main objective of this section.

Proposition 2.1. *When* $1.82 \le \kappa \le 1.85$, *we have*

$$\dot{z}_{\tilde{\Pi}}(\kappa) < 2.19.$$

We differentiate

$$\tilde{\Pi}_\kappa(z_{\tilde{\Pi}}(\kappa)) - 2 = 0$$

with respect to κ to obtain

$$\dot{z}_{\tilde{\Pi}}(\kappa) = -\frac{\dot{\tilde{\Pi}}_\kappa(u)}{\tilde{\Pi}'_\kappa(u)}\bigg|_{u=z_{\tilde{\Pi}}(\kappa)}. \tag{2.22}$$

By [DHR 5], Lemma Π2, $\dot{z}_{\tilde{\Pi}}(\kappa) > 0$, that is, $z_{\tilde{\Pi}}(\kappa)$ is strictly increasing with κ.

Indeed, we showed there that $z_{\tilde{\Pi}}(\kappa) > 2\kappa$, $\dot{\tilde{\Pi}}_\kappa(u) > 0$ when $u \ge \kappa+1$ and $\tilde{\Pi}'_\kappa(u) < 0$ when $u > 2$. In the short κ-range under consideration we deduce from the monotonicity of $z_{\tilde{\Pi}}(\kappa)$ and numerical data that

$$4.84 \le z_{\tilde{\Pi}}(\kappa) \le 4.93; \tag{2.23}$$

here we err on the side of caution, for $z_{\tilde{\Pi}}(1.82) = 4.8506\ldots$ and $z_{\tilde{\Pi}}(1.85) = 4.9156\ldots$.

We have only to estimate $\dot{\tilde{\Pi}}_\kappa(u)$ from above, and $-\tilde{\Pi}'_\kappa(u)$ from below, on the rectangle

$$1.82 \le \kappa \le 1.85, \quad 4.84 \le u \le 4.93.$$

To this end we prove

Lemma 2.7. *On the above rectangle,* $\tilde{\Pi}''(u) > 0$, $\dot{\tilde{\Pi}}'(u) < 0$ *and* $\ddot{\tilde{\Pi}}_\kappa(u) > 0$.

Proof. From [DHR 5], (2.17),

$$\tilde{\Pi}'_\kappa(u) = \kappa\left\{\frac{p_\kappa(u)\sigma_\kappa(u-2)}{\sigma_\kappa^2(u)} - \frac{p_\kappa(u-1)}{\sigma_\kappa(u-2)}\right\}, \quad u > 2. \tag{2.24}$$

Hence, after a simple computation,

$$\begin{aligned}\frac{1}{\kappa}\tilde{\Pi}''(u) &= \frac{p(u)\sigma(u-2)}{\sigma^2(u)}\left\{\frac{p'}{p}(u) + \frac{\sigma'}{\sigma}(u-2) - 2\frac{\sigma'}{\sigma}(u)\right\}\\ &\quad + \frac{p(u-1)}{\sigma(u-2)}\left\{\frac{\sigma'}{\sigma}(u-2) - \frac{p'}{p}(u-1)\right\}.\end{aligned}$$

Because $p'(t) < 0$ (by (2.3)), the second expression on the right is positive. For the same reason $p(u-1) > p(u)$ and therefore

$$\frac{p(u-1)}{\sigma(u-2)} > \frac{p(u)\sigma(u-2)}{\sigma^2(u)},$$

whence

$$\frac{1}{\kappa}\tilde{\Pi}''(u) > \frac{p(u)\sigma(u-2)}{\sigma^2(u)}\left\{\frac{p'}{p}(u) - \frac{p'}{p}(u-1) + 2\left(\frac{\sigma'}{\sigma}(u-2) - \frac{\sigma'}{\sigma}(u)\right)\right\} > 0,$$

because $(p'/p)(t)$ is increasing in t by (2.8) and $(\sigma'/\sigma)(t)$ is decreasing by (2.17). This proves the first assertion of the Lemma for $u > 2$ and $\kappa > 1$.

Next, after a simple computation based on (2.24) and then some rearrangement,

$$\begin{aligned}\frac{\sigma(u)}{\kappa}\dot{\tilde{\Pi}}'(u) &= \frac{\sigma(u-2)}{\sigma(u)}\left\{\dot{p}_\kappa(u) + \frac{1}{\kappa}p(u)\right\}\\ &\quad - \frac{\sigma(u)}{\sigma(u-2)}\left\{\dot{p}_\kappa(u-1) + \frac{1}{\kappa}p(u-1)\right\}\\ &\quad + 2p(u)\frac{\sigma(u-2)}{\sigma(u)}\left\{\frac{\dot{\sigma}_\kappa}{\sigma}(u-2) - \frac{\dot{\sigma}_\kappa}{\sigma}(u)\right\}\\ &\quad - \frac{\sigma(u)}{\sigma(u-2)}\frac{\dot{\sigma}_\kappa}{\sigma}(u-2)\left\{\left(\frac{\sigma(u-2)}{\sigma(u)}\right)^2 p(u) - p(u-1)\right\}\\ &< \frac{\sigma(u-2)}{\sigma(u)}\left\{\dot{p}_\kappa(u) + \frac{1}{\kappa}p(u)\right\} - \frac{\sigma(u)}{\sigma(u-2)}\left\{\dot{p}_\kappa(u-1) + \frac{1}{\kappa}p(u-1)\right\}\\ &< \left\{\frac{\sigma(u-2)}{\sigma(u)} - \frac{\sigma(u)}{\sigma(u-2)}\right\}\left\{\dot{p}_\kappa(u-1) + \frac{1}{\kappa}p(u-1)\right\}.\end{aligned}$$

The first inequality follows from (2.18), (2.19) and (2.3), and the second from Lemma 2.3. Since $0 < \sigma(u-2)/\sigma(u) < 1$ for $u > 2$, $\sigma(u-2)/\sigma(u) - \sigma(u)/\sigma(u-2) < 0$, and the second assertion of the Lemma will hold as soon as we show that $\dot{p}(u-1) + p(u-1)/\kappa > 0$. But this follows immediately from Lemma 2.3 and the fact that this function vanishes as $u \to \infty$.

Finally, from the definition of $\widetilde{\Pi}_\kappa(u)$ we get

$$\ddot{\widetilde{\Pi}}_\kappa(u) = \frac{u}{\sigma(u)}\left\{\ddot{p}_\kappa(u) + 2\dot{p}_\kappa(u)\left(-\frac{\dot{\sigma}_\kappa}{\sigma}(u)\right) + p(u)\left(2\left(\frac{\dot{\sigma}_\kappa}{\sigma}(u)\right)^2 - \frac{\ddot{\sigma}_\kappa}{\sigma}(u)\right)\right\}$$
$$+ \int_{u-2}^{u}\left\{\kappa\ddot{p}_\kappa(t+1) + 2\dot{p}_\kappa(t+1)\left(1 - \kappa\frac{\dot{\sigma}_\kappa}{\sigma}(t)\right)\right.$$
$$\left. + 2p(t+1)\left(-\frac{\dot{\sigma}_\kappa}{\sigma}(t) + \left(\frac{\dot{\sigma}_\kappa}{\sigma}(t)\right)^2\kappa - \frac{\ddot{\sigma}_\kappa(t)}{2\sigma(t)}\kappa\right)\right\}\frac{dt}{\sigma(t)}.$$

By (2.4) $\ddot{p}_\kappa(t) > 0$, and by (2.20) $(\dot{\sigma}_\kappa/\sigma)^2(u) \geq (\ddot{\sigma}_\kappa/\sigma)(u)$. Hence

$$\ddot{\widetilde{\Pi}}_\kappa(u) > \frac{up(u)}{\sigma(u)}\left(-\frac{\dot{\sigma}_\kappa}{\sigma}(u)\right)\left\{2\frac{\dot{p}_\kappa}{p}(u) + \left(-\frac{\dot{\sigma}_\kappa}{\sigma}(u)\right)\right\}$$
$$+ \int_{u-2}^{u}\left\{1 + \kappa\left(-\frac{\dot{\sigma}_\kappa}{\sigma}(t)\right)\right\}\left\{2\frac{\dot{p}_\kappa}{p}(t+1) + \left(-\frac{\dot{\sigma}_\kappa}{\sigma}(t)\right)\right\}\frac{p(t+1)}{\sigma(t)}\,dt$$
$$> 0$$

by (2.19), provided

$$2\frac{\dot{p}_\kappa}{p}(u) + \left(-\frac{\dot{\sigma}_\kappa}{\sigma}(u)\right) > 0 \tag{2.25}$$

and

$$2\frac{\dot{p}_\kappa}{p}(t+1) + \left(-\frac{\dot{\sigma}_\kappa}{\sigma}(t)\right) > 0, \quad u-2 \leq t \leq u. \tag{2.26}$$

Now $(\dot{p}/p)_\kappa(t)$ is increasing in both κ and t by (2.10), while $-(\dot{\sigma}/\sigma)_\kappa(t)$ is increasing in κ but decreasing in t, by (2.20) and (2.19). Hence the expression in (2.26) is at least

$$2\left(\frac{\dot{p}}{p}\right)_{1.82}(3.84) + \left(-\frac{\dot{\sigma}}{\sigma}\right)_{1.82}(4.93) > -0.3504 + 0.3565 = 0.0061 > 0$$

and (2.25) follows by the monotonicity of $\dot{p}/p$. ■

Corollary. *When* $\underline{\kappa} \le \kappa \le \overline{\kappa}$ *and* $\underline{u} \le u \le \overline{u}$, *then*

$$\frac{\dot{\tilde{\Pi}}_{\kappa}(u)}{-\tilde{\Pi}'_{\kappa}(u)} \le \frac{\dot{\tilde{\Pi}}_{\overline{\kappa}}(\underline{u})}{-\tilde{\Pi}'_{\underline{\kappa}}(\bar{u})} .$$

We are now in a position to prove Proposition 2.1. Let

$$\kappa_r = 1.82 + 2r \cdot 10^{-3}, \quad 0 \le r \le 14,$$

write $z(\kappa)$ for $z_{\tilde{\Pi}}(\kappa)$ and z_r for $z(\kappa_r)$.

Since $\dot{z}(\kappa) > 0$ it follows that if

$$\kappa_r \le \kappa \le \kappa_{r+1} = \kappa_r + 2 \cdot 10^{-3},$$

then

$$z_r \le z(\kappa) \le z_{r+1};$$

hence, by the Corollary above,

$$\dot{z}(\kappa) \le \frac{\dot{\tilde{\Pi}}_{\kappa_{r+1}}(z_r)}{-\tilde{\Pi}'_{\kappa_r}(z_{r+1})} =: Z_r ,$$

say. The upper bounds $Z_r (0 \le r \le 14)$ are computed using Lemma 2.6 and (2.24): thus, for each r,

$$Z_r < \frac{\epsilon^{-1}\left\{\tilde{\Pi}_{\kappa_{r+1}+\epsilon}(z_r) - \tilde{\Pi}_{\kappa_{r+1}}(z_r)\right\}}{\kappa_r \left\{\frac{p_{\kappa_r}(z_{r+1}-1)}{\sigma_{\kappa_r}(z_{r+1}-2)} - \frac{p_{\kappa_r}(z_{r+1})\sigma_{\kappa_r}(z_{r+1}-2)}{\sigma^2_{\kappa_r}(z_{r+1})}\right\}} .$$

The following Table 1 gives upper estimates for Z_r.

Table 1

κ_r	Z_r
1.820	2.1849
1.822	2.1848
1.824	2.1848
1.826	2.1847
1.828	2.1847
1.830	2.1846
1.832	2.1846
1.834	2.1846
1.836	2.1845
1.838	2.1845
1.840	2.1845
1.842	2.1844
1.844	2.1844
1.846	2.1843
1.848	2.1843

Thus $\dot{z}(\kappa) < 2.19$ for $1.82 \leq \kappa \leq 1.85$. ■

3. Narrowing the gap

We proved in [DHR 8], Theorem 2.3, that

$$z_{\tilde{\Pi}}(\kappa) > z_{\tilde{\chi}}(\kappa) \quad \text{for} \quad 1.5 \leq \kappa \leq 1.82$$

and

$$z_{\tilde{\chi}}(\kappa) > z_{\tilde{\Pi}}(\kappa) \quad \text{for} \quad 1.85 \leq \kappa \leq 2.$$

We proceed to narrow the gap between 1.82 and 1.85 in the following

Proposition 3.1. *We have*

$$z_{\tilde{\Pi}}(\kappa) > z_{\tilde{\chi}}(\kappa) \quad \textit{for} \quad 1.82 \leq \kappa \leq 1.834$$

and

$$z_{\tilde{\chi}}(\kappa) > z_{\tilde{\Pi}}(\kappa) \quad \textit{for} \quad 1.835 \leq \kappa \leq 1.85.$$

Proof. Let

$$f(\kappa) := z_{\tilde{\Pi}}(\kappa) - z_{\tilde{\chi}}(\kappa).$$

We showed in [DHR 8], Theorem 2.2, that $\dot{z}_{\tilde{\chi}}(\kappa) > 0$ (for $1.5 < \kappa < 2$ even), and in [DHR 5], Lemma Π2, that $\dot{z}_{\tilde{\Pi}}(\kappa) > 0$ (even for $\kappa \geq 1$); also, by Proposition 2.1 above, $\dot{z}_{\tilde{\Pi}}(\kappa) < 2.2$ for $1.82 \leq \kappa \leq 1.85$. Thus

$$\dot{f}(\kappa) < 2.2 \quad \text{for} \quad 1.82 \leq \kappa \leq 1.85. \tag{3.1}$$

Numerical evidence suggests that, in fact, $\dot{f}(\kappa)$ is about -0.436 in this κ-range, but the crude upper bound (3.1) suffices for our argument. Let $f_c(\kappa)$ denote the *computed* value of $f(\kappa)$. Our computed data of $z_{\tilde{\Pi}}(\kappa)$ and $z_{\tilde{\chi}}(\kappa)$ are guaranteed each to within 10^{-5}, and hence

$$|f_c(\kappa) - f(\kappa)| < 2 \times 10^{-5}. \tag{3.2}$$

We proceed as follows: We shall define a sequence

$$x_1 = 1.834 > x_2 > \cdots > x_{21} > 1.82 > x_{22}$$

in such a way that we shall be able to show that $f(\kappa) > 0$ on each interval $x_n \geq \kappa \geq x_{n+1}$.

This is how the step-by-step definition of the sequence $\{x_j\}$ is effected. Suppose that $f_c(x_n)$ has been calculated. By the mean value theorem and both (3.1) and (3.2) we obtain

$$\begin{aligned} f(\kappa) &> f(x_n) + 2.2(\kappa - x_n), \quad \kappa < x_n \\ &> f_c(x_n) - 2 \times 10^{-5} - 2.2(x_n - \kappa). \end{aligned}$$

If, for the moment, we write $y_n = f_c(x_n) - 2\times 10^{-5}$, this inequality tells us that, in the $\kappa - y$ plane, the point $(\kappa, f(\kappa))$ lies *above* the line $y - y_n = 2.2(\kappa - x_n)$ when $\kappa < x_n$. This line meets the κ-axis at

$$\bar{x}_{n+1} := x_n - y_n/2.2 = x_n - \frac{1}{2.2}(f_c(x_n) - 2 \times 10^{-5}),$$

and we may assert that

$$f(\kappa) > 0 \quad \text{for} \quad \bar{x}_{n+1} \leq \kappa \leq x_n.$$

Now define x_{n+1} to be the least integer multiple of 10^{-5} that exceeds $\bar{x}_{n+1}$; and repeat the procedure. The first part of Table 2 shows that the successive values of $f_c(x_n)$ are large enough to keep the process going, and to deal with the interval $1.82 \leq \kappa \leq 1.834$ in 22 steps.

This completes the proof of the first inequality in our Proposition, and the second is handle in an analogous manner.

Define a sequence

$$\xi_0 = 1.835 < \xi_1 < \cdots < \xi_{19} < 1.85 < \xi_{20}$$

so that $f(\kappa) < 0$ on each interval $\xi_n \le \kappa \le \xi_{n+1}$ as follows: Suppose that $f_c(\xi_n)$ has been calculated. By the mean value theorem again, together with (3.1) and (3.2),

$$\begin{aligned} f(\kappa) &< f(\xi_n) + 2.2(\kappa - \xi_n), \quad \kappa > \xi_n \\ &< f_c(\xi_n) + 2 \times 10^{-5} + 2.2(\kappa - \xi_n). \end{aligned}$$

Writing $\eta_n = f_c(\xi_n) + 2 \times 10^{-5}$, the point $(\kappa, f(\kappa))$ lies *below* the line $y - \eta_n = 2.2(x - \xi_n)$ when $\kappa > \xi_n$. We let

$$\bar{\xi}_{n+1} := \xi_n - \frac{\eta_n}{2.2} = \xi_n - \frac{1}{2.2}(f_c(\xi_n) + 2 \times 10^{-5}),$$

and then choose ξ_{n+1} to be the largest multiple of 10^{-5} less than $\bar{\xi}_{n+1}$. Once again the successive values of $f_c(\xi_n)$ are small enough (large enough, numerically) to keep the process going and to terminate it in 21 steps, as the second part of Table 2 shows.

Table 2

κ	$f(\kappa) \ge$	κ	$f(\kappa) \le$
1.83400	0.00018	1.83500	-0.00024
1.83393	0.00021	1.83510	-0.00029
1.83385	0.00025	1.83522	-0.00034
1.83375	0.00029	1.83536	-0.00040
1.83363	0.00034	1.83553	-0.00047
1.83349	0.00040	1.83573	-0.00056
1.83332	0.00048	1.83597	-0.00067
1.83312	0.00057	1.83626	-0.00079
1.83287	0.00067	1.83661	-0.00094
1.83258	0.00080	1.83702	-0.00112

Table 2, continued

κ	$f(\kappa) \geq$	κ	$f(\kappa) \leq$
1.83223	0.00095	1.83752	-0.00134
1.83181	0.00113	1.83812	-0.00160
1.83131	0.00135	1.83883	-0.00191
1.83071	0.00161	1.83968	-0.00229
1.82999	0.00192	1.84071	-0.00274
1.82913	0.00230	1.84194	-0.00328
1.82810	0.00274	1.84342	-0.00393
1.82687	0.00327	1.84519	-0.00471
1.82540	0.00391	1.84732	-0.00565
1.82364	0.00466	1.84987	-0.00677
1.82154	0.00556	1.85293	-0.00813
1.81903	0.00664		

4. A lower bound for $\dot{z}_{\tilde{\chi}}(\kappa)$: Estimating $\dot{\tilde{\chi}}_\kappa(u)$ near $u = z_{\tilde{\chi}}(\kappa)$

We have the formula (1.14) for $\dot{z}_{\tilde{\chi}}(\kappa)$, and we proceed to deal with the numerator in this section. After section 3 we can confine ourselves to the range $1.834 \leq \kappa \leq 1.835$, and since $\dot{z}_{\tilde{\chi}}(\kappa) > 0$ for $1.5 < \kappa < 2$ by [DHR 8], Theorem 2.2, we know that the points $z_{\tilde{\chi}}(\kappa)$ corresponding to the short κ-range lie in $[z_{\tilde{\chi}}(1.834), z_{\tilde{\chi}}(1.835)]$. Thanks to Dr. Bradley's secured computations we can restrict attention to the range

$$4.88 < z_{\tilde{\chi}}(\kappa) < 4.884, \quad \kappa \in [1.834, 1.835].$$

(In fact, $z_{\tilde{\chi}}(1.834) = 4.88077\ldots$ and $z_{\tilde{\chi}}(1.835) = 4.88337\ldots$.)

From [DHR 8], (4.3) et seq., we have

$$\begin{aligned}\frac{\partial}{\partial\kappa}(\sigma_\kappa(u)\tilde{\chi}_\kappa(u)) &= (u-1)\dot{q}_\kappa(u-1) \\ &\quad - \kappa\int_{u-2}^{u}\frac{\sigma(u)}{\sigma(t)}\left\{\frac{\dot{\sigma}_\kappa}{\sigma}(u) - \frac{\dot{\sigma}_\kappa}{\sigma}(t)\right\}q(t+1)dt \\ &\quad - \int_{u-2}^{u}\left(\frac{\sigma(u)}{\sigma(t)} - 1\right)\{\kappa\dot{q}_\kappa(t+1) + q(t+1)\}dt \\ &= -A_0 - B_0 - C_0, \end{aligned} \tag{4.1}$$

say, where

$$A_0 := -(u-1)\dot{q}_\kappa(u-1), \tag{4.2}$$

$$B_0 := \kappa\sigma(u)\int_{u-2}^{u}\left\{\frac{\sigma_\kappa}{\sigma}(u)-\frac{\sigma_\kappa}{\sigma}(t)\right\}\frac{q(t+1)}{\sigma(t)}\,dt \tag{4.3}$$

and

$$C_0 := \int_{u-2}^{u}\left(\frac{\sigma(u)}{\sigma(t)}-1\right)\{\kappa\dot{q}_\kappa(t+1)+q(t+1)\}dt. \tag{4.4}$$

Let $\mathcal{R}$ denote from now on the small two-dimensional range $[\underline{\kappa},\overline{\kappa}]\times[\underline{u},\bar{u}]$, where $\underline{\kappa}=1.834$, $\overline{\kappa}=1.835$, $\underline{u}=4.88$, $\bar{u}=4.884$.

We shall prove that

Proposition 4.1. *When (κ,u) lies in $\mathcal{R}$, then*

$$A_0 > 48.95, \quad B_0 > 11.74 \ \textit{and}\ \ C_0 > 1.52; \tag{4.5}$$

and therefore

$$\frac{\partial}{\partial\kappa}\left(\sigma_\kappa(u)\tilde{\chi}_\kappa(u)\right) < -62.21. \tag{4.6}$$

The expression on the left of (4.6) is

$$\dot{\sigma}_\kappa(u)\tilde{\chi}_\kappa(u)+\sigma_\kappa(u)\dot{\tilde{\chi}}_\kappa(u)=\sigma_\kappa(z_{\tilde{\chi}}(\kappa))\dot{\tilde{\chi}}_\kappa(z_{\tilde{\chi}}(\kappa))$$

at $u = z_{\tilde{\chi}}(\kappa)$ since $\tilde{\chi}_\kappa(z_{\tilde{\chi}}(\kappa)) = 0$, and we deduce at once, since $\sigma_\kappa(u)$ is a positive function, that

$$\dot{\tilde{\chi}}_\kappa(z_{\tilde{\chi}}(\kappa)) < -\frac{62.21}{\sigma_\kappa(z_{\tilde{\chi}}(\kappa))}, \quad \underline{\kappa}\le\kappa\le\overline{\kappa}.$$

But $\sigma_\kappa(u)$ increases with u and decreases in κ (the latter by (2.18)), whence, on $\mathcal{R}$, $\sigma_\kappa(u) < \sigma_{1.834}(4.884) < 0.7574$ (computer calculation), and hence we shall obtain

$$\dot{\tilde{\chi}}_\kappa(z_{\tilde{\chi}}(\kappa)) < -82.13, \quad \underline{\kappa}\le\kappa\le\overline{\kappa}. \tag{4.7}$$

Our method for dealing with $\dot{\tilde{\chi}}_\kappa$ is evidently quite different from that used to estimate $\dot{\tilde{\Pi}}_\kappa$ in section 2. The reason is that to imitate section 2 would require us to deal with $\ddot{q}_\kappa$ and $\ddot{\tilde{\chi}}_\kappa$, and just a glance at the Appendix suffices to convince one that it would be better to go back to [DHR8] and sharpen the method developed there in Section 4. Indeed, we showed then that $A_0 > 0$ on a much more extensive (κ, u)-range ([DHR8], Prop. 4.2), and also that $B_0 > 0$ (loc. cit., (4.5) and Prop. 4.3); and it seems reasonable that we could obtain more accurate lower bounds on the small range $\mathcal{R}$. We isolated the term A_0 because numerical experimentation showed A_0 to be by far the most important. The last term, C_0, is the price we had to pay for separating off A_0 (cf. [DHR8], (4.4) and (4.6)); C_0 is positive too, but to prove this is delicate because the integrand changes sign within the range of integration.

The inequality for A_0 in (4.5) is proved in the Appendix, Proposition A5.2, Cor.

Here we deal first with C_0, and then with B_0.

We begin by writing C_0 in the form

$$C_0 = \int_0^2 f(t)g(t)\,dt = \sum_{n=1}^{8} C_{0,n}$$

where

$$f(t) = f(t,u,\kappa) := \frac{\sigma_\kappa(u)}{\sigma_\kappa(u+t-2)} - 1 =: Y_\kappa(u, 2-t)^{-1} - 1,$$
$$g(t) = g(t,u,\kappa) := \kappa\dot{q}_\kappa(u+t-1) + q_\kappa(u+t-1)$$

and

$$C_{0,n} = \int_{(n-1)/4}^{n/4} f(t)g(t)\,dt, \quad n = 1, 2, \ldots, 8.$$

For $0 \le t \le 2$, $f(t)$ is positive, decreasing and convex in t, but increasing in κ and decreasing in u ([DHR5], Lemmas $\sigma 4$, $\sigma 5$), so that

$$\underline{f}(t) := f(t, \bar{u}, \underline{\kappa}) \le f(t) \le f(t, \underline{u}, \bar{\kappa}) =: \overline{f}(t), \quad 0 \le t \le 2,$$

when $\underline{u} \le u \le \bar{u}$ and $\underline{\kappa} \le \kappa \le \bar{\kappa}$. The function $g(t)$ is increasing and convex in t (also in u) (Propositions A3 and A4) and we denote by $\gamma(t)$ a lower bound for $g(t)$ that is valid for $\underline{u} \le u \le \bar{u}$ and $\underline{\kappa} \le \kappa \le \bar{\kappa}$. We shall construct such a function $\gamma(t)$ by first estimating $\dot{q}$ via Propositions A5.1 and A5.2; according to the sign of $\dot{q}(u+t-1)$ we shall bound q_κ by $q_{\bar{\kappa}}$ or $q_{\underline{\kappa}}$ and we use $\bar{\kappa}$ or $\underline{\kappa}$ as the coefficient of $\dot{q}(u+t-1)$. We have shown [Proposition A5.3] that the change of sign of $\dot{q}_\kappa(x)$ occurs in the interval $4.70 < x < 4.72$ for all $\kappa \in [\underline{\kappa}, \bar{\kappa}]$. Since our nearest γ evaluations occur at the arguments 4.63 and 4.88, we know whether to use $\underline{\kappa}$ or $\bar{\kappa}$ in forming γ at these points.

We have

$$C_{0,n} = \gamma(\tfrac{n-1}{4}) \int_{(n-1)/4}^{n/4} f + \int_{(n-1)/4}^{n/4} f(g - \gamma(\tfrac{n-1}{4})),$$

$$\text{(4.8)} \qquad \geq \gamma(\tfrac{n-1}{4}) \int_{(n-1)/4}^{n/4} f + f(\tfrac{n}{4}) \int_{(n-1)/4}^{n/4} (g - \gamma(\tfrac{n-1}{4})),$$

since $g(t) \geq \gamma(\frac{n-1}{4})$ on $\frac{1}{4}(n-1) \leq t \leq \frac{1}{4}n$ and f is decreasing.

When $n = 1$, 2 and 3, $\gamma(\frac{n-1}{4}) < 0$, and in each of these three cases we first rewrite (4.8) as

$$C_{0,n} \geq \gamma(\tfrac{n-1}{4}) \int_{(n-1)/4}^{n/4} \left(f - f\left(\tfrac{n}{4}\right)\right) + f\left(\tfrac{n}{4}\right) \int_{(n-1)/4}^{n/4} g,$$

and then deduce by monotonicity and convexity that

$$\begin{aligned} C_{0,n} &\geq \tfrac{1}{4}\gamma(\tfrac{n-1}{4}) \cdot \tfrac{1}{2}(f(\tfrac{n-1}{4}) - f(\tfrac{n}{4})) + \tfrac{1}{4} f(\tfrac{n}{4}) g(\tfrac{n-1/2}{4}) \\ &\geq \tfrac{1}{4}\left\{\gamma(\tfrac{n-1}{4}) \cdot \tfrac{1}{2}(f(\tfrac{n-1}{4}) - f(\tfrac{n}{4})) + f(\tfrac{n}{4})\gamma(\tfrac{n-1/2}{4})\right\} \\ &= \tfrac{1}{4}\left\{\tfrac{1}{2}\gamma(\tfrac{n-1}{4}) f(\tfrac{n-1}{4}) + f(\tfrac{n}{4})(\gamma(\tfrac{n-1/2}{4}) - \tfrac{1}{2}\gamma(\tfrac{n-1}{4}))\right\} \end{aligned}$$

$$\text{(4.9)} \qquad \geq \tfrac{1}{4}\left\{\tfrac{1}{2}\gamma(\tfrac{n-1}{4}) \overline{f}(\tfrac{n-1}{4}) + f^*(\tfrac{n}{4})(\gamma(\tfrac{n-1/2}{4}) - \tfrac{1}{2}\gamma(\tfrac{n-1}{4}))\right\}, \quad n = 1, 2, 3,$$

where

$$f^*(\tfrac{n}{4}) = \begin{cases} \underline{f}(\frac{n}{4}) & \text{if } \gamma(\frac{n-1/2}{4}) > \frac{1}{2}\gamma(\frac{n-1}{4}) \\ \overline{f}(\frac{n}{4}) & \text{if } \gamma(\frac{n-1/2}{4}) < \frac{1}{2}\gamma(\frac{n-1}{4}). \end{cases}$$

For each of $n = 4, 5, \ldots, 8$, $\gamma(\frac{n-1}{4}) > 0$ and then, by (4.8),

$$\begin{aligned} C_{0,n} &\geq \tfrac{1}{4}\gamma(\tfrac{n-1}{4}) f(\tfrac{n-1/2}{4}) + \tfrac{1}{4} f(\tfrac{n}{4})(g(\tfrac{n-1/2}{4}) - \gamma(\tfrac{n-1}{4})) \\ &\geq \tfrac{1}{4}\gamma(\tfrac{n-1}{4}) f(\tfrac{n-1/2}{4}) + \tfrac{1}{4} f(\tfrac{n}{4})(\gamma(\tfrac{n-1/2}{4}) - \gamma(\tfrac{n-1}{4})) \end{aligned}$$

$$\text{(4.10)} \qquad \geq \tfrac{1}{4}\left\{\gamma(\tfrac{n-1}{4}) \underline{f}(\tfrac{n-1/2}{4}) + \underline{f}(\tfrac{n}{4})(\gamma(\tfrac{n-1/2}{4}) - \gamma(\tfrac{n-1}{4}))\right\}, \quad n = 4, \ldots, 8.$$

The outcome of the numerical estimations is set out below, with lines $n = 1, 2, 3$ deriving from (4.9) and the rest from (4.10).

Table 3

n	$C_{0,n} \geq$	n	$C_{0,n} \geq$
1	-4.0216	4	1.1631
2	-1.8847	5	1.9238
3	-0.1267	6	2.1156
		7	1.7017
		8	0.6541

We find that $\sum_{n=1}^{8} C_{0,n} > 1.5253$, which establishes the bound for C_0 stated in Proposition 4.1.

In the last part of this section we derive the inequality (4.5) for B_0. Turning to the definition (4.3) of B_0, the integrand is non-negative; this is clear for q since $u - 1 > \rho_\kappa$ in $\mathcal{R}$, and by (2.19), the difference in curly brackets is also positive. Hence, on $\mathcal{R}$,

$$B_0 \geq 1.834\sigma_{1.835}(4.88)B^*(\kappa, u) > 1.3873B^*(\kappa, u)$$

using $\sigma_{1.835}(4.88) > 0.75646$ (from our data), where

$$B^*(\kappa, u) := \int_{u-2}^{u} \left\{ \frac{\dot{\sigma}_\kappa}{\sigma}(u) - \frac{\dot{\sigma}_\kappa}{\sigma}(t) \right\} \frac{q(t+1)}{\sigma(t)}\, dt.$$

We set out to bound $B^*(\kappa, u)$ from below on $\mathcal{R}$. We shall prove that

$$B^*(\kappa, u) \geq 8.466 \quad \text{when} \quad (\kappa, u) \in \mathcal{R}; \tag{4.11}$$

from the discussion above, this implies (cf. (4.5)) that $B_0 > 11.74$.

We prepare the ground by discussing separately the two components $q(t+1)/\sigma(t)$ and $(\dot{\sigma}_\kappa/\sigma)(u) - (\dot{\sigma}_\kappa/\sigma)(t)$ that make up the integrand. Actually we know that $q(t+1)/\sigma(t)$ is convex on a much larger κ-range, but we make do here, at little cost, with simpler considerations.

Lemma 4.1. *Suppose that* $\underline{u} - 2 \leq a < b \leq \bar{u}$. *Let* $f(t) = q(t+1)/\sigma(t)$ *and define*

$$c = \min_{a \leq t \leq b} \frac{f'}{f}(t).$$

Then

$$f(t) \geq f(a) \exp c(t-a), \quad a \leq t \leq b,$$

$$\int_a^b f(t)\,dt \geq (b-a)f(a)\exp\{c(b-a)/2\},$$

and, if $\underline{\kappa} \leq \kappa \leq \overline{\kappa}$,

$$c \geq \frac{2\underline{\kappa}-1}{b+1-\underline{\kappa}} - \frac{\overline{\kappa}}{a}(1 - Y_{\overline{\kappa}}(a)),$$

where $Y_\kappa(t) := \sigma_\kappa(t-2)/\sigma_\kappa(t)$.

Proof. The first inequality comes from integrating f'/f and the second from estimating from below the convex function $\exp c(t-a)$. For the third inequality we argue as follows.

In $\mathcal{R}$, $q(x)$, $q'(x)$ and $q''(x)$ are positive. It follows from (1.3), the delay differential equation for q, and by Taylor's Theorem (along lines we have often followed before) that

$$xq'(x) = (\kappa-1)q(x) + \kappa q(x+1) > (\kappa-1)q(x) + \kappa(q(x)+q'(x)),$$

whence

$$\frac{q'}{q}(t+1) > \frac{2\kappa-1}{t+1-\kappa} \geq \frac{2\underline{\kappa}-1}{b+1-\underline{\kappa}}. \tag{4.12}$$

Also, from the delay differential equation for σ,

$$t\frac{\sigma'_\kappa}{\sigma_\kappa}(t) = \kappa\left(1 - \frac{\sigma_\kappa(t-2)}{\sigma_\kappa(t)}\right) = \kappa(1 - Y_\kappa(t)),$$

and the function $Y_\kappa(t)$ is increasing in t and decreasing in κ ([DHR 5], Lemmas $\sigma 4$ and $\sigma 5$). Hence

$$\frac{\sigma'}{\sigma}(t) \leq \frac{\overline{\kappa}}{a}(1 - Y_{\overline{\kappa}}(a)).$$

Together with (4.12) this yields the lower bound for c. ∎

Corollary. *Under the hypotheses of Lemma 4.1,* $q(t+1)/\sigma(t)$ *increases with* t.

Proof. It suffices to show that $c > 0$. We have

$$c \geq \frac{2\underline{\kappa}-1}{\bar{u}+1-\underline{\kappa}} - \frac{\bar{\kappa}}{\underline{u}-2}(1 - Y_{\bar{\kappa}}(\underline{u}-2)) \geq .658 - .563 > 0. \quad ∎$$

Lemma 4.2. *For* $\underline{\kappa} \leq \kappa \leq \bar{\kappa}$ *we have*

$$\int_{2.88}^{4.88} \frac{q_\kappa(t+1)}{\sigma_\kappa(t)}\,dt > 51.15.$$

Proof. We express the integral as a sum of ten integrals, each taken over an interval of length 0.2 and apply Lemma 4.1 on each interval. The functions in each integrand are positive and satisfy $\sigma_{\underline{\kappa}}(t) \geq \sigma_\kappa(t)$ by (2.18) and $q_\kappa(t+1) \geq q_*(t+1)$ by (9) of the Appendix and Lemma A.5.1. Thus we have

$$\int_{2.88}^{4.88} \frac{q_\kappa(t+1)}{\sigma_\kappa(t)}\, dt \geq \sum_{r=0}^{9} 0.2 \frac{q_*(3.88+0.2r)}{\sigma_{\underline{\kappa}}(2.88+0.2r)} \exp\{\frac{1}{10}c_r\},$$

where c_r is the value of c in Lemma 4.1 for the interval $[2.88+0.2r, 2.88+0.2r+.2]$. The claimed result follows from totaling up the entries in the second column of Table 4 below.

Lemma 4.3. *For $\kappa > 1$ and $x > 2$ we have*

$$\left(\frac{\dot{\sigma}_\kappa}{\sigma}\right)'(x) > \frac{\kappa}{x} Y(x) \left\{\frac{\dot{\sigma}_\kappa}{\sigma}(x) - \frac{\dot{\sigma}_\kappa}{\sigma}(x-2)\right\}.$$

Proof. From (2.18)

$$\begin{aligned}
\frac{\dot{\sigma}_\kappa}{\sigma}(x) - \frac{\dot{\sigma}_\kappa}{\sigma}(x-2) &= \int_0^2 \left\{\frac{\sigma(x-t)}{\sigma(x)} - \frac{\sigma(x-2-t)}{\sigma(x-2)}\right\} \frac{dt}{t} \\
&= \int_0^2 \frac{\sigma(x-t)}{\sigma(x-2)} \{Y(x) - Y(x-t)\} \frac{dt}{t} \\
&= \frac{1}{Y(x)} \int_0^2 \frac{\sigma(x-t)}{\sigma(x)} \left\{\frac{x-t}{\kappa}\frac{\sigma'}{\sigma}(x-t) - \frac{x}{\kappa}\frac{\sigma'}{\sigma}(x)\right\} \frac{dt}{t} \\
&< \frac{1}{Y(x)} \frac{x}{\kappa} \int_0^2 \frac{\sigma(x-t)}{\sigma(x)} \left\{\frac{\sigma'}{\sigma}(x-t) - \frac{\sigma'}{\sigma}(x)\right\} \frac{dt}{t} \\
&= \frac{1}{Y(x)} \cdot \frac{x}{\kappa} \left(\frac{\dot{\sigma}_\kappa}{\sigma}\right)'(x)
\end{aligned}$$

after differentiating $(\dot{\sigma}_\kappa/\sigma)(x)$ with respect to x from (2.18). ■

We are now in a position to tackle $B^*(\kappa, u) = B^*(u)$.

Lemma 4.4. *We have $B^{*\prime}(u) > 0$ if $4.88 \leq u \leq 4.884$. In particular,*

$$B^*(\kappa, u) \geq B^*(\kappa, 4.88).$$

Proof. From its definition,

$$B^{*\prime}(u) = \left(\frac{\dot{\sigma}}{\sigma}\right)'(u) \int_{u-2}^{u} \frac{q(t+1)}{\sigma(t)}\, dt - \left\{\frac{\dot{\sigma}_\kappa}{\sigma}(u) - \frac{\dot{\sigma}_\kappa}{\sigma}(u-2)\right\} \frac{q(u-1)}{\sigma(u-2)}.$$

Since the integral is positive, we may apply Lemma 4.3, and we obtain then

$$B^{*\prime}(u) > \left\{\frac{\dot\sigma_\kappa}{\sigma}(u) - \frac{\dot\sigma_\kappa}{\sigma}(u-2)\right\}\left(\kappa\frac{Y(u)}{u}\int_{u-2}^{u}\frac{q(t+1)}{\sigma(t)}\,dt - \frac{q(u-1)}{\sigma(u-2)}\right).$$

By (2.19) the difference $(\dot\sigma_\kappa/\sigma)(u) - (\dot\sigma_\kappa/\sigma)(u-2)$ is positive, and we have therefore only to show that

$$\kappa\frac{Y(u)}{u}\int_{u-2}^{u}\frac{q(t+1)}{\sigma(t)}\,dt > \frac{q(u-1)}{\sigma(u-2)}. \tag{4.13}$$

In view of the monotonicities of $Y_\kappa(u)$ cited in the proof of Lemma 4.1, and of the monotonicity of $q(t+1)/\sigma(t)$ in t established in Lemma 4.1, Cor., the expression on the left of (4.13) is of least

$$1.834\frac{Y_{1.835}(4.88)}{4.884}\int_{2.88}^{4.88}\frac{q(t+1)}{\sigma(t)}\,dt.$$

We have shown in Lemma 4.3 that the last integral is at least 51.15. Thus the left side of (4.13) is at least

$$(1.834)(0.503)(51.15)/4.884 > 9.66.$$

We estimate $q_\kappa(u-1)$ on the right side of (4.13) as follows:

$$\begin{aligned} q_\kappa(u-1) &\le q_\kappa(3.884) \quad (q' > 0 \quad \text{on} \quad \mathcal{R}) \\ &\le q_{\underline{\kappa}}(3.884) \quad (\text{by Proposition A5.3}) \\ &\le \Sigma(3.884, \underline{\kappa}, 4) \quad (\text{by [DHR 7], Theorem 0.1}). \end{aligned}$$

Thus $q_\kappa(u-1)/\sigma_\kappa(u-2) \le 3.2364/0.38057 < 8.51$.

This establishes (4.13) and the Lemma is proved. ■

By Lemma 4.4 we have now to estimate $B^*(\kappa, 4.88)$ from below. By (2.19), $(\dot\sigma/\sigma)(u) - (\dot\sigma/\sigma)(t)$ is positive for $u-2 \le t < u$ and decreasing in t. Thus we have

$$\begin{aligned} &B^*(\kappa, 4.88) \\ &\ge \sum_{j=1}^{9}\left\{\left(\frac{\dot\sigma}{\sigma}\right)_\kappa(4.88) - \left(\frac{\dot\sigma}{\sigma}\right)_\kappa\left(4.88 - \frac{j}{5}\right)\right\}\int_{4.88-(j+1)/5}^{4.88-j/5}\frac{q(t+1)}{\sigma(t)}\,dt \\ &\ge \sum_{j=1}^{9}\left\{\left(\frac{\dot\sigma}{\sigma}\right)_{\bar\kappa}(4.88) - \left(\frac{\dot\sigma}{\sigma}\right)_{\underline{\kappa}}\left(4.88 - \frac{j}{5}\right)\right\}\int_{4.88-(j+1)/5}^{4.88-j/5}\frac{q(t+1)}{\sigma(t)}\,dt \end{aligned} \tag{4.14}$$

since $(\dot{\sigma}/\sigma)^{\cdot}_{\kappa}(u) < 0$ by (2.20). The nine integrals on the right have been tabulated in Table 4. We have found a method to estimate rather simply and accurately the $\dot{\sigma}/\sigma$ expressions as follows.

Lemma 4.5. *Let $\delta > 0$, $\kappa > 1 + \delta$ and $t > 2$. Then*

$$\frac{1}{\delta}\{\log \sigma_{\kappa+\delta}(t) - \log \sigma_{\kappa}(t)\} < \left(\frac{\dot{\sigma}}{\sigma}\right)_{\kappa}(t) < -\frac{1}{\delta}\{\log \sigma_{\kappa-\delta}(t) - \log \sigma_{\kappa}(t)\}.$$

Proof. We showed in (2.20) that

$$(\log \sigma_{\kappa}(t))^{\cdot\cdot} = (\dot{\sigma}/\sigma)^{\cdot}_{\kappa}(t) < 0$$

in the specified κ and t ranges. The stated inequalities follow by concavity. ■

For actual calculations, our numerical information about σ-values is (in our ranges) accurate to within 10^{-10}. Thus we estimate $(\dot{\sigma}_{\kappa}/\sigma_{\kappa})(t)$ from above by

$$\delta^{-1}\{\log(\sigma^*_{\kappa}(t) + 10^{-10}) - \log(\sigma^*_{\kappa-\delta}(t) - 10^{-10})\}$$

with $\delta = 10^{-4}$ and σ^* a machine-generated table value of σ. A similar procedure leads to lower estimates of $(\dot{\sigma}_{\kappa}/\sigma_{\kappa})(t)$. The outcomes of such calculations for $\Delta(\dot{\sigma}/\sigma) = (\dot{\sigma}/\sigma)_{\bar{\kappa}}(4.88) - (\dot{\sigma}/\sigma)_{\underline{\kappa}}(4.88 - j/5)$ are listed in column 3 of Table 4. The products of columns 2 and 3 are listed in column 4 of Table 4 for use in (4.14).

Totaling the entries of column 4, we obtain

$$B^*(\kappa, 4.88) \geq 8.4664, \quad 1.834 \leq \kappa \leq 1.835,$$

which proves (4.11).

Table 4

Interval I	$\int_I q/\sigma \geq$	$\Delta\dot{\sigma}/\sigma \geq$	$B^*_I \geq$
[2.88,3.08]	1.7733	0.5192	0.9206
[3.08,3.28]	2.4395	0.4466	1.0894
[3.28,3.48]	3.1226	0.3782	1.1809
[3.48,3.68]	3.8299	0.3139	1.2022
[3.68,3.88]	4.5679	0.2532	1.1565
[3.88,4.08]	5.3425	0.1961	1.0476
[4.08,4.28]	6.1612	0.1422	0.8761
[4.28,4.48]	7.0296	0.0915	0.6432
[4.48,4.68]	7.9534	0.0440	0.3499
[4.68,4.88]	8.9394	0.0000	0.0000
	51.1593		8.4664

5. A lower bound for $\dot{z}_{\tilde{\chi}}(\kappa)$: estimating $\tilde{\chi}'_\kappa(u)$ near $u = z_{\tilde{\chi}}(\kappa)$

By reference to the formula (1.14) for $\dot{z}_{\tilde{\chi}}(\kappa)$, it remains after (4.7) to estimate the positive quantity $\tilde{\chi}'_\kappa(z_{\tilde{\chi}}(\kappa))$ from above for $1.834 \le \kappa \le 1.835$. It suffices to prove

Lemma 5.1. *When (κ, u) lies in $\mathcal{R}$, then*

$$0 < \tilde{\chi}'_\kappa(u) < 34.01.$$

Assuming this result for the moment, we deduce from (1.14), (4.7) and Lemma 5.1 that

$$\dot{z}_{\tilde{\chi}}(\kappa) > 2.41, \quad 1.834 \le \kappa \le 1.835$$

and this suffices, following the discussion in the introduction, to complete the proof of Theorem 0.

We turn to Lemma 5.1 and prepare the ground with:

Lemma 5.2. *$\tilde{\chi}_\kappa(u)$ is increasing and convex in $u \ge \rho_\kappa + 1$, $\kappa > 1$.*

Proof. From (1.8) and (1.6),

$$\kappa^{-1}\tilde{\chi}'(u) = \frac{q(u)\sigma(u-2)}{\sigma^2(u)} + \frac{q(u-1)}{\sigma(u-2)} > 0, \quad u \ge \rho + 1.$$

Next,

$$\begin{aligned}\kappa^{-1}\tilde{\chi}''(u) &= \frac{q'(u)\sigma(u-2)}{\sigma^2(u)} + \left(\frac{\sigma'(u-2)}{\sigma^2(u)} - 2\frac{\sigma(u-2)\sigma'(u)}{\sigma^3(u)}\right) q(u) \\ &\quad + \frac{q'(u-1)}{\sigma(u-2)} - \frac{\sigma'(u-2)}{\sigma^2(u-2)} q(u-1) \\ &= \frac{\sigma(u-2)}{\sigma^2(u)}\left\{\frac{q'}{q}(u) + \frac{\sigma'}{\sigma}(u-2) - 2\frac{\sigma'}{\sigma}(u)\right\} q(u) \\ &\quad + \left\{\frac{q'}{q}(u-1) - \frac{\sigma'}{\sigma}(u-2)\right\}\frac{q(u-1)}{\sigma(u-2)}.\end{aligned}$$

It suffices to show that each of the two expressions on the right is positive.

The second expression is equal to $q'(\rho)/\sigma(\rho-1) > 0$ at $u = \rho + 1$; and if $u > \rho + 1$ we have, by (4.12) and (2.16) (with $u - 2$ in place of u),

$$\frac{q'}{q}(u-1) - \frac{\sigma'}{\sigma}(u-2) \ge \frac{2\kappa - 1}{u - 1 - \kappa} - \frac{\kappa}{u-2} > \frac{2\kappa-1}{u-2} - \frac{\kappa}{u-2} = \frac{\kappa - 1}{u-2} > 0,$$

so that the second expression is positive when $u > \rho + 1$.

Next, consider the terms from the curly brackets in the first expression on the right. From [DHR 2], Lemma $\sigma 1$, $t(\sigma'/\sigma)(t)$ is decreasing in t and therefore $\frac{\sigma'}{\sigma}(u-2) \geq \frac{u}{u-2}\frac{\sigma'}{\sigma}(u)$; hence $\frac{\sigma'}{\sigma}(u-2) - 2\frac{\sigma'}{\sigma}(u) \geq \frac{4-u}{u-2} \cdot \frac{\sigma'}{\sigma}(u) \geq 0$ if $2 < u \leq 4$. If $u > 4$, using (4.12) and (2.16) again, these terms have a sum at least as large as

$$\frac{2\kappa-1}{u-\kappa} - \frac{u-4}{u-2}\frac{\sigma'}{\sigma}(u) \geq \frac{2\kappa-1}{u-\kappa} - \frac{(u-4)\kappa}{(u-2)u}$$
$$= \frac{(\kappa-1)u^2 + (\kappa^2+2)u - 4\kappa^2}{(u-\kappa)(u-2)u} > 0. \quad \blacksquare$$

We are now ready to prove Lemma 5.1. We have just shown in Lemma 5.2 that $\tilde{\chi}''_\kappa(u) > 0$, so that it suffices to give an upper estimate of $\tilde{\chi}'_\kappa(4.884)$. It is known that $Y_\kappa(u) = \sigma_\kappa(u-2)/\sigma_\kappa(u)$ is a decreasing function of κ [DHR 5, Lemma $\sigma 4$], and the same is true of each of $\sigma_\kappa(u)$, $\sigma_\kappa(u-2)$ [GR, Theorem 1]. Since each of $q_\kappa(4.884)$, $q_\kappa(3.884)$ is positive, we have

$$\begin{aligned}\tilde{\chi}'_\kappa(4.884) &\leq \frac{1.835}{\sigma_{1.835}(4.884)} Y_{1.834}(4.884) q_\kappa(4.884) + 1.835 \frac{q_\kappa(3.884)}{\sigma_{1.835}(2.884)} \\ &< 1.835 \left\{ \frac{0.50405}{0.75694} q_\kappa(4.884) + \frac{q_\kappa(3.884)}{0.38130} \right\} \\ &< 1.222 q_{1.835}(4.884) + 4.8125 q_{1.834}(3.884)\end{aligned}$$

by Proposition A5.3. By [DHR 7, Theorem 0.1], $q_\kappa(t) \leq \Sigma(t, \kappa, 4)$ and from this inequality

$$q_{1.834}(3.884) < 3.23635, \quad q_{1.835}(4.884) < 15.0812.$$

Hence

$$\tilde{\chi}'(u) < 34.005.$$

Appendix

Let

$$\text{(1)} \qquad \operatorname{Ein} z = \int_0^z (1 - e^{-t}) t^{-1}\, dt, \qquad z \in \mathbb{C}$$

and for $\kappa > 0, u > 0$ define

$$\text{(2)} \qquad q_\kappa(u) = \frac{\Gamma(2\kappa)}{2\pi i} \int_C z^{-2\kappa} e^{\kappa \operatorname{Ein}(-z)} e^{uz} dz,$$

where $\mathcal{C}$ is the Hankel contour that surrounds the negative real axis in the positive sense.

We showed in [DHR 7] how to approximate q and its derivatives with respect to u and κ by reasonably simple expressions that can be used for calculations and estimations. Here we carry out an analysis of $\dot{q}_\kappa$ and its derivatives, to be used in calculations on a highly restricted range of values of u and κ. We shall deal separately, in different sections below, with $\dot{q}''_\kappa$, $\dot{q}'_\kappa$ and $\dot{q}_\kappa$.

A1. Preliminaries. In this section we establish an identity for functions $\dot{q}_\kappa^{(\nu)}$ for $\nu = 1, 2, \ldots$ that will be useful for making approximations. We start by introducing the notation we use.

Let $a_j = a_j(\kappa)$ $(j = 0, 1, 2, 3, 4)$ be defined as follows: $a_0 = 1$, $a_1 = \kappa$, $a_2 = \frac{1}{2}\kappa(\kappa - \frac{1}{2})$, $a_3 = (\kappa/36)(6\kappa^2 - 9\kappa + 2)$, and $a_4 = -\kappa^3/95$. For $r = 3$ or 4 and $u > 0$ set

$$\Sigma(u, \kappa, r) = \sum_{n=0}^{r} (-1)^n a_n \frac{\Gamma(2\kappa)}{\Gamma(2\kappa - n)} u^{2\kappa - n - 1}, \tag{3}$$

$$\Delta(u, \kappa, r) = \exp(\kappa \operatorname{Ein} u) - \sum_{n=0}^{r} a_n u^n, \tag{4}$$

and

$$\delta(u, \kappa, r) = \frac{1}{\Gamma(1 - 2\kappa)} \int_0^\infty e^{-ux} x^{-2\kappa} \Delta(x, \kappa, r)\, dx. \tag{5}$$

We have shown that for $r = 3$ or 4 and $u > 0$

$$q_\kappa(u) = \Sigma(u, \kappa, r) + \delta(u, \kappa, r) \tag{6}$$

and that for $3/2 < \kappa < 2$ and all $x > 0$,

$$\frac{-\kappa^3}{95} x^4 \le \Delta(x, \kappa, 3) \le 0 \le \Delta(x, \kappa, 4) \le \frac{\kappa^3}{95} x^4. \tag{7}$$

It follows that, for $r = 3$ or 4; $3/2 < \kappa < 2$; $\nu = 0, 1, 2, \ldots$; and all $u > 0$

$$(-1)^{r+\nu+1} \delta^{(\nu)}(u, \kappa, r) > 0 \tag{8}$$

and hence, for $3/2 < \kappa < 2$, all $u > 0$, and $\nu = 0, 1, 2, \ldots$

$$(-1)^\nu \Sigma^{(\nu)}(u, \kappa, 3) \le (-1)^\nu q_\kappa^{(\nu)}(u) \le (-1)^\nu \Sigma^{(\nu)}(u, \kappa, 4). \tag{9}$$

For $r = 3$ or 4, $\nu = 0, 1, 2, \ldots$ and $u > 0$ set

$$(10)\quad S_\nu(u,\kappa,r) = \sum_{n=1}^{r}(-1)^n a_n \frac{\Gamma(2\kappa)}{\Gamma(2\kappa-n-\nu)}\left(\sum_{k=\nu+1}^{\nu+n}\frac{1}{2\kappa-k}\right)u^{2\kappa-\nu-n-1}.$$

We now have the following extension of Proposition 5.1 of [DHR7].

Proposition A1. *Suppose that* $1.5 < \kappa < 2$ *and* $u > 0$. *Then for* $\nu = 0, 1, 2, \ldots$ *we have*

$$\begin{aligned}\frac{1}{2}\dot q_\kappa^{(\nu)}(u) = {} & \left\{\log u + \sum_{k=1}^{\nu}\frac{1}{2\kappa-k}\right\}\Sigma^{(\nu)}(u,\kappa,r)\\ & -\frac{1}{2}\int_0^1\{q^{(\nu)}(u+s)-q^{(\nu)}(u)\}\frac{ds}{s}\\ & + S_\nu(u,\kappa,r) + \{\log u + \psi(1-2\kappa)\}\delta^{(\nu)}(u,\kappa,r)\\ & + (-1)^{\nu+1}\frac{u^{2\kappa-\nu-1}}{\Gamma(1-2\kappa)}\int_0^\infty x^{\nu-2\kappa}e^{-x}\Delta(\frac{x}{u},\kappa,r)\log x\,dx.\end{aligned}$$

Here ψ *denotes the logarithmic derivative of* Γ.

Proof. From the definition of q and after some manipulation (as was done in the case $\nu = 0$ in [DHR 7, Prop 5.1]) we obtain

$$\begin{aligned}\frac{1}{2}\dot q_\kappa^{(\nu)}(u) = {} & \{\psi(2\kappa)+\log u\}q^{(\nu)}(u) - \frac{1}{2}\int_0^1\{q^{(\nu)}(u+s)-q^{(\nu)}(u)\}\frac{ds}{s}\\ & - u^{2\kappa-\nu-1}\frac{\Gamma(2\kappa)}{2\pi i}\int_C w^{\nu-2\kappa}(\log w)e^{\kappa\,\mathrm{Ein}(-w/u)}e^w\,dw,\end{aligned}$$

where C is the aforementioned Hankel contour.

By substituting (4) in the last expression we obtain

$$(11)\quad\begin{aligned}\frac{1}{2}\dot q_\kappa^{(\nu)}(u) = {} & \{\psi(2\kappa)+\log u\}q^{(\nu)}(u)\\ & -\frac{1}{2}\int_0^1\{q^{(\nu)}(u+s)-q^{(\nu)}(u)\}\frac{ds}{s} + A + B,\end{aligned}$$

where

$$A := -u^{2\kappa-\nu-1}\frac{\Gamma(2\kappa)}{2\pi i}\sum_{n=0}^{r}a_n\left(\frac{-1}{u}\right)^n\int_C w^{n+\nu-2\kappa}(\log w)e^w\,dw$$

and

$$B := -u^{2\kappa-\nu-1}\frac{\Gamma(2\kappa)}{2\pi i}\int_C w^{\nu-2\kappa}(\log w)e^w\Delta(\frac{-w}{u},\kappa,r)\,dw.$$

We shall rewrite the last two integrals, using the Hankel identity in the first one. This yields, by (9) and (10),

$$\begin{aligned} A &= -\sum_{n=0}^{r}(-1)^n a_n u^{2\kappa-\nu-n-1}\frac{\Gamma(2\kappa)}{\Gamma(2\kappa-\nu-n)}\psi(2\kappa-\nu-n) \\ &= -\sum_{n=0}^{r}(-1)^n a_n u^{2\kappa-\nu-n-1}\frac{\Gamma(2\kappa)}{\Gamma(2\kappa-\nu-n)}\left\{\psi(2\kappa)-\sum_{k=1}^{n+\nu}\frac{1}{2\kappa-k}\right\} \\ &= \left\{\sum_{k=1}^{\nu}\frac{1}{2\kappa-k}-\psi(2\kappa)\right\}\Sigma^{(\nu)}(u,\kappa,r)+S_\nu(u,\kappa,r). \end{aligned}$$

Turning to B, we collapse $\mathcal{C}$ onto the negative real axis and change the sign of the integration variable. We get

$$\begin{aligned} B &= \{\psi(1-2\kappa)-\psi(2\kappa)\}\delta^{(\nu)}(u,\kappa,r) \\ &\quad + (-1)^{\nu+1}\frac{u^{2\kappa-\nu-1}}{\Gamma(1-2\kappa)}\int_0^\infty x^{\nu-2\kappa}e^{-x}\Delta\left(\frac{x}{u},\kappa,r\right)\log x\,dx. \end{aligned}$$

Finally, we combine the last expressions for A and B with (6) and (11) to obtain the desired formula for $\dot{q}^{(\nu)}$. ■

A2. Convexity estimates of some Hellinger Integrals. Let

$$H(u) := \int_0^1 \{q(u+s)-q(u)\}\frac{ds}{s}.$$

We estimate $H^{(\nu)}(u)$. We have

$$\{q^{(\nu)}(u+s)-q^{(\nu)}(u)\}/s = \int_0^1 \left(\frac{d}{du}\right)^\nu q'(u+st)dt$$

and hence

$$\left(\frac{\partial}{\partial s}\right)^2\{q^{(\nu)}(u+s)-q^{(\nu)}(u)\}/s = \int_0^1 q^{(\nu+3)}(u+st)t^2\,dt = \frac{1}{3}q^{(\nu+3)}(\xi)$$

for some $\xi \in (u, u+s)$. The sign of this expression is $(-1)^{[2\kappa]+\nu+3}$ for all $u > 0$ [DHR3, Prop. 10] which is $(-1)^\nu$ in our κ range. Thus the integrand of $H^{(\nu)}$

is convex for ν even and concave for ν odd, and we have

$$\begin{aligned}
&2\{q^{(\nu)}(u+\tfrac{1}{2}) - q^{(\nu)}(u)\}\\
&\qquad \le H^{(\nu)}(u) \le \frac{1}{2}\{q^{(\nu)}(u+1) - q^{(\nu)}(u) + q^{(\nu+1)}(u)\}, \quad 2 \mid \nu,\\
&\frac{1}{2}\{q^{(\nu)}(u+1) - q^{(\nu)}(u) + q^{(\nu+1)}(u)\}\\
&\qquad \le H^{(\nu)}(u) \le 2\{q^{(\nu)}(u+\tfrac{1}{2}) - q^{(\nu)}(u)\}, \quad 2 \nmid \nu.
\end{aligned} \tag{12}$$

A3. Convexity of $\dot{q}$. For $\dot{q}''$ we estimate H'' from above by (12), using Taylor's Theorem and noting that $q^{(4)} < 0$; we obtain $H''(u) < q'''(u)$. Then by the difference differential equation for q''' and Taylor's formula we get

$$\begin{aligned}
uq'''(u) &= (\kappa - 3)q''(u) + \kappa q''(u+1)\\
&\le (\kappa - 3)q''(u) + \kappa\{q''(u) + q'''(u)\}
\end{aligned}$$

and hence

$$H''(u) < q'''(u) \le (2\kappa - 3)q''(u)/(u - \kappa), \quad u > \kappa.$$

If we replace $q''(u)$ by $\Sigma''(u,\kappa,3) + \delta''(u,\kappa,3)$ and H'' by the above estimate in Prop. A1 we get

$$\begin{aligned}
\frac{1}{2}\dot{q}''(u) \ge &\left\{\log u + \frac{1}{2\kappa - 1} + \frac{1}{2\kappa - 2} - \frac{\kappa - 3/2}{u - \kappa}\right\}\Sigma''(u,\kappa,3)\\
&+ S_2(u,\kappa,3) + \left\{\log u + \psi(1 - 2\kappa) - \frac{\kappa - 3/2}{u - \kappa}\right\}\delta''(u,\kappa,3)\\
&- \frac{u^{2\kappa-3}}{\Gamma(1-2\kappa)}\int_0^\infty x^{2-2\kappa}e^{-x}\Delta\left(\frac{x}{u},\kappa,3\right)\log x\,dx.
\end{aligned} \tag{13}$$

We use this relation to show

Proposition A3. *Suppose that* $u \ge 2.87$ *and* $1.834 \le \kappa \le 1.835$. *Then* $\dot{q}''(u) > 0$.

Proof. We give lower estimates of each of the terms in (13), starting with the last. For $1.5 < \kappa < 2$ we have $\Gamma(1 - 2\kappa) < 0$ and $\Delta(x/u,\kappa,3)$ satisfies (7). Thus

$$\begin{aligned}
&\frac{-u^{2\kappa-3}}{\Gamma(1-2\kappa)}\int_0^\infty x^{2-2\kappa}e^{-x}\Delta(\frac{x}{u},\kappa,3)\log x\,dx\\
&\qquad \ge \frac{\kappa^3}{95}\frac{u^{2\kappa-7}}{\Gamma(1-2\kappa)}\int_1^\infty x^{6-2\kappa}e^{-x}\log x\,dx.
\end{aligned}$$

In our range $1/\Gamma(1-2\kappa) \geq 1/\Gamma(-2.67) \geq -1.1036$ and $x^{6-2\kappa} \leq x^{2.332}$ $(x \geq 1)$, and thus the last term in (13) is at least $-0.0718u^{2\kappa-7}\int_1^\infty x^{2.332}e^{-x}\log x\,dx$.

We have

$$(14) \qquad \int_1^\infty x^{2.332}e^{-x}\log x\,dx = \Gamma'(3.332) - \int_0^1 x^{2.332}e^{-x}\log x\,dx.$$

If we expand e^{-x} in its Maclaurin series in the last integral and integrate termwise, the integral becomes the alternating series

$$-\sum_{n=0}^{\infty}\frac{(-1)^n}{n!}(n+3.332)^{-2}.$$

It follows that the right side of (14) is at most

$$2.90211 + 0.05089 = 2.953$$

and the last term in (13) is at least $-0.212u^{2\kappa-3}$.

We have noted in (8) that $\delta''(u,\kappa,3) > 0$, and we show that its coefficient in (13) is positive in our κ and u ranges. This coefficient,

$$\log u + \psi(1-2\kappa) - (\kappa - 3/2)/(u-\kappa),$$

is increasing in u and decreasing in κ in our range, and therefore is at least

$$\log 2.87 + \psi(-2.67) - 0.335/(2.87 - 1.835) > 0.018 > 0.$$

Thus we drop the δ'' term in (13).

The coefficient of Σ'' in (13),

$$(15) \qquad \log u - (\kappa - 3/2)/(u-\kappa) + 1/(2\kappa-1) + 1/(2\kappa-2),$$

is increasing in u and decreasing in κ in our range. Thus (15) is minimal at $u = 2.87$ and $\kappa = 1.835$ and has value exceeding $17/10$. Assuming for the while that $\Sigma''(u,\kappa,3) > 0$ in our range, we replace (15) by $17/10$ in (13) and obtain

$$\frac{1}{2}\dot{q}''(u) > \frac{17}{10}\Sigma''(u,\kappa,3) + S_2(u,\kappa,3) - 0.212u^{2\kappa-7}.$$

We now multiply the last inequality by $90u^{6-2\kappa}$ to create reasonably simple polynomials on the right side. We obtain

$$(16) \quad 45u^{6-2\kappa}\dot{q}''(u) > 153u^{6-2\kappa}\Sigma''(u,\kappa,3) + 90u^{6-2\kappa}S_2(u,\kappa,3) - 19.1u^{-1}.$$

The first two terms on the right-hand side of (16) equal

$$(\kappa-1)(2\kappa-1)\{306u^3+(-612\kappa^2+738\kappa)u^2+(612\kappa^4-2088\kappa^3 \\ +2097\kappa^2-603\kappa)u-408\kappa^6+2700\kappa^5-6622\kappa^4+7377\kappa^3-3593\kappa^2+550\kappa\} \\ =(\kappa-1)(2\kappa-1)f(u,\kappa),$$

say. This expression is increasing in u in our range. Indeed,

$$\frac{\partial f}{\partial u}=918u^2+(-1224\kappa^2+1476\kappa)u \\ +612\kappa^4-2088\kappa^3+2097\kappa^2-603\kappa,$$

and if we replace κ in the positive terms by 1.834 and in the negative terms by 1.835, we see that

$$\frac{\partial f}{\partial u}>918u^2-1415u-31>0.$$

Thus, for $u\geq 2.87$, the first two terms on the right side of (16) are at least $(\kappa-1)(2\kappa-1)f(2.87,\kappa)$, a polynomial in κ of degree 8. This polynomial is decreasing for $1.834\leq\kappa\leq 1.835$ (as shown by applying Sturm's Theorem to the derivative, via computer algebra), and we conclude that $(\kappa-1)(2\kappa-1)f(u,\kappa)\geq 3122$ by evaluation at $\kappa=1.835$. The last term in (16) has magnitude at most $19.1/2.87<6.66$ and thus the right side of (13) is positive.

Finally, we show that $\Sigma''(u,\kappa,3)>0$ for $u\geq 2.87$ and $1.834\leq\kappa\leq 1.835$. We have

$$9u^{6-2\kappa}\Sigma''(u,\kappa,3)/\{1-3\kappa+2\kappa^2\} \\ =18u^3+(-36\kappa^2+54\kappa)u^2+(36\kappa^4-144\kappa^3+171\kappa^2-54\kappa)u \\ -24\kappa^6+180\kappa^5-506\kappa^4+651\kappa^3-364\kappa^2+60\kappa \\ =: a(u,\kappa).$$

Now, for $0\leq e\leq 0.001$,

$$a'(u,1.835-e)=54u^2-44.2602u-4.87471+(156.24u-8.68433)e \\ +(105.601-72u)e^2-120.24e^3+36e^4,$$

which is positive for $u\geq 2.87$. Thus

$$a(u,\kappa)\geq a(2.87,\kappa),$$

a 6th degree polynomial in κ which is positive throughout the interval $[1.834,1.835]$ by Sturm's Root Algorithm and an evaluation at a single point.

This concludes the proof that $\dot{q}_k''(u)>0$ for $u\geq 2.87$ and $1.834\leq\kappa\leq 1.835$.

A4. Monotonicity of $\dot{q}$.

Proposition A4. *Suppose that* $u \geq 3.8$ *and* $1.834 \leq \kappa \leq 1.835$. *Then* $\dot{q}'(u) > 0$.

Proof. Since $\dot{q}''(u) > 0$ for all $u \geq 2.87$, it suffices to show that $\dot{q}'(3.8) > 0$. We start with the representation

$$\begin{aligned}\frac{1}{2}\dot{q}'(u) &= \left\{\log u + \frac{1}{2\kappa - 1}\right\}\Sigma'(u, \kappa, 4) - \frac{1}{2}H'(u)\\ &+ S_1(u, \kappa, 4) + \{\log u + \psi(1 - 2\kappa)\}\delta'(u, \kappa, 4)\\ &+ \frac{u^{2\kappa-2}}{\Gamma(1 - 2\kappa)}\int_0^\infty x^{1-2\kappa}e^{-x}\Delta\left(\frac{x}{u}, \kappa, 4\right)\log x \, dx\end{aligned} \tag{17}$$

from Proposition A1 with $\nu = 1$ and $r = 4$. We shall give a lower estimate for each term. We chose $r = 4$ because $\delta'(u, \kappa, 4) > 0$, and since $\log u + \psi(1 - 2\kappa) \geq \log 3.8 + \psi(-2.67) > 0.6348 > 0$, we simply drop the term containing $\delta'(u, \kappa, 4)$.

Let L denote the last term in (17). Here $\Gamma(1 - 2\kappa) < 0$, $\Delta(x/u, \kappa, 4) > 0$ and $\log x > 0$ for $x > 1$. Thus, by (7),

$$\begin{aligned}L &\geq \frac{u^{2\kappa-2}}{\Gamma(1 - 2\kappa)}\int_1^\infty x^{1-2\kappa}e^{-x}\Delta(\frac{x}{u}, \kappa, 4)\log x \, dx\\ &\geq \frac{\kappa^3 u^{2\kappa-6}}{95\Gamma(1 - 2\kappa)}\int_1^\infty x^{5-2\kappa}e^{-x}\log x \, dx.\end{aligned}$$

Using essentially the same method we applied to estimate the last term in (13), we find that

$$L \geq \frac{(1.835)^3 3.8^{-2.33}}{95\Gamma(-2.67)}\int_1^\infty x^{1.332}e^{-x}\log x \, dx \geq -0.0028.$$

By (12) and then (9)

$$\begin{aligned}\frac{1}{2}H'(u) \leq q'(u + \frac{1}{2}) - q'(u) &= \int_u^{u+\frac{1}{2}} q''(x)dx\\ &< \int_u^{u+\frac{1}{2}} \Sigma''(x, \kappa, 4)dx\\ &= \Sigma'(u + \frac{1}{2}, \kappa, 4) - \Sigma'(u, \kappa, 4).\end{aligned}$$

It remains to give lower estimates of $\Sigma'(3.8, \kappa, 4)$ and $S_1(3.8, \kappa, 4)$ and an upper estimate of $\Sigma'(4.3, \kappa, 4)$. Each of these expressions has the form $w^{2\kappa-6}P(w)$ for P a polynomial in w and κ, with $w = 3.8$ or 4.3. In each case we factor out $w^{2\kappa-6}$, bound the resulting polynomial in κ on the range $1.834 \le \kappa \le 1.835$, and restore the factor $w^{2\kappa-6}$ to get our result.

We have

$$\begin{aligned}(\frac{38}{10})^{6-2\kappa}\Sigma'(\frac{38}{10},\kappa,4) = &-\frac{32}{95}\kappa^8 - \frac{2168}{285}\kappa^7 + \frac{16732}{285}\kappa^6 \\ &- \frac{430966}{4275}\kappa^5 + \frac{22631}{4275}\kappa^4 - \frac{2404696}{21375}\kappa^3 \\ &+ \frac{563407}{2250}\kappa^2 + \frac{1214537}{3750}\kappa - \frac{130321}{625},\end{aligned}$$

a decreasing function for $1.834 \le \kappa \le 1.835$, which is at least 161.6043. Thus

$$\Sigma'(3.8,\kappa,4) \ge 3.8^{-2.332}\cdot 161.6043 > 7.1845.$$

Arguing similarly, we find that

$$\Sigma'(4.3,\kappa,4) < 11.0825 \quad \text{and} \quad S_1(3.8,\kappa,4) \ge -7.0990.$$

If we combine the estimates we have found for the terms in (17) we obtain $\dot{q}'(3.8) > 2.56 > 0$. ■

Proposition A4, Corollary. *Let* $1.834 \le \kappa \le 1.835$ *and* $u \ge 3.8$. *Then*

$$H_\kappa(u) := \int_0^1 \{q_\kappa(u+s) - q_\kappa(u)\}ds/s$$

is increasing in κ.

Proof. We have

$$H_\kappa(u) = \int_{s=0}^1 \int_{t=0}^1 q_k'(u+st)dtds.$$

If we differentiate the last expression with respect to κ and use the result of Prop. A4, we find that $\dot{H}_\kappa(u) > 0$. ■

By the Corollary to Proposition A4, (12), and (9) we get bounds

$$\begin{aligned}H_\kappa(v) \ge H_{\underline{\kappa}}(v) &\ge 2\{q_{\underline{\kappa}}(v+\frac{1}{2}) - q_{\underline{\kappa}}(v)\} \\ &= 2\int_v^{v+\frac12} q_{\underline{\kappa}}' dt \ge 2\int_v^{v+\frac12} \Sigma'(t,\underline{\kappa},4)dt \\ &= 2\{\Sigma(v+\frac{1}{2},\underline{\kappa},4) - \Sigma(v,\underline{\kappa},4)\},\end{aligned} \tag{18}$$

and similarly

$$
\begin{aligned}
H_\kappa(v) \le H_{\bar\kappa}(v) &= \frac{1}{2}\{q_{\bar\kappa}(v+1) - q_{\bar\kappa}(v) + q'_{\bar\kappa}(v)\} \\
&\le \frac{1}{2}\{\Sigma(v+1,\bar\kappa,3) - \Sigma(v,\bar\kappa,3) + \Sigma'(v,\bar\kappa,3)\}.
\end{aligned}
\tag{19}
$$

A5. Upper and lower estimates of $\dot{q}_\kappa$. Our aim in this last section is to prove the following three Propositions.

Proposition A5.1. *Let* $3.8 \le v \le 10$ *and* $1.834 =: \underline{\kappa} \le \kappa \le \bar\kappa := 1.835$. *Then*

$$
\begin{aligned}
\dot{q}_\kappa(v) \ge \{&\frac{1}{2} + 2v^{-0.002}\log v\}\Sigma(v,\bar\kappa,3) - \frac{1}{2}\Sigma(v+1,\bar\kappa,3) \\
&- \frac{1}{2}\Sigma'(v,\bar\kappa,3) + 2v^{0.002}S(v,\bar\kappa,3) - 0.043v^{-1.33}.
\end{aligned}
$$

Proposition A5.2. *Let* $3.8 \le v \le 10$ *and* $\underline{\kappa} \le \kappa \le \bar\kappa$. *Then*

$$
\begin{aligned}
\dot{q}_\kappa(v) \le \{&2v^{0.002}\Sigma(v,\underline{\kappa},3) + 0.1284v^{1.33}\}\log v \\
&- 2\Sigma(v+\frac{1}{2},\underline{\kappa},4) + 2\Sigma(v,\underline{\kappa},4) + 2v^{-0.002}S(v,\bar\kappa,3).
\end{aligned}
$$

Corollary. *When* $4.88 \le u \le 4.884$ *and* $\underline{\kappa} \le \kappa \le \bar\kappa$, *the quantity (cf. (4.2) and (4.5))*

$$A_0 := -(u-1)\dot{q}_\kappa(u-1) > 48.95$$

Proof. By Proposition A4, and then Proposition A5.2,

$$\dot{q}_\kappa(u-1) < \dot{q}_\kappa(3.884) < -12.6168,$$

whence

$$(u-1)\{-\dot{q}_\kappa(u-1)\} > (3.88)(12.6168) > 48.9531. \quad \blacksquare$$

Proposition A5.3. *Let* $\underline{\kappa} \le \kappa \le \bar\kappa$. *Then*

$$\dot{q}_\kappa(v) < 0 \quad \textit{for} \quad 3.8 \le v \le 4.70$$

and

$$\dot{q}_\kappa(v) > 0 \quad \textit{for} \quad 4.72 \le v < \infty.$$

The proofs of the first two Propositions require some preparation, but then Proposition A5.3 follows as a Corollary, after some numerical computation. For, by Proposition A4, $\dot{q}'_\kappa(v) > 0$ for all $v \geq 3.8$. It suffices, therefore, to check via Proposition A5.2, that $\dot{q}_\kappa(4.70) < -0.008 < 0$; and, via Proposition A5.1, that $\dot{q}_\kappa(4.72) \geq 0.071 > 0$.

To prove Propositions A5.1 and A5.2 we start from the representation of $\dot{q}_\kappa$ in Proposition A1(with $\nu = 0$, $r = 3$). In both cases we require estimates of the functions of type S and Σ.

Lemma A5.1. *Let* $\underline{\kappa} \leq \kappa \leq \bar{\kappa}$ *and* $v \geq 3.35$. *Then*

$$q_*(v) := v^{-0.002}\Sigma(v, \bar{\kappa}, 3) \leq \Sigma(v, \kappa, 3) \leq v^{0.002}\Sigma(v, \underline{\kappa}, 3).$$

Proof. Let

$$P(v, \kappa, 3) = v^{4-2\kappa}\Sigma(v, \kappa, 3),$$

a polynomial in v and κ. We show that this polynomial is decreasing in κ in our range. We have

$$\begin{aligned}\dot{P}(v, \kappa, 3) = {} & (-4\kappa + 1)v^2 + (8\kappa^3 - 12\kappa^2 + 5\kappa - 1/2)v \\ & - 8\kappa^5 + 30\kappa^4 - 364/9\ \kappa^3 + 47/2\ \kappa^2 - 49/9\ \kappa + 1/3.\end{aligned}$$

Setting $\kappa = \underline{\kappa} + e, \quad 0 \leq e \leq 0.001$, and dropping all negative terms in expressions containing a power of e, we get the upper bound

$$\dot{P}(v, \kappa, 3) \leq -6.336v^2 + 17.6994v - 6.6868,$$

which is negative for all $v \geq 2.35$.

Also, we have

$$P(v, \kappa, 3) \geq P(v, \bar{\kappa}, 3) > v^3 - 4.8995v^2 + 5.4615v - 0.8663,$$

and this expression is positive for all $v \geq 3.35$. For all $v > 1$, $v^{2\kappa-4}$ is increasing in κ. Thus we obtain

$$v^{2\underline{\kappa}-4}P(v, \bar{\kappa}, 3) \leq \Sigma(v, \kappa, 3) \leq v^{2\bar{\kappa}-4}P(v, \underline{\kappa}, 3)$$

and hence

$$v^{-0.002}\Sigma(v, \bar{\kappa}, 3) \leq \Sigma(v, \kappa, 3) \leq v^{0.002}\Sigma(v, \underline{\kappa}, 3). \quad \blacksquare$$

Lemma A5.2. *Let* $2.42 \le v \le 10$ *and* $\underline{\kappa} \le \kappa \le \bar{\kappa}$. *Then*

$$v^{0.002} S(v, \underline{\kappa}, 3) \le S(v, \kappa, 3) \le v^{-.002} S(v, \bar{\kappa}, 3).$$

Proof. Let

$$R(v, \kappa, 3) = v^{4-2\kappa} S(v, \kappa, 3),$$

a polynomial in v and κ. We claim that R is increasing in κ in our range. We have

$$\begin{aligned}\dot{R}(v, \kappa, 3) = &- v^2 + (6\kappa^2 - 5\kappa)v + (3/4)v \\ &- 10\kappa^4 + 28\kappa^3 - (51/2)\kappa^2 + (49/6)\kappa - 11/18.\end{aligned}$$

Setting $\kappa = \underline{\kappa} + e, 0 \le e \le 0.001$, we have

$$\begin{aligned}\dot{R}(v, \kappa, 3) \ge &- v^2 + 11.7613v - 11.8139 \\ &+ e(17v - 49.6) + e^2(6v - 74) - 46e^3 - 10e^4.\end{aligned}$$

For $v \ge 2.4$ we have

$$\dot{R}(v, \kappa, 3) \ge -v^2 + 11.7613v - 11.8228,$$

which is positive for $2.4 \le v \le 10.65$.

Also, we have

$$R(v, \kappa, 3) \le R(v, \bar{\kappa}, 3) \le -1.835v^2 + 5.3160v - 2.1359,$$

which is negative for all $v \ge 2.42$. Thus

$$\begin{aligned}&- v^{2\kappa-4} R(v, \bar{\kappa}, 3) \le -S(v, \kappa, 3) \le -v^{2\kappa-4} R(v, \underline{\kappa}, 3), \\ &- v^{0.002} S(v, \bar{\kappa}, 3) \le -S(v, \kappa, 3) \le -v^{0.002} S(v, \underline{\kappa}, 3). \quad \blacksquare\end{aligned}$$

We are now able to prove Propositions A5.1 and A5.2. Begin with Proposition A5.1. Here we argue, on the basis of Proposition A1 (with $\nu = 0$ and $r = 3$) much as we did in Proposition A3 when deriving a lower bound for $\overset{\bullet}{q}''_k$. The term $\{\log v + \psi(1 - 2\kappa)\}\delta(v, \kappa, 3)$ is positive and can be omitted. The last term on the right of the identity for $\dot{q}_\kappa$ is

$$
\begin{aligned}
&-2\frac{v^{2\kappa-1}}{\Gamma(1-2\kappa)}\int_0^\infty x^{-2\kappa}e^{-x}\Delta(\frac{x}{v},\kappa,3)\log x\;dx \\
&\geq \frac{2}{95}\frac{\kappa^3 v^{2\kappa-5}}{\Gamma(1-2\kappa)}\int_1^\infty x^{4-2\kappa}e^{-x}\log x\;dx \\
&\geq -0.1436v^{2\kappa-5}\int_1^\infty x^{0.332}e^{-x}\log x\;dx \\
&\geq -0.1436v^{2\kappa-5}\{\Gamma'(1.332)-\sum_{n=0}^{\infty}\frac{(-1)^n}{n!}(n+1.332)^{-2}\} \\
&\geq -0.043v^{-1.33}.
\end{aligned}
$$

To complete the proof we apply the upper bounds for $H_\kappa(v)$, $\Sigma(v,\kappa,3)$ and $S(v,\kappa,3)$ from (19), Lemmas A5.1 and A5.2.

Finally we turn to Proposition A5.2. The choice $r=3$ in Proposition A.1 (with $\nu=0$) here causes a minor complication in that we shall have to estimate $\delta(v,\kappa,3)$ from above. We give upper estimates for the terms occurring on the right in Proposition A.1 (with $\nu=0, r=3$). We use the upper estimate of $\Sigma(v,\kappa,3)$ from Lemma A5.1. For H we apply the lower estimate (18), and for $S(v,\kappa,3)$ we apply the upper estimate of Lemma A5.2.

Next, $\log v+\psi(1-2\kappa)$ is positive in our range and we give an upper estimate of $2\delta(v,\kappa,3)$. Recall that $\Gamma(1-2\kappa)$ and $\Delta(x,\kappa,3)$ are each negative. We have

$$
\begin{aligned}
2\delta(v,\kappa,3) &\leq \frac{-2}{\Gamma(1-2\kappa)}\frac{\kappa^3}{95}v^{2\kappa-5}\int_0^\infty x^{4-2\kappa}e^{-x}dx \\
&\leq \frac{-2}{\Gamma(-2.668)}\frac{\bar{\kappa}^3}{95}v^{2\bar{\kappa}-5}\int_0^\infty x^{4-2\underline{\kappa}}e^{-x}dx \\
&\leq 0.14375v^{-1.33}\Gamma(1.332) \\
&\leq 0.1284v^{-1.33}.
\end{aligned}
$$

The last term in the formula for $\dot{q}$ is the integral involving $\Delta(x/v,\kappa,3)$ and $\log x$. An upper estimate is

$$
\begin{aligned}
&\frac{-2}{\Gamma(1-2\kappa)}\frac{\kappa^3}{95}v^{2\kappa-5}\int_0^1 x^{4-2\kappa}e^{-x}\log\frac{1}{x}dx \\
&\leq 0.14375v^{-1.33}\int_0^1 x^{0.332}e^{-x}\log\frac{1}{x}\;dx. \qquad (20)
\end{aligned}
$$

We expand e^{-x} in its Maclaurin series and integrate (by parts) termwise the resulting series. We find the last integral equals

$$\sum_{n=0}^{\infty} \frac{(-1)^n}{n!}(n+1.332)^{-2} = 0.41717891\ldots$$

It follows that the right hand side of (20) is at most $0.05997v^{-1.33}$.

Combining the terms we have obtained we get the claimed upper estimate for $\dot{q}$.

References

[AO] N. C. Ankeny and H. Onishi, *The general sieve*, Acta Arith. **10** (1964), 31–62.

[AS] M. Abramowitz and I. Stegun, *Handbook of Mathematical Functions*, Dover, New York, 1965.

[Br] D. Bradley, *A sieve auxiliary function*, These Proceedings, 173–210.

[Brü-F] J. Brüdern and E. Fouvry, *Le crible à vecteurs* (to appear).

[DHR1] H. G. Diamond, H. Halberstam, and H.–E. Richert, *Combinatorial sieves of dimension exceeding one*, J. Number Theory **28** (1988), 306–346.

[DHR2] ———, *A boundary value problem for a pair of differential delay equations related to sieve theory, I*, in Analytic Number Theory: Proceedings of a Conference in honor of P. T. Bateman, B. Berndt, et al. eds., Birkhauser, 1990, 133–157.

[DHR3] ———, *Sieve auxiliary functions I*, in Number Theory, Proceedings of First Conf. of the Canadian Number Theory Assn., R. Mollin, ed., W. de Gruyter, 1990, 99–113.

[DHR4] ———, *Sieve auxiliary functions II*, in *A tribute to Emil Grosswald: Number theory and related analysis*, M. Knopp and M. Sheingorn, eds., Contemporary Math. **143** (1993), AMS, Providence, RI, 247–253.

[DHR5] ———, *A boundary value problem for a pair of differential delay equations related to sieve theory II*, J. Number Theory **45** (1993), 129–185.

[DHR6] ———, *A boundary value problem for a pair of differential delay equations related to sieve theory III*, J. Number Theory **47** (1994), 300–328.

[DHR7] ———, *Estimation of the sieve auxiliary functions q_κ in the range* $1 < \kappa < 2$, Analysis **14** (1994), 75–102.

[DHR8] ———, *Monotonicity of the zero of the sieve auxiliary function $\tilde{\chi}_\kappa$ in the range* $3/2 < \kappa < 2$, Analysis **15** (1995), 1-16.

[Grp] F. Grupp, *On zeros of functions satisfying certain difference–*

differential equations, Acta Arith. **51** (1988), 247–268.

[GR] F. Grupp and H.–E. Richert, *Notes on functions connected with the sieve*, Analysis **8** (1988), 1–23.

[Iw] H. Iwaniec, *Rosser's sieve*, Acta Arith. **36** (1980), 171–202.

[ILR] H. Iwaniec, J. van de Lune, and H. J. J. te Riele, *The limits of Buchstab's iteration sieve*, Indag. Math. **42** (1980), 409–417.

[Raw] D. A. Rawsthorne, *Improvements in the small sieve estimate of Selberg by iteration*, Ph.D. Thesis, University of Illinois (1980).

[teR] H. J. J. te Riele, *Numerical solution of two coupled nonlinear equations related to the limits of Buchstab's iteration sieve*, Math. Centrum Amsterdam Afd. Numerieke Wisk. NW 86/80, 15 pp.

[Sel] A. Selberg, *Sifting problems, sifting density and sieves*, Number Theory, Trace Formulas and Discrete Groups, Acad. Press (1988), 467–484.

[VDW] B. L. Van der Waerden, *Modern Algebra*, I, F. Unger, New York, 1949.

[Whe1] F. S. Wheeler, *On two differential-difference equations arising in analytic number theory*, Ph.D. Thesis, 110 pages, University of Illinois, 1988.

[Whe2] F. S. Wheeler, *Two differential difference equations arising in number theory*, Trans. Amer. Math. Soc. **318** (1990), 491–523.

[Wol] S. Wolfram et al., *Mathematica*, Addison–Wesley, Redwood City, CA, 1988.

H. Diamond and H. Halberstam
Department of Mathematics
University of Illinois
1409 West Green Street
Urbana, IL 61801
diamond@math.uiuc.edu

Congruences for partition functions

Dennis Eichhorn and Ken Ono[1]

Dedicated to Heini Halberstam on the occasion of his retirement

Abstract. The ordinary partition function $p(n)$ and some of its generalizations satisfy some beautiful congruence properties. For instance, Ramanujan proved that for every integer n

$$p(5n+4) \equiv 0 \pmod{5},$$
$$p(7n+5) \equiv 0 \pmod{7},$$
$$p(11n+6) \equiv 0 \pmod{11}.$$

Here we consider congruences for $p(n)$ and $c_h(n)$, the number of partitions of n into h colors. If l is prime and s is a positive integer, then, using a result of Sturm, we compute a constant $C(h,t,r,l^s)$ such that $c_h(tn+r) \equiv 0 \pmod{l^s}$ for all n if and only if the congruence holds for every $n \le C(h,t,r,l^s)$. If $h = 1$, these results pertain to $p(n)$. In many cases, $C(h,t,r,l^s)$ is small enough that one obtains an effective method of determining the truth of alleged congruences. For example, Ramanujan's congruences are easily verified because $C(1,5,4,5) = 2, C(1,7,5,7) = 4$, and $C(1,11,6,11) = 10$.

1. Preliminaries

An ordinary partition of a non-negative integer n is a non-increasing sequence of positive integers with sum n. Let $p(n)$ denote the ordinary partition function; that is, let $p(n)$ denote the number of partitions of n. Ordinary partitions have many interesting generalizations, including partitions of n into h colors. A partition of a non-negative integer n into h colors is a non-increasing sequence of positive integers in which each positive integer is assigned one of h distinct colors, the order of the colors is not considered, and the sum of all of these positive integers is n. If $c_h(n)$ denotes the number of partitions of n into h colors, then it is easy to see that the generating function for $c_h(n)$ is:

$$\sum_{n=0}^{\infty} c_h(n) q^n = \prod_{n=1}^{\infty} \left(\frac{1}{1-q^n} \right)^h .$$

[1]The second author is supported by grants DMS-9508976 and DMS-9304580 from the National Science Foundation.

Of course, if $h = 1$, then $c_1(n) = p(n)$.

These partition functions possess a number of outstanding congruence properties. The most well known examples of these congruences are the Ramanujan Congruences. Some examples of colored partition congruences were given by Gandhi [4], who exhibited infinite families of congruences using identities of Euler and Jacobi. For instance, he proved that

$$c_2(5n+3) \equiv 0 \pmod{5},$$
$$c_8(11n+4) \equiv 0 \pmod{11}.$$

Other colored partition congruences were also found by Newman [13].

There are a number of methods that prove such congruences. Often these proofs depend on the manipulation of generating functions, the theory of modular forms, or beautiful combinatorial arguments where the partitions are equally distributed into various equivalence classes (see [1, 2, 3, 4, 5, 6, 7, 9, 10, 13, 16, 18]). The purpose of this paper is to define a simple algorithm for which the truth of an alleged congruence for $c_h(n)$ is easily verified by computing *enough* values.

We shall use the theory of modular forms to obtain our results. We begin by recalling some essential preliminaries. If N is a positive integer, then $\Gamma_0(N)$ and $\Gamma_1(N)$ denote subgroups of $SL_2(\mathbb{Z})$ that are defined by:

$$\Gamma_0(N) = \left\{ \begin{pmatrix} a & b \\ c & d \end{pmatrix} \mid ad - bc = 1,\ c \equiv 0 \pmod{N} \right\}$$

$$\Gamma_1(N) = \left\{ \begin{pmatrix} a & b \\ c & d \end{pmatrix} \mid ad - bc = 1,\ a \equiv d \equiv 1 \pmod{N}, \text{and}\, c \equiv 0 \pmod{N} \right\}.$$

It is well known [11, p. 107] that

$$[SL_2(\mathbb{Z}) : \Gamma_0(N)] = N \prod_{p|N} \left(1 + \frac{1}{p}\right)$$

and

$$[SL_2(\mathbb{Z}) : \Gamma_1(N)] = N^2 \prod_{p|N} \left(1 - \frac{1}{p^2}\right).$$

Let $SL_2(\mathbb{Z})$ act on $\mathfrak{H}$, the upper half of the complex plane, in the usual way. That is, if $A = \begin{pmatrix} a & b \\ c & d \end{pmatrix} \in SL_2(\mathbb{Z})$, then $Az = \frac{az+b}{cz+d}$. If χ is a Dirichlet character modulo N and k is a positive integer, then any meromorphic function

$f(z)$ on $\mathfrak{H}$ satisfying

$$f(Az) = \chi(d)(cz+d)^k f(z)$$

for all $A \in \Gamma_0(N)$, $z \in \mathfrak{H}$ is called a *modular form* of weight k and *Nebentypus character* χ with respect to $\Gamma_0(N)$. If $f(z)$ is holomorphic on $\mathfrak{H}$ and at the cusps (that is, the rationals) of $\Gamma_0(N)$, then $f(z)$ is called a *holomorphic modular form* of type (k,χ) on $\Gamma_0(N)$. If $f(z)$ is such a form that also vanishes at the cusps, then it is called a *cusp form* of type (k,χ) on $\Gamma_0(N)$. The set of weight k holomorphic modular (resp. cusp) forms of type (k,χ) on $\Gamma_0(N)$ forms a finite dimensional vector space over $\mathbb{C}$ which we denote by $M_k(N,\chi)$ (resp. $S_k(N,\chi)$).

We shall also require other modular forms. If $f(z)$ is a meromorphic function on $\mathfrak{H}$ and

$$f(Az) = (cz+d)^k f(z)$$

for all $A = \begin{pmatrix} a & b \\ c & d \end{pmatrix} \in \Gamma_1(N)$, then $f(z)$ is called a *modular form* of weight k with respect to $\Gamma_1(N)$. If $f(z)$ is holomorphic on $\mathfrak{H}$ and at the cusps of $\Gamma_1(N)$, then $f(z)$ is called a *holomorphic modular form* with respect to $\Gamma_1(N)$. Moreover, if $f(z)$ vanishes at the cusps, then $f(z)$ is called a *cusp form* of weight k with respect to $\Gamma_1(N)$. The set of holomorphic modular (resp. cusp) forms of weight k forms a finite dimensional vector space over $\mathbb{C}$ which we denote by $M_k(N)$ (resp. $S_k(N)$). It is easy to see that if $f(z) \in M_k(N,\chi)$ (resp. $S_k(N,\chi)$) for any χ, then $f(z) \in M_k(N)$ (resp. $S_k(N)$).

If $f(z)$ is a holomorphic modular form with respect to $\Gamma_1(N)$, then it admits a Fourier expansion in the variable $q = e^{2\pi i z}$ of the form $f(z) = \sum_{n=0}^{\infty} a(n)q^n$.

A crucial function in the study of modular forms is Dedekind's eta-function. It is defined by

$$\eta(z) := q^{\frac{1}{24}} \prod_{n=1}^{\infty} (1-q^n).$$

It is important to note that $\eta(z)$ is a weight $\frac{1}{2}$ holomorphic modular form that is non-vanishing on $\mathfrak{H}$. From this, we now see that the generating function for $p(n)$ may be written as $q^{\frac{1}{24}}\eta^{-1}(z)$, and that the generating function for $c_h(n)$ may be written as $q^{\frac{h}{24}}\eta^{-h}(z)$.

By [8], it is known that if $f(z) = \prod_{\delta|N} \eta^{r_\delta}(\delta z)$ is an eta-product where

$$\sum_{\delta|N} \delta r_\delta \equiv 0 \pmod{24} \qquad (1)$$

and

$$N\sum_{\delta\mid N}\frac{r_\delta}{\delta}\equiv 0 \pmod{24}, \tag{2}$$

then $f(z)$ is a modular form with respect to $\Gamma_0(N)$ with Nebentypus character χ where $k = \frac{1}{2}\sum_{\delta\mid N} r_\delta$ and χ is defined by $\chi(d) = \left(\frac{(-1)^k w}{d}\right)$ where $w = \prod_{\delta\mid N}\delta^{r_\delta}$. Note that when χ is defined in this manner, it is a Dirichlet character modulo N. Notice that if k is a positive integer and $f(z)$ is holomorphic at all of the cusps, then $f(z)\in M_k(N,\chi)$.

Fortunately, it is fairly easy to deduce whether or not $f(z)$ is holomorphic at the cusps of $\Gamma_0(N)$ [12]. Since every cusp of $\Gamma_0(N)$ has a representative of the form $\frac{c}{d}$ where $d\mid N$, c and d are positive integers, and $\gcd(c,d)=1$, it suffices to know the orders of $f(z)$ at all such cusps. By [12], the order of $f(z)$ at such a cusp $\frac{c}{d}$ is

$$\frac{N}{24d\cdot\gcd(d,\frac{N}{d})}\sum_{\delta\mid N}\frac{\gcd(d,\delta)^2 r_\delta}{\delta}. \tag{3}$$

Now we recall the *Hecke operators* T_m. The Hecke operators are linear transformations on the vector space $M_k(N,\chi)$ that preserve both $M_k(N,\chi)$ and $S_k(N,\chi)$. If $f(z)=\sum_{n=0}^{\infty}a(n)q^n\in M_k(N,\chi)$, then the modular form $f(z)\mid T_m\in M_k(N,\chi)$ is defined by

$$f(z)\mid T_m=\sum_{n\geq 0}\sum_{d\mid\gcd(m,n)}\chi(d)d^{k-1}a(nm/d^2)q^n. \tag{4}$$

In this paper, it is important to know when two holomorphic modular forms $f(z)$ and $g(z)$ are congruent modulo a positive integer j. Let j be a positive integer and let $F(q)=\sum_{n\geq N_0}A(n)q^n$ be a formal power series with integer coefficients. Then the function $\mathrm{Ord}_j(F(q))$ is defined by

$$\mathrm{Ord}_j(F(q)) := \min\{n \mid A(n)\not\equiv 0 \pmod{j}\}.$$

A theorem of Sturm [17] says that if $f(z)=\sum_{n=0}^{\infty}a(n)q^n$ and $g(z)=\sum_{n=0}^{\infty}b(n)q^n$ are holomorphic modular forms of weight k with respect to some congruence subgroup Γ of $SL_2(\mathbb{Z})$ with integer coefficients, then $f(z)\equiv g(z) \pmod{l}$ where l is prime if and only if

$$\mathrm{Ord}_l(f(z)-g(z)) > \frac{k}{12}[SL_2(\mathbb{Z}):\Gamma].$$

Sturm's theorem can be easily extended from l to an arbitrary integer j. Without loss of generality, we may assume that j is a power of a prime, say $j =$

l^s. Suppose that $f(z) \equiv g(z) \pmod{l^s}$ where $s > 1$. Let $F_1(z) := f(z) - g(z)$ and let $G_1(z) := 0$. Since both $F_1(z)$ and $G_1(z)$ are modular forms with integer coefficients, Sturm's criterion allows us to deduce that $F_1(z) \equiv G_1(z) \pmod{l}$. Then let $F_2(z) := \frac{1}{l}F_1(z)$ and $G_2(z) := 0$. Since both $F_2(z)$ and $G_2(z)$ are weight k modular forms with integer coefficients, Sturm's criterion again allows us to deduce that $F_2(z) \equiv G_2(z) \pmod{l}$ which is equivalent to the statement that $f(z) \equiv g(z) \pmod{l^2}$. By iterating this process, it is easy to see that Sturm's criterion holds in generality.

It will be convenient to use certain modular forms whose Fourier expansions are congruent to the constant 1 modulo powers of any given fixed prime. We shall use the following eta-product.

Lemma 1. *If l is a prime and $s \geq 1$, then the eta-product* $\frac{\eta^{l^s}(z)}{\eta^{l^{s-1}}(lz)}$ *satisfies*

$$\frac{\eta^{l^s}(z)}{\eta^{l^{s-1}}(lz)} \equiv 1 \pmod{l^s}.$$

Proof. We first note that if $f(q) = 1 + \sum_{n=1}^{\infty} a(n)q^n$ is a power series with integer coefficients such that $a(n) \equiv 0 \pmod{l}$ for all $n \geq 1$, then $f^{l^s}(q)$, the l^s power of $f(q)$, satisfies

$$f^{l^s}(q) \equiv 1 \pmod{l^{s+1}}.$$

This holds by hypothesis for $s = 0$, and if $f^{l^s}(q) = 1 + l^{s+1}g(q)$, then

$$f^{l^{s+1}}(q) = [1 + l^{s+1}g(q)]^l = 1 + l^{s+2}h(q),$$

completing the induction.

Since $1 - X^l \equiv (1 - X)^l \pmod{l}$, satisfies (1) and (2), then $f(z) \in M_k(N, \chi)$. it is easy to see that $f(q) := \frac{\eta^l(z)}{\eta(lz)} \equiv 1 \pmod{l}$. □

2. The main results

In this section, we compute $C(h, t, r, l^s)$. The strategy is to construct a holomorphic integer weight modular form whose Fourier coefficients are all multiples of l^s if and only if $c_h(tn + r) \equiv 0 \pmod{l^s}$ for every integer n. Once such a form is constructed, Sturm's theorem provides the explicit constant $C(h, t, r, l^s)$.

The first two theorems cover all possible cases; however, to achieve such generality we often obtain constants that are far from optimal. The other theorems in this section illustrate cases where the results of the first two theorems are sharpened dramatically. Theorem 1 gives us a result if $l \geq 5$ is prime.

Theorem 1. *Let $l \geq 5$ be prime and let $0 \leq r < t$. Then we have*

$$c_h(tn+r) \equiv 0 \pmod{l^s}$$

for every integer n if and only if the congruence holds for every $n \leq C(h,t,r,l^s)$. The constant $C(h,t,r,l^s)$ is defined as follows:

1. *Let*

$$m := \begin{cases} 8, & \text{if } h \equiv 1 \pmod 2 \text{ and } t \equiv 0 \pmod 2; \\ 1, & \text{otherwise.} \end{cases}$$

2. *Let b be any integer such that $b \geq ht/\gcd(l,t)^2$ and $b \geq t[h - \frac{1}{24}l^{s-2}(l^2 - 1)]$ for which $b \equiv h \pmod 2$ and $mb \equiv mht \pmod{24}$.*

Then define $C(h,t,r,l^s)$ by

$$C(h,t,r,l^s) := \frac{2^{12}3^4m^2t^3\mathrm{lcm}[t,l]^2[l^{s-1}(l-1)+b-h]}{\gcd(24t, 24r+bt-h)^2}\prod_{p|6tl}\left(1-\frac{1}{p^2}\right).$$

Proof. Let

$$f(z) = \frac{\eta^b(24tz)}{\eta^h(24z)}\left(\frac{\eta^l(z)}{\eta(lz)}\right)^{l^{s-1}} = \sum_{n=0}^{\infty} a(n)q^n.$$

Using the notation from our discussion prior to the statement of Theorem 1, we see that $r_1 = l^s, r_l = -l^{s-1}, r_{24} = -h$, and $r_{24t} = b$. We can see that conditions (1) and (2) hold where $N = 24m \cdot \mathrm{lcm}[t,l]$, and hence $f(z)$ is a modular form with respect to $\Gamma_1(N)$.

We would like to show that the orders at all of the cusps are non-negative. Since the scalar in front of the sum in (3) is always positive, it suffices to show that

$$\frac{l^s}{1} - \frac{l^{s-1}\gcd(d,l)^2}{l} - \frac{h\cdot\gcd(d,24)^2}{24} + \frac{b\cdot\gcd(d,24t)^2}{24t} \geq 0.$$

If $\gcd(d,l) = l$, we have that $\frac{l^s}{1} - \frac{l^{s-1}\gcd(d,l)^2}{l} = 0$, and so it suffices to show that $\frac{b\cdot\gcd(d,24t)^2}{24t} \geq \frac{h\cdot\gcd(d,24)^2}{24}$. This is equivalent to $b \geq \frac{ht\cdot\gcd(d,24)^2}{\gcd(d,24t)^2}$, but since $l \geq 5$ is prime and $l|d$, we have that $\frac{ht}{\gcd(l,t)^2} \geq \frac{ht\cdot\gcd(d,24)^2}{\gcd(d,24t)^2}$, and hence our inequality holds by our selection of b in 2.

If $\gcd(d,l)=1$, it suffices to show that $b\cdot\gcd(d,24t)^2 \geq t[h\cdot\gcd(d,24)^2 - 24l^{s-2}(l^2-1)]$, but since $\gcd(d,24t)\geq\gcd(d,24)$ and since $b\geq ht$ (see 2.) the result follows.

Hence $f(z)$ is holomorphic at the cusps, and since $\eta(z)$ is non-vanishing and holomorphic on $\mathfrak{H}$, $f(z)\in M_k(N)$ where $k=\frac{1}{2}[l^{s-1}(l-1)+b-h]$. There are other holomorphic modular forms that could be used in place of $f(z)$, but we have defined $f(z)$ as above to efficiently cancel poles while keeping the level and the weight of the modular form to a minimum in an effort to achieve the best possible bound.

By Lemma 1, if $\eta^b(24tz)=\sum_{n=0}^{\infty} d(24tn)q^{24tn+bt}$, then

$$(5)\qquad a(24tn+24r+bt-h)\equiv\sum_{m=0}^{\infty} d(24tm)c_h(tn+r-tm) \pmod{l^s}.$$

Since $d(0)=1$, (5) becomes

$$a(24tn+24r+bt-h)\equiv c_h(tn+r)+\sum_{m=1}^{\infty} d(24tm)c_h(tn+r-tm) \pmod{l^s}.$$

By induction, it is easy to see that $c_h(tn+r)\equiv 0 \pmod{l^s}$ for all n if $a(24tn+24r+bt-h)\equiv 0 \pmod{l^s}$ for all n. Hence we have that $c_h(tn+r)\equiv 0 \pmod{l^s}$ for all n if and only if $a(24tn+24r+bt-h)\equiv 0 \pmod{l^s}$ for all n.

Now define

$$f_1(z) := \sum_{n\equiv 24r+bt-h \bmod 24t} a(n)q^n.$$

By [Lemma 2, 14, 15], it is known that $f_1(z)\in M_k(\frac{576Nt^2}{g})$ where $g=\gcd(24t,24r+bt-h)$. Hence, by Sturm's theorem, we find that $f_1(z)\equiv 0 \pmod{l^s}$ if and only if $a(24tn+24r+bt-h)\equiv 0 \pmod{l^s}$ for all $24tn+24r+bt-h\leq\frac{27648kN^2t^4}{g^2}\prod_{p\mid 6tl}\left(1-\frac{1}{p^2}\right)$. Therefore $c_h(tn+r)\equiv 0 \pmod{l^s}$ for every integer n if and only if the congruence holds for every $n\leq C(h,t,r,l^s)$. □

The next theorem is the analog of Theorem 1 where $l=2$ or 3.

Theorem 2. *Let $l=2$ or 3 and let $0\leq r<t$. Then we have*

$$c_h(tn+r)\equiv 0 \pmod{l^s}$$

for every integer n if and only if the congruence holds for every $n\leq C(h,t,r,l^s)$. The constant $C(h,t,r,l^s)$ is defined as follows:

1. *Let*

$$m:=\begin{cases}8, & \text{if } h\equiv 1 \pmod 2 \text{ and } t\equiv 0 \pmod 2;\\ 1, & \text{otherwise.}\end{cases}$$

2. *Let b be any integer such that $b \geq ht$ for which $b \equiv h \pmod{2}$ and $mb \equiv mht \pmod{24}$.*
3. *Let*

$$N := \begin{cases} 24mt, & \text{if } s \geq 2; \\ 48mt, & \text{if } s = 1 \text{ and } l = 2; \\ 72mt, & \text{if } s = 1 \text{ and } l = 3. \end{cases}$$

Then define $C(h, t, r, l^s)$ by

$$C(h, t, r, l^s) := \frac{2^6 3^2 N^2 t^3 [l^{s-1}(l-1) + b - h]}{\gcd(24t, 24r + bt - h)^2} \prod_{p|N} \left(1 - \frac{1}{p^2}\right).$$

The proof of Theorem 2 is similar to the proof of Theorem 1.

Now, using Theorem 1 and Theorem 2, we can verify any alleged congruence for $c_h(n)$ in any arithmetic progression. However, in many instances the bounds established by Theorem 1 and Theorem 2 are not optimal. Next we consider convenient cases which appear to cover most non-trivial congruences of this type. In these cases we can construct a modular form with trivial Nebentypus character with respect to $\Gamma_0(N)$.

Theorem 3. *Let $l \geq 5$ be prime, let $0 \leq r < t$, and suppose h is odd. If t and h satisfy either*

- $\gcd(t, 6) = 1$
- $\gcd(t, 6) = 3$ and $h \equiv 0 \mod 3$,

then define N and b by:

1. *Let N be the smallest integer multiple of $\mathrm{lcm}[t, l]$ such that N/l is a perfect square.*
2. *Let b be any integer such that $b \geq hN/l^2$ and $b \geq N[h - l^{s-2}(l^2 - 1)]$ for which $b \equiv hN \pmod{24}$.*

If $r \equiv -\frac{bN-h}{24} \pmod{t}$, then

$$c_h(tn + r) \equiv 0 \pmod{l^s}$$

for every integer n if and only if the congruence holds for every $n \leq C(h, t, r, l^s)$. Here $C(h, t, r, l^s)$ is defined by

$$C(h, t, r, l^s) := \frac{N[l^{s-1}(l-1) + b - h]}{24} \prod_{p|N} \left(1 + \frac{1}{p}\right).$$

Proof. Let

$$f(z) = \frac{\eta^b(Nz)}{\eta^h(z)}\left(\frac{\eta^l(z)}{\eta(lz)}\right)^{l^{s-1}} = \sum_{n=0}^{\infty} a(n)q^n.$$

Using the notation from our discussion prior to the statement of Theorem 1, we see that $r_1 = l^s - h, r_l = -l^{s-1}$, and $r_N = b$. We can see that conditions (1) and (2) hold where N is defined as above. We can also see that $w = \prod_{\delta|N}\delta^{r_\delta}$ is a square and that $k = \frac{1}{2}[l^{s-1}(l-1)+b-h]$ is even, and hence $f(z)$ is a modular form of weight k and character χ_0, the trivial character, with respect to $\Gamma_0(N)$.

We would like to show that the orders at all of the cusps are non-negative. Since the scalar in front of the sum in (3) is always positive, it suffices to show that

$$\frac{l^s}{1} - \frac{l^{s-1}\gcd(d,l)^2}{l} - \frac{h}{1} + \frac{b\cdot\gcd(d,N)^2}{N} \geq 0.$$

If $\gcd(d,l) = l$, we have that $\frac{l^s}{1} - \frac{l^{s-1}\gcd(d,l)^2}{l} = 0$, and so it suffices to show that $\frac{b\cdot\gcd(d,N)^2}{N} \geq h$. This is equivalent to $b \geq \frac{hN}{\gcd(d,N)^2}$, but since $l|d$ and $l|N$, we have that $\frac{hN}{l^2} \geq \frac{hN}{\gcd(d,N)^2}$, and hence our inequality holds by our selection of b above.

If $\gcd(d,l) = 1$, it suffices to show that $b\cdot\gcd(d,N)^2 \geq N[h - l^{s-2}(l^2-1)]$, but since $\gcd(d,N) \geq 1$ our inequality again holds by our selection of b above. Hence $f(z)$ is holomorphic at the cusps, and since $\eta(z)$ is non-vanishing and holomorphic on $\mathfrak{H}$, $f(z) \in M_k(N,\chi_0)$. There are other holomorphic modular forms that could be used in place of $f(z)$, but we have defined $f(z)$ as above to efficiently cancel poles while keeping the level and the weight of the modular form to a minimum in an effort to achieve the best possible bound.

By Lemma 1, we may conclude that if $q^{-\frac{bN}{24}}\eta^b(Nz) = \sum_{n=0}^{\infty} d(Nn)q^{Nn}$, then

$$a(tn + r + \frac{bN-h}{24}) \equiv \sum_{m=0}^{\infty} d(Nm)c_h(tn+r-Nm) \pmod{l^s}. \tag{6}$$

Since $d(0) = 1$, (6) becomes

$$a(tn + r + \frac{bN-h}{24}) \equiv c_h(tn+r) + \sum_{m=1}^{\infty} d(Nm)c_h(tn+r-Nm) \pmod{l^s}.$$

By induction, it is easy to see that $c_h(tn+r) \equiv 0 \pmod{l^s}$ for all n if

$a(tn+r+\frac{bN-h}{24}) \equiv 0 \pmod{l^s}$ for all n. Hence we have that $c_h(tn+r) \equiv 0 \pmod{l^s}$ for all n if and only if $a(tn+r+\frac{bN-h}{24}) \equiv 0 \pmod{l^s}$ for all n.

Now notice that $r+\frac{bN-h}{24} \equiv 0 \pmod{t}$ by hypothesis, so let us consider

$$f_1(z) = f(z)|T_t = \sum_{n=0}^{\infty} a(tn)q^n.$$

Notice that $f_1(z) \in M_k(N,\chi_0)$. Hence, by Sturm's theorem, we find that $f_1(z) \equiv 0 \pmod{l^s}$ if and only if $a(tn) \equiv 0 \pmod{l^s}$ for all $n \le \frac{kN}{12}\prod_{p|N}\left(1+\frac{1}{p}\right)$. Therefore $c_h(tn+r) \equiv 0 \pmod{l^s}$ for every integer n if and only if the congruence holds for every $n \le C(h,t,r,l^s)$. □

The next theorem is the analog of Theorem 3 where $l = 2$ or 3.

Theorem 4. *Let $l = 2$ or 3, let $0 \le r < t$, and suppose h is odd. If t and h satisfy either*

- $\gcd(t,6) = 1$
- $\gcd(t,6) = 3$ and $h \equiv 0 \mod 3$,

then define J, N, and b by:

1. *Let J be the smallest integer square such that $t|J$.*
2. *Let b be any integer such that $b \ge hJ/\gcd(l,t)^2$ and $b \ge J[h-2l^{s-2}(l^2-1)]$ for which $b \equiv hJ \pmod{24}$.*
3. *Let*

$$N := \begin{cases} J, & \text{if } s \ge 4; \\ J, & \text{if } s = 3 \text{ and } l = 3; \\ 2^{4-s}J, & \text{if } s < 4 \text{ and } l = 2; \\ 3^{3-s}J, & \text{if } s < 3 \text{ and } l = 3. \end{cases}$$

If $r \equiv -\frac{bJ-h}{24} \pmod{t}$, then

$$c_h(tn+r) \equiv 0 \pmod{l^s}$$

for every integer n if and only if the congruence holds for every $n \le C(h,t,r,l^s)$. Here $C(h,t,r,l^s)$ is defined by

$$C(h,t,r,l^s) := \frac{N[2l^{s-1}(l-1)+b-h]}{24}\prod_{p|N}\left(1+\frac{1}{p}\right).$$

The proof of Theorem 4 is similar to the proof of Theorem 3 with

$$f(z) = \frac{\eta^b(Jz)}{\eta^h(z)}\left(\frac{\eta^l(z)}{\eta(lz)}\right)^{2l^{s-1}}$$

being the appropriate modular form.

By carefully selecting some of the techniques introduced in the proof of Theorem 3, one can prove theorems that are more general than Theorems 3 and 4 which produce bounds smaller than those in Theorems 1 and 2.

3. Examples

Now we use these theorems to obtain *new* proofs of some well known congruences.

Corollary 1. *If $p(n)$ denotes the number of partitions of n, then*

$$p(5n+4) \equiv 0 \pmod{5},$$
$$p(7n+5) \equiv 0 \pmod{7},$$
$$p(11n+6) \equiv 0 \pmod{11}$$

for all n.

Proof. For the first congruence, we find that the hypotheses of Theorem 3 are met with $J = b = l = 5$, and hence $C(1,5,4,5) = 2$. Since $p(5n+4) \equiv 0 \pmod{5}$ for $0 \le n \le 2$, we find that $p(5n+4) \equiv 0 \pmod{5}$ for all non-negative integers n.

For the other two congruences, the hypotheses are met with $J = b = l = 7$ and $J = b = l = 11$ respectively. In these cases we find that $C(1,7,5,7) = 4$ and $C(1,11,6,11) = 10$. □

Corollary 2. *For all n,*

$$c_2(5n+3) \equiv 0 \pmod{5}$$

and

$$c_8(11n+4) \equiv 0 \pmod{11}.$$

Proof. For the first congruence, we find that the hypotheses of Theorem 3 are met with $J = l = 5$ and $b = 10$, and hence $C(2,5,3,5) = 3$. Since $c_2(5n+3) \equiv 0 \pmod{5}$ for all $0 \le n \le 3$, we find that $c_2(5n+3) \equiv 0 \pmod{5}$ for all non-negative integers n.

We find that the hypotheses of Theorem 3 are met with $J = l = 11$ and $b = 20$, and hence $C(8,11,4,11) = 11$. □

The authors would like to thank the referee for his/her helpful corrections and suggestions.

References

[1] G. Andrews, *The theory of partitions*, Addison-Wesley, 1976.

[2] G. Andrews and F. Garvan, *Dyson's crank of a partition*, Bull. Amer. Math. Soc. **18** (1988), 167-171.

[3] A.O.L. Atkin, *Proof of a conjecture of Ramanujan*, Glasgow Math. J. **8** (1967), 14-32.

[4] J. M. Gandhi, *Congruences for $p_r(n)$ and Ramanujan's $\tau-$function*, Amer. Math. Monthly **70** (1963), 265-274.

[5] F. Garvan, *New combinatorial interpretations of Ramanujan's partition congruences* mod *5, 7 and* 11, Trans. Amer. Math. Soc. **305** (1988), 47-77.

[6] F. Garvan and D. Stanton, *Sieved partition functions and $q-$binomial coefficients*, Math. Comp. **55 191** (1990), 299-311.

[7] F. Garvan and D. Stanton, *Cranks and $t-$cores*, Invent. Math. **101** (1990), 1-17.

[8] B. Gordon and K. Hughes, *Multiplicative properties of $\eta-$products II*, A tribute to Emil Grosswald: Number Theory and Related Analysis, Cont. Math. **143** (1993), Amer. Math. Soc., 415-430.

[9] M. Hirschhorn, *Ramanujan's partition congruences*, Discrete Math. **131** (1994), 351-355.

[10] M. Hirschhorn and D.C. Hunt, *A simple proof of the Ramanujan conjecture for powers of* 5, J. Reine Angew. Math. **336** (1981), 1-17.

[11] N. Koblitz, *Introduction to elliptic curves and modular forms*, Springer-Verlag, 1984.

[12] G. Ligozat, *Courbes modulaires de genre 1*, Bull. Math. Soc. France [Memoire 43] (1972), 1-80.

[13] M. Newman, *Some theorems about $p_r(n)$*, Canad. J. Math. **9** (1957), 68-70.

[14] K. Ono, *Parity of the partition function*, Elec. Res. Ann. Amer. Math. Soc. **1, 1** (1995), 35-42.

[15] K. Ono, *Parity of the partition function in arithmetic progressions*, J. Reine Agnew. Math. to appear.

[16] S. Ramanujan, *Congruence properties of partitions*, Proc. London Math. Soc.(2) **18** (1920), xix-xx.

[17] J. Sturm, *On the congruence of modular forms*, Springer Lect. Notes Math. **1240** (1984).

[18] G.N. Watson, *Ramanujan's vermutung über zerfällungsanzahlen*, J. Reine Angew. Math. **179** (1938), 97-128.

Dennis Eichhorn
Department of Mathematics
The University of Illinois,
Urbana, IL 61801
eichhorn@math.uiuc.edu

Ken Ono
School of Mathematics
Institute for Advanced Study
Princeton, NJ 08540
ono@math.ias.edu

Fractional power large sieves

P.D.T.A. Elliott

In honor of the retirement of Professor Heini Halberstam

1. Introduction

In the form given it by Davenport and Halberstam, [2], the inequality of the Large Sieve asserts that for points δ_j, $j = 1, 2, \ldots$, spaced according to $\|\delta_j - \delta_k\| \geq \delta > 0$, $j \neq k$, the function

$$S(\alpha) = \sum_{n=-N}^{N} a_n e^{2\pi i n\alpha}, \qquad \alpha \text{ real},$$

satisfies

$$\sum_j |S(\alpha_j)|^2 \leq 2.2 \max(\delta^{-1}, 2N) \sum_{n=-N}^{N} |a_n|^2. \tag{1}$$

Here N denotes a positive integer, $\|y\|$ the distance of the real y from a nearest integer.

Choosing δ_j of the form bp^{-1}, for primes p, and appealing to the identity of Parseval (or Plancherel, as shown below), allows us to derive an inequality of a more arithmetic nature:

$$\sum_{p\leq Q} p \sum_{r=0}^{p-1} \left| \sum_{\substack{n=-N \\ n\equiv r (\operatorname{mod} p)}}^{N} a_n - \frac{1}{p}\sum_{n=-N}^{N} a_n \right|^2 \leq 2.2 \max(2N, Q^2) \sum_{n=-N}^{N} |a_n|^2. \tag{2}$$

Is there an analogue of the second inequality with the powers 2 replaced by α, $1 < \alpha < 2$?

Considerations of averages suggest that for some positive β

$$\sum_{p\leq Q} p^{\alpha-1} \sum_{r=0}^{p-1} \left| \sum_{\substack{n=1 \\ n\equiv r (\operatorname{mod} p)}}^{N} a_n - \frac{1}{p}\sum_{n=1}^{N} a_n \right|^\alpha \ll (N^{\alpha-1} + Q^\beta) \sum_{n=1}^{N} |a_n|^\alpha, \tag{3}$$

the implied constant depending at most upon α. Impeding the establishment of such an inequality is our limited belief in its validity, in particular, in its form.

Why would we wish such a generalisation?

2. Background

According to Theorem (10.1) of my book [4], there is a positive constant c so that every additive function f satisfies

$$\sum_{q\le x} q^{-1}|f(q)-F\log q|^2 \ll \sup_{x\le y\le x^c} y^{-1}\sum_{x<n\le y}|f(n+1)-f(n)|^2, \tag{4}$$

where q denotes a prime power, F a certain function of f and x. Is there a version of this inequality with the powers 2 replaced by α, $1<\alpha<2$?

An important ingredient in the proof of (4) is an abstract analog for additive functions of the Bombieri–Vinogrdov theorem on primes in arithmetic progression. The Bombieri–Vinogradov theorem asserts that for $A>0$,

$$\sum_{D\le Q}\max_{(r,D)=1}\max_{y\le x}\left|\pi(y,D,r)-\frac{Li(x)}{\phi(D)}\right| \ll \frac{x}{(\log x)^A} \tag{5}$$

provided Q does not exceed $x^{\frac{1}{2}}(\log x)^{-B}$ for a suitable B depending at most upon A. In a recent form, [5], the analogue asserts that for each positive δ not exceeding 1/2,

$$\sum_{q\le x^\delta}\phi(q)\max_{(r,q)=1}\max_{y\le x}\left|\sum_{\substack{n\le y\\ n\equiv r(\operatorname{mod} q)}} f(n)-\frac{1}{\phi(q)}\sum_{\substack{n\le y\\ (n,q)=1}} f(n)\right|^2 \ll \frac{x^2}{\log x}\sum_{q\le x} q^{-1}|f(q)|^2. \tag{6}$$

It is essential that the prime power moduli be allowed to reach a fixed power of x, and that a fixed power of $\log x$ be saved over the factor x^2 in the upper bound. Unlike the situation in the Bombieri–Vinogradov theorem, we cannot gain an arbitrary high power of $\log x$; the factor $x^2/\log x$ in the upper bound is best possible.

We are led to ask for an analogue of (6), again replacing exponents 2 by α, $1<\alpha<2$. Of course, by analogy with (2), we would expect to replace the weight $\phi(q)$ by $\phi(q)^{\alpha-1}$. However, doubts about the form of inequality (3) induce similar doubts concerning the generalisations of (6).

As a heuristic we appeal to the definition of an additive function:

$$\sum_{\substack{n\le x\\ n\equiv r(\operatorname{mod} p)}} f(n)=\sum_{n\le x}\sum_{q\| n} f(q)=\sum_{\substack{mq\le x\\ (m,q)=1\\ mq\equiv r(\operatorname{mod} p)}} f(q),\quad (r,p)=1.$$

Consider a typical section of bilinear form

$$\sum_{m=M}^{2M} \sum_{\substack{N<q\le 2N \\ mq\equiv r(\bmod p)}} f(q) = \sum_{t=1}^{p-1} \sum_{\substack{N<q\le 2N \\ q\equiv t(\bmod p)}} f(q) \sum_{\substack{m=M \\ m\equiv r\bar{t}(\bmod p)}}^{2M} 1,$$

where $t\bar{t} \equiv 1(\bmod p)$ The innermost of the three right hand sums contributes $Mp^{-1} + O(1)$ and we may view the (section of) bilinear form reasonably represented by

$$\frac{M}{p} \sum_{t=1}^{p-1} \sum_{\substack{N<q\le 2N \\ q\equiv t(\bmod p)}} f(q).$$

We expect to raise this to power α, apply Hölder's inequality and appeal to a fractional power version (3) of the large sieve inequality (2). The form of an α-power generalisation of (2) will control the form of an α-power generalisation of (6), and ultimately of an α-power generalisation of the inequality (4) involving differences of additive functions. How then to proceed?

According to a dictum of Hecke, whenever a Fourier expansion is available one should employ it. In the step between analogues of (1) and (2), what is to play the rôle of Parseval's inequality?

3. Fourier analysis on groups

Let G be a finite abelian group, of order $|G|$, $\widehat{G}$ its dual group, χ a typical character.

The Fourier transform $\hat{h}$ of a function $h : G \to \mathbb{C}$ is given by

$$\hat{h}(\chi) = |G|^{-\frac{1}{2}} \sum_{g\in G} h(g)\overline{\chi(g)}.$$

The Parseval/Plancherel relation asserts that

$$\sum_{\chi\in\widehat{G}} |\hat{h}(\chi)|^2 = \sum_{g\in G} |h(g)|^2.$$

This amounts to an isometry between the spaces $L^2(G)$, $L^2(\widehat{G})$ of complex-valued functions on $G, \widehat{G}$, equipped with norms

$$\left(|G|^{-1} \sum_{g\in G} |h(g)|^2 \right)^{\frac{1}{2}}, \quad \left(|\widehat{G}|^{-1} \sum_{\chi\in\widehat{G}} |k(\chi)|^2 \right)^{\frac{1}{2}},$$

respectively.

Replacing the exponents 2, 1/2, in these norms by α, $1/\alpha$, we may likewise consider spaces $L^\alpha(G)$, $L^\alpha(\widehat{G})$, $\alpha \geq 1$ including, suitably interpreted, $\alpha = \infty$. The Fourier transform is a linear map between $L^2(G)$, $L^2(\widehat{G})$, of norm 1. Since

$$\max_{\chi} |\hat{h}(\chi)| \leq |G|^{\frac{1}{2}} \sum_{g\in G} |h(g)|,$$

we may also estimate its norm between $L^1(G)$ and $L^\infty(\widehat{G})$. Application of the Riesz–Thorin interpolation theorem gives at once a Young–Hausdorff inequality:

$$\left(\sum_{\chi\in\widehat{G}} \left| \sum_{g\in G} b_g \overline{\chi(g)} \right|^{\alpha'} \right)^{\frac{1}{\alpha'}} \leq |G|^{\frac{1}{\alpha'}} \left(\sum_{g\in G} |b_g|^\alpha \right)^{\frac{1}{\alpha}},$$

where $1/\alpha + 1/\alpha' = 1$, $1 < \alpha \leq 2$, and the b_g are arbitrary complex numbers. But for our purposes this is in the wrong direction!

We dualise and use the inverse Fourier transform. Actually the inverse of the Fourier transform is its adjoint, but the upshot is here the same:

$$\left(\sum_{g\in G} |d_g|^{\alpha'} \right)^{\frac{1}{\alpha'}} \leq |G|^{-\frac{1}{\alpha}} \left(\sum_{\chi\in\widehat{G}} \left| \sum_{w\in G} \chi(w) d_w \right|^{\alpha} \right)^{\frac{1}{\alpha}},$$

(again) for arbitrary complex d_g.

Let G be the group of reduced residue classes (mod D) for a positive integer D, and set

$$d_r = \sum_{\substack{n\leq x \\ n\equiv r (\mathrm{mod}\, D)}} a_n - \frac{1}{\phi(D)} \sum_{\substack{n\leq x \\ (n,D)=1}} a_n.$$

Then

$$\left(\phi(D)^{\alpha'-1} \sum_{\substack{r=1 \\ (r,D)=1}}^{D} \left| \sum_{\substack{n\leq x \\ n\equiv r (\mathrm{mod}\, D)}} a_n - \frac{1}{\phi(D)} \sum_{\substack{n\leq x \\ (n,D)=1}} a_n \right|^{\alpha'} \right)^{\frac{1}{\alpha'}}$$

$$\leq \left(\sum_{\chi\neq\chi_0 (\mathrm{mod}\, D)} \left| \sum_{n\leq x} a_n \chi(n) \right|^{\alpha} \right)^{\frac{1}{\alpha}},$$

$1 \le \alpha \le 2$, with χ_0 the principal (Dirichlet) character (mod D). Looking back at inequalities (2) and (3) we see that we have gained an inequality in the right direction, but for $\alpha' > 2$, the wrong range!

It becomes clear that for the range $1 < \alpha < 2$, mean-power character sums lie below an arithmetic object which we wish to make small.

We don't give up yet.

We might believe that nevertheless, the Young–Hausdorff inequalities are essentially equalities, or that the underlying functional analytic argument by which they are established can be readily retrieved to act successfully in our particular case. An example rapidly disabuses us.

To seek an upper bound for

$$\left(\sum_j \left|\sum_{n\le x} a_n \chi_j(n)\right|^{\frac{4}{3}}\right)^{\frac{3}{4}}, \tag{7}$$

where χ_j runs over the $p-1$ Dirichlet characters $(\bmod\, p)$, we consider the dual expression

$$S^{\frac{1}{4}} = \left(\sum_{n\le x} \left|\sum_j c_j \chi_j(n)\right|^4\right)^{\frac{1}{4}}.$$

Direct computation, employing the Pólya–Vinogradov theorem, shows that

$$S = x\left(1-\frac{1}{p}\right)\widetilde{\sum} c_{j_1} c_{j_2} \bar{c}_{j_3} \bar{c}_{j_4} + O\left(p^{\frac{1}{2}} \log p \left(\sum_j |c_j|\right)^4\right),$$

where $\widetilde{}$ indicates that summation is over those products $\chi_{j_1}\chi_{j_2}\bar{\chi}_{j_3}\bar{\chi}_{j_4}$ of characters that are principal. Here

$$\left(\widetilde{\sum} c_{j_1} c_{j_2} \bar{c}_{j_3} \bar{c}_{j_4}\right)^{\frac{1}{4}}$$

is in fact a norm. We can thus estimate $S^{\frac{1}{4}}$ in terms of the maximum of this norm and an L^1 norm on $\mathbb{C}^{p-1}$. To derive an equivalent version with which to bound our sum (7) we need the dual norm of this maximum norm. What is it? Apparently it is bound up with the arithmetic properties of the group of reduced residue classes $(\bmod\, p)$; and something of a fright. As I show elsewhere, simplifying the argument by applying Hölder's inequality at some stage introduces superfluous powers of p into the upper bound and destroys its canonical nature.

For the problems considered in this paper I abandon the dictum of Hecke and the use of characters, and seek another understanding.

4. The fractional power large sieve

In one of its variant forms the Turán–Kubilius inequality asserts that every additive function f satisfies

$$\sum_{n\le x} |f(n)-A|^2 \ll x\sum_{q\le x} q^{-1}|f(q)|^2,$$

with

$$A=\sum_{q\le x} f(q)q^{-1}(1-q_0^{-1}),$$

where q denotes a prime power, q_0 the prime of which it is a power, and the implied constant is absolute.

Taking (non-negative) square roots of both sides, we may view the inequality as a bound for an operator norm, the operator between appropriate spaces endowed with norms of mean-square type. This allows the immediate deduction of a dual inequality:

$$\sum_{q\le x} q\left|\sum_{\substack{n\le x\\ n\cong 0(\mathrm{mod}\, q)}} a_n - \frac{1}{q}\left(1-\frac{1}{q}\right)\sum_{n\le x} a_n\right|^2 \ll x\sum_{n\le x}|a_n|^2. \tag{8}$$

Here $n\cong 0(\mathrm{mod}\, q)$ denotes that q divides n, but qq_0 does not.

As I show in [6], this interpretation suggests natural generalisations of the Turán–Kubilius inequality, involving powers other than squares. For present purposes I note the similarity between the form of the dual Turán–Kubilius inequality (8) and the arithmetic version of the large sieve inequality, (2). This suggests that we

(i) View the Large Sieve as the dual of something.
(ii) Find the something (i.e., an appropriate operator).
(iii) Choose a suitable set of spaces.
(iv) Estimate the norm of the something.

To exemplify this procedure, consider the space J^β of vectors h, one coordinate $h(p,r)$ for each pair $(p,r), p$ a prime not exceeding Q, r a positive integer not exceeding p, endowed with norm

$$\|h\|_\beta=\left(\sum_{p\le Q}\sum_{r}\frac{1}{p}\left(1-\frac{1}{p}\right)|h(p,r)|^\beta\right)^{\frac{1}{\beta}}, \quad \beta\ge 1.$$

Note that the weight $p^{-1}(1-p^{-1})$ does not involve r.

Let K^β be the space of vectors a, one coordinate a_n for each positive integer n in the range $x-\rho < n \le x$, where $2 \le \rho \le x$; endowed with norm

$$\|a\|_\beta = \left(\frac{1}{[x]-[x-\rho]} \sum_{x-\rho<n\le x} |a_n|^\beta \right)^{\frac{1}{\beta}}, \quad \beta \ge 1.$$

An appropriate dual operator associated with an inequality of (hopefully) large sieve type would then be given by

$$(W'a)(p,r) = \frac{p(1-p^{-1})^{-1}}{[x]-[x-\rho]} \left(\sum_{\substack{x-\rho<n\le x \\ n-r\cong 0 (\mathrm{mod}\, p)}} a_n - \frac{1}{p}\left(1-\frac{1}{p}\right) \sum_{x-\rho<n\le x} a_n \right).$$

The form of this operator is partially controlled by the need to reproduce the original mean-square large sieve, (2). There is scope to vary the definitions of the various norms, and thus of the operator W'. In the present formulation, having followed (i) and (iii), I offer towards (ii) the operator W given by

$$(Wh)(n) = \sum_{\substack{r,p \\ p\|(n-r)}} h(p,r) - \sum_{r,p} \frac{1}{p}\left(1-\frac{1}{p}\right) h(p,r).$$

Very roughly, Wh at n is the sum of differing additive functions f_r, taken at shifted values $n-k$, $1 \le k \le Q$.

To establish a fractional power sieve I must carry out (iv). Omitting the details I state a finished result.

Theorem 1 (Fractional Power Large Sieve). *Let* $1 \le \alpha \le 2$. *The operator*

$$W' : (K^{\alpha'})' \to (J^2 \cap J^{\alpha'})'$$

has norm $\ll (1+\rho^{-1}Q^{2\alpha'+3})^{\frac{1}{\alpha'}}$. *For* $\alpha \ge 2$, *the operator*

$$W' : (K^{\alpha'})' \to (J^2 \cap J^{\alpha})'' \simeq J^2 \cap J^\alpha$$

has norm $\ll (1+\rho^{-1}Q^7)^{\frac{1}{\alpha'}}$.

Here $J^2 \cap J^\beta$ denotes the space of vectors h with norm $\max(\|h\|_2, \|h\|_\beta)$, and M' denotes the dual space of the space M. The (otherwise arcane) notation in this theorem reflects the derivation of the results. The treatment of the operator W ultimately determines the spaces upon which we operate.

By identifying the appropriate dual norms we gain a practical version of the theorem. I give the case $1 < \alpha \le 2$, originally sought.

Theorem 2. *Let $1 < \alpha \le 2$. In the notation of Theorem 1,*

$$\sum_{p \le Q} p^{\alpha-1} \sum_{r=0}^{p-1} \left| \sum_{\substack{x-\rho<n\le x \\ n-r \cong 0 (\mathrm{mod}\, p)}} a_n - \frac{1}{p}\left(1-\frac{1}{p}\right) \sum_{x-\rho<n\le x} a_n \right|^{\alpha}$$
$$\ll (\rho^{\alpha-1} + Q^{5\alpha-3}) \sum_{x-\rho<n\le x} |a_n|^{\alpha}$$

provided the pairs (p, r) are restricted to those for which

$$\frac{p}{\rho} \left| \sum_{\substack{x-\rho<n\le x \\ n-r \cong 0 (\mathrm{mod}\, p)}} a_n - \frac{1}{p}\left(1-\frac{1}{p}\right) \sum_{x-\rho<n\le x} a_n \right| > \|W'\| \left(\frac{1}{\rho} \sum_{x-\rho<n\le x} |a_n|^{\alpha} \right)^{\frac{1}{\alpha}}.$$

For the remaining pairs

$$\left(\sum_{p \le Q} p \sum_{r=0}^{p-1} \left| \sum_{\substack{x-\rho<n\le x \\ n-r \cong 0 (\mathrm{mod}\, p)}} a_n - \frac{1}{p}\left(1-\frac{1}{p}\right) \sum_{x-\rho<n\le x} a_n \right|^2 \right)^{\frac{1}{2}}$$
$$\ll \rho(1 + \rho^{-1} Q^{2\alpha'+3})^{\frac{1}{\alpha'}} \left(\rho^{-1} \sum_{x-\rho<n\le x} |a_n|^{\alpha} \right)^{\frac{1}{\alpha}}.$$

Moreover, if $Q^{2\alpha'+3} \le \rho$, then $\|W'\|$ may be replaced by 1. For this range the inequalities have 'the correct form' and are best possible.

The form of fractional large sieve surmised in inequality (3) is neither quite right, nor quite wrong. A main difference is that the underlying operator acts upon the intersection of two spaces, and not upon a single space.

5. Conjectures

The validity of Theorem 1 ensures analogues of (6) and analogues of (4), for every $\alpha > 1$. Announcing these results in 1987, [7], I made a conjecture, the validity of which would guarantee the following version of (6).

Conjecture. *Given an integer $r \neq 0$, and a real $\alpha > 1$, there is a positive δ*

so that

$$(9)\quad \sum_{q\le x}\phi(q)^{\alpha-1}\left|\sum_{\substack{n\le x\\ n\equiv r(\mathrm{mod}\, q)}} f(n)-\frac{1}{\phi(q)}\sum_{\substack{n\le x\\ (n,q)=1}} f(n)\right|^{\alpha} \ll \frac{x^\alpha}{(\log x)^\delta}\sum_{q\le x}\frac{|f(q)|^\alpha}{q},$$

with prime powers q, holds for every additive function f and real $x \ge 2$.

A comparable version of the Bombieri–Vinogradov theorem would allow Q, in (5), up to x. That Q may be allowed almost up to x is a conjecture of Elliott and Halberstam, [8]. Bombieri, Friedlander and Iwaniec, [1], showed that at least for an $A > 1$ one may allow the Q to reach $x^{\frac{1}{2}}\exp((\log x)^c)$ for a certain c, $0 < c < 1$. However, the work of Friedlander, Granville, Hildebrand and Maier, [9], shows that (for example) $Q = x\exp(-\log\log x)^2(\log\log\log x)^{-1})$ is not possible.

This appears to cast doubt upon the likelihood of the conjecture (9). However, a form of the Elliott–Halberstam conjecture more closely analogous to (9) would assert the existence of a positive δ so that

$$\sum_{q\le x}\phi(q)\left|\pi(x,q,r)-\frac{Li(x)}{\phi(q)}\right|^2 \ll \frac{x^2}{(\log x)^{2+\delta}}.$$

The Bombieri–Vinogradov theorem guarantees a range $q \le x^{\frac{1}{2}}(\log x)^{-c}$, and the Brun–Titchmarsh theorem the range $x^{\frac{1}{2}}(\log x)^{-c} < q \le x^{\frac{1}{2}}\exp((\log x)^\varepsilon)$, any fixed ε, $0 < \varepsilon < 1$. Because r is fixed and the moduli q are confined to prime-powers, an application of the Selberg sieve (cf. [3], Chapter 4) shows the range $x^{1-\beta} < q \le x$ to contribute $\ll \beta x^2(\log x)^{-1}$. With β a negative power of $\log x$ we can finesse the limitations (for composite moduli) expressed in the work of Friedlander, Granville, Hildebrand and Maier.

Detailed proofs of the various theorems on additive functions mentioned in the present paper, including the derivation of fractional power large sieves and a discussion of the final example in §**3**, may be found in my book: *Duality in Analytic Number Theory*, to be published by Cambridge University Press.

On the final page of their paper, [2], Davenport and Halberstam modify their argument to deliver

$$(10)\qquad \sum_j |S(\alpha_j)|^k \ll k^{1/2}\max(\delta^{-1},N)\left(\sum_{n=-N}^{N}|a_n|^\ell\right)^{k/\ell},$$

valid uniformly for $k \ge 2$, $k^{-1}+\ell^{-1} = 1$. In particular, they apply the standard Young–Hausdorff inequality, on the additive group of reals (mod 1). Whilst feeling it important to give a detailed argument, they make no comment concerning the result itself. I suspect that this is because they had no application in view. As an inequality in analysis, (10) is of immediate interest. Its implications for number theory (cf. §**3**) are more problematic.

The present paper is a slightly elaborated version of the lecture which I gave at the International Conference on Analytic Number Theory, held at Allerton Park in Illinois, May 16–20, 1995, to celebrate the forthcoming retirement of Professor Heini Halberstam. I thank the organisers Bruce Berndt, Harold Diamond and Adolph Hildebrand for their invitation to speak, and for their support.

References

[1] Bombieri, E., Friedlander, J. B., and Iwaniec, H., Primes in arithmetic progression to large moduli, III, Journ. of the American Math. Soc. **2** (1989), 215–224.

[2] Davenport, H. and Halberstam, H., The values of a trigonometrical polynomial at well spaced points, Mathematika **13** (1966), 91–96; Corrigendum and addendum, Mathematika **14** (1967), 229–232.

[3] Elliott, P.D.T.A., Probabilistic Number Theory, I: Mean Value Theorems., Grund. der math. Wiss. **239**, Springer Verlag, New York, Heidelberg, Berlin, 1979.

[4] ______________, Arithmetic Functions and Integer Products, Grund. der math. Wiss. **272**, Springer Verlag, New York, Berlin, Heidelberg, Tokyo, 1985.

[5] ______________, Additive arithmetic functions on arithmetic progressions, Proc. Lond. Math. Soc. **54** (1987), 15–37.

[6] ______________, Functional analysis and additive arithmetic functions, Bull. Amer. Math. Soc., **16** (2) (1987), 179–223.

[7] ______________, Application of elementary functional analysis to the study of arithmetic functions, Colloquia Math. Soc. János Bolyai **51**, Number Theory, Budapest (1987), 35–43.

[8] Elliott, P.D.T.A. and Halberstam, H., A conjecture in prime number theory, Symposia Mathematica IV, Academic Press, London and New York, 1970, 59–72.

[9] Friedlander, J., Granville, A., Hildebrand, A. and Maier, H., Oscillation theorems for primes in arithmetic progressions and for sifting functions, J. Amer. Math. Soc., **4** (1) (1991), 25–86.

P.D.T.A. Elliott
Department of Mathematics
University of Colorado
Boulder, CO 80309

Some problems I presented or planned to present in my short talk

Paul Erdös

To my old friend Heini Halberstam

I begin with a problem on sieves which is old but has been neglected: Let x and y be large and consider the primes $p_i < y$. Determine or estimate the smallest $y = y(x)$ for which it is possible to find a residue $a_{p_i} \pmod{p_i}$, $p_i < y$, for which every integer in the interval $(1, x]$ satisfies at least one of the congruences $a_{p_i} \pmod{p_i}$. In other words, if we sieve with the primes $p_i < y$, then all the integers in $(1, x]$ disappear. It is known that $y > x^{1/2}$, but as far as I know the estimate $y < x^{1-\epsilon}$ has never been proved and would perhaps be important. The same question can of course be asked if we sieve with r residues mod p_i; the answer will probably be very different and the exponent of x will strongly depend on r.

Over fifteen years ago, Straus, Selfridge, and I considered the following question. Let n be large and represent $n!$ as the product of n integers, the smallest of which is as large as possible. In other words,

$$n! = a_1 a_2 \cdots a_n, \quad a_1 \leq a_2 \leq \ldots \leq a_n. \tag{1}$$

Clearly $a_1 \leq (n/e)(1 + o(1))$. We observed that this is best possible and I believed that Straus had written up our proof. Unfortunately Straus suddenly died and no trace was ever found of his notes. Furthermore, we never could reconstruct our proof, so our assertion now can be called only a conjecture. Guy and Selfridge had difficulties in proving even $a_1 > n/3$ for $n > 300000$. I believe that it would be of some interest to determine $\min a_1(n)$ as accurately as possible and to try some computation.

Let $f(n)$ denote the maximum possible value of a_1 which satisfies (1). It is not true that $f(n+1) - f(n) \leq 1$; see the report in Guy's "Unsolved Problems in Number Theory" for a counter-example.

Gordon called my attention to the following problem which we had discussed many years ago and which I had completely forgotten. Is it true that for every positive integer t there is an integer a for which

$$\binom{n}{k} = a, \quad 1 \leq k \leq n/2, \quad 1 < n < \infty \tag{2}$$

has t solutions? We found that for $a = 120$, (2) has three solutions and Selfridge told me that also for $a = 3003$ (2) has three solutions. We have never found an integer a for which (2) has four or more solutions, and perhaps such an a does not exist.

Balasubramanian called to my attention another old problem of mine which I had forgotten. Can there ever be two distinct integers x and y for which for $0 \leq i \leq 2$ the integers $x+i$ and $y+i$ have the same prime factors? The answer is Yes for a more restricted problem. For x, y and $x+1, y+1$ we can take $x = 2(2^r - 1)$, $y + 1 = (x+1)^2$. Then x and y also have the same prime factors. Are there other x, y and $x+1, y+1$ with the same factors?

One of my favorite conjectures, which goes back to 1934, is as follows: A system of congruences

$$a_i \pmod{n_i},\ 1 < n_1 < n_2 < \ldots < n_k \tag{3}$$

is called a covering system if every integer satisfies at least one of the congruences (3). The simplest such system is 0 (mod 2), 0 (mod 3), 1 (mod 4), 5 (mod 6), and 7 (mod 12). I conjectured that there is such a system for which n_1 is arbitrarily large. I offered and offer \$1000 for a proof or a disproof.

Selfridge, Schinzel, and I asked: Is there a covering system (3) in which all moduli are odd or where none of the moduli divides another?

Selfridge recently asked: Is it true that if we divide the integers into two (or more generally into k) classes, then one of the classes contain a covering system (3)? Perhaps the following stronger conjecture also holds: Let

$$\sum_{n_i < x} \frac{1}{n_i} > c \log x, \quad x > x_0(c).$$

Is it then true that our set $\{n_i\}$ contains a covering system (3)? Perhaps the answer is negative.

R. L. Graham and I asked: Is there an integer m_k such that if we divide the integers $1 \leq t \leq m_k$ into k classes, then the equation

$$\sum \frac{1}{x_i} = 1$$

is monochromatically solvable? The answer must surely be positive, but we do not have a proof even for $k = 2$. Very likely, for every $\epsilon > 0$ there is an x_0 such that if

$$\sum_{n_i < x_0} \frac{1}{n_i} > \epsilon \log x_0,$$

then

$$\sum_{n_i < x_0} \frac{\epsilon_i}{n_i} = 1, \quad \epsilon_i = 0 \text{ or } 1$$

is solvable. In fact, $\epsilon \log x_0$ can perhaps be replaced by $(\log\log x)^c$. In other words, our Ramsey type conjecture can be strengthened to a Turán type conjecture. Unfortunately, so far we have no results of any kind.

Is it true that for every k there are integers n_k such that if we split the proper divisors of n_k into k classes, n_k can always be expressed as a sum of distinct elements from the same class? I cannot even prove that n_2 exists. It would be of interest to estimate the smallest such n_k. I expect that for large enough $m_k, m_k!$ has our property.

Benkoski and I [1] have the following old problem. Let $\sigma(n)$ denote the sum of the divisors of n. Is it true that if $\sigma(n)/n$ is large enough then n is always the sum of distinct proper divisors of n? In fact we expected that if $\sigma(n)/n$ is large enough, only divisors smaller than ϵn have to be used.

Is it true that there is a c so that there are arbitrarily large integers n for which every $m < n$ is the sum of fewer than $(\log\log n)^c$ distinct divisors of n? Vose proved that for large n every $m < n!$ is the sum of $n^{1/2}$ divisors of $n!$ Probably this should hold with n^ϵ or even $(\log n)^c$ distinct divisors of $n!$

References

[1] S. J. Benkoski and P. Erdös, *On weird and pseudoperfect numbers*, Mathematics of Computation **28** (1979), 617 - 623.

[2] Michael D. Vose, *Integers with consecutive divisors in small ratio*, J. Number Theory **19** (1984), 233 - 238.

Paul Erdös
Mathematical Institute
Hungarian Academy of Sciences
P.O. Box 127
1364 Budapest
HUNGARY

On the number of divisors of n!

Paul Erdös, S. W. Graham, Aleksandar Ivić and Carl Pomerance

For Heini Halberstam on his retirement

Abstract. Several results involving $d(n!)$ are obtained, where $d(m)$ denotes the number of positive divisors of m. These include estimates for $d(n!)/d((n-1)!)$, $d(n!) - d((n-1)!)$, as well as the least number K with $d((n+K)!)/d(n!) \geq 2$.

1. Introduction

Let as usual $d(m)$ denote the number of positive divisors of m. We are interested here in problems involving $d(n!)$. There seem to exist only a few results in the literature on this subject, one of which is the result of G. Tenenbaum [13] that

$$\sum_{1 \leq j \leq d(n!)} \left(\frac{d_{j+1}}{d_j} - 1 \right)^{\alpha} \ll_{\alpha} 1$$

for any fixed $\alpha > 1$, where $1 = d_1 < d_2 < \cdots < d_m = n!$ are the divisors of $n!$ (and so $m = d(n!)$).

We note that the divisor function can at least occasionally be large, since by a classical result of Wigert, one has

$$\log d(m) \leq \frac{\log 2 \log m}{\log \log m} + O\left(\frac{\log m}{(\log \log m)^2} \right), \tag{1}$$

with equality holding if $m = p_1 p_2 \cdots p_r$, where $r \to \infty$, and p_r is the r-th prime number. Our first goal is to obtain an asymptotic formula for $\log d(n!)$, showing that it is $\sim c_0 \log(n!)/(\log\log(n!))^2$, where c_0 is an explicit constant approximately equal to 1.25775.

We next are concerned with the function $d(n!)/d((n-1)!)$. We show that this is well approximated by $1 + P(n)/n$, where $P(n)$ is the largest prime

---

Research of the second author was partially supported by the National Security Agency; research of the third author financed by the Mathematical Insititute of Belgrade; research of the fourth author partially supported by the National Science Foundation.

factor of n. From this we are able to find the limit points of the sequence $d(n!)/d((n-1)!)$.

Let $K = K(n)$ denote the least number with $d((n+K)!)/d(n!) \geq 2$. We are able to show that infinitely often $K(n)$ is abnormally large, namely that $K(n)/\log n$ is unbounded. The method of proof involves the Erdös-Rankin method for showing that there are sometimes abnormally large gaps between the primes. On the other hand we show that $K(n)$ cannot be too large: it is $< n^{4/9}$ for all sufficiently large numbers n. The proof comes from the circle of ideas, begun by Ramachandra, for showing that short intervals contain integers with large prime factors.

Finally we consider the function $D(n) = d(n!) - d((n-1)!)$. In particular we consider "champs" for this sequence, namely numbers n with $D(n) > D(m)$ for all numbers m smaller than n. We show that primes and the doubles of primes are such champs, we show on the prime k-tuples conjecture that there are infinitely many other champs, and we show on the Riemann Hypothesis that the set of champs has asymptotic density zero.

2. An approximate formula for $d(n!)$

Our first aim is to express $d(n!)$ in terms of elementary functions. If $[x]$ denotes the integer part of x, then

$$n! = \prod_{p\leq n} p^{w_p(n)}, \text{ where } w_p(n) := \left[\frac{n}{p}\right] + \left[\frac{n}{p^2}\right] + \left[\frac{n}{p^3}\right] + \cdots,$$

hence

$$\begin{aligned} \log d(n!) &= \sum_{p\leq n} \log(w_p(n)+1) \\ &= \sum_{p\leq n^{3/4}} \log(w_p(n)+1) + \sum_{n^{3/4}<p\leq n} \log(w_p(n)+1) = \sum_1 + \sum_2, \end{aligned} \tag{2}$$

say. Since $w_p(n) < n\sum_j p^{-j} = n/(p-1)$, we have

$$\sum_1 < \sum_{p\leq n^{3/4}} \log\left(1+\frac{n}{p-1}\right) \ll \log n \sum_{p\leq n^{3/4}} 1 \ll n^{3/4}.$$

Now note that in Σ_2 we have $w_p(n) = [n/p]$, since $[n/p^k] = 0$ for $k \geq 2$.

Therefore by the prime number theorem we have

$$\sum_2 = \int_{n^{3/4}}^{n} \log\left(\left[\frac{n}{x}\right]+1\right) d\pi(x)$$
$$= \int_{n^{3/4}}^{n} \frac{\log\left(\left[\frac{n}{x}\right]+1\right)}{\log x}\,dx + \int_{n^{3/4}}^{n} \frac{\log\left(\left[\frac{n}{x}\right]+1\right)}{\log x}\,dR(x) = I_1 + I_2,$$

say, where (see e.g. [2] or [7]) $R(x) = O(xe^{-\sqrt{\log x}})$. Since $\log\left(\left[\frac{n}{x}\right]+1\right)$ is a non-increasing function of x in $[1, n]$ we obtain, on integrating by parts,

$$I_2 \ll ne^{(-1/2)\sqrt{\log n}}.$$

The main contribution to $\log d(n!)$ comes from

$$I_1 = \int_{n^{3/4}}^{n} \frac{\log\left(\left[\frac{n}{x}\right]+1\right)}{\log x}\,dx = n\int_{1}^{n^{1/4}} \frac{\log([t]+1)}{t^2\log\left(\frac{n}{t}\right)}\,dt$$
$$= \frac{n}{\log n}\sum_{k=0}^{K}\frac{1}{\log^k n}\int_1^{n^{1/4}}\frac{\log([t]+1)}{t^2}\log^k t\,dt + O\left(\frac{n}{\log^{K+2} n}\right)$$
$$= \frac{n}{\log n}\sum_{k=0}^{K}\frac{c_k}{\log^k n} + O\left(\frac{n}{\log^{K+2} n}\right)$$

for any fixed integer $K \geq 0$, where

$$c_k = \int_1^{\infty}\frac{\log([t]+1)}{t^2}\log^k t\,dt. \tag{3}$$

In particular,

$$c_0 = \int_1^{\infty}\log([t]+1)\frac{dt}{t^2} = \sum_{k=2}^{\infty}\int_{k-1}^{k}\log k\frac{dt}{t^2} = \sum_{k=2}^{\infty}\frac{\log k}{k(k-1)} \approx 1.25775.$$

Thus we have proved

Theorem 1. *For any fixed integer $K \geq 0$ and c_k given by (3) we have*

$$d(n!) = \exp\left\{\frac{n}{\log n}\sum_{k=0}^{K}\frac{c_k}{\log^k n} + O\left(\frac{n}{\log^{K+2} n}\right)\right\}.$$

Let $m = n!$. Then by Stirling's formula one has $\log m = n\log n - n + O(\log n)$, which gives

$$\log n = \log\log m + O(\log\log\log m), \quad n = \frac{\log m}{\log\log m}\left(1 + O\left(\frac{\log\log\log m}{\log\log m}\right)\right).$$

Therefore we obtain from Theorem 1 that

$$\log d(m) = \frac{c_0 \log m}{(\log\log m)^2}\left(1 + O\left(\frac{\log\log\log m}{\log\log m}\right)\right),$$

which may be compared with (1).

3. The function $d(n!)/d((n-1)!)$

We now note another consequence of Theorem 1. It implies that $d(n!)/d((n-1)!) \to 1$ as $n \to \infty$ on a set of asymptotic density 1. We can show more. We begin with the following lemma.

Lemma 1. *Let $S(n)$ denote the sum of the prime factors of n where they are summed with multiplicity. Then for every integer $n \ge 1$,*

$$1 + \frac{S(n)}{2n} \le \frac{d(n!)}{d((n-1)!)} \le 1 + \frac{2S(n)}{n}.$$

Proof. For the upper bound, we have

$$\frac{d(n!)}{d((n-1)!)} = \prod_{p^a \| n} \frac{w_p(n-1)+1+a}{w_p(n-1)+1}$$

$$= \prod_{p^a \| n}\left(1 + \frac{a}{w_p(n-1)+1}\right) \le \exp\left(\sum_{p^a \| n} \frac{a}{w_p(n-1)+1}\right). \tag{4}$$

We have for any prime p dividing n that $[(n-1)/p] = n/p - 1$. Thus, $w_p(n-1)+1 \ge [(n-1)/p]+1 = n/p$, and so

$$\frac{d(n!)}{d((n-1)!)} \le \exp\left(\sum_{p^a \| n} \frac{a}{n/p}\right) = \exp(S(n)/n). \tag{5}$$

But clearly $S(n) \le n$ for every natural number n, so that $\exp(S(n)/n) \le 1 + 2S(n)/n$, which completes our proof of the upper bound in the lemma.

For the lower bound, we have

$$\frac{d(n!)}{d((n-1)!)} = \prod_{p^a \| n}\left(1 + \frac{a}{w_p(n-1)+1}\right) \ge 1 + \sum_{p^a \| n} \frac{a}{w_p(n-1)+1}.$$

But $w_p(n-1)+1 < 1 + (n-1)\sum_j p^{-j} = 1 + (n-1)/(p-1) \le 2n/p$ (the last inequality easily proved for all integers p, n with $2 \le p \le n$). Thus we have the lower bound in the lemma.

We are now ready to prove the following theorem.

Theorem 2. *Let $P(n)$ denote the largest prime factor of n. Then*

$$\frac{d(n!)}{d((n-1)!)} = 1 + \frac{P(n)}{n} + O\left(\frac{1}{n^{1/2}}\right).$$

Proof. Let $\Omega(n)$ denote the number of prime factors of n, counted with multiplicity. Then $S(n) \le P(n)\Omega(n) \le P(n)\log_2 n$. First suppose that $p = P(n) \le n^{1/2}$ and $q = P(n/p) \le n^{1/3}$. Then $S(n) = p + S(n/p) \le p + q\log_2 n \ll n^{1/2}$. Next suppose that $p \le n^{1/2}$ and $n^{1/3} < q \le p$. Then $S(n) = p + q + S(n/pq) \le p + q + n/pq \le 3p \ll n^{1/2}$. Thus, if $P(n) \le n^{1/2}$, then $S(n) \ll n^{1/2}$, and so the theorem follows from Lemma 1 in this case.

Now suppose that $p = P(n) > n^{1/2}$. Writing $n = mp$, we have $w_p(n) = [n/p] = m$, $w_p(n-1) = m - 1$, so that from (4),

$$\frac{d(n!)}{d((n-1)!)} = \frac{m+1}{m} \prod_{q^b \| m} \left(1 + \frac{b}{w_q(n-1)+1}\right).$$

By the calculations in the proof of the lemma we have

$$\begin{aligned} 1 &\le \prod_{q^b\|m}\left(1+\frac{b}{w_q(n-1)+1}\right) \le \prod_{q^b\|m}\left(1+\frac{b}{n/q}\right) \le \exp\left(\sum_{q^b\|m}\frac{bq}{n}\right) \\ &= \exp\left(\frac{S(m)}{n}\right) \le \exp\left(\frac{m}{n}\right) \le 1+\frac{2m}{n} \le 1+\frac{2}{n^{1/2}}. \end{aligned}$$

This completes the proof of the theorem.

Corollary 1. *The set of limit points of the sequence $d(n!)/d((n-1)!)$ consists of the number 1 and the numbers $1 + 1/m$, where m is a natural number.*

From Theorem 2 it is straightforward to get asymptotic estimates, or even asymptotic expansions, for the average order of $d(n!)/d((n-1)!)$, or any fixed positive or negative power of this fraction. For example, such results follow from the circle of papers that includes [3].

For a positive integer k, let $F_k(n) = d((n+k)!)/d(n!)$. It follows from Theorem 2 that for fixed k, the average order of $F_k(n)$ is 1. Indeed, from Theorem 2,

$$1 \le F_k(n) \le \exp\left(\frac{1}{n}\sum_{i=1}^{k} P(n+i) + O\left(\frac{k}{n^{1/2}}\right)\right). \tag{6}$$

But

$$\sum_{n\le x}\sum_{i=1}^{k} P(n+i) \le k \sum_{m\le x+k} P(m) \ll kx^2/\log x = o(x^2).$$

Thus, on average, $\sum_{i=1}^{k} P(n+i) = o(n)$, and for all n, we have this sum $\ll kn$. Hence our assertion follows from (6). This argument can be used to show that if $k = k(n)$ varies with n in such a way that $k(n) = o(\log n)$, then the normal order of $F_k(n)$ is 1. That is, $F_k(n) \sim 1$ as n tends to ∞ through a set of integers of asymptotic density 1. It is also possible to show via (6) that if $c > 0$ is fixed and $k = k(n) \sim c\log n$, then the normal order of $F_k(n)$ is $g(c)$, where $g(c) > 1$ is a number that depends on c. These results may well be true too for the average order, but the proofs are likely to be harder.

In the spirit of Corollary 1, it may also be of interest to consider the set of limit points of the set of numbers $F_k(n)$, where both k and n vary. It is likely possible to show that if the prime k-tuples conjecture is true, then these numbers are dense in the interval $[1, \infty)$. Without assuming this hypothesis, the most we can actually prove is contained in Corollary 1 and the additional assertion that if $x \ge 1$ is arbitrary, then there is a limit point for the set of $F_k(n)$ in the interval $[x, 2x]$. This follows from the fact that $d(n!)/d((n-1)!) \le 2$ for all n. We know no specific finite limit points of the set of $F_k(n)$ other than the numbers specified in Corollary 1.

Let p_n denote the nth prime. We conjecture that the sequence $d(p_{n+1}!)/d(p_n!)$ is dense in $[1, \infty)$. At present we cannot prove that the liminf is 1, the limsup is ∞, nor can we identify any specific limit points. It follows from Theorem 1 that there is at least one finite limit point.

4. The least K with $d((n+K)!)/d(n!) \ge 2$

Let $K = K(n)$ denote the least positive integer with $F_K(n) \ge 2$. That is, $d((n+K)!) \ge 2d(n!)$. If $n+1$ is prime, then $K(n) = 1$. From Theorem 1 it seems that one should compare $K(n)$ with $\log n$. In fact, this theorem immediately implies that the average order of $K(n)$ is $\asymp \log n$. One might ask about the maximal order of $K(n)$. The following two results show that $K(n) < n^{4/9}$ for all large numbers n and that $K(n)/\log n$ is unbounded.

Theorem 3. *Recall that $S(n)$ denotes the sum of the prime factors of n, with multiplicity. Let $f(n)$ denote the least number such that*

$$\sum_{i=1}^{f(n)} S(n+i) > n.$$

For each number $\varepsilon > 0$ there are infinitely many integers n for which

$$f(n) \geq (1/4 - \varepsilon) \log n \log\log n \log\log\log\log n / (\log\log\log n)^3.$$

Corollary 2. *For infinitely many natural numbers n we have*

$$K(n) > \log n \frac{\log\log n \log\log\log\log n}{9(\log\log\log n)^3}.$$

In particular, $K(n)/\log n$ is unbounded.

Proof of the Corollary. The sum of $S(m)$ over the interval $(n, n + f(n))$ is $\leq n$. Thus, either over the first half of this interval the sum is $\leq \frac{1}{2}n$, or over the second half the sum is $\leq \frac{1}{2}n$. From the theorem with a choice of ε such that $2/9 < 1/4 - \varepsilon$, there are infinitely many pairs n, K with $K > (1/9) \log n \log\log n \log\log\log\log n / (\log\log\log n)^3$ and with $\sum_{i=1}^{K} S(n+i) \leq \frac{1}{2}n$. It follows from (5) that

$$\frac{d((n+K)!)}{d(n!)} < \exp\left(\sum_{i=1}^{K} \frac{S(n+i)}{n+i}\right) < \exp\left(\frac{1}{n}\sum_{i=1}^{K} S(n+i)\right) \leq \exp\left(\frac{1}{2}\right) < 2.$$

Thus, $K(n) > K$ and the corollary is proved.

Proof of Theorem 3. Let u denote a large number and let $M = M(u)$ denote the product of the primes in the interval $[\log^2 u, u]$. Let $\varepsilon > 0$ be arbitrary and fixed. By the Erdös–Rankin argument, for all sufficiently large numbers u, depending on the choice of ε, there is a residue class $A \bmod M$, such that $(A + i, M) > 1$ for each integer i with

$$1 \leq i \leq L := (1/2 - \varepsilon/4) u \log u \log\log\log u / (\log\log u)^3.$$

Indeed, it suffices to show that for each prime $p|M$ there is a residue class $a_p \bmod p$, such that for each integer i with $1 \leq i \leq L$, there is a prime $p|M$ with $i \equiv a_p \bmod p$. For then we may choose A via the Chinese remainder theorem to be that residue class mod M satisfying $A \equiv -a_p \bmod p$ for each prime factor p of M. The numbers a_p can be chosen as follows: for $y := u^{(1-\varepsilon/4)\log\log\log u/\log\log u} < p \leq u/\log\log u$, we choose $a_p = 0$. The number of integers i in $[1, L]$ that are not congruent to 0 modulo any of these primes p is $\sim L \log\log u / \log u$ (using de Bruijn [1]). Next, for the primes p with $\log^2 u \leq p \leq y$, we choose a_p sequentially in such a way that for each p we have as many as possible remaining integers in $[1, L]$ congruent to $a_p \bmod p$. The number of remaining integers i in $[1, L]$ that are still not covered by any of the residue classes a_p mod p for $\log^2 u \le p \le u/\log\log u$ is, by Mertens' theorem, less than or asymptotically equal to

$$\frac{2\log\log u}{\log y}\cdot\frac{L\log\log u}{\log u}=\frac{2}{1-\varepsilon/4}\cdot\frac{L(\log\log u)^3}{(\log u)^2\log\log\log u}=\frac{1-\varepsilon/2}{1-\varepsilon/4}\cdot\frac{u}{\log u}.$$

It follows from the prime number theorem that for u sufficiently large, there are fewer residual values of i left unsieved in $[1,L]$ than there are primes p in the remaining interval $u/\log\log u < p \le u$, so even if we use these primes to remove just one value of i each, we will succeed in covering the entire interval $[1,L]$ as claimed.

For any integer $l \le u^3$, let $N(l)$ denote the number of pairs i,j of integers with $1\le i\le L$, $M/2\le j\le M$, and $(jM+A+i)/l$ is prime.

Lemma 2. *We have uniformly for $l \le u^3$,*

$$N(l) \ll \frac{LM\log u}{\varphi(l)u\log\log u}+\frac{(l,M)M\log u}{\varphi(l)u\log\log u}.$$

Proof. Let $k=l/(l,M)$. First consider the case $(l,M)\le L$. For l to divide $jM+A+i$, we must have $i\equiv -A \bmod (l,M)$. There are $\ll L/(l,M)$ values of $i\in[1,L]$ for which this holds. Fixing one of these, for l to divide $jM+A+i$, we must have j in a fixed residue class mod k. So the number of such j values in $[M/2,M]$ for which $jM+A+i$ is a prime, is, by the sieve,

$$\ll \frac{M/k}{\log(M/k)}\cdot\frac{Mk}{\varphi(Mk)} \ll \frac{M}{\varphi(k)\log M}\cdot\frac{M}{\varphi(M)} \ll \frac{M}{\varphi(k)u}\cdot\frac{M}{\varphi(M)},$$

where we have used $\log(M/k)\sim\log M\sim u$ and $\varphi(Mk)\ge\varphi(M)\varphi(k)$. Now $M/\varphi(M)\asymp \log u/\log\log u$, and multiplying by the number of possible i values, we get

$$N(l)\ll\frac{L}{(l,M)}\cdot\frac{M\log u}{\varphi(k)u\log\log u}\le\frac{LM\log u}{\varphi(l)u\log\log u}.$$

Now consider the case $(l,M)>L$. Then there is at most one value of $i\in[1,L]$ for which l can divide $jM+A+i$. Thus again using the sieve, we have

$$N(l)\ll\frac{M/k}{\log(M/k)}\cdot\frac{Mk}{\varphi(Mk)}\ll\frac{M\log u}{\varphi(k)u\log\log u}\le\frac{(l,M)M\log u}{\varphi(l)u\log\log u}.$$

This completes the proof of the lemma.

We now return to the proof of the theorem. We are going to show that

$$\sum_{M/2\le j\le M}\ \sum_{1\le i\le L} S(jM+A+i)=o(M^3) \tag{7}$$

as $u \to \infty$. Note that for $j \in [M/2, M]$ and $i \in [1, L]$, we have $jM+A+i \asymp M^2$. If we show (7) it will follow that for u large there is some number $j \in [M/2, M]$ with $\sum_{1 \le i \le L} S(jM+A+i) < jM+A$, so that $f(jM+A) > L$, and the theorem follows.

Let $S(n) = S_1(n) + S_2(n)$, where $S_1(n)$ is the sum of the prime factors of n that are bigger than n/u^3, while $S_2(n)$ is the sum of the smaller prime factors of n. Note that $S_2(n) \ll (n/u^3) \log n$. Thus

$$\sum_{M/2 \le j \le M} \sum_{1 \le i \le L} S_2(jM + A + i) \ll \sum_{M/2 \le j \le M} \sum_{1 \le i \le L} \frac{M^2}{u^3} \log(M^2) \ll \frac{LM^3}{u^2} = o(M^3),$$

as $u \to \infty$. Thus it suffices to show

$$\sum_{M/2 \le j \le M} \sum_{1 \le i \le L} S_1(jM + A + i) = o(M^3) \tag{8}$$

as $u \to \infty$. If $S_1(jM + A + i) > 0$, then $jM + A + i = Pl$, where P is prime and $l < u^3$. By the choice of A, we have $(l, M) > 1$, so that $l \ge \log^2 u$. Thus, by the lemma, we have

$$\begin{aligned}
&\sum_{M/2 \le j \le M} \sum_{1 \le i \le L} S_1(jM + A + i) \\
&= \sum_{\log^2 u \le l < u^3} \sum_{M/2 < j < M} \sum_{\substack{1 \le i \le L \\ (jM+A+i)/l \text{ is prime}}} \frac{jM + A + i}{l} \\
&\ll \sum_{\log^2 u \le l < u^3} \sum_{M/2 < j < M} \sum_{\substack{1 \le i \le L \\ (jM+A+i)/l \text{ is prime}}} M^2/l \\
&= \sum_{\log^2 u \le l < u^3} N(l) M^2/l \\
&\ll \sum_{\log^2 u \le l} \left(\frac{LM^3 \log u}{l\varphi(l) u \log\log u} + \frac{(l, M) M^3 \log u}{l \varphi(l) u \log\log u} \right) \\
&\ll \frac{LM^3 \log u}{u \log^2 u \log\log u} + \sum_{d|M} \sum_{v=1}^{\infty} \frac{dM^3 \log u}{dv\varphi(dv) u \log\log u} \\
&\ll \frac{LM^3}{u \log u \log\log u} + \sum_{d|M} \frac{M^3 \log u}{\varphi(d) u \log\log u} \\
&\ll \frac{LM^3}{u \log u \log\log u} + \frac{M^3 \log^2 u}{u (\log\log u)^2} = o(M^3),
\end{aligned}$$

as $u \to \infty$. This proves (8) and the theorem.

Theorem 3 shows that sometimes $f(n)$, and hence $K(n)$, can be abnormally large. We now show that they cannot be too large.

Theorem 4. *Let $f(n)$ be as in Theorem 3. If n is sufficiently large, then $f(n) < n^{4/9}$.*

Corollary 3. *For all sufficiently large numbers n, we have $K(n) < n^{4/9}$.*

Theorem 4 and Corollary 3 are immediate consequences of the following lemma.

Lemma 3. *Let x be a sufficiently large positive real number, let $c = 4/9$, and $\delta = 1/1000$. Then the number of primes p such that p divides some m in the interval $(x, x + x^c]$ and $p > x^{1-c+\delta}$ is $\gg x^c$.*

To see how Theorem 4 follows from Lemma 3, note that if we take $g(n) = [n^{4/9}]$, then

$$\sum_{i=1}^{g(n)} S(n+i) \geq n^{5/9+\delta} R,$$

where R is the number of primes $p > n^{5/9+\delta}$ that divide some integer in the interval $(n, n + g(n)]$. By the Lemma, $R \gg n^{4/9}$, so

$$\sum_{i=1}^{g(n)} S(n+i) \gg n^{1+\delta}.$$

Therefore, $f(n) \leq g(n)$ for n sufficiently large.

To see how Corollary 3 follows from Lemma 3, we have using Lemma 1 and the same notation as above that

$$\frac{d((n+g(n))!)}{d(n!)} > \prod_{i=1}^{g(n)} \left(1 + \frac{S(n+i)}{2(n+i)}\right) > 1 + \frac{1}{2}\sum_{i=1}^{g(n)} \frac{S(n+i)}{n+i}$$
$$> 1 + \frac{1}{3n}\sum_{i=1}^{g(n)} S(n+i) \gg n^{\delta}.$$

Thus, for n sufficiently large, we have $K(n) \leq g(n)$, and so we have the corollary.

Proof of Lemma 3. For most of the argument, we shall assume only that c is a real number with $c < 1/2$; only at the last step will we specialize to $c = 4/9$. Our proof is an adaptation of K. Ramachandra's proof [10] that the greatest prime factor of the product of the integers in the interval $(x, x + x^{1/2}]$ exceeds $x^{15/26}$, for x sufficiently large. Ramachandra's exponent has been improved by a series of authors; see the references in [9]. The current best exponent is 0.723, and it is due independently to Hong-Quan Liu [9] and Jia Chaohua [8]. It is very likely that our exponent of 4/9 can be improved by using ideas from these papers, but, for reasons of brevity, we do not consider these possible improvements here.

In the initial stage, we follow an argument due to Chebyshev (cf. [6], Chapter 2). We begin by observing that

$$\begin{aligned} x^c \log x + O(x^c) &= \sum_{x<m\le x+x^c} \log m \\ &= \sum_{x<m\le x+x^c} \sum_{d\mid m} \Lambda(d) = \sum_{d} \Lambda(d) N(d) \end{aligned} \tag{9}$$

where

$$N(d) = \sum_{\substack{d\mid m \\ x<m\le x+x^c}} 1, \tag{10}$$

and Λ is von Mangoldt's function; i.e., $\Lambda(d) = \log p$ if $d = p^a$ for some prime p and 0 otherwise. We decompose the last sum in (9) as

$$\begin{aligned} \sum_{d} \Lambda(d)N(d) &= \sum_{d\le x^c} \Lambda(d)N(d) + \sum_{p\le x^c<p^a} N(p^a)\log p + \sum_{p>x^c}\sum_{a\ge 1} N(p^a)\log p \\ &= \Sigma_1 + \Sigma_2 + \Sigma_3, \end{aligned} \tag{11}$$

say. From the trivial estimate $N(d) = x^c/d + O(1)$, we get

$$\Sigma_1 = x^c \sum_{d\le x^c} \Lambda(d)/d + O(x^c) = cx^c \log x + O(x^c).$$

If $p^a > x^c$, then $N(p^a) \le 1$, and $N(p^a) = 0$ if $p^a > 2x > x + x^c$. Therefore

$$\Sigma_2 \le \sum_{p\le x^c} \sum_{a\le \log(2x)/\log p} 1 \ll \sum_{p\le x^c} \frac{\log x}{\log p} \ll x^c.$$

Combining (9) and (11) together with the above estimates for Σ_1 and Σ_2 gives

$$\Sigma_3 = (1-c)x^c \log x + O(x^c). \tag{12}$$

Since we wish to get prime factors p with $p > x^{1-c+\delta}$, we write

$$\Sigma_3 = \sum_{x^c<p\le x^{1-c+\delta}} \sum_{a\ge 1} N(p^a)\log p + \sum_{x^{1-c+\delta}<p} N(p)\log p = \Sigma_4 + \Sigma_5, \tag{13}$$

say. Here, δ is a small positive constant that will be chosen later. In our notation for Σ_5, we have used the fact that if $c < 1/2$ then $(1-c) > 1/2$; consequently, if $p > x^{1-c}$ and $N(p^a) \neq 0$, then we must have $a = 1$.

Our objective is to show that $\Sigma_5 \gg x^c \log x$ for some value of c. We do this by using Selberg's upper bound sieve to give an upper bound for Σ_4. We will split the range of summation in this sum into subintervals of the form $v < p^a \leq ev$; accordingly, it is convenient to define

$$T(v) = \sum_{\substack{x^c < p \\ v < p^a \leq ev}} N(p^a).$$

Now we let $z = z(v, x)$ be a parameter to be chosen later; for now, we assume only that $z \leq x^c$. Suppose that we have real numbers λ_d with $\lambda_1 = 1$ and $\lambda_d = 0$ if $d > z$ or if d is not squarefree. Then

$$T(v) \leq \sum_{v < m \leq ev} N(m) \left(\sum_{d|m} \lambda_d \right)^2.$$

Upon expanding the square, we obtain

$$T(v) \leq \sum_{k,l} \lambda_k \lambda_l \sum_{v < m[k,l] \leq ev} N(m[k,l]). \tag{14}$$

We expect that the inner sum is about $x^c/[k,l]$. Accordingly, we wish to choose the λ_k's to minimize the bilinear form $\sum_{k,l} \lambda_k \lambda_l [k,l]^{-1}$ subject to the restraints $\lambda_1 = 1$ and $\lambda_k = 0$ if $k > z$. From the theory of the Selberg sieve (cf. [6], pp. 7–9), it is known that this conditional minimum is $\leq (\log z)^{-1}$. Using this choice of λ_k and writing $\rho_d = \sum_{[k,l]=d} \lambda_k \lambda_l$, we see that (14) may be re-written as

$$T(v) \leq \sum_{d < z^2} \rho_d \sum_{v/d < m \leq ev/d} N(md). \tag{15}$$

For future reference, we note also that the minimizing choice of λ_k satisfies $|\lambda_k| \leq \mu^2(k)$ ([6], equation 18); therefore

$$\rho_d \leq \mu^2(d) \tau_3(d), \tag{16}$$

where $\tau_3(d)$ denotes the number of ways of writing d as a product of three ordered factors.

Now let $\psi(w) = w - [w] - 1/2$. The inner sum in (15) may be written as

$$\sum_{v/d<m\le ev/d} N(md) = \sum_{v/d<m\le ev/d} \left(\frac{x^c}{md} - \psi\left(\frac{x+x^c}{md}\right) + \psi\left(\frac{x}{md}\right)\right)$$
$$= \frac{x^c}{d} + O\left(\frac{x^c}{v}\right) + \sum_{v/d<m\le ev/d} \left(\psi\left(\frac{x}{md}\right) - \psi\left(\frac{x+x^c}{md}\right)\right).$$

Using $\sum \rho(d)/d = \sum_{k,l} \lambda_k\lambda_l[k,l]^{-1} \le (\log z)^{-1}$ for the main term and (16) for the first error term gives

$$T(v) \le \frac{x^c}{\log z} + O\left(x^c v^{-1} z^2 \log^2 z\right) + R_1 - R_2, \tag{17}$$

where

$$R_1 = \sum_{d<z^2} \rho_d \sum_{v/d<m\le ev/d} \psi\left(\frac{x}{md}\right),$$

and R_2 is the corresponding sum with x replaced by $x + x^c$.

We use the theory of exponent pairs to estimate R_1 and R_2. Assume that (k,l) is an exponent pair with $k \ne 0$ and $l \ne 1$. (In our application, we will use only the exponent pairs $B(0,1) = (1/2, 1/2)$, $AB(0,1) = (1/6, 2/3)$ and $A^2B(0,1) = (1/14, 11/14)$.) From Lemma 4.3 of [5], we see that

$$R_i \ll \sum_{d<z^2} \tau_3(d)\left(x^{k/(k+1)} d^{-l/(k+1)} v^{(l-k)/(k+1)} + v^2 x^{-1} d^{-1}\right)$$
$$\ll x^{k/(k+1)} v^{(l-k)/(k+1)} z^{2(1+k-l)/(k+1)} \log^2 z + v^2 x^{-1} \log^3 z$$

for $i = 1$ and 2. Now we choose z to make the first term in the above estimate $< x^c/(\log x)^2$; in other words, we choose z to satisfy

$$z^{2+2k-2l} = v^{k-l} x^{c(k+1)-k} (\log x)^{-4(k+1)}.$$

With this choice of z, the above estimates give

$$T(v) \le \frac{\theta(u,k,l)x^c}{\log x} + O\left(\frac{x^c \log\log x}{\log^2 x}\right) + O\left(\frac{v^2 \log^3 x}{x}\right)$$
$$+ O\left(\frac{x^{(2c+2ck-cl-k)/(1+k-l)}}{v^{1/(1+k-l)} (\log x)^{(2+2k+2l)/(1+k-l)}}\right), \tag{18}$$

where $u = \log v / \log x$ and

$$\theta(u,k,l) = \frac{2(1+k-l)}{c+ck-k+ku-lu}.$$

Now if $v \leq x^{(1+c)/2}(\log x)^{-5/2}$, then the second error term in (18) is $\ll x^c/\log^2 x$. We also note that if $v \geq x^c$, then the third error term is

$$\ll x^c x^{(c-1)k/(1+k-l)}(\log x)^{-(2+2k+2l)/(1+k-l)} \ll x^c(\log x)^{-2}.$$

Therefore, if

$$x^c \leq v \leq x^{(1+c)/2}(\log x)^{-5/2},$$

then our bound for $T(v)$ may be simplified to

$$T(v) \leq \frac{\theta(u,k,l)x^c}{\log x} \cdot (1 + O(\epsilon(x))), \tag{19}$$

where, for brevity, we have written $\epsilon(x)$ for $(\log\log x)/\log x$.

Now we are ready to bound Σ_4. Let

$$\mathcal{V} = \{e^j x^c : 0 \leq j \text{ and } e^j x^c \leq x^{1-c+\delta}\},$$

so that

$$\Sigma_4 \leq \sum_{v\in\mathcal{V}} \log(ev)T(v).$$

We now set

$$\alpha = \frac{1+17c}{18} \text{ and } \beta = \frac{1+5c}{6}.$$

For $x^c \leq v \leq x^\alpha$, we will use the exponent pair $(1/14, 11/14)$. For $x^\alpha < v \leq x^\beta$, we use $(1/6, 2/3)$, and for $x^\beta < v \leq x^{1-c+\delta}$, we use $(1/2, 1/2)$. We assume that $1-c+\delta < (1+c)/2$, so that for x sufficiently large, $x^{1-c+\delta} \leq x^{(1+c)/2}(\log x)^{-5/2}$ and (19) holds. We find that

$$\Sigma_4 \leq I(c)x^c(\log x)(1 + O(\epsilon(x))),$$

where

$$I(c) = \int_c^\alpha u\theta(u, 1/14, 11/14)du + \int_\alpha^\beta u\theta(u, 1/6, 2/3)du + \int_\beta^{1-c+\delta} u\theta(u, 1/2, 1/2)du.$$

Henceforth, we specialize to the choice of $c = 4/9$ and $\delta = .001$. Now a lengthy but straightforward computation shows that $4/9 + I(4/9) < 0.9998\ldots.$

We conclude from (12) and (13) that if x is sufficiently large, then

$$\sum_{x^{5/9+\delta}<p} N(p)\log p > 0.0002x^{4/9}\log x.$$

This completes the proof of Lemma 3.

5. The differences $d(n!) - d((n-1)!)$

We now turn our attention to the function

$$D(n) := d(n!) - d((n-1)!).$$

Thus, $D(n)$ is the number of divisors of $n!$ that are not divisors of $(n-1)!$. It is easy to assess the average order of $D(n)$, since the sum of $D(n)$ for $n \le x$ is exactly $d([x]!)$. We may also find the average order of $\log D(n)$. Indeed, from Theorem 2, we have

$$D(n) = d((n-1)!)\left(\frac{d(n!)}{d((n-1)!)} - 1\right) = d((n-1)!)\left(\frac{P(n)}{n} + O\left(\frac{1}{n^{1/2}}\right)\right).$$

Thus, $\log D(n) = \log d((n-1)!) + O(\log n)$, and so the average order of $\log D(n)$ is essentially the same as the average order of $\log d(n!)$. In fact, we have immediately from Theorem 1 that there are numbers $d_0, d_1, \ldots$ such that

$$\sum_{2\le n\le x} \log D(n) = \frac{x^2}{\log x}\sum_{k=0}^{K}\frac{d_k}{\log^k x} + O\left(\frac{x^2}{\log^{K+2} x}\right)$$

for any fixed positive integer K. Note that the coefficient of the main term, d_0, is $c_0/2 \approx 0.6289$.

We shall call a natural number n a *champ* if $D(n) > D(m)$ for all natural numbers $m < n$. Thus a champ is entirely analogous to the concept, introduced by Ramanujan [11], of a highly composite number. This is an integer n with $d(n) > d(m)$ for all natural numbers m smaller than n.

We have the following elementary result.

Theorem 5. *For each prime p, both p and $2p$ are champs.*

Proof. We have

$$D(p) = d(p!) - d((p-1)!) = d((p-1)!) \ge d(m!) > D(m)$$

for every $m < p$. Thus, p is a champ.

Note that for every positive integer m we have $D(m) \leq d(m!)/2$. Indeed, we note that there are two kinds of divisors of $m!$: those that divide $(m-1)!$ and those that don't. There are at least as many divisors of the first kind as of the second kind, since if d is a divisor of the second kind, then d/m is a divisor of the first kind. Thus, $D(m)$, the number of divisors of the second kind, is at most $d(m!)/2$. Now say p is an odd prime. Since $p^2 || (2p)!$ and $p || (2p-1)!$, we have $d((2p)!) > \frac{3}{2} d((2p-1)!)$. Thus,

$$D(2p) = d((2p)!) - d((2p-1)!) > \frac{1}{2} d((2p-1)!) \geq D(m),$$

for every positive integer $m < 2p$. Thus, $2p$ is a champ. To conclude the proof of the theorem, it remains to note that 4 is a champ.

Theorem 5 suggest two natural questions. The first is if there are any, or infinitely many, champs n which are not of the form p or $2p$. The second is if, in some sense, *most* of the champs are of the form p or $2p$, or at least to decide if the set of champs has asymptotic density 0.

On the first question, we first note that yes indeed, there are champs other than the prescribed forms of Theorem 5. The least such champ is 8. Marc Deleglise computed all of the champs up to 500 and found that there are 30 of them that are not of the form p or $2p$. These exceptional champs are all of the form mp where p is a prime $\geq P(m)$ and m is 3, 4, 5, 6 or 7. These calculations were continued by Jean-Pierre Massias up to $200,000$. All the champs up to this level are of the form mp with $m \leq 7$. There are 3273 champs of the form $3p$, 911 of the form $4p$, 196 of the form $5p$, 49 of the form $6p$ and 17 champs of the form $7p$. The largest champ of the form $7p$ below $200,000$ happens to be $194,411$. It is not trivial to compute these numbers, since the values of $D(n)$ get large quickly. For example, for the champs n near $200,000$, $D(n)$ has about $10,000$ decimal digits. We warmly thank Dr. Deleglise and Dr. Massias for permitting us to include this summary of their very interesting calculations.

It is reasonable to conjecture that there are infinitely many champs not of the form p, $2p$. In fact, this conjecture follows from the prime k-tuples conjecture. For example, it is relatively easy to show that if q, r are primes with $2q+1 = 3r$, then $3r$ is a champ. And of course, the prime k-tuples conjecture implies that there are infinitely many such pairs of primes q, r. (In fact, whenever q and r are primes with $2q < 3r$ and such that there are no primes in the interval $[2q, 3r]$, then $3r$ is a champ. It may be not hopeless to show *unconditionally* that such pairs of primes q, r occur infinitely often.) Similar arguments can likely show that for each fixed positive integer m, there are infinitely many champs that are m times a prime. In fact it may be that for each m there is a positive density of primes p with mp a champ. If there are arbitrarily large numbers m, p with mp a champ, it would follow that for each $\varepsilon > 0$ there is a champ n with $d(n!) < (1+\varepsilon) d((n-1)!)$. In fact, the converse holds as well.

We also conjecture that the set of champs has asymptotic density zero. It is very annoying that we have not been able to prove this. We can at least show the following conditional result.

Theorem 6. *If the Riemann Hypothesis is true, then the set of champs has asymptotic density zero.*

Proof. We first show the following unconditional result: If $P(n) \le n/\log^3 n$ and there is a prime in the interval $(n - \frac{1}{3}\log^3 n, n]$, then n is not a champ. Indeed, suppose n is a champ and $P(n) \le n/\log^3 n$. Let $m = [n - \frac{1}{3}\log^3 n]$. Since n is a champ,

$$\begin{aligned} d((n-1)!) &= d(m!) + D(m+1) + D(m+2) + \cdots + D(n-1) \\ &< d(m!) + (n-1-m)D(n) \le d(m!) + \frac{1}{3}D(n)\log^3 n. \end{aligned}$$

By Theorem 2, $d(n!) \le (1 + \log^{-3} n + O(n^{-1/2}))d((n-1)!)$, so that $D(n) \le (\log^{-3} n + O(n^{-1/2}))d((n-1)!)$. It follows that if n is sufficiently large, $D(n) \le \frac{3}{2}(\log^{-3} n)d((n-1)!)$, and so

$$d((n-1)!) < d(m!) + \frac{1}{3} \cdot \frac{3}{2} d((n-1)!),$$

that is, $d((n-1)!) < 2d(m!)$. Since $d(p!) = 2d((p-1)!)$ for a prime p, it follows that there are no primes in the interval $(m, n]$. This proves our assertion.

Thus, the set of non-champs contains the intersection of the set of (sufficiently large) n with $P(n) \le n/\log^3 n$ and the set of n for which there is a prime in the interval $(n - \frac{1}{3}\log^3 n, n]$. The number of n up to x not in the first set is $O(x \log\log x/\log x)$, unconditionally, so that the first set has asymptotic density 1. It follows from a theorem of Selberg [12] that if the Riemann Hypothesis holds, then the second set also has asymptotic density 1. This concludes the proof of the theorem.

From a well known conjecture of H. Cramér we have that for n sufficiently large, the interval $(n - \frac{1}{3}\log^3 n, n]$ always contains a prime. If this holds, then the above argument gives that the number of champs up to x is $\ll x \log\log x/\log x$. On the other hand, from Theorem 5, the number of champs up to x is $\gg x/\log x$. We are unsure what function to suggest for the true order of magnitude for the counting function of the champs.

We close finally with the following conjecture. Show that the asymptotic density of the set of integers n with $D(n+1) > D(n)$ is equal to $1/2$. We can show that this conjecture is equivalent to the conjecture that the asymptotic density of the set of integers n with $P(n+1) > P(n)$ is $1/2$. It follows by the method of [4] that the set of integers n with $D(n) > D(n+1)$ has positive lower density and an upper density that is less than 1. In particular, the upper density of the set of champs is < 1.

Added later. Some of the problems in this paper were previously considered by the first author and R. L. Graham in their monograph *Old and new problems and results in combinatorial number theory*, L'Enseignement Mathématique, Geneva, 1980 (see page 83).

We have recently learned that R. C. Baker and G. Harman (*Numbers with a large prime factor*, Acta Arithmetica, to appear) have shown that for all sufficiently large numbers x there is an integer in the interval $[x, x+x^{1/2}]$ with a prime factor exceeding $x^{0.732}$. The exponent 0.732 improves on the prior record of 0.723 in [8] and [9].

References

[1] N. G. de Bruijn, *On the number of positive integers $\leq x$ and free of prime factors $> y$*, Nederl. Akad. Wet. **54** (1951), 50-60.

[2] H. Davenport, *Multiplicative Number Theory (2nd ed.)*, Springer-Verlag, Berlin-Heidelberg-New York, 1980.

[3] P. Erdös, A. Ivić and C. Pomerance, *On sums involving reciprocals of the largest prime factor of an integer*, Glasnik Mat. **21** (1986), 283-300.

[4] P. Erdös and C. Pomerance, *On the largest prime factors of n and $n+1$*, Aequationes Math. **17** (1978), 311-321.

[5] S. W. Graham and G. Kolesnik, *Van der Corput's method of exponential sums*, Cambridge University Press, Cambridge, 1991.

[6] C. Hooley, *Applications of sieve methods to the theory of numbers*, Cambridge University Press, Cambridge, 1976.

[7] A. Ivić, *The Riemann zeta-function*, John Wiley & Sons, New York, 1985.

[8] Jia Chaohua, *The greatest prime factor of the integers in a short interval (III)*, Acta Mathematica Sinica (New Series) **9** (1993), 321-336.

[9] Hong-Quan Liu, *The greatest prime factor of the integers in an interval*, Acta Arith. **65** (1993), 301-328.

[10] K. Ramachandra, *A note on numbers with a large prime factor*, J. London Math. Soc. (2) **1** (1969), 303-306.

[11] S. Ramanujan, *Collected Works*, Chelsea, New York, 1962.

[12] A. Selberg, *On the normal density of primes in small intervals, and the difference between consecutive primes*, Arch. Math. Naturvid. **47** (1943), 87-105.

[13] G. Tenenbaum, *Sur un problème extrémal en arithmétique*, Ann. Inst. Fourier **27** (1987), 1-18.

Paul Erdös
Matematikai Kutáto Intézete
Reáltanoda utca 13-15
H-1364 Budapest V
HUNGARY

S. W. Graham
Mathematical Sciences Department
Michigan Technological University
Houghton, MI 49931
Current address:
National Science Foundation
Arlington, VA 22230
swgraham@mtu.edu

Aleksandar Ivić
Katedra Matematike RGF-a
Universiteta u Beogradu
Djušina 7, 11000 Beograd
SERBIA (YUGOSLAVIA)
aleks@ivic.matf.bg.ac.yu

Carl Pomerance
Department of Mathematics
University of Georgia
Athens, GA 30602
carl@ada.math.uga.edu

Generalized Lambert series

Ronald Evans

Dedicated to Heini Halberstam on the occasion of his retirement

Abstract. Let $q = \exp(2\pi i\tau)$ with Im $\tau > 0$, so $0 < |q| < 1$. For any positive integer n, define

$$S_n = \sum_{k \neq 0} \frac{(-1)^k q^{(k^2+k)/2}}{(1-q^k)^n},$$

where the sum is over all nonzero integers k. Malcolm Perry needed to know the modular properties of S_2, which arose in his work in quantum string theory. Ramanujan evaluated S_2 in terms of Eisenstein series. We prove a general transformation formula that enables us to evaluate each sum S_n in terms of Eisenstein series, and to thus determine the modular properties of S_n. Moreover, our formula yields systematic proofs of related q-series identities of Ramanujan, proofs which are considerably simpler than those in the literature.

1. Introduction

Let $q = \exp(2\pi i\tau)$ with Im $\tau > 0$, so $0 < |q| < 1$. For $a, n \in \mathbb{C}$, define

(1.1) $$(a)_\infty = (a;q)_\infty = \prod_{k=0}^{\infty}(1-aq^k),$$

(1.2) $$(a)_n = (a)_\infty/(aq^n)_\infty.$$

Ramanujan [3, (17.6), p. 34] proved that for $|b/a| < |z| < 1$,

(1.3) $$\sum_{n=-\infty}^{\infty} \frac{(a)_n}{(b)_n} z^n = \frac{(q)_\infty (b/a)_\infty (az)_\infty (q/az)_\infty}{(b)_\infty (q/a)_\infty (z)_\infty (b/az)_\infty}.$$

(This is his famous summation formula for the bilateral ${}_1\psi_1$ basic hypergeometric series.) The q-binomial theorem

(1.4) $$\sum_{n=0}^{\infty} \frac{(a)_n}{(q)_n} z^n = \frac{(az)_\infty}{(z)_\infty}$$

is the case $b = q$ of (1.3), since $(q)_{-n} = \infty$ for positive integers n. Replace b by 0 and z by z/a in (1.3) to deduce that

$$\sum_{n=-\infty}^{\infty} \left\{ \frac{(a)_n}{a^n} \right\} z^n = \frac{(q)_\infty (z)_\infty (q/z)_\infty}{(q/a)_\infty (z/a)_\infty}; \tag{1.5}$$

then since

$$\lim_{a \to \infty} \frac{(a)_n}{a^n} = (-1)^n q^{(n^2-n)/2}, \tag{1.6}$$

(1.5) yields (with $a \to \infty$)

$$\theta(z) := \sum_{n=-\infty}^{\infty} (-1)^n q^{(n^2-n)/2} z^n = (q)_\infty (z)_\infty (q/z)_\infty. \tag{1.7}$$

The Jacobi triple product formula (1.7) for the theta function $\theta(z)$ thus turns out to be a limiting special case of Ramanujan's formula (1.3). In Section 2, we use (1.3) to prove the following transformation formula.

Theorem. *Let $xy \neq 0$. For any integer n, define*

$$T_n(x, y) := (qx)_\infty (q/x)_\infty \sum_{k=-\infty}^{\infty} (-1)^k \frac{q^{(k^2+k)/2}(x-1)y^k}{(xq^n - q^k)}; \tag{1.8}$$

then

$$T_n(x, y) = T_n(y, x). \tag{1.9}$$

As an application, we show in Section 3 how to evaluate the sums

$$S_n := \sum_{k \neq 0} \frac{(-1)^k q^{(k^2+k)/2}}{(1-q^k)^n}, \qquad n = 1, 2, 3, \ldots \tag{1.10}$$

as polynomials over $\mathbb{Q}$ in the Lambert series L_1, L_3, and L_5, where

$$L_m := \sum_{k=1}^{\infty} \frac{k^m q^k}{1-q^k}. \tag{1.11}$$

For example,

$$S_2 = -L_1, \tag{1.12}$$

$$S_3 = -L_1, \tag{1.13}$$

(1.14) $$S_4 = -\frac{1}{2}L_1^2 - \frac{11}{12}L_1 - \frac{1}{12}L_3,$$

(1.15) $$S_5 = -L_1^2 - \frac{5}{6}L_1 - \frac{1}{6}L_3,$$

and

(1.16) $$S_6 = -\frac{17}{12}L_1^2 - \frac{1}{12}L_1L_3 - \frac{1}{6}L_1^3 - \frac{137}{180}L_1 - \frac{17}{72}L_3 - \frac{1}{360}L_5.$$

Identity (1.12) is due to Ramanujan [3, Entry 8(viii), p. 114], [4, Entry 21, p. 153].

Identities of the type (1.12) - (1.16) yield modular transformation formulas for the sums S_n, because the Lambert series can be expressed in terms of Eisenstein series $E_{2m}(\tau)$, $m = 1, 2, 3, \dots$ [2, p. 318], [7, pp. 194-195]. For $m > 1, E_{2m}(\tau)$ satisfies the modular transformation formula [2, (10.2), p. 320]

(1.17) $$E_{2m}(V\tau) = (c\tau + d)^{2m} E_{2m}(\tau),$$

where

(1.18) $$V\tau = \frac{a\tau + b}{c\tau + d}, \quad ad - bc = 1, \quad a, b, c, d \in \mathbb{Z}.$$

For $m = 1$, we have [2, (10.2), p. 320]

(1.19) $$E_2^*(V\tau) = (c\tau + d)^2 E_2^*(\tau) - \frac{6ic}{\pi}(c\tau + d),$$

where

(1.20) $$E_2^*(\tau) := E_2(\tau) + \frac{3}{\pi \mathrm{Im}\, \tau}.$$

The connection with the Lambert series L_{2m-1} is given by [2, p. 318]

(1.21) $$1 - 24L_1 = E_2^*(\tau),$$

(1.22) $$1 - \frac{4m}{B_{2m}} L_{2m-1} = E_{2m}(\tau), \qquad m \geq 2,$$

where the B_{2m} are Bernoulli numbers. From (1.12), (1.19), and (1.21), we determine the modular behavior of the sum S_2. M. Perry [6] needed to know the modular behavior of S_2 in connection with his work in quantum string theory.

As another application of the Theorem, we present in Section 4 the general identities (4.6), (4.12), and (4.15), and then we specialize them to produce systematic proofs of the nine Lambert series identities recorded on p. 264 of Ramanujan's Second Notebook [4, Entries 18 - 26, pp. 152-155]. We could derive new identities of similar type by specializing in (4.10), for example. Our proofs are considerably simpler and shorter than those in [4], [1], which rely on heavy machinery and an elaborate network of prior results. Section 4 may be read independently from Section 3.

Acknowledgment. I am very grateful to George Andrews, who on short notice at the Conference on Analytic Number Theory at Allerton Park, May 1995, presented my talk on these results and showed how they fit into a bigger picture.

2. Proof of the Theorem

Here we show that $T_n(x,y) = T_n(y,x)$, where $T_n(x,y)$ is defined by (1.8). The idea is to show that the generating function

$$S(x,y,v) := \sum_{n=-\infty}^{\infty} T_n(x,y)v^n \tag{2.1}$$

equals

$$(q)_\infty^2 \theta(xv)\theta(yv)/\theta(v), \tag{2.2}$$

for $|q| < |v| < 1$. This will prove that $S(x,y,v)$ and hence each Laurent coefficient $T_n(x,y)$ is symmetric in x and y.

By definitions (1.8) and (2.1),

$$S(x,y,v) = (qx)_\infty (q/x)_\infty (x-1) \sum_k \sum_n (-1)^k q^{(k^2+k)/2} y^k \frac{v^n}{xq^n - q^k} . \tag{2.3}$$

Replace n by $n+k$ to get

$$S(x,y,v) = (qx)_\infty (q/x)_\infty (x-1) \sum_k (-1)^k q^{(k^2-k)/2} (vy)^k \sum_n \frac{v^n}{xq^n - 1}. \tag{2.4}$$

For $|q| < |v| < 1$,

$$\sum_n \frac{v^n}{xq^n - 1} = \frac{1}{x-1} \sum_n \frac{(x)_n}{(xq)_n} v^n = \frac{1}{x-1} \frac{(xv)_\infty (q/xv)_\infty (q)_\infty^2}{(v)_\infty (q/v)_\infty (qx)_\infty (q/x)_\infty} , \tag{2.5}$$

by (1.3). By (2.4) and (2.5),

$$S(x,y,v) = \frac{(xv)_\infty (q/xv)_\infty (q)_\infty^2}{(v)_\infty (q/v)_\infty} \sum_k (-1)^k q^{(k^2-k)/2} (vy)^k . \tag{2.6}$$

Hence, by (1.7),

$$S(x,y,v) = (q)_\infty^2 \theta(vx)\theta(vy)/\theta(v). \tag{2.7}$$

Remark. Another way to prove that $T_0(x,y) = T_0(y,x)$ is to use Bailey's transformation formula for ${}_2\psi_2\left(\begin{matrix} a & b \\ c & d \end{matrix}\middle| z\right)$. In [5, 5.20(i), p. 137], set $b = 1/y$, $c = q/a$, $d = q/y$, $z = q/ax$, and let $a \to \infty$.

3. Evaluation of S_n in terms of L_1, L_3, L_5

In this section, we apply the Theorem to show that the sum S_n defined by (1.10) is a polynomial in the Lambert series L_1, L_3, L_5 over $\mathbb{Q}$.

By taking $n = 0, y = 1$ in (1.9), we obtain, for all $x \neq 0$,

$$S(x) := \sum_{k=-\infty}^{\infty} (-1)^k q^{(k^2+k)/2} \frac{(x-1)}{x-q^k} = \frac{P(1)}{P(x)}, \tag{3.1}$$

where

$$P(x) := (qx)_\infty (q/x)_\infty, \qquad x \neq 0. \tag{3.2}$$

After taking the nth derivative with respect to x on both sides of (3.1), where $n \geq 1$, we obtain

$$S_n(x) := \sum_{k\neq 0} \frac{(-1)^k q^{(k^2+k)/2}(1-q^k)}{(x-q^k)^{n+1}} = \frac{(-1)^{n+1}}{n!} P(1) \left(\frac{1}{P(x)}\right)^{(n)}. \tag{3.3}$$

With $x = 1$, this becomes

$$S_n = S_n(1) = \sum_{k\neq 0} \frac{(-1)^k q^{(k^2+k)/2}}{(1-q^k)^n} = \frac{(-1)^{n+1}}{n!} P(1) \left(\frac{1}{P(x)}\right)^{(n)}\Bigg|_{x=1}. \tag{3.4}$$

By grouping the k and $-k$ terms, we could express S_n as a unilateral series

$$S_n = \sum_{k=1}^{\infty} \frac{(-1)^k q^{(k^2+k)/2} \left(1 + (-1)^n q^{(n-1)k}\right)}{(1-q^k)^n}. \tag{3.5}$$

For example (cf. [4, Entry 21, p. 153]),

$$S_2 = \sum_{k=1}^{\infty} (-1)^k q^{(k^2+k)/2} \frac{(1+q^k)}{(1-q^k)^2}. \tag{3.6}$$

From (3.5), we see that

$$S_1 = 0 \,. \tag{3.7}$$

Thus, by (3.4) with $n = 1$,

$$P'(1) = 0 \,. \tag{3.8}$$

By induction on n, $\left(\frac{1}{P(x)}\right)^{(n)}$ equals $\frac{1}{P(x)}$ times a polynomial (over $\mathbb{Q}$) in the variables $\frac{P^{(m)}(x)}{P(x)}$, $1 \le m \le n$. Thus, by (3.4), S_n is a polynomial in the variables $\frac{P^{(m)}(1)}{P(1)}$, $2 \le m \le n$. It therefore suffices to show that each $\frac{P^{(m)}(1)}{P(1)}$ is a polynomial in L_1, L_3, L_5. Define

$$Q(x) := P'(x)/P(x), \qquad x \neq 0 \,. \tag{3.9}$$

We have $P'(x) = P(x)Q(x)$ and thus

$$P^{(m)}(1) = \sum_{j=0}^{m-1} \binom{m-1}{j} P^{(m-1-j)}(1) Q^{(j)}(1) \,. \tag{3.10}$$

By (3.10) and induction on m, it suffices to show that each $Q^{(j)}(1)$ is a polynomial in L_1, L_3, L_5.

Logarithmically differentiating in (3.2), we obtain, by (3.9),

$$Q(x) = -\sum_{k=1}^{\infty} \frac{q^k}{1 - q^k x} + \frac{1}{x^2} \sum_{k=1}^{\infty} \frac{q^k}{1 - q^k x^{-1}} \,. \tag{3.11}$$

Hence , for $|q| < |x| < |q|^{-1}$,

$$\begin{aligned} Q(x) &= -\sum_{k=1}^{\infty} \sum_{r=0}^{\infty} q^k q^{kr} x^r + \sum_{k=1}^{\infty} \sum_{r=0}^{\infty} q^k q^{kr} x^{-r-2} \\ &= \sum_{r=1}^{\infty} \frac{q^r}{1 - q^r} (x^{-r-1} - x^{r-1}) \,. \end{aligned} \tag{3.12}$$

Therefore,

$$Q^{(j)}(1) = \sum_{r=1}^{\infty} \frac{q^r}{1 - q^r} \{(-r-1) \cdots (-r-j) - (r-1) \cdots (r-j)\} \,. \tag{3.13}$$

The expression in braces is an integral linear combination of odd powers of r, so $Q^{(j)}(1)$ is an integer linear combination of $L_1, L_3, L_5, L_7, \ldots, L_{2m-1}$ with $2m - 1 \le j$, by definition (1.11). The Eisenstein series $E_{2m}(\tau)$ for $m \ge 2$ are all polynomials in $E_4(\tau)$ and $E_6(\tau)$ [7, p. 199], and hence by (1.22), the Lambert series L_{2m-1} for $m \geq 2$ are all polynomials in L_3 and L_5. This completes the proof that the $Q^{(j)}(1)$ are all polynomials in L_1, L_3, L_5.

The first few values of $Q^{(j)}(1)$ are, by (3.13),

$$Q^{(0)}(1) = 0, \quad Q^{(1)}(1) = -2L_1, \quad Q^{(2)}(1) = 6L_1, \quad Q^{(3)}(1) = -22L_1 - 2L_3,$$
$$Q^{(4)}(1) = 100L_1 + 20L_3, \quad Q^{(5)}(1) = -548L_1 - 170L_3 - 2L_5,$$
$$Q^{(6)}(1) = 3528L_1 + 1470L_3 + 42L_5.$$

Thus, from (3.10), the first few values of $A_m := P^{(m)}(1)/P(1)$ are

$$A_0 = 1, \quad A_1 = 0, \quad A_2 = -2L_1, \quad A_3 = 6L_1, \quad A_4 = 12L_1^2 - 22L_1 - 2L_3,$$
$$A_5 = -120L_1^2 + 100L_1 + 20L_3,$$
$$A_6 = -120L_1^3 + 1020L_1^2 + 60L_1L_3 - 548L_1 - 170L_3 - 2L_5 \ .$$

From (3.4),

$$S_2 = \frac{1}{2}A_2, \quad S_3 = -\frac{1}{6}A_3, \quad S_4 = -\frac{1}{4}A_2^2 + \frac{1}{24}A_4,$$
$$S_5 = \frac{1}{6}A_2A_3 - \frac{1}{120}A_5, \; S_6 = \frac{1}{8}A_2^3 - \frac{1}{36}A_3^2 - \frac{1}{24}A_2A_4 + \frac{1}{720}A_6,$$

and so formulas (1.12) - (1.16) follow.

4. Lambert series identities

This section is independent of Section 3. Write, for brevity,

$$a_k = (-1)^k q^{(k^2+k)/2} \tag{4.1}$$

and

$$P(x) = (qx)_\infty (q/x)_\infty, \qquad x \neq 0 \ . \tag{4.2}$$

We have

$$\frac{P'(x)}{P(x)} = \sum_{k=1}^{\infty} \frac{q^k}{x^2 - xq^k} - \sum_{k=1}^{\infty} \frac{q^k}{1 - xq^k} \ , \tag{4.3}$$

so that in particular,

$$P'(1) = P'(-1) = 0 \ . \tag{4.4}$$

The case $n = 0$ of (1.9) yields, for $xy \neq 0$,

$$P(x)\sum_k a_k y^k \frac{(x-1)}{(x-q^k)} = P(y)\sum_k a_k x^k \frac{(y-1)}{(y-q^k)} \,. \tag{4.5}$$

In particular, for $y = 1$,

$$\sum_k a_k \frac{(x-1)}{(x-q^k)} = \frac{P(1)}{P(x)} \,. \tag{4.6}$$

One can derive many interesting identities by differentiating in (4.5). Take the derivative with respect to x to obtain

$$\begin{aligned} P(y)\sum_k k a_k x^{k-1} \frac{(y-1)}{(y-q^k)} &= P'(x)\sum_k a_k y^k \frac{(x-1)}{(x-q^k)} \\ &\quad + P(x)\sum_{k\neq 0} a_k y^k \frac{(1-q^k)}{(x-q^k)^2} \,. \end{aligned} \tag{4.7}$$

Take the derivative once more with respect to x to obtain

$$\begin{aligned} P(y)\sum_k k(k-1) a_k x^{k-2} \frac{(y-1)}{(y-q^k)} &= P''(x)\sum_k a_k y^k \frac{(x-1)}{(x-q^k)} \\ + 2P'(x)\sum_{k\neq 0} a_k y^k \frac{(1-q^k)}{(x-q^k)^2} &- 2P(x)\sum_{k\neq 0} a_k y^k \frac{(1-q^k)}{(x-q^k)^3} \,. \end{aligned} \tag{4.8}$$

Take the derivative with respect to y in (4.7) to obtain

$$\begin{aligned} &P'(y)\sum_k k a_k x^{k-1} \frac{(y-1)}{(y-q^k)} + P(y)\sum_k k a_k x^{k-1} \frac{(1-q^k)}{(y-q^k)^2} \\ &\quad = P'(x)\sum_k k a_k y^{k-1} \frac{(x-1)}{(x-q^k)} + P(x)\sum_k k a_k y^{k-1} \frac{(1-q^k)}{(x-q^k)^2} \,. \end{aligned} \tag{4.9}$$

Setting $x = 1$ in (4.7) - (4.9), we obtain with the aid of (4.4) the interesting identities

$$P(y)\sum_k k a_k \frac{(y-1)}{(y-q^k)} = P(1)\sum_{k\neq 0} \frac{a_k y^k}{(1-q^k)} \,, \tag{4.10}$$

$$P(y)\sum_k k(k-1) a_k \frac{(y-1)}{(y-q^k)} = P''(1) - 2P(1)\sum_{k\neq 0} \frac{a_k y^k}{(1-q^k)^2} \,, \tag{4.11}$$

and

$$(4.12)\quad P'(y)\sum_k ka_k\frac{(y-1)}{(y-q^k)}+P(y)\sum_k ka_k\frac{(1-q^k)}{(y-q^k)^2}=P(1)\sum_k\frac{ka_k y^{k-1}}{(1-q^k)}\,.$$

We proceed to give systematic new proofs of Entries 18-26 in [4, pp.152-155]. We will express each of these entries in bilateral form; to convert to the corresponding unilateral form given in [4], simply group appropriate pairs of terms (e.g., the k and $-k$ terms in Entry 18, the k and $-1-k$ terms in Entry 19, etc.)

Entry 18.
$$\sum_k\frac{(-1)^k q^{(k^2+k)/2}}{1+q^k}=\frac{(q)_\infty^2}{2(-q)_\infty^2}$$

Proof. Put $x=-1$ in (4.6).

Entry 19.
$$\sum_k\frac{(-1)^k q^{(k^2+k)/2}}{1-q^{2k+1}}=\frac{(q^4;q^4)_\infty(-q^2;q^4)_\infty^2(q^2;q^2)_\infty}{(q;q^2)_\infty}$$

Proof. By (4.6),

$$(4.13)\qquad \sum_k\frac{a_k}{x-q^k}=-\frac{(q)_\infty^2}{(x)_\infty(q/x)_\infty}\,.$$

Replace x by $-x$ and combine with (4.13) to get

$$\begin{aligned}-\sum_k\frac{a_kq^k}{x^2-q^{2k}}&=\frac12\sum_k\left(\frac{a_k}{x+q^k}-\frac{a_k}{x-q^k}\right)\\&=\frac{(q)_\infty^2}{2}\left\{\frac{1}{(-x)_\infty(-q/x)_\infty}+\frac{1}{(x)_\infty(q/x)_\infty}\right\}.\end{aligned}$$

Replacing x by $\sqrt{q}$, we obtain

$$-\sum_k\frac{a_kq^{k-1}}{1-q^{2k-1}}=\frac{(q)_\infty^2\{(\sqrt{q})_\infty^2+(-\sqrt{q})_\infty^2\}}{2(-\sqrt{q})_\infty^2(\sqrt{q})_\infty^2}\,.$$

Replace k by $-k$ on the left side and simplify the right side using (1.7) to get

$$\begin{aligned}\sum_k\frac{(-1)^kq^{(k^2+k)/2}}{1-q^{2k+1}}&=\frac{(q)_\infty^2\{(\sqrt{q})_\infty^2+(-\sqrt{q})_\infty^2\}}{2(\sqrt{q})_\infty^2(-\sqrt{q})_\infty^2}\\&=\frac{(q)_\infty\{\theta(\sqrt{q})+\theta(-\sqrt{q})\}}{2(q;q^2)_\infty^2}=\frac{(q)_\infty}{(q;q^2)_\infty^2}\sum_k q^{2k^2}\,.\end{aligned}$$

By (1.7) once more,

$$\sum_k q^{2k^2} = \left(q^4;q^4\right)_\infty \left(-q^2;q^4\right)_\infty^2,$$

and the result follows.

Entry 20.
$$\sum_k \frac{(-1)^k q^{(k^2+k)}}{1-q^{2k+1}} = \frac{(q^2;q^2)_\infty^2}{(q;q^2)_\infty^2}$$

Proof. By (4.13) with $x = \sqrt{q}$,

$$-\sum_k \frac{a_k}{\sqrt{q}\left(1-q^{k-\frac{1}{2}}\right)} = \frac{(q)_\infty^2}{(\sqrt{q})_\infty^2}\,. \tag{4.14}$$

Replace k by $-k$ on the left and then replace q everywhere by q^2.

Entry 21.
$$\sum_{k\neq 0} \frac{(-1)^k q^{(k^2+k)/2}}{(1-q^k)^2} = -L_1$$

Proof. This is (1.12). To prove it without reference to Section 3, first differentiate twice with respect to x in (4.6) and set $x = 1$. It remains to show that $P''(1)/(2P(1))$ equals $-L_1$. Differentiate in (4.3) with respect to x and set $x = 1$. In view of (4.4), this yields

$$\begin{aligned} -P''(1)/(2P(1)) &= \sum_{k=1}^\infty \frac{q^k}{(1-q^k)^2} = \sum_{k=1}^\infty \sum_{m=1}^\infty m q^{km} \\ &= \sum_{m=1}^\infty \frac{mq^m}{1-q^m} = L_1\,. \end{aligned}$$

Remark. Since the right side of Entry 21 is $P''(1)/(2P(1))$, (4.11) can be written in the form

$$P(y)\sum_k (k^2-k)a_k \frac{(y-1)}{(y-q^k)} = 2P(1)\sum_{k\neq 0} \frac{a_k(1-y^k)}{(1-q^k)^2}\,. \tag{4.15}$$

Entry 22.
$$\sum_k \frac{(-1)^k q^{(k^2+k)/2}}{1-q^{2k-1}} = -2q\frac{(q^2;q^2)_\infty (q^8;q^8)_\infty}{(q;q^2)_\infty (q^4;q^8)_\infty}$$

Proof. In (4.13), replace x by $-x$ and then combine with (4.13) to get

$$\sum_k \frac{a_k}{x^2-q^{2k}} = \frac{1}{2x}\sum_k\left(\frac{a_k}{x+q^k}+\frac{a_k}{x-q^k}\right)$$
$$= \frac{(q)_\infty^2}{2x}\left\{\frac{1}{(-x)_\infty(-q/x)_\infty} - \frac{1}{(x)_\infty(q/x)_\infty}\right\}.$$

Replacing x by $\sqrt{q}$ and multiplying by q, we obtain

$$\sum_k \frac{(-1)^k q^{(k^2+k)/2}}{1-q^{2k-1}} = \frac{(q)_\infty^2\sqrt{q}\,\{(\sqrt{q})_\infty^2-(-\sqrt{q})_\infty^2\}}{2(-\sqrt{q})_\infty^2(\sqrt{q})_\infty^2}$$
$$= \frac{\sqrt{q}(q)_\infty\{\theta(\sqrt{q})-\theta(-\sqrt{q})\}}{2(q;q^2)_\infty^2} = \frac{-(q)_\infty q}{(q;q^2)_\infty^2}\sum_k q^{2(k^2+k)}$$
$$= \frac{-2q(q)_\infty\left(q^8;q^8\right)_\infty}{(q;q^2)_\infty^2\left(q^4;q^8\right)_\infty},$$

by (1.7), and the result follows.

Entry 23.

$$\frac{(-q)_\infty^2}{2(q)_\infty^2}\sum_k \frac{k^2(-1)^k q^{(k^2+k)/2}}{1+q^k} = -\sum_k \frac{q^{2k^2-k}}{(1-q^{2k-1})^2}$$

Proof. By (4.15) with $y=-1$,

$$\frac{P(-1)}{P(1)}\sum_k \frac{(k^2-k)a_k}{1+q^k} = 2\sum_{k\text{ odd}}\frac{a_k}{(1-q^k)^2} = -2\sum_k \frac{q^{2k^2-k}}{(1-q^{2k-1})^2}. \tag{4.16}$$

Now,

$$\sum_k \frac{ka_k}{1+q^k} = 0 \tag{4.17}$$

since the sum in (4.17) changes sign when k is replaced by $-k$. The result follows from (4.16) and (4.17).

Entry 24.

$$\frac{-(q;q^2)_\infty^2}{4\,(q^2;q^2)_\infty^2}\sum_k \frac{(-1)^k(2k-1)^2 q^{k^2-k}}{1-q^{2k-1}} = \sum_k \frac{(-1)^k q^{k^2+k}}{(1+q^k)^2}$$

Proof. Put $y=\sqrt{q}$ in (4.15) and divide by $P(1)$ to obtain

$$\frac{-(\sqrt{q})_\infty^2}{(q)_\infty^2}\sum_k \frac{(k^2-k)a_k}{\sqrt{q}(1-q^{k-1/2})} = R, \tag{4.18}$$

where

$$R = 2\sum_{k\neq 0} \frac{a_k(1-q^{k/2})}{(1-q^k)^2}. \tag{4.19}$$

Replace k by $-k$ in (4.19) to see that

$$R = -2\sum_{k\neq 0} \frac{a_k(1-q^{k/2})q^{k/2}}{(1-q^k)^2}. \tag{4.20}$$

Adding (4.19) and (4.20), we see that

$$R = \sum_{k\neq 0} \frac{a_k}{(1+q^{k/2})^2}. \tag{4.21}$$

By (4.14), (4.18), and (4.21), we have

$$\frac{-(\sqrt{q})_\infty^2}{(q)_\infty^2} \sum_k \frac{(k-\frac{1}{2})^2 a_k}{\sqrt{q}(1-q^{k-\frac{1}{2}})} = \sum_k \frac{a_k}{(1+q^{k/2})^2}. \tag{4.22}$$

The result now follows after replacing k by $1-k$ on the left side of (4.22), and then replacing q everywhere by q^2.

Entry 25. $$\frac{(q;q^2)_\infty^2}{2(q^2;q^2)_\infty^2} \sum_k \frac{(-1)^k(2k-1)q^{k^2+k-1}}{(1-q^{2k-1})^2} = \sum_k \frac{k(-1)^k q^{k^2}}{1-q^{2k}}$$

Proof. The right side of (4.3) telescopes for $x=\sqrt{q}$, and so (4.3) yields

$$\frac{P'(\sqrt{q})}{P(\sqrt{q})} = \frac{1}{1-\sqrt{q}}. \tag{4.23}$$

Thus with $y=\sqrt{q}$, (4.12) becomes

$$\sum_k \frac{ka_k(1-\sqrt{q})}{(\sqrt{q}-q^k)^2} = \frac{P(1)}{P(\sqrt{q})} \sum_k \frac{ka_k q^{(k-1)/2}}{(1-q^k)}.$$

Thus

$$\begin{aligned} \frac{(\sqrt{q})_\infty^2}{(q)_\infty^2} \sum_k \frac{k(-1)^k q^{(k^2+k-1)/2}}{(1-q^{k-1/2})^2} &= \sum_k \frac{k(-1)^k q^{(k^2+2k)/2}}{(1-q^k)} \\ &= \sum_k \frac{k(-1)^k q^{k^2/2}}{(1-q^k)}, \end{aligned} \tag{4.24}$$

where the last equality results from replacing k by $-k$. Now,

$$\sum_k \frac{(-1)^k q^{(k^2+k-1)/2}}{(1-q^{k-\frac{1}{2}})^2} = 0$$

since the sum changes sign when k is replaced by $1-k$. Thus the first factor k on the left of (4.24) may be replaced by $(k-\frac{1}{2})$, and the result then follows after replacing q by q^2.

Entry 26.
$$\frac{-2(-q)_\infty^2}{(q)_\infty^2} \sum_k \frac{k(-1)^k q^{(k^2+k)/2}}{(1+q^k)^2} = \sum_k \frac{kq^{(k^2+k)/2}}{(1-q^k)}$$

Proof. By (4.4) and (4.12) with $y=-1$,

$$\frac{-P(-1)}{P(1)} \sum_k \frac{ka_k(1-q^k)}{(1+q^k)^2} = \sum_k \frac{kq^{(k^2+k)/2}}{(1-q^k)} .$$

The result now follows, since

$$\sum_k \frac{ka_k}{(1+q^k)^2} = -\sum_k \frac{ka_k q^k}{(1+q^k)^2} ,$$

which we can see by replacing k by $-k$.

References

[1] G.E. Andrews, *Bailey chains and generalized Lambert series: I. Four identities of Ramanujan*, Illinois J. Math **36** (1992), 251-274.
[2] B.C. Berndt, *Ramanujan's Notebooks, Part II*, Springer, N.Y., 1989.
[3] B.C. Berndt, *Ramanujan's Notebooks, Part III*, Springer, N.Y., 1991.
[4] B.C. Berndt, *Ramanujan's Notebooks, Part IV*, Springer, N.Y., 1994.
[5] G. Gasper and M. Rahman, *Basic hypergeometric series*, Cambridge Univ. Press, Cambridge, 1990.
[6] M. Perry, manuscript in preparation.
[7] R.A. Rankin, *Modular forms and functions*, Cambridge Univ. Press, Cambridge, 1977.

Ronald Evans
Department of Mathematics
University of California at San Diego
La Jolla, CA 92093-0112
revans@euclid.ucsd.edu

A generalization of an irreducibility theorem of I. Schur

Michael Filaseta[1]

Dedicated to my Ph.D. advisor, Heini Halberstam, on the occasion of his retirement

1. Introduction

In 1929, I. Schur [19] established the following result:

Theorem 1 (I. Schur). *Let n be a positive integer, and let $a_0, a_1, \cdots, a_n$ denote arbitrary integers with $|a_0| = |a_n| = 1$. Then*

$$a_n \frac{x^n}{n!} + a_{n-1} \frac{x^{n-1}}{(n-1)!} + \cdots + a_1 x + a_0$$

is irreducible.

Irreducibility here and throughout this paper refers to irreducibility over the rationals. Some condition, such as $|a_0| = |a_n| = 1$, on the integers a_j is necessary; otherwise, the irreducibility of all polynomials of the form above would imply every polynomial in $\mathbb{Z}[x]$ is irreducible (which is clearly not the case). In this paper, we will mainly be interested in relaxing the condition $|a_n| = 1$. Specifically, we will show:

Theorem 2. *Let $a_0, a_1, \ldots, a_n$ denote arbitrary integers with $|a_0| = 1$, and let $f(x) = \sum_{j=0}^{n} a_j x^j / j!$. If $0 < |a_n| < n$, then $f(x)$ is irreducible unless*

$$a_n = \pm 5 \quad \textit{and} \quad n = 6$$

or

$$a_n = \pm 7 \quad \textit{and} \quad n = 10$$

in which cases either $f(x)$ is irreducible or $f(x)$ is the product of two irreducible polynomials of equal degree. If $|a_n| = n$, then either $f(x)$ is irreducible or $f(x)$ is $x \pm 1$ times an irreducible polynomial of degree $n - 1$.

[1]Research was supported by NSF Grant DMS-9400937.

In the cases that the pair (a_n, n) is $(\pm 5, 6)$ or $(\pm 7, 10)$, one can obtain reducible $f(x)$. In the way of examples, we note

$$5 \times \frac{x^6}{6!} - 1 = \left(\frac{x^3}{12} + 1\right)\left(\frac{x^3}{12} - 1\right)$$

and

$$7 \times \frac{x^{10}}{10!} - 1 = \left(\frac{x^5}{720} + 1\right)\left(\frac{x^5}{720} - 1\right).$$

These particular examples arise partially because $6!/5$ and $10!/7$ are squares. It is possible to show that if n is a positive integer and p is a prime for which $n!/p$ is a square, then $(n, p) \in \{(2,2), (6,5), (10,7)\}$.

If one chooses $a_0 = \pm 1$, the integers $a_1, a_2, \ldots, a_{n-2}$ arbitrarily, and $a_n = \pm n$ (or any multiple of n will do), it is easy to see that there is an integer a_{n-1} for which $f(-1) = 0$ and there is an integer a_{n-1} for which $f(1) = 0$. Thus, $x \pm 1$ may be a factor of $f(x)$ when $|a_n| = n$. The last statement of the theorem implies that the remaining factor of degree $n-1$ will necessarily be irreducible.

In the next three sections, we establish Theorem 2. It is of some interest to know to what extent one can further extend the range on a_n in Theorem 2. In Section 5, we will establish

Theorem 3. *Let C be a positive number $< 1/\sqrt{2}$. Let n be a positive integer, and let $a_0, a_1, \ldots, a_n$ be arbitrary integers with $|a_0| = 1$ and $0 < |a_n| \le Cn^{3/2}$. Let $f(x) = \sum_{j=0}^{n} a_j x^j / j!$. Then for all but finitely many pairs of integers (a_n, n), either $f(x)$ is irreducible or $f(x)$ factors in $\mathbb{Q}[x]$ as the product of a linear polynomial and an irreducible polynomial of degree $n-1$. In the case that $f(x)$ has a linear factor, one necessarily has $n | a_n$.*

The proof we will give of Theorem 3 is non-effective; the exceptional pairs (a_n, n) cannot be determined. A weaker inequality such as

$$0 < |a_n| \le n \exp\left(\sqrt{\frac{\log n}{(\log\log n)^3}}\right)$$

would make Theorem 3 effective, and we describe how this can be done in Section 6.

From Theorem 3, it follows that if $|a_n|$ is not too large, then either $f(x)$ is irreducible or it has a linear factor. Suppose instead we wish to conclude that either $f(x)$ is irreducible or it has a factor of degree ≤ 2. Then a stronger result, with a larger upper bound on $|a_n|$, can be obtained directly from our methods. In fact, an even stronger result can be obtained if for some $k \ge 3$ one allows for $f(x)$ to have a factor of degree $\le k$. We will not elaborate further on this, but these remarks are easily deduced from the proof we give of Theorem 3.

The proof of Theorem 3 will be based on finding an upper bound on n for the exceptional pairs (a_n, n) in the statement of the theorem. The condition $0 < |a_n| \le Cn^{3/2}$ then implies that this list of exceptional pairs is finite. The author has considered specifically the case that $0 < |a_n| \le 2n$. In particular, if $f(x)$ is as above with $n < |a_n| < 2n$, then either $f(x)$ is irreducible or $(a_n, n) = (\pm 6, 4)$. If $|a_n| = 2n$ and $f(x)$ is reducible, then $f(x)$ is the product of two irreducible polynomials. One of these two polynomials will be linear, and in fact will be $x \pm 1$ or $2x \pm 1$, unless $(a_n, n) = (\pm 10, 5)$ and $f(x)$ factors as the product of an irreducible quadratic and an irreducible cubic. In this regard, the examples

$$6 \times \frac{x^4}{4!} - 1 = \left(\frac{x^2}{2} + 1\right)\left(\frac{x^2}{2} - 1\right)$$

and

$$10 \times \frac{x^5}{5!} + \frac{x^3}{3!} + \frac{x^2}{2!} + 1 = \frac{1}{12}\left(x^2 + 2\right)\left(x^3 + 6\right)$$

are worth noting. This result with $0 < |a_n| \le 2n$ can be obtained in a very similar manner to the proof of Theorem 2, and we do not elaborate on the details.

In the sixth section of this paper, we will discuss other results related to the irreducibility of $f(x)$. In particular, we will give a proof of the following nice generalization of Schur's theorem due to T.Y. Lam (private communication):

Theorem 4 (T.Y. Lam). *Let n be an integer ≥ 2, and let $a_0, a_1, \dots, a_n$ be arbitrary integers with* $\gcd(a_0 a_n, n!) = 1$. *Then $\sum_{j=0}^{n} a_j x^j / j!$ is irreducible.*

We observe that, as a consequence of Lam's theorem, if p is a prime, $|a_0| = 1$, and $|a_n| = p$, then $f(x)$ is irreducible for $n < p$. Together with Theorem 2, this implies that if $|a_n| = p$, then the condition $|a_n| < n$ in Theorem 2 may be replaced by $|a_n| \ne n$. In other words, if $|a_n| = p$, then $f(x)$ irreducible unless $(a_n, n) \in \{(\pm 5, 6), (\pm 7, 10)\}$ or $n = p$.

2. Background and sketch of the proof of Theorem 2

Recently, the author [9] established that all but finitely many Bessel polynomials are irreducible. The proof of Theorem 2 will be based on the same techniques. However, there are really two basic elements of the proof and these both have a long history. The first is the use of Newton polygons to deduce the irreducibility of the polynomials in Theorem 2. In 1906, G. Dumas [6] obtained an important result, discussed below, which has been the basis of many irreducibility theorems since then (cf. [9,11,13,24]), so it is not surprising that Newton polygons should play a role in the proof of Theorem 2. In fact, R.F. Coleman [4] has already observed that Newton polygons can be used to establish the irreducibility of the polynomials in Theorem 1 in the case that $a_n = a_{n-1} = \cdots = a_1 = a_0 = 1$; and the author [9] recently gave a proof of Theorem 1 in its full generality using Newton polygons. The second basic element of our proof of Theorem 2 is the use of information on the distribution of primes. In particular, a result of E.F. Ecklund, Jr., R.B. Eggleton, P. Erdős, and J.L. Selfridge [7] on prime divisors of binomial coefficients will play an important role.

Newton polygons can be described as follows. Let

$$f(x) = \sum_{j=0}^{n} a_j x^j \in \mathbb{Z}[x]$$

with $a_0 a_n \neq 0$. Let p be a prime, and let m be an integer. We use the p-adic notation

$$\nu(m) = \nu_p(m) = r \text{ if } p^r | m \text{ and } p^{r+1} \nmid m.$$

If $m = 0$, then we will understand this to mean $\nu(m) = +\infty$. For $j \in \{0, \ldots, n\}$, we define a set of points

$$S = \{(0, \nu(a_n)), (1, \nu(a_{n-1})), \ldots, (n, \nu(a_0))\}$$

in the extended plane. The elements of S we refer to as spots. We consider the lower edges along the convex hull of these spots. The left-most edge has one endpoint being $(0, \nu(a_n))$ and the right-most edge has $(n, \nu(a_0))$ as an endpoint. The endpoints of every edge belong to the set S. We emphasize that, for our purposes, when referring to the "edges" of a Newton polygon, we shall not allow 2 different edges to have the same slope. The polygonal path formed by these edges is called the Newton polygon of $f(x)$ with respect to the prime p. Observe that the slope of the edges are always increasing when calculated from the left-most edge to the right-most edge. The following theorem is due to Dumas [6].

Theorem 5. *Let $g(x)$ and $h(x)$ be in $\mathbb{Z}[x]$ with $g(0)h(0) \neq 0$, and let p be a prime. Let k be a non-negative integer such that p^k divides the leading coefficient of $g(x)h(x)$ but p^{k+1} does not. Then the edges of the Newton polygon for $g(x)h(x)$ with respect to p can be formed by constructing a polygonal path beginning at $(0, k)$ and using translates of the edges in the Newton polygons for $g(x)$ and $h(x)$ with respect to the prime p, using exactly one translate for each edge of the Newton polygons for $g(x)$ and $h(x)$. Necessarily, the translated* *edges are translated in such a way as to form a polygonal path with the slopes of the edges increasing from left to right.*

Although we will stick here to the use of Newton polygons as just described, we note that there are other contexts in which Newton polygons occur. In particular, one may describe Newton polygons, or more appropriately Newton polytopes, in several variables as follows. Associate with a polynomial $f(x_1, \dots, x_m)$ a set T of points $(e_1, e_2, \dots, e_m)$ in $\mathbb{R}^m$ corresponding to the terms $ax_1^{e_1}x_2^{e_2}\cdots x_m^{e_m}$ of f with $a \neq 0$. Denote by C_f the convex hull of the set T in $\mathbb{R}^m$. Then C_f is called the Newton polytope of f. We mention two results associated with Newton polytopes. D.N. Bernstein [3] has shown that the number of intersection points (counted with multiplicity) of m algebraic curves in m variables can be determined from the use of Newton polytopes. If f factors as the product of polynomials g and h in $\mathbb{Z}[x_1, \dots, x_m]$, A.M. Ostrowski [16] has established that $C_f = C_g + C_h$. If one replaces the prime p in our definition of Newton polygons above with a variable, then Ostrowski's theorem comes close to directly implying Theorem 5 (but note the example $f(x) = g(x)h(x)$ with $g(x) = h(x) = x + 1$ and $p = 2$ in this context).

Our main use of Theorem 5 is summarized in our first lemma. We note that the proof is similar to the proof of Lemma 2 in [9].

Lemma 1. *Let $a_0, a_1, \dots, a_n$ denote arbitrary integers with $|a_0| = 1$, and let*

$$f(x) = \sum_{j=0}^{n} a_j \frac{x^j}{j!}.$$

Let k be a positive integer $\leq n/2$. Suppose there exists a prime $p \geq k + 1$ and a positive integer r for which

$$p^r | n(n-1)\cdots(n-k+1) \quad \textit{and} \quad p^r \nmid a_n.$$

Then $f(x)$ cannot have a factor of degree k.

Proof. It suffices to show that

$$F(x) = n!f(x) = \sum_{j=0}^{n} a_j\,(n!/j!)\,x^j$$

cannot have a factor of degree k. We set $b_j = a_j n!/j!$ for $0 \leq j \leq n$. The condition $p^r | n(n-1)\cdots(n-k+1)$ implies that $p^r | b_j$ for $j \in \{0, 1, \dots, n-k\}$. Thus, the $n - k + 1$ right-most spots, $(k, \nu(a_{n-k}n!/(n-k)!)), \dots, (n, \nu(a_0 n!))$, associated with the Newton polygon of $F(x)$ with respect to p have y-coordinates $\geq r$. Since $p^r \nmid a_n$ and $b_n = a_n$, we have $p^r \nmid b_n$, and the left-most spot $(0, \nu(a_n))$ has y-coordinate $< r$. Recall that the slopes of the edges of the

Newton of $F(x)$ with respect to p increase from left to right. This is sufficient to imply that the spots $(j, \nu(b_{n-j}))$ for $j \in \{k-1, k, k+1, \ldots, n\}$ all lie on or above edges of the Newton polygon of $F(x)$ which have positive slope.

Next, we show that each of these positive slopes is $< 1/k$. Since the slopes of the edges of the Newton polygon increase from left to right, it suffices to show that the right-most edge has slope $< 1/k$. The slope of that edge is

$$\max_{1 \le j \le n} \left\{ \frac{\nu(b_0) - \nu(b_j)}{j} \right\}.$$

For $1 \le j \le n$, we have

$$\begin{aligned} \nu(b_0) - \nu(b_j) &= \nu(a_0 n!) - \nu(a_j n!/j!) \\ &\le \nu(n!) - \nu(n!/j!) \\ &= \nu(j!) = \left[\frac{j}{p}\right] + \left[\frac{j}{p^2}\right] + \left[\frac{j}{p^3}\right] + \cdots \\ &< j\left(\frac{1}{p} + \frac{1}{p^2} + \cdots\right) = \frac{j}{p-1}. \end{aligned}$$

Since $p \ge k+1$, we obtain

$$\max_{1 \le j \le n} \left\{ \frac{\nu(b_0) - \nu(b_j)}{j} \right\} < \frac{1}{p-1} \le \frac{1}{k}.$$

Thus, each edge of the Newton polygon of $F(x)$ has slope $< 1/k$.

Now, assume $F(x)$ has a factor $g(x) \in \mathbb{Z}[x]$ of degree k. We obtain a contradiction to Theorem 5 by showing that translates of all the edges of the Newton polygon of $g(x)$ with respect to p cannot be found among the edges of the Newton polygon of $F(x)$ with respect to p. First, we show that no translate of an edge of the Newton polygon of $g(x)$ can be found among those edges in the Newton polygon of $F(x)$ having positive slope. Suppose (a, b) and (c, d), with $a < c$, are two lattices points on an edge of the Newton polygon of $F(x)$ having positive slope. Since we have already shown such a slope is $< 1/k$, we deduce

$$\frac{1}{k} > \frac{d-b}{c-a} \ge \frac{1}{c-a}.$$

Therefore, $c - a > k = \deg g(x)$. It follows that (a, b) and (c, d) cannot be endpoints of a translated edge of the Newton polygon of $g(x)$. Therefore, the translates of the edges of the Newton polygon of $g(x)$ with respect to p must be among the edges of the Newton polygon of $F(x)$ having zero or negative slope. We have already observed that all the spots $(j, \nu(b_{n-j}))$ for $j \in \{k-1, k, \cdots, n\}$ lie on or above the edges of the Newton polygon of $F(x)$

having positive slope. Therefore, the spots forming the endpoints of the edges of the Newton polygon of $F(x)$ having zero or negative slope must be among the spots $(j, \nu(b_{n-j}))$ where $j \in \{0, 1, \ldots, k-1\}$. Since $k - 1 < \deg g(x)$, these edges by themselves cannot consist of a complete collection of translated edges of the Newton polygon of $g(x)$. Therefore, we obtain a contradiction, completing the proof. ■

As a consequence, we deduce

Lemma 2. *If* $0 < |a_n| < n$, *then* $f(x)$ *cannot have a linear factor.*

Proof. If $0 < |a_n| < n$, then there must be a prime power p^r such that $p^r | n$ and $p^r \nmid a_n$. The lemma follows by taking $k = 1$ in Lemma 1. ■

In the next two sections, we will show that if $0 < |a_n| \leq n$ and $f(x)$ has a factor of degree $k \in [2, n/2]$, then either $a_n = \pm 5$, $n = 6$, and $k = 3$ or $a_n = \pm 7$, $n = 10$, and $k = 5$. We explain here why Theorem 2 will then follow. Suppose $f(x)$ is reducible. Then $f(x)$ and likewise $F(x)$ has a factor of degree $k \in [1, n/2]$. If $0 < |a_n| < n$, then Lemma 2 and the results just mentioned that we will be establishing in the next two sections easily imply this case of Theorem 2. Now, consider the possibility that $|a_n| = n$. In this case, we still know that $F(x)$ cannot have a factor of degree $k \in [2, n/2]$. For $n > 3$, it follows that $F(x)$ has a linear factor and the remaining factor is irreducible; for $n = 3$, considering the Newton polygon of $F(x)$ with respect to 2 allows us to make the same conclusion (i.e., in this case, $F(x)$ factors as a linear polynomial times an irreducible quadratic in $\mathbb{Z}[x]$). We want to show that either $F(x) = (x+1)g(x)$ or $F(x) = (x-1)g(x)$ where $g(x) \in \mathbb{Z}[x]$. Since $|a_n| = n$, we deduce that n divides each coefficient of $F(x)$. We write $F(x) = nh(x)$ where $h(x)$ is a monic polynomial in $\mathbb{Z}[x]$ having constant term $\pm(n-1)!$. We deduce from the rational root test applied to $h(x)$ that $F(x)$ has an integer root m which divides $(n-1)!$. We now only need to show that no prime p divides m. Assume some prime p divides m, and let $r = \nu_p((n-1)!)$. As in the proof of Lemma 1, we have $\nu_p(j!) < j$ for $j \geq 1$. Thus,

$$\nu_p\left(a_j \frac{(n-1)!}{j!} m^j\right) \geq \nu_p((n-1)!) - \nu_p(j!) + j \geq r + 1 \quad \text{for } 1 \leq j \leq n.$$

Hence,

$$h(m) = \sum_{j=0}^{n} a_j \frac{(n-1)!}{j!} m^j \equiv \pm(n-1)! \not\equiv 0 \pmod{p^{r+1}}.$$

This contradicts that $F(m)$, and hence $h(m)$, is zero.

The above discussion explains the main role of Newton polygons in the proof of Theorem 2. We now turn to the role of results from the distribution of primes. In proving Theorem 1, Schur used the following:

Theorem 6 (J.J. Sylvester [22]). *Let k be a positive integer. Then at least one of any k consecutive integers $> k$ is divisible by a prime $> k$.*

Theorem 6 was rediscovered by Schur in [19]. The theorem implies immediately that for any positive integer k, one of $k+1, k+2, \dots, 2k$ is a prime (since one of these integers must be divisible by a prime $\geq k+1$). Thus, Bertrand's Postulate is a consequence of Theorem 6.

As mentioned earlier, if $f(x)$ is as in Theorem 2 (so its degree is n) and $f(x)$ is reducible, then $f(x)$ has a factor of degree $k \in [1, n/2]$. This implies $n-k+1 > k$, so Theorem 6 implies that one of the k numbers $n, n-1, \dots, n-k+1$ is divisible by a prime $p \geq k+1$. Observe that if $0 < |a_n| \leq k$, then $p \nmid a_n$. Hence, from Lemma 1, we obtain

Lemma 3. *If $0 < |a_n| \leq k$, then $f(x)$ cannot have a factor of degree k.*

Observe that if $a_n = \pm 1$, then we deduce immediately that $f(x)$ is irreducible. In other words, we have just established Theorem 1.

To prove Theorem 2, we will make use of a generalization of Theorem 6 obtained by Ecklund, Eggleton, Erdős, and Selfridge [7]. We note the important related work by P. Erdős [8], K. Ramachandra [17], R. Tijdeman [23], M. Jutila [14], K. Ramachandra and T.N. Shorey [18], and T.N. Shorey [20] involving estimates for the largest prime factor of the product of k consecutive positive integers, say $n(n-1)\cdots(n-k+1)$. Also, see A. Granville [10] for a result concerning $p \geq k+1$ for which $p \| n(n-1)\cdots(n-k+1)$.

With $F(x)$ as before and $0 < |a_n| \leq n$, we want to show for $(a_n, n) \notin \{(\pm 5, 6), (\pm 7, 10)\}$ that $F(x)$ cannot have a factor of degree $k \in [2, n/2]$. Assume otherwise. Then Lemma 1 implies

$$\prod_{\substack{p^r \| n(n-1)\cdots(n-k+1) \\ p \geq k+1}} p^r$$

divides a_n. Since $0 < |a_n| \leq n$, we will obtain a contradiction if the above product is $> n$. In other words, we would like to know not just that $n(n-1)\cdots(n-k+1)$ is divisible by a prime $> k$ (as in Theorem 6) but further that the contribution of all the prime factors of this product which are $> k$ is $> n$. This is not in fact always the case, but it usually is. We will obtain the following lemma as a consequence of the work in [7], but we note here that the work in [7] contains considerably stronger estimates (see the next section).

Lemma 4. *Let k be an integer $\in [2, n/2]$. Then*

$$\prod_{\substack{p^r \| n(n-1)\cdots(n-k+1) \\ p \geq k+1}} p^r > n \tag{1}$$

unless one of the following holds:

$$\begin{array}{rcl} n = 12 & \text{and} & k = 5 \\ n = 10 & \text{and} & k = 5 \\ n = 9 & \text{and} & k = 4 \\ n = 18 & \text{and} & k = 3 \\ n = 10 & \text{and} & k = 3 \\ n = 9 & \text{and} & k = 3 \\ n = 8 & \text{and} & k = 3 \\ n = 6 & \text{and} & k = 3 \\ n = 2^\ell + 1 & \text{and} & k = 2 \\ n = 2^\ell & \text{and} & k = 2, \end{array}$$

where ℓ *represents an arbitrary positive integer.*

Although the proof of Theorem 2 does not require an inequality stronger than (1), we note that Lemma 4 holds if the right-hand side of (1) is replaced by $2n$. One easily checks that each value for the pair (n, k) in the list above gives rise to a choice of n and k for which (1) does not hold. After establishing Lemma 4, the proof of the irreducibility of the polynomials $f(x)$ in Theorem 2 will almost be complete. To handle the remaining cases of n and k given above, we will appeal once again to the use of Newton polygons. This will be done in Section 4.

3. The proof of Lemma 4

Throughout this section, we let n and k denote positive integers with $2 \le k \le n/2$. Our goal in this section is to establish Lemma 4. As mentioned in the previous section, Lemma 4 will be a fairly direct consequence of the work in [7]. There we find the following result:

Lemma 5. *Set* $\binom{n}{k} = UV$ *where the prime factors of* U *are all* $\le k$ *and the prime factors of* V *are all* $\ge k+1$. *If* $k \notin \{3,5,7\}$ *and* $U > V$, *then* $(n,k) \in S$ *where*

$$S = \{(9,4), (21,8), (33,13), (33,14), (36,13), (36,17), (56,13)\}\,.$$

Now, consider $k \ge 4$. Observe that since $n \ge 2k$, $\binom{n}{k} \ge \binom{n}{4} > n^2$ provided

$$n(n-1)(n-2)(n-3) - 24n^2 > 0.$$

The left-hand side above can be written as

$$n(n+1)\big(n(n-7)-6\big)$$

which is clearly positive since $n \geq 2k \geq 8$. Hence, we obtain

Lemma 6. *For $k \geq 4$, $\binom{n}{k} > n^2$.*

For $k \geq 4$ and $k \notin \{5,7\}$, we deduce from the above two lemmas that if $(n,k) \notin S$, then

$$\prod_{\substack{p^r \| n(n-1)\cdots(n-k+1) \\ p \geq k+1}} p^r = \prod_{\substack{p^r \| \binom{n}{k} \\ p \geq k+1}} p^r \geq \sqrt{\binom{n}{k}} > n.$$

In other words, for such (n,k), (1) holds. One checks directly that each pair $(n,k) \in S$ also satisfies (1) with the exception of the pair $(9,4)$. Thus, we have

Lemma 7. *If $k \geq 4$ and $k \neq 5$ and $k \neq 7$, then (1) holds unless $(n,k) = (9,4)$.*

We are left now with determining when (1) holds with $k \in \{2,3,5,7\}$. One can deal with $k=5$ and $k=7$ in a fairly simple manner. We explain the argument for $k=5$ and leave the analogous argument for $k=7$ to the reader. We remove from the set $T = \{n, n-1, n-2, n-3, n-4\}$ the integer divisible by the largest power of 2, the integer divisible by the largest power of 3, and the integer divisible by 5. Two of these may be the same; but in any case, at least two, say a and b, of the five integers will remain in T. Since the integer divisible by the largest power of 2 was removed from T, we deduce that $16 \nmid ab$. Similarly, $9 \nmid ab$ and $5 \nmid ab$. Hence,

$$\prod_{\substack{p^r \| n(n-1)\cdots(n-4) \\ p \geq 6}} p^r \geq \frac{ab}{24} \geq \frac{(n-3)(n-4)}{24}.$$

It is easy to verify that this last expression is $> n$ for $n \geq 32$. It follows that (1) holds for $k=5$ and $n \geq 32$. One checks directly that (1) holds for $k=5$ and $n \in \{11\} \cup \{13, 14, \ldots, 31\}$.

Next, we consider $k=3$. We will use the following result which was established by G.C. Gerono in 1857 (cf. [5, p. 744]).

Lemma 8. *The only solutions to the equation $|p^r - q^s| = 1$, where p and q are primes and r and s are integers greater than one, are $(p,r,q,s) = (3,2,2,3)$ and $(2,3,3,2)$.*

The next lemma is well-known and follows immediately from Lemma 8.

Lemma 9. *If $n > 9$, then there exists a prime $p > 3$ such that $p|n(n-1)$.*

Lemma 10. *If $k = 3 \leq n/2$, then (1) holds unless $n \in \{6, 8, 9, 10, 18\}$.*

Proof. One checks (1) directly for $k = 3$ and $6 \leq n \leq 18$. We therefore only consider $n \geq 19$. Let u be the greatest positive integer such that 2^u divides one of n, $n-1$, and $n-2$. Let v be the greatest positive integer such that 3^v divides one of n, $n-1$, and $n-2$. Note that exactly one member of $\{n, n-1, n-2\}$, say m_1, is divisible by 2^u, and one of these quantities, say m_2, is divisible by 3^v. The numbers m_1 and m_2 may not be distinct, but there must be at least one number, say m_3, among n, $n-1$, and $n-2$ which is different from m_1 and m_2. Observe that either $m_3 = m$ or $m_3 = 2m$ where m is a positive integer having each of its prime divisors > 3.

Next, we show that $n(n-1)(n-2)/m_3$ is divisible by a prime > 3. If $m_3 = n-2$, then we deduce from Lemma 9 that $n(n-1)$ is divisible by a prime > 3. Similarly, if $m_3 = n$, we deduce such a prime exists dividing $(n-1)(n-2)$. If $m_3 = n-1$ and m_3 is odd, then n and $n-2$ are even and Lemma 9 implies that $(n/2)(n-2)/2$ is divisible by some prime > 3. If $m_3 = n-1$ and m_3 is even, then n and $n-2$ are odd and only one is divisible by 3; hence, there exists a prime > 3 dividing one of n and $n-2$. Thus, we have shown that $n(n-1)(n-2)/m_3$ is divisible by a prime > 3. It now follows that

$$\prod_{p>3} p^{\nu_p(n(n-1)(n-2))} \geq 5m_3/2 \geq 5(n-2)/2 = (5n-10)/2 > n.$$

Thus, (1) holds. ■

The proof of Lemma 4 now follows from

Lemma 11. *If $k = 2 \leq n/2$, then (1) holds unless $n = 2^\ell$ or $2^\ell + 1$ for some positive integer ℓ.*

Proof. For $k = 2$, the inequality (1) is simply

$$\prod_{\substack{p^r || n(n-1) \\ p \geq 3}} p^r > n.$$

One of n or $n-1$ is odd and, hence, divides the product on the left. If n is not of the form 2^ℓ or $2^\ell + 1$, then neither n nor $n-1$ is a power of 2. Therefore,

$$\prod_{\substack{p^r || n(n-1) \\ p \geq 3}} p^r \geq 3(n-1) > n,$$

completing the proof. ■

4. The remaining cases for Theorem 2

We set

$$F(x) = n!f(x) = \sum_{j=0}^{n} b_j x^j \qquad \text{where } b_j = a_j\,(n!/j!)\,.$$

From Section 2, Theorem 2 holds provided we can show that if $0 < |a_n| \le n$ and $F(x)$ has a factor of degree $k \in [2, n/2]$, then either $a_n = \pm 5$, $n = 6$, and $k = 3$ or $a_n = \pm 7$, $n = 10$, and $k = 5$. We therefore assume $F(x)$ has a factor of degree $k \in [2, n/2]$. Since $0 < |a_n| \le n$, Lemma 1 implies that (1) does not hold. Hence, Lemma 4 implies that the pair (n, k) belongs to a short list of possible values. In this section, we complete the proof of Theorem 2 by examining each of these possibilities for (n, k).

For $n = 12$ and $k = 5$, we observe that Lemma 1 implies that $11 | a_n$. Since $|a_n| \le n = 12$, we deduce $a_n = \pm 11$. We consider the Newton polygon of $F(x)$ with respect to 3. Since we do not know what the values of a_j are, we cannot determine precisely what this Newton polygon looks like. Since $a_n = \pm 11$ and $a_0 = \pm 1$, $\nu_3(b_n) = 0$ and $\nu_3(b_0) = 5$. This means that the left-most spot of the Newton polygon is $(0, 0)$ and the right-most spot is $(12, 5)$. If we first consider the case where $a_j = 1$ for $1 \le j \le 11$, we see that the Newton polygon consists of the line segment joining $(0, 0)$ to $(3, 1)$ together with the line segment joining $(3, 1)$ to $(12, 5)$. Since $\nu_3(b_j)$ can only increase by choosing the numbers a_j in a different way, in general if we let the values of a_j vary over the integers, the Newton polygon of $F(x)$ with respect to 3 will consist of precisely these 2 line segments unless $3 | a_{n-3}$. If $3 | a_{n-3}$, then the Newton polygon is simply the segment joining $(0, 0)$ to $(12, 5)$. In either case, Theorem 5 implies that $F(x)$ cannot have a factor of degree 5 in $\mathbb{Z}[x]$, and we obtain a contradiction.

Now, suppose $n = 10$ and $k = 5$. Here, Lemma 1 implies $7 | a_n$. Since $|a_n| \le n = 10$, we obtain $a_n = \pm 7$. One checks that the Newton polygon of $F(x)$ with respect to 5 is the segment joining $(0, 0)$ to $(10, 2)$. The lattice points along this Newton polygon are $(0, 0)$, $(5, 1)$, and $(10, 2)$. Hence, Theorem 5 implies that if $F(x)$ is reducible, then it is the product of two irreducible polynomials of degree 5. Recall that in Theorem 2, we allow the possibility that $a_n = \pm 7$, $n = 10$, and $f(x)$ (or equivalently $F(x)$) factors as a product of two irreducible quintics. Thus, we are done in this case.

If $n = 9$ and $k = 4$, then Lemma 1 implies $7 | a_n$ so that $a_n = \pm 7$. The Newton polygon of $F(x)$ with respect to 3 is simply the line segment joining $(0, 0)$ to $(9, 4)$. We deduce from Theorem 5 that $F(x)$ cannot have a factor of degree 4 in $\mathbb{Z}[x]$, obtaining a contradiction.

If $n = 18$ and $k = 3$, then Lemma 1 implies $17 | a_n$ so that $a_n = \pm 17$. The Newton polygon of $F(x)$ with respect to 3 is the line segment joining $(0, 0)$ to $(18, 8)$. There are only three lattice points on the Newton polygon, namely

(0, 0), (9, 4), and (18, 8). Hence, $F(x)$ cannot have a factor of degree 3 in $\mathbb{Z}[x]$.

If $n = 10$ and $k = 3$, we use Lemma 1 to deduce that $5|a_n$. Since $0 < |a_n| \leq 10$, we deduce here that $a_n = \pm 5$ or $a_n = \pm 10$. In either case, we consider the Newton polygon of $F(x)$ with respect to 3. If $3 \nmid a_{n-1}$, then the Newton polygon consists of 2 edges, one joining (0, 0) to (1, 0) and the other joining (1, 0) to (10, 4). If $3|a_{n-1}$, then the Newton polygon consists of exactly 1 edge joining (0, 0) to (10, 4). In either case, we again deduce that $F(x)$ cannot have a cubic factor in $\mathbb{Z}[x]$.

If $n = 9$ and $k = 3$, Lemma 1 implies that $a_n = \pm 7$. As before, one can show that $F(x)$ cannot have a cubic factor in $\mathbb{Z}[x]$ by considering the Newton polygon of $F(x)$ with respect to 3.

If $n = 8$ and $k = 3$, Lemma 1 implies that $a_n = \pm 7$, and one shows that $F(x)$ cannot have a cubic factor in $\mathbb{Z}[x]$ by considering the Newton polygon of $F(x)$ with respect to 2.

For $n = 6$ and $k = 3$, Lemma 1 implies that $a_n = \pm 5$. Theorem 2 allows for the possibility that $a_n = \pm 5$, $n = 6$, and $k = 3$, so we are done in this case.

Now, suppose that $n = 2^\ell + 1$ for some positive integer ℓ and that $k = 2$. Lemma 1 implies that n divides a_n. Since $0 < |a_n| \leq n$, we deduce that $a_n = \pm n$. We consider

$$g(x) = a_n x^n + \sum_{j=0}^{n-1} (n!/j!)\, x^j,$$

so that $g(x)$ is the polynomial $F(x)$ with $a_j = 1$ for $0 \leq j \leq n-1$. As we have already seen in the cases above, the Newton polygon of $g(x)$ with respect to a prime is related to the Newton polygon of $F(x)$ with respect to that prime. We consider the Newton polygon of $g(x)$ with respect to 2. We now justify that this Newton polygon consists of two line segments, one joining (0, 0) to (1, 0) and one joining (1, 0) to $(n, n-2)$. Clearly (0, 0) and (1, 0) are spots obtained in the construction of the Newton polygon of $g(x)$ with respect to 2. The right-most spot is $(n, \nu_2(n!))$, and

$$\nu_2(n!) = \left[\frac{n}{2}\right] + \left[\frac{n}{4}\right] + \cdots = 2^{l-1} + 2^{l-2} + \cdots + 2 + 1 = 2^l - 1 = n - 2.$$

Since (1, 0) and $(n, n-2)$ are spots, the slope of the right-most edge of the Newton polygon of $g(x)$ with respect to 2 is $\geq (n-2)/(n-1)$. To show that the line segment joining (1, 0) and $(n, n-2)$ is an edge of this Newton polygon, it suffices to show that the slope of the right-most edge is $\leq (n-2)/(n-1)$. Since (0, 0) and (1, 0) are spots, the right-most edge has slope

$$\max_{1 \leq j \leq n-1} \left\{ \frac{\nu_2(n!) - \nu_2(n!/j!)}{j} \right\} = \max_{1 \leq j \leq n-1} \left\{ \frac{\nu_2(j!)}{j} \right\}.$$

For any positive integer j, $\nu_2(j!) = [j/2] + [j/4] + \cdots < j$. Thus, $\nu_2(j!) \le j - 1$, and

$$\max_{1 \le j \le n-1} \left\{ \frac{\nu_2(j!)}{j} \right\} \le \max_{1 \le j \le n-1} \left\{ \frac{j-1}{j} \right\} = \frac{n-2}{n-1}.$$

This completes the justification that the Newton polygon of $g(x)$ with respect to 2 consists of a line segment joining $(0,0)$ to $(1,0)$ together with a line segment joining $(1,0)$ to $(n, n-2)$.

We now concern ourselves with the Newton polygon of $F(x)$ with respect to 2. Note that the spots associated with the Newton polygon of $F(x)$ with respect to 2 lie on or above the edges of the Newton polygon of $g(x)$ with respect to 2. The interior of the triangle T with vertices $(0,0)$, $(1,0)$, and $(n, n-2)$ will in general contain many lattice points. We will show that none of these lattice points can be a spot associated with the Newton polygon of $F(x)$ with respect to 2. It will then follow that if $(1,0)$ is not a spot associated with the Newton polygon of $F(x)$ with respect to 2, then this Newton polygon is simply the line segment joining $(0,0)$ to $(n, n-2)$. On the other hand, if $(1,0)$ is a spot associated with the Newton polygon of $F(x)$ with respect to 2, then the Newton polygon of $F(x)$ with respect to 2 is the same as the Newton polygon of $g(x)$ with respect to 2. Since n is odd, $\gcd(n, n-2)$ and $\gcd(n-2, n-1)$ are both 1 so that in any case the only lattice points on the edges of the Newton polygon of $F(x)$ with respect to 2 are the endpoints of its edges. This will easily imply that $F(x)$ cannot have a quadratic factor.

Now, we show that the lattice points strictly inside the triangle T are not spots associated with the Newton polygon of $F(x)$ with respect to 2. The point $(n, n-2)$ is the right endpoint of the right-most edge of this Newton polygon. The line passing through $(n, n-2)$ and a point interior to T has slope $> (n-2)/n$. In particular, this implies $(1,1)$ is not inside T. Since the spots associated with the Newton polygon of $F(x)$ with respect to 2 are the points of the form $(n-j, \nu_2(b_j))$, it suffices to show

$$\max_{1 \le j \le n-2} \left\{ \frac{\nu_2(b_0) - \nu_2(b_j)}{j} \right\} \le \frac{n-2}{n}.$$

Since $\nu_2(a_0) = \nu_2(\pm 1) = 0$ and $\nu_2(a_j) \ge 0$, we need only show

$$\max_{1 \le j \le n-2} \left\{ \frac{\nu_2(n!) - \nu_2(n!/j!)}{j} \right\} \le \frac{n-2}{n}. \tag{2}$$

Observe that this is stronger than the earlier inequality we obtained with $(n-2)/n$ replaced by $(n-2)/(n-1)$. Since $n-2 = 2^\ell - 1$, we obtain for $1 \le j \le n-2$

that

$$\begin{aligned}\nu_2(j!) &= \left[\frac{j}{2}\right] + \left[\frac{j}{4}\right] + \cdots + \left[\frac{j}{2^{\ell-1}}\right] \\ &\le j\left(\frac{1}{2} + \frac{1}{4} + \cdots + \frac{1}{2^{\ell-1}}\right) = \frac{j\left(2^{\ell-1}-1\right)}{2^{\ell-1}} = \frac{j(n-3)}{n-1}.\end{aligned}$$

Thus,

$$\frac{\nu_2(n!) - \nu_2(n!/j!)}{j} = \frac{\nu_2(j!)}{j} \le \frac{n-3}{n-1} < \frac{n-2}{n} \qquad \text{for } 1 \le j \le n-2.$$

Hence, (2) follows.

Now, suppose that $n = 2^\ell$ for some positive integer ℓ and that $k = 2$. Here, Lemma 1 implies $a_n = \pm(n-1)$. We define $g(x)$ as above. Since a_n is odd, the left-most endpoint of the Newton polygon of $g(x)$ with respect to 2 is $(0,0)$. Since

$$\nu_2(n!) = [n/2] + [n/4] + \cdots = 2^{\ell-1} + 2^{\ell-2} + \cdots + 2 + 1 = 2^\ell - 1,$$

the right-most endpoint of the Newton polygon of $g(x)$ with respect to 2 is $(n, 2^\ell - 1) = (n, n-1)$. Analogous to the previous case, the slope of the right-most edge of the Newton polygon of $g(x)$ with respect to 2 is

$$\max_{1\le j\le n}\left\{\frac{\nu_2(n!) - \nu_2(n!/j!)}{j}\right\} = \max_{1\le j\le n}\left\{\frac{\nu_2(j!)}{j}\right\} \le \max_{1\le j\le n}\left\{\frac{j-1}{j}\right\} = \frac{n-1}{n}.$$

We deduce that the Newton polygon of $g(x)$, and hence of $F(x)$, with respect to 2 consists of the line segment joining $(0,0)$ to $(n, n-1)$. Since $\gcd(n, n-1) = 1$, we conclude $F(x)$ cannot have a quadratic factor.

This completes the proof of Theorem 2.

5. The proof of Theorem 3

The proof of Theorem 3 will mainly be based on the ideas already presented in the proof of Theorem 2. The possibility that $f(x)$ has a linear factor or has a factor of degree $k \ge 8$ will not involve any new ideas, and we shall not elaborate much on these details. There are however two aspects of the arguments here that will be different, one which deals with the possibility that $f(x)$ has a factor of degree $k \in [3,7]$ and one which deals with the possibility that $f(x)$ has a quadratic factor.

As before we set $F(x) = \sum_{j=0}^{n} b_j x^j$ with $b_j = n!a_j/j!$, and assume that $F(x)$ has a factor of degree $k \in [1, n/2]$. Lemma 2 was an immediate consequence of Lemma 1. A stronger result than Lemma 2 that also follows immediately from Lemma 1 is the following.

Lemma 12. *If $F(x)$ has a linear factor in $\mathbb{Z}[x]$, then $n|a_n$.*

The above lemma clarifies the situation when $k = 1$. To establish Theorem 3, we only need to show now that if n is sufficiently large, then any $f(x)$ as in the statement of the theorem cannot have a factor of degree $k \in [2, n/2]$. To deal with large values of k, we use a variant of (1). Specifically, Lemma 1 implies that if $F(x)$ is as in the statement of Theorem 3, then $F(x)$ cannot have a factor of degree k if

$$\prod_{\substack{p^r || n(n-1)\cdots(n-k+1) \\ p \geq k+1}} p^r > Cn^{3/2}. \tag{3}$$

We consider $k \in [8, n/2]$. Lemma 5 (with n sufficiently large) implies

$$\prod_{\substack{p^r || n(n-1)\cdots(n-k+1) \\ p \geq k+1}} p^r \geq \sqrt{\binom{n}{k}} \geq \sqrt{\binom{n}{8}},$$

and it easily follows that (3) holds in this case.

Next, we consider $k \in [3, 7]$. We will make use of a result of K. Mahler which we state in the form given in [15].

Lemma 13. *Let c and v be two positive constants, and let*

$$p_1, \ldots, p_r, p_{r+1}, \ldots, p_{r+r'}, p_{r+r'+1}, \ldots, p_{r+r'+r''}$$

be finitely many distinct primes. Denote by Σ an infinite sequence of distinct triples

$$\{P^{(k)}, Q^{(k)}, R^{(k)}\} \qquad (k = 1, 2, 3, \ldots)$$

where $P^{(k)}$, $Q^{(k)}$, and $R^{(k)}$ are integers as follows,

$$P^{(k)} \neq 0, \quad Q^{(k)} \neq 0, \quad R^{(k)} \neq 0, \quad P^{(k)} + Q^{(k)} + R^{(k)} = 0,$$

$$\left(P^{(k)}, Q^{(k)}\right) = \left(P^{(k)}, R^{(k)}\right) = \left(Q^{(k)}, R^{(k)}\right) = 1.$$

Put

$$H^{(k)} = \max\left(|P^{(k)}|, |Q^{(k)}|, |R^{(k)}|\right),$$

and write $P^{(k)}$, $Q^{(k)}$, and $R^{(k)}$ as products of integers,

$$P^{(k)} = P_1^{(k)} P_2^{(k)}, \quad Q^{(k)} = Q_1^{(k)} Q_2^{(k)}, \quad R^{(k)} = R_1^{(k)} R_2^{(k)},$$

where $P_1^{(k)}$ *has no prime factors distinct from* $p_{r+1}, \dots, p_{r+r'}$, $Q_1^{(k)}$ *has no prime factors distinct from* $p_{r+r'+1}, \dots, p_{r+r'+r''}$, *and* $R_1^{(k)}$ *has no prime factors distinct from* $p_1, \dots, p_r$. *If*

$$\left|P_2^{(k)} Q_2^{(k)} R_2^{(k)}\right| \leq cH^{(k)v} \qquad (k = 1, 2, 3, \dots),$$

then $v \geq 1$.

We make use of the above lemma to establish the following.

Lemma 14. *Let a be a fixed non-zero integer, and let N be a fixed positive integer. Let* $\varepsilon > 0$. *If n is sufficiently large (depending on a, N, and* ε*), then the largest divisor of* $n(n+a)$ *which is relatively prime to N is* $\geq n^{1-\varepsilon}$.

Proof. We show that there are only finitely many positive integers n satisfying $\mathcal{P}$ where $\mathcal{P} = \mathcal{P}(n)$ denotes the property that the largest divisor of $n(n+a)$ relatively prime to N is $< n^{1-\varepsilon}$. For each positive integer n satisfying $\mathcal{P}$, we consider $d = \gcd(a, n)$. Thus, n is associated with a pairwise relatively prime triple $\{P, Q, R\}$ as in Lemma 13 with

$$P = (n+a)/d, \quad Q = -a/d, \quad \text{and} \quad R = -n/d.$$

Observe that d is uniquely determined by Q and hence n is uniquely determined by Q and R. In other words, there is a one-to-one correspondence between n satisfying $\mathcal{P}$ and the triples $\{P, Q, R\}$ arising from such n. For each triple $\{P, Q, R\}$, we consider $H = \max\{|P|, |Q|, |R|\}$. Let $p_1, \dots, p_s$ be the complete list of distinct prime divisors of N. We momentarily fix an ordering of these primes and integers r and r' with $0 \leq r \leq r + r' \leq s$. With the ordering and r and r' fixed, we consider triples $\{P, Q, R\}$ arising from n as above and write

$$P = P_1 P_2, \quad Q = Q_1 Q_2, \quad \text{and} \quad R = R_1 R_2$$

where P_1 has no prime factors distinct from $p_{r+1}, \dots, p_{r+r'}$, Q_1 has no prime factors distinct from $p_{r+r'+1}, \dots, p_s$, and R_1 has no prime factors distinct from $p_1, \dots, p_r$. Setting $r'' = s - r - r'$, $c = |a|^2$, and $v = 1 - \varepsilon$, we obtain from Lemma 13 that there can be only finitely many triples $\{P, Q, R\}$ satisfying

$$|P_2 Q_2 R_2| \leq |a|^2 H^{1-\varepsilon}.$$

Since $d \leq |a|$, we deduce

$$|a|^2 H^{1-\varepsilon} \geq |a|^2 \left(\frac{n}{d}\right)^{1-\varepsilon} \geq |a|\, n^{1-\varepsilon}.$$

Thus, there are only finitely many n for which $|P_2 Q_2 R_2| \leq |a|\, n^{1-\varepsilon}$. We now let the ordering on the primes $p_1, \dots, p_s$ vary as well as the values of r

and r' noting that there are only a finite number (depending on a and N) of possible orderings and values of r and r' to consider. Observe that if n satisfies property $\mathcal{P}$, then the largest divisor of PQR relatively prime to N is $< |a|n^{1-\varepsilon}$. In particular, there is an ordering of the primes $p_1, \dots, p_s$ and values of r and r' for which $|P_2 Q_2 R_2| \le |a|n^{1-\varepsilon}$. As we have just seen, there can be only finitely many such n. ∎

Now, we make an argument similar to that used by Erdős in [8]. For each prime $p \le k$, consider s_p equal to an element from $\{n, n-1, \dots, n-k+1\}$ with $\nu_p(s_p)$ as large as possible. Let

$$S = \{n, n-1, \dots, n-k+1\} - \{s_p : p \le k\}.$$

Since $k \ge 3$, the set S contains at least one element, say s, and there are at least two additional integers, say m and $m+a$, among the numbers $n, n-1, \dots, n-k+1$. For each $p \le k$, we have removed a multiple of p, namely s_p, from $\{n, n-1, \dots, n-k+1\}$ in obtaining S so that there are at most $[k/p]$ multiples of p that can be in S. The definition of s_p in fact implies that there are at most $[k/p^j]$ multiples of p^j in S for each $j \ge 1$. Thus,

$$\nu_p\left(\prod_{u \in S} u\right) \le \sum_{j=1}^{\infty} \left[\frac{k}{p^j}\right] = \nu_p(k!) \qquad \text{for } p \le k.$$

In particular, for such p, $\nu_p(s) \le \nu_p(k!)$. We deduce

$$\prod_{\substack{p^r \| s \\ p \ge k+1}} p^r \ge \frac{s}{k!}.$$

Let $N = \prod_{p \le k} p$ and recall $k \le 7$. By Lemma 14, provided n (and hence $n-k+1$) is sufficiently large,

$$\prod_{\substack{p^r \| m(m+a) \\ p \ge k+1}} p^r \ge n^{3/4} \ge 2(7!)n^{1/2}.$$

Since $s \ge n-k+1 > n/2$, it follows that (3) holds. Hence, (3) holds for n sufficiently large and for all $k \in [3, n/2]$.

In the above arguments, it is clear that we could in fact establish a considerably stronger inequality than (3). The more delicate case which we now consider is when $k = 2 \le n/2$. To clarify how the bound in (3) arises and hence our bound on $|a_n|$ in Theorem 3, we define $\psi = \psi(n) = Cn^{1/2}$, the ratio of the right-hand side of (3) to n. We write $g(x) = a_n x^n + \sum_{j=0}^{n-1} (n!/j!)x^j$. As before (in dealing with the possibility that $f(x)$ has a quadratic factor in Theorem 2), we want to relate the Newton polygon of $g(x)$ with respect to 2 with the Newton polygon of $F(x)$ with respect to 2. In this case, the Newton polygon of $g(x)$ with respect to 2 cannot be described as precisely. Observe that Lemma 1 implies that the largest odd factor of $n(n-1)$ divides a_n. Define non-negative integers r, s, m, and m' by

$$n(n-1) = 2^r m \quad \text{and} \quad |a_n| = 2^s m'm \quad \text{where } 2 \nmid m'm.$$

From

$$n(n-1) = 2^r m = 2^{r-s}(2^s m) \leq 2^{r-s}|a_n| \leq 2^{r-s} n\psi(n),$$

we deduce

$$r - s > \log_2 \left(\frac{n-1}{\psi} \right) = \log_2 \left(\frac{n}{\psi} \right) + O(1/n).$$

In particular, $r > s$ and r is greater than the right-hand side above.

We consider two cases depending on whether n is odd or even. First, suppose n is odd and sufficiently large. The definition of s implies that $(0, s)$ is the left-most spot of the Newton polygon of $g(x)$ with respect to 2. Since n is odd, $(1, 0)$ is also a spot obtained in the construction of this Newton polygon. The definition of r implies we can write $n - 1 = 2^r w$ for some positive odd integer w. Observe that $m = wn$ so that $2^s m \leq |a_n| \leq n\psi$ implies

$$w \leq \psi \quad \text{and} \quad s \leq \log_2 \left(\frac{\psi}{w} \right).$$

Let D denote the number of times the digit 1 occurs in the binary expansion of w. It is an easy exercise to show that

$$D = w - \left[\frac{w}{2} \right] - \left[\frac{w}{4} \right] - \cdots.$$

Let γ denote the greatest integer $\leq \log_2 w$. If $D = \gamma + 1$, then necessarily $w = 2^{\gamma+1} - 1$ and we deduce $D = \log_2(w+1)$. On the other hand, if $D \neq \gamma + 1$, then $D \leq \gamma \leq \log_2 w$. Thus, in any case, we deduce

$$w - \left[\frac{w}{2} \right] - \left[\frac{w}{4} \right] - \cdots \leq \log_2(w+1).$$

Thus,

$$\begin{aligned}
\nu_2(n!) = \nu_2((n-1)!) &= \left[\frac{2^r w}{2}\right] + \left[\frac{2^r w}{4}\right] + \cdots + \left[\frac{2^r w}{2^r}\right] + \left[\frac{2^r w}{2^{r+1}}\right] + \cdots \\
&= 2^{r-1}w + 2^{r-2}w + \cdots + 2w + w + \left[\frac{w}{2}\right] + \left[\frac{w}{4}\right] + \cdots \\
&= 2^r w - w + \left[\frac{w}{2}\right] + \left[\frac{w}{4}\right] + \cdots \\
&= n - 1 - w + \left[\frac{w}{2}\right] + \left[\frac{w}{4}\right] + \cdots \\
&= n - E,
\end{aligned}$$

where

$$2 \le E = 1 + w - \left[\frac{w}{2}\right] - \left[\frac{w}{4}\right] - \cdots \le 1 + \log_2(w+1).$$

Note that the bound on s given above implies

$$s + E \le 1 + \log_2 \psi + \log_2\left(1 + \frac{1}{w}\right) \le 2 + \log_2 \psi.$$

From this we obtain from the previous bound on $r - s$ that

$$r - 2s - E = (r - s) - (s + E) \ge \log_2 n - 2\log_2 \psi - 2 + O(1/n).$$

The definition of ψ now implies

$$\text{(4)} \qquad r - 2s - E \ge -2 - 2\log_2 C + O(1/n).$$

We will make use of this inequality momentarily but note here that the condition $C < 1/\sqrt{2}$ in Theorem 3 implies that the right-hand side of (4) is > -1.

The right-most spot of the Newton polygon of $g(x)$ with respect to 2 is $(n, n - E)$. In dealing with the analogous situation in the proof of Theorem 2, we had $n = 2^\ell + 1$, $s = 0$, and $E = 2$. We were able to describe precisely the Newton polygon of $g(x)$ with respect to 2. In particular, we showed that the right-most edge had slope $\le (n-2)/(n-1)$. That same argument works here. Let ℓ be the line passing through the spot $(n, n - E)$ and having slope $(n-2)/(n-1)$. Then every spot obtained in the construction of the Newton polygon of $g(x)$ with respect to 2 is on or to the left of ℓ. Since $n - 1 = 2^r w$ and $n!/(n-j)!$ has $n-1$ as a factor for $j \in \{2, 3, \ldots, n\}$, each spot $(j, \nu_2(n!/(n-j)!))$ for such j is on or above the line $y = r$. Since n is sufficiently large, $r > s$. Thus, each spot other than the two left-most spots $(0, s)$ and $(1, 0)$ is in the closed region R in the plane bounded above $y = r$ and to the left of ℓ. Observe that $E \ge 2$ and the definition of ℓ imply that both $(0, s)$ and $(1, 0)$ are on or above ℓ. The points $(0, s)$ and $(n, n - E)$ are also spots obtained in constructing the Newton polygon of $F(x)$ with respect to 2, and the remaining spots in obtaining this Newton polygon, other than possibly the second spot from the left, must also lie in the region R. Let $(1, u)$ denote the second spot from the left. The left-most edge of this Newton polygon will have a negative slope if $u < s$, but since $r > s$, the remaining edges will have positive slopes. In particular, the assumption that $F(x)$ has a factor of degree $k = 2$ implies that there are two lattice points, say (a, b) and (c, d), along an edge of the Newton polygon of $F(x)$ with respect to 2 such that $d - b \geq 1$ and either $c - a = 1$ or $c - a = 2$. The right-most edge of the Newton polygon of $g(x)$ with respect to 2, and hence the right-most edge of the Newton polygon of $F(x)$ with respect to 2, has slope $\leq (n-2)/(n-1) < 1$. Since the slopes of the edges increase from left to right, we obtain $(d-b)/(c-a) < 1$. Since $d - b \geq 1$, we deduce $c - a \neq 1$. Thus, $c - a = 2$. Now, $(d-b)/(c-a) < 1$ and $d - b \geq 1$ imply $d - b = 1$. Therefore, there must be an edge of the Newton polygon of $F(x)$ with respect to 2 that has slope $1/2$.

We show that the above is impossible by showing that each edge with a positive slope has slope $> 1/2$. Let P be the point of intersection of the lines $y = r$ and ℓ. Let ℓ' denote the line passing through $(0, s)$ and P, and let t denote its slope. We claim that if an edge of the Newton polygon of $F(x)$ with respect to 2 has positive slope, then that slope must be $\geq \min\{t, 1\}$. To see this, suppose first that $(1, u)$ is on or above ℓ'. Then all the spots in obtaining the Newton polygon of $F(x)$ with respect to 2 lie on or above ℓ', and we can deduce that the left-most edge must have slope at least t. Since the slopes of the edges increase from left to right, we get in this case all the edges have slope $\geq t$. Now, suppose $(1, u)$ is below ℓ'. Then the left-most edge has slope $u - s$ which is either ≤ 0 or ≥ 1. Every spot lies on or above the line passing through $(1, u)$ and P, and it follows that the slope of the second left-most edge is at least as large as the slope of the line through $(1, u)$ and P. But the slope of that line is $\geq t$. Hence, every edge with a positive slope has slope $\geq \min\{t, 1\}$.

It now suffices to show that $t > 1/2$. The coordinates of P and, hence, the slope of ℓ' can be computed directly. We obtain

$$t = \frac{(r-s)(n-2)}{(r+E)(n-1)-n}.$$

The inequality $t > 1/2$ is equivalent to

$$(r - 2s - E)(n-1) > r - s - n.$$

Recall that n is sufficiently large and $C < 1/\sqrt{2}$. We deduce from (4) that $r - 2s - E$ is greater than -1 plus a positive constant (depending on C). On the other hand, it is easy to see that both r and s must be $\ll \log n$. Hence, $t > 1/2$.

We have still to consider the possibility that n is even and sufficiently large. Here, $n = 2^r w$ for some odd integer w. The left-most spot on the Newton polygon of $g(x)$ with respect to 2 is $(s, 0)$ and the right-most spot is $(n, n - E)$ for some $E \in [1, \log_2(w + 1)]$. Here, the right-most edge of the Newton polygon has slope $\leq (n-1)/n$. Letting ℓ now denote the line passing through $(n, n - E)$ with slope $(n - 1)/n$ and considering R to be the set of points on or above $y = r$ and on or to the left of ℓ, we continue as in the case that n is odd. Here, the situation is somewhat easier since every edge will have a positive slope. We omit further details.

6. Miscellaneous remarks

We begin with a proof of Theorem 4. We consider $r = 1$ in Lemma 1 and observe that Lemma 1 holds with the condition $p \nmid a_n$ replaced by $p \nmid a_n a_0$; in fact, no change (other than taking $r = 1$) is required in the proof of Lemma 1 as given. By Theorem 6, for any integer $k \in [1, n/2]$, there is a prime $p \geq k+1$ dividing $n(n-1)\cdots(n-k+1)$. The condition $\gcd(a_n a_0, n!) = 1$ implies that such a p does not divide $a_n a_0$. Hence, by Lemma 1 so revised, $f(x)$ cannot have a factor of degree k for any $k \in [1, n/2]$. It follows that $f(x)$ must be irreducible, and hence Theorem 4 is established.

Since Theorem 2 is a generalization of Theorem 1 with the condition $|a_n| = 1$ being relaxed, it is reasonable to ask whether an analogous result to Theorem 2 holds with instead the condition $|a_0| = 1$ in Theorem 1 relaxed. There is certainly irreducibility results that can be obtained along this line, but nothing as strong as the analog to Theorem 2 can hold in this case. For example, in Theorem 2, if $a_n = \pm 2$ and $a_0 = 1$, we can deduce that $f(x)$ is irreducible unless $f(x)$ is one of the following:

$$\pm(x+1)^2, \quad \pm(x-1)^2, \quad \text{or} \quad \pm(x+1)(x-1).$$

On the other hand, if $a_0 = 2$ and a_n is fixed, then there are arbitrarily large n for which $f(x)$ can be reducible. Specifically, we consider $n = 2^\ell$ with ℓ a positive integer, and we take $a_2 = a_3 = \cdots = a_{n-2} = 0$. Observe that $\nu_2(n!) = n - 1$, and write $n! = 2^{n-1}m$ where m is an odd integer. We deduce that

$$\frac{n! f(2)}{2^n} = a_n + 2^{\ell-1} a_{n-1} + m a_1 + m.$$

Since m is odd, there exist integers a_{n-1} and a_1 for which this last expression above is 0. Hence, for some $a_1, a_2, \ldots, a_{n-1}$, $f(x)$ has $x - 2$ as a factor.

As mentioned in the introduction, an effective version of Theorem 3 can be obtained if we require

$$0 < |a_n| \leq n \exp\left(\sqrt{\frac{\log n}{(\log\log n)^3}}\right).$$

We sketch how this can be done. It will be clear from the argument that a slight improvement on this bound is possible, but we do not bother with such details here.

The more difficult case in Section 5 with $k = 2$ does not need to be modified (the estimates there are already effective). In fact, the only change that needs to be made is in the case that $k \in [3, 7]$ (and now the worse estimate comes from $k = 3$). To obtain the above version of Theorem 3, we replace the role of Lemmas 13 and 14 with the following (the remaining arguments are essentially the same).

Lemma 13′. *Let ℓ be a positive integer. Let $a_1, \ldots, a_\ell$ be positive integers with $a_j \leq A_1$ for $1 \leq j \leq \ell - 2$ and with each of $a_{\ell-1}$ and $a_\ell \leq A_2$ where A_1 and A_2 are both ≥ 3. Let $b_1, \ldots, b_\ell$ be integers with each b_j having absolute value $\leq B$ where $B \geq 2$. Set*

$$\Lambda = b_1 \log a_1 + \cdots + b_\ell \log a_\ell.$$

There is a positive constant c depending only on ℓ and A_1 such that either $\Lambda = 0$ or

$$|\Lambda| > \exp\left(-c \log^2 A_2 \log B \log\log A_2\right).$$

Lemma 13′ is due to A. Baker [1]. Baker actually proved substantially more, but the above will suffice for our purposes. In particular, c can be made explicit. For a discussion of some other related work, see the section on "New Developments" in [2] and the notes on page 31 of [21].

Lemma 14′. *Let a be a non-zero integer, and let N be a fixed positive integer. There is an effectively computable constant $n_0 = n_0(a, N)$ such that if $n \geq n_0$, then the largest divisor of $n(n + a)$ which is relatively prime to N is*

$$\geq 2(7!) \exp\left(\sqrt{\frac{\log n}{(\log\log n)^3}}\right).$$

Proof. Let a be as in the lemma, and let $p_1, \ldots, p_r$ be the complete list of prime divisors of N. Let n be large, and let m_1 and m_2 be the largest divisors of n and $n+a$, respectively, which are relatively prime to N. Thus, we can find non-negative integers $e_1, \dots, e_r$ and $f_1, \dots, f_r$ such that

$$n = p_1^{e_1} p_2^{e_2} \cdots p_r^{e_r} m_1 \quad \text{and} \quad n + a = p_1^{f_1} p_2^{f_2} \cdots p_r^{f_r} m_2.$$

Set

$$\Lambda = (f_1 - e_1)\log p_1 + (f_2 - e_2)\log p_2 + \cdots + (f_r - e_r)\log p_r + \log m_2 - \log m_1.$$

Thus, $\Lambda = \log((n+a)/n) \asymp |a|/n \ll 1/n$. In particular, $\Lambda \neq 0$. Since N is fixed, each p_j is $\ll 1$. Also, each e_j and f_j (and, hence, $f_j - e_j$) is $\ll \log n$. Letting $A = \max\{3, m_1, m_2\}$, we deduce from Lemma 13 that for some constant c depending only on N,

$$\frac{1}{n} \gg \exp\left(-c \log^2 A \log\log A \log\log n\right).$$

This inequality implies that

$$A \geq 2(7!) \exp\left(\sqrt{\frac{\log n}{(\log\log n)^3}}\right),$$

and the lemma follows. ■

Finally, we comment that in [12], Patrick Harley obtained a complete list of the reducible polynomials that exist when $2 \leq a_n \leq 10$. The case that a_n is prime has already been addressed at the end of the introduction. We summarize some of Harley's findings by noting that when $a_n = 4$, there are reducible $f(x)$ only of degrees $n = 2$ and 4; when $a_n = 6$, there are reducible $f(x)$ only of degrees $n = 2$, 3, 4, and 6; when $a_n = 8$, there are reducible $f(x)$ only of degrees $n = 2$, 4, and 8; when $a_n = 9$, there are reducible $f(x)$ only of degrees $n = 3$ and 9; and when $a_n = 10$, there are reducible $f(x)$ only of degrees $n = 2$, 5, and 10.

Acknowledgments: The author thanks P. Moree, A. Schinzel, and the referee for several of the references below. In particular, the referee pointed out reference [7] and Schinzel reference [15], and each of these played a crucial role in the final version of this paper.

References

[1] A. Baker, *Transcendence Theory: Advances and Applications*, Academic Press, London, 1977, pp. 1–27.

[2] A. Baker, *Transcendental Number Theory*, Cambridge University Press, Cambridge, 1979.

[3] D.N. Bernstein, *The number of integral points in integral polyhedra*, Funct. Anal. Appl. **10** (1976), 223-224.

[4] R.F. Coleman, *On the Galois groups of the exponential Taylor polynomials*, L'Enseignement Math. **33** (1987), 183–189.

[5] L.E. Dickson, *History of the Theory of Numbers, Vol. II*, Chelsea, N.Y., 1971.

[6] G. Dumas, *Sur quelques cas d'irréductibilité des polynômes à coefficients rationnels*, Journal de Math. Pures et Appl. **2** (1906), 191–258.

[7] E.F. Ecklund, Jr., R.B. Eggleton, P. Erdős, and J.L. Selfridge, *On the prime factorization of binomial coefficients*, J. Austral. Math. Soc. (Series A) **26** (1978), 257–269.

[8] P. Erdős, *On consecutive integers*, Nieuw Archief voor Wiskunde III **3** (1955), 124–128.

[9] M. Filaseta, *The irreducibility of all but finitely many Bessel polynomials*, Acta Math. **174** (1995), 383–397.

[10] A. Granville, *On the scarcity of powerful binomial coefficients*, preprint.

[11] E. Grosswald, *Bessel Polynomials*, Lecture Notes in Mathematics, Vol. 698, SpringerVerlag, Berlin, 1978.

[12] P. Harley, *A generalization of an irreducibility theorem of I. Schur*, Masters' Thesis, University of South Carolina (1995).

[13] H. Ille, *Zur Irreduzibilität der Kugelfunktionen*, Jahrbuch der Dissertationen der Universität Berlin (1924).

[14] M. Jutila, *On numbers with a large prime factor II*, J. Indian Math. Soc. (N.S.) **38** (1974/75), 124–130.

[15] K. Mahler, *Lectures on Diophantine Approximations*, Cushing-Malloy, Ann Arbor, 1961.

[16] A.M. Ostrowski, *On multiplication and factorization of polynomials, I. Lexicographic orderings and extreme aggregates of terms*, Aequationes Math. **13** (1975), 201–228.

[17] K. Ramachandra, *A note on numbers with a large prime factor III*, Acta Arith. **19** (1971), 77–90.

[18] K. Ramachandra and T.N. Shorey, *On gaps between numbers with a large prime factor*, Acta Arith. **24** (1973), 99–111.

[19] I. Schur, *Einige Sätze über Primzahlen mit Anwendungen auf Irreduzibilitätsfragen, I*, Sitzungsberichte der Preussischen Akademie der Wissenschaften, Physikalisch-Mathematische Klasse **14** (1929), 125–136.

[20] T.N. Shorey, *On gaps between numbers with a large prime factor II*, Acta Arith. **25** (1973/74), 365–373.

[21] T.N. Shorey and R. Tijdeman, *Exponential Diophantine Equations*, Cambridge Univ. Press, Cambridge, 1986.

[22] J.J. Sylvester, *On arithmetical series*, Messenger of Math. **21** (1892), 1–19.

[23] R. Tijdeman, *On the maximal distance of numbers with a large prime factor*, J. London Math. Soc. (2) **5** (1972), 313–320.

[24] J.H. Wahab, *New cases of irreducibility of Legendre polynomials*, Duke Math J. **19** (1952), 165–176.

Michael Filaseta
Mathematics Department
University of South Carolina
Columbia, SC 29208
filaseta@math.sc.edu

Small values of indefinite binary quadratic forms

M. E. Flahive[1] *and A. C. Woods*

Dedicated to Professor Halberstam on the occasion of his retirement

Abstract. We develop the notion of convergents to a full lattice in two dimensions. We apply the idea to investigate the numerically small values of indefinite binary quadratic forms.

1.

Let Λ be a full lattice in R^2 (that is, its determinant $d(\Lambda)$ is nonzero); Λ^* will denote the nonzero elements of Λ. For any $(a,b) \in \Lambda$ with $b \geq 0$, we define $B(a,b) = \{(x,y) \neq (a,b) \; : \; |x| \leq |a|; 0 \leq y \leq b\}$, and $S(a,b) = B(a,b) \cap \Lambda^*$.

Definition. A lattice point $(a,b) \in \Lambda$ will be called a $\Lambda-$convergent if $b \geq 0$ and $S(a,b) \subseteq \{(-a,b)\}$.

Note. Every lattice has at least one convergent. Indeed, let (p,q) be an arbitrary element of Λ^* with $q \geq 0$. The discreteness of Λ implies that $S(p,q)$ is finite. If there exists $(a,b) \in S(p,q)$, $(a,b) \neq (-p,q)$, then we consider such $(a,b) \in S(p,q)$ with minimal non-negative b, and then minimize such $|a|$. By construction, we have that $S(a,b) \subseteq \{(-a,b)\}$, proving that (a,b) is a $\Lambda-$convergent.

Theorem 1.1. *Let (p,q) be a primitive point of Λ^* with $q > 0$.*
(i) If $|pq| > d(\Lambda)$, then (p,q) is not a $\Lambda-$convergent.
(ii) If $0 < |pq| < \frac{d(\Lambda)}{2}$, then (p,q) is a $\Lambda-$convergent.

Proof. First we consider a general primitive point $P = (p,q) \in \Lambda^*$ with $|pq| \neq 0$ and $q > 0$. Replacing Λ by $-\Lambda$ if necessary, we may assume that $p > 0$ holds. Then the line $\{(x,y) \; : \; y = \frac{q}{p}x\}$ contains elements of Λ^*, spaced at a distance of $\sqrt{p^2+q^2}$. Let $k > 0$ be minimal such that the line $L_0 = \{(x,y) \; : \; y = \frac{q}{p}x + k\}$ contains elements of Λ^*. If $Q = (a,b) \in \Lambda$ is on this line, then the minimality of k implies that $\{P,Q\}$ is a basis for Λ, and

$$d(\Lambda) = |pb - qa| = |p(\frac{q}{p}a + k) - qa| = pk \; .$$

[1]We are grateful for the financial support given to the first author by the Mathematics Institute at Ohio State University

If $pq > d(\Lambda)$ holds, then $0 < k < q$, and the line L_0 intersects $B(p,q)$ in a line segment of length $\sqrt{p^2+q^2}$. Therefore, $B(p,q)$ contains at least one element $Q = (a,b) \in \Lambda^*$ with $|a| < |p|$, which implies that (p,q) is not a $\Lambda-$convergent.

If $0 < pq < \frac{d(\Lambda)}{2}$ holds, then $k > 2q$ and the line L_0 does not intersect $B(p,q)$. This implies that $S(p,q) \subseteq \{(-p,q)\}$, proving (p,q) is a $\Lambda-$convergent.

Before continuing with the general theory of convergents, we apply Theorem 1.1 to obtain small values of binary quadratic forms.

In what follows, f will denote an indefinite binary quadratic form with nonzero discriminant $D(f)$. Since $D(f) > 0$, f decomposes into linear factors over R; that is, there exist $a,b,c,d \in R$ such that for every $k \neq 0$, $f(x,y) = (akx + cky)(\frac{b}{k}x + \frac{d}{k}y)$. Defining the lattices $\Lambda_k = \{(k(am+bn), \frac{cm+dn}{k}) : m,n \textit{ integers }\}$, we observe that $d(\Lambda_k)^2 = D(f)$ and $f(m,n)$ is the product of the coordinates of the lattice points $\pm((am+cn)k, \frac{bm+dn}{k})$. Moreover, for each $k > 0$ we know that (p,q) is a Λ_1-convergent if and only if $(pk, \frac{q}{k})$ is a Λ_k-convergent.

Definition. Let $f(x,y) = (ax+cy)(bx+dy)$, and let Λ be the lattice generated by $\{(a,b),(c,d)\}$. Then (m,n) will be called a convergent of f if at least one of $\pm(am+cn, bm+dn)$ is a $\Lambda-$convergent, where the sign is chosen so that $\pm(bm+dn) \geq 0$ holds.

Definition. The indefinite binary quadratic form f will be called Hermite reduced if $f(x,y) = \mu(x-\gamma y)(x-\delta y)$ with $\gamma > 1$; $-1 < \delta < 0$.

Notes. 1. The definition of convergent is independent of the particular factorization: It's independent of the order of the factors, and in the previous paragraph we showed that it's independent of multiplying the respective factors by $k, \frac{1}{k}$.

2. In Section 3 we obtain an explicit description of the convergents of Hermite reduced forms. Hermite reduction was used by A. Markoff in [4,5], where he shows that every indefinite binary quadratic form has a chain of equivalent reduced forms.

Theorem 1.2. *Let $f(x,y)$ be a form with nonzero $D(f)$, and let m,n be relatively prime integers.*

(i) If $|f(m,n)| > \sqrt{D(f)}$, then (m,n) is not a convergent of f.

(ii) If $|f(m,n)| < \frac{\sqrt{D(f)}}{2}$, then (m,n) is a convergent of f.

Proof. We set $f(x,y) = (ax+cy)(bx+dy)$ and consider the lattice Λ generated by (a,b) and (c,d). Since m,n are relatively prime, then $P = m(a,b)+n(c,d) = (p,q)$ is a primitive point of Λ, and

$$f(m,n) = (am+cn)(bm+dn) = pq.$$

Replacing (m,n) by $-(m,n)$ if necessary, we may assume that $q > 0$. We also recall that $d(\Lambda) = \sqrt{D(f)}$.

If $|f(m,n)| > \sqrt{D(f)}$, then $|pq| > d(\Lambda)$, and Theorem 1.1(i) implies that P is not a Λ−convergent; (m,n) is therefore not a convergent of f. In a similar manner, we obtain (ii) from Theorem 1.1(ii).

Theorem 1.2 (ii) was proved by E.S.Barnes in [1,Lemma 16], where it is pointed out that A. M. Legendre proved the result for rational forms in [2 or 3]. A. Rockett and P. Szűsz [7, Theorem 1] used work of A. Ostrowski [6] to generalize Theorem 1.2 (ii) for the forms associated with Pell's equation.

2.

Definition. Let Λ be a lattice in R^2. Then Λ is an anisotropic lattice if Λ is a full lattice with no point of Λ^* on either coordinate axis.

We note that in an anisotropic lattice every line which is parallel to either coordinate axis contains at most one nonzero lattice point. After some detailed analysis of special cases, much of what follows can be generalized to arbitrary full lattices. For simplicity, we've restricted to anistropic lattices, in which case the associated forms are nonzero forms; that is, f does not represent zero nontrivally.

Theorem 2.1. *Let Λ be an anisotropic lattice in R^2. If $P_0 = (p_0, q_0)$ is any convergent of Λ, then the convergents of Λ can be ordered in a sequence, $\{P_k = (p_k, q_k)\}$, with the following properties:*

(i) for each k, $q_k < q_{k+1}$;

(ii) $\lim_{k\to-\infty} q_k = 0 = \lim_{k\to\infty} p_k$; $\lim_{k\to\infty} q_k = \infty = \lim_{k\to-\infty} |p_k|$*;*

(iii) for each k, $|p_k| > |p_{k+1}|$;

(iv) for any k, there are no elements of Λ^* in $|x| < |p_k|$, $0 < y < q_{k+1}$;

(v) for each k, $p_k p_{k+1} < 0$.

Proof. From the definition of convergent and the discreteness of Λ, we can order the convergents of Λ according to increasing second coordinate, say $\{P_k = (p_k, q_k)\}$. The fact that Λ is anisotropic further ensures that (i) holds. By Minkowski's Little Theorem, for every $\epsilon > 0$ each of the strips $|x| < \epsilon$ and $|y| < \epsilon$ contains an element of Λ^*. Therefore, (ii) holds for this sequence. The fact that P_{k+1} is a Λ−convergent implies (iii).

By way of contradiction to (iv), we suppose that there exists $(p,q) \in \Lambda^*$ such that $|p| < |p_k|, 0 < q < q_{k+1}$. By the discreteness of Λ the number of such elements is finite. We take q minimal, and then minimize $|p|$. The fact that P_k is a convergent further implies that $q_k < q < q_{k+1}$, and P cannot be in our sequence of convergents. However, since $S(P_k) \cup S(P_{k+1}) = \emptyset$, by construction we must have $S(P) = \emptyset$, a contradiction which proves (iv).

If, for some choice of k, $p_k p_{k+1} > 0$ were to hold, then $|p_{k+1} - p_k| < |p_k|$,

$0 < q_{k+1} - q_k < q_{k+1}$, which violates (*iv*). This completes the proof of the theorem.

Theorem 2.2. *Let Λ be an anisotropic lattice in R^2. For any fixed anchor convergent P_0, let $\{P_k = (p_k, q_k)\}$ be the ordering of convergents given in Theorem 2.1. Then for each integer k there exists a positive integer a_k such that*

$$P_{k+1} = a_k P_k + P_{k-1}. \tag{2.1}$$

Proof. Property (*iv*) implies that there are no lattice points in the triangle formed by $(0,0)$, P_k , P_{k+1}. Therefore, $\{P_k, P_{k+1}\}$ is a basis for Λ. Similarly, $\{P_k, P_{k-1}\}$ is also a basis for Λ. Hence, there exist integers a, b, c, d such that

$$P_{k+1} = aP_k + bP_{k-1} \ ; \ P_{k-1} = cP_k + dP_{k+1}.$$

Also, $d(\Lambda) = |p_k q_{k+1} - p_{k+1} q_k| = |b||p_k q_{k-1} - p_{k-1} q_k| = |b| d(\Lambda)$, implying that $b = \pm 1$. If $b = -1$ were to hold, then from $q_{k+1} = aq_k - q_{k-1} > q_k$ we would obtain $a > 0$. However, since p_k and p_{k-1} have opposite signs, $p_{k+1} = ap_k - p_{k-1}$ would then imply that p_k and p_{k+1} have the same sign, contrary to Theorem 2.1(v). Therefore, $b = 1$ and

$$0 < q_{k+1} = aq_k + q_{k-1} < (a+1)q_k;$$

$a > -1$. Since $P_{k+1} \neq P_{k-1}$, then a is a positive integer, proving (2.1).

Definition. Let Λ be an anisotropic lattice in R^2, with $\{P_k = (p_k, q_k)\}$ its sequence of convergents. A lattice point P of the form $P = aP_k + P_{k-1}$ with $1 \le a < a_k$ will be called an intermediate $\Lambda-$convergent. An intermediate $\Lambda-$convergent P will be called a restricted intermediate $\Lambda-$convergent if $a = 1$ or $a = a_k - 1$.

Definition. Let $f(x, y) = (ax + cy)(bx + dy)$ be a nonzero form, and let Λ be the lattice generated by $\{(a, b), (c, d)\}$. Then (m, n) will be called a restricted intermediate convergent of f if one of $\pm(am + cn, bm + dn)$ is a restricted intermediate $\Lambda-$convergent.

In the following theorems we continue the analysis begun in Theorem 1.1.

Theorem 2.3. *Let Λ be an anisotropic lattice in R^2. If $P = (p, q) \in \Lambda$, $q > 0$ is a primitive point with $\frac{d(\Lambda)}{2} \le |pq| \le d(\Lambda)$, then P is either a $\Lambda-$convergent or a restricted intermediate $\Lambda-$convergent.*

Proof. As in the proof of Theorem 1.1, we may assume that $p > 0$. If $k > 0$ be minimal such that the line $L_0 = \{(x, y) \ : \ y = \frac{q}{p}x + k\}$ contains lattice points, then $d(\Lambda) = kp$ holds. Therefore, $\frac{d(\Lambda)}{2} \leq pq \leq d(\Lambda)$ implies that $q \leq k \leq 2q$, and from this we obtain that there is at most one element in $L_0 \cap S(p,q)$. If P is not a $\Lambda-$convergent, then there is exactly one element in $L_0 \cap S(p,q)$, and that point is a $\Lambda-$convergent, say P_j.

We also observe that when $q_j \geq \frac{q}{2}$ we have $P - P_j = P_{j-1}$, and $P = P_j + (P - P_j) = P_j + P_{j-1}$. Otherwise, $q_j < \frac{q}{2}$ and $P_{j+1} = P + P_j$, implying $P = P_{j+1} - P_j = (a_j - 1)P_j + P_{j-1}$. This completes the proof of the theorem.

Note. This proof of Theorem 2.3 also identifies when each type of restricted intermediate occurs.

Theorem 2.4. *Let $f(x,y) = (ax - cy)(bx - dy)$ be a nonzero form, and let m, n be relatively prime integers. If $|f(m,n)| \leq \sqrt{D(f)}$, then (m,n) is either a convergent or a restricted intermediate convergent of f.*

Proof. As in the proof of Theorem 1.2, we consider the lattice Λ generated by (a,b) and (c,d), and set $P = m(a,b) - n(c,d) = (p,q)$. Since f is a nonzero form, then Λ is anisotropic, and the result follows from the previous theorem.

In [7] A. Rockett and P. Szűsz have obtained this result for the forms associated with Pell's equation. Their method of proof is phrased in terms of approximation, and their argument can be extended to obtain the results for arbitrary nonzero forms.

3.

In this section we describe the convergents of f for the case when f is Hermite reduced.

Theorem 3.1. *Let $f(x,y) = \mu(x - \gamma y)(x - \delta y)$; $|\gamma| > |\delta|$ be a nonzero indefinite binary quadratic form. If Λ is the lattice generated by the points $(1,1), (-\gamma, -\delta)$, then f is Hermite reduced if and only if $(1,1)$ is the $\Lambda-$convergent following $(-\gamma, -\delta)$.*

Proof. The fact that f is nonzero implies that the lattice Λ is anisotropic, and the results of Sections 1 and 2 apply.

We first assume that $P_0 = (1,1)$ and $P_{-1} = (-\gamma, -\delta)$, consecutive convergents of Λ. Then $-\delta > 0$, and parts (i), (iii), (v) of Theorem 2.1 respectively imply that $0 < -\delta < 1$; $|\gamma| > 1$; $\gamma\delta < 0$, as claimed.

On the other hand, when f is Hermite reduced with $|\gamma| > |\delta|$, then $\gamma > 1$ and $-1 < \delta < 0$. We consider integers m, n such that the lattice point $P = m(1,1) + n(-\gamma, -\delta) = (p,q)$ lies inside

$$|p| < \gamma \,, \; 0 < q < 1; \tag{3.1}$$

$$(3.2) \qquad |m - n\gamma| < \gamma \ , \ 0 < m - n\delta < 1.$$

We note that m, n must be nonzero. Adding the inequalities in (3.2) we obtain

$$-\gamma < n(\gamma - \delta) < \gamma + 1.$$

Under our assumptions on γ and δ, this implies that $n = 1$, and $0 < m - \delta < 1$, which contradicts $m \neq 0$. Therefore, the open rectangle given by (3.1) contains no lattice point, and we obtain that $(1,1)$ is the Λ−convergent following $(-\gamma, -\delta)$, completing the proof of the theorem.

Theorem 3.2. *Let f be a nonzero Hermite reduced indefinite binary quadratic form, and let m, n be relatively prime integers. Then (m, n) is a convergent of f if and only if $\frac{m}{n}$ is a convergent of one of the roots of $f(x, 1) = 0$.*

Proof. Let $f(x, y) = \mu(x - \gamma y)(x - \delta y)$ with $|\gamma| > |\delta|$ and let Λ denote the lattice generated by $P = (1,1), Q = (-\gamma, -\delta)$. From Theorem 3.1 we obtain that P and Q are successive Λ−convergents, say $P = P_0, Q = P_{-1}$. Since Λ is anisotropic, Theorem 2.2 implies the existence of a sequence $\{a_k\}$ of positive integers for which (2.1) holds.

Identity (2.1) allows us to explicitly write each P_k as a linear combination of P_{-1} and P_0: We define $R_0 = 1 = S_{-1}$, $R_{-1} = 0 = S_0$, and for any $k \geq 1$, we iteratively define

$$R_k = a_{k-1}R_{k-1} + R_{k-2} \ , \ S_k = a_{k-1}S_{k-1} + S_{k-2}.$$

Then, for $k \geq 1$, the continued fraction of $\frac{R_k}{S_k}$ is $[a_0; a_1, \dots, a_{k-1}]$; using (2.1), induction on k yields

$$(3.3) \qquad P_k = R_k P_0 + S_k P_{-1}.$$

For $k < 0$ we iteratively define two sequences R_k, S_k which alternate in sign (with $R_{-2} > 0, S_{-2} < 0$) for which

$$|R_k| = a_{k+1}|R_{k+1}| + |R_{k+2}| \ , \ |S_k| = a_{k+1}|S_{k+1}| + |S_{k+2}|.$$

Then, for $k \leq -2$, the continued fraction of $\frac{R_k}{S_k}$ is $-[0; a_{-1}, a_{-2}, \dots, a_{k+1}]$, and (3.3) also follows by induction on $k \leq -2$.

Comparing the first coordinates of the points in (3.3) gives $p_k = R_k p_0 + S_k p_{-1}$. For all $k > 0$, $S_k \neq 0$ and $p_0 \neq 0$, implying that $\frac{p_k}{p_0 S_k} = \frac{R_k}{S_k} + \frac{p_{-1}}{p_0}$. Therefore, Theorem 2.1(ii) implies that $0 = \lim_{k\to\infty}\left(\frac{R_k}{S_k}\right) + \frac{p_{-1}}{p_0}$; that is,

$$(3.4) \qquad [a_0; a_1, a_2, \dots] = -\frac{p_{-1}}{p_0} = \gamma.$$

From a similar analysis of the second coordinates in (3.3) we obtain

$$-[0; a_{-1}, a_{-2}, \dots] = \frac{q_{-1}}{q_0} = -\delta. \tag{3.5}$$

For relatively prime integers m, n we consider the lattice point $P = mP_0 + nP_{-1}$. By definition, (m, n) is a convergent of f if and only if one of $\pm P$ is a Λ−convergent. Identity (3.3) thus yields that (m, n) is a convergent of f if and only if $(m, n) = (R_k, S_k)$, for some k. The conclusion follows from (3.4) and (3.5).

4.

In this section we give a finer analysis of the structure of restricted intermediate convergents. This analysis yields results on small values of forms in terms of the minimum of the form, rather than its discriminant. The minimum of a form f is

$$M(f) = inf\{|f(m, n)| > 0 \ : \ integers\ m, n \},$$

and we restrict to forms with nonzero minimum. Any associated lattice Λ has a nonzero homogeneous minimum,

$$\mu(\Lambda) = inf\{ \ |pq| > 0 \ : \ (p, q) \in \Lambda^*\}.$$

A lattice that satisfies this property must be anisotropic, and the theory of convergents applies.

Proposition 4.1. *Let $P = (p, q) \in \Lambda^*$ and j be an integer such that $pp_j > 0$, $|p| > |p_j|$ and $q > q_j$. If $0 < |pq| \leq 4\mu(\Lambda)$, then $P = 2P_j$ and $|pq| = 4\mu(\Lambda)$.*

Proof. We set $\mu = \mu(\Lambda)$. Since $pp_j > 0$, $|p| > |p_j|$, and $P - P_j \in \Lambda^*$, then

$$|p - p_j| = ||p| - |p_j|| = |p| - |p_j| > 0;$$

$$\mu \leq |p - p_j||q - q_j| = (|p| - |p_j|)(q - q_j) = |p|q(1 - \gamma)(1 - \delta), \tag{4.1}$$

where $\gamma = \left|\frac{p_j}{p}\right|, \delta = \frac{q_j}{q}$. The hypotheses imply that $0 < \gamma, \delta < 1$ and

$$\gamma\delta = \frac{|p_j|q_j}{|p|q} \geq \frac{\mu}{4\mu} = \frac{1}{4}.$$

From this, (4.1), and $|pq| \leq 4\mu$ we obtain

$$(1-\gamma)(1-\frac{1}{4\gamma}) \geq (1-\gamma)(1-\delta) \geq \frac{\mu}{|p|q} \geq \frac{1}{4};$$

$$(1-\gamma)(4\gamma-1) \geq \gamma;$$

$$4\gamma^2 - 4\gamma + 1 = (2\gamma-1)^2 \leq 0.$$

Therefore, equality holds throughout: $\gamma = \frac{1}{2}$; $\gamma\delta = \frac{1}{4}$. This implies that $P = 2P_j$ and $|p|q = 4\mu$, as claimed.

Proposition 4.2. *Suppose $P = (p, q) \in \Lambda$, $q > 0$ is not a $\Lambda-$convergent. If j is an integer with $q_{j-1} < q < q_{j+1}$ and $pp_j < 0$, then either $|p| > |p_{j-1}| + |p_{j+1}|$ or there exists an integer $1 \leq a < a_j$ such that $P = aP_j + P_{j-1}$.*

Proof. From $q_{j-1} < q < q_{j+1} = a_j q_j + q_{j-1}$ we obtain an integer a, $1 \leq a < a_j$, such that

$$(a-1)q_j + q_{j-1} < q \leq aq_j + q_{j-1};$$

$$0 \leq aq_j + q_{j-1} - q < q_j. \tag{4.2}$$

If the equality holds, then $P = aP_j + P_{j-1}$ and the result is proved. Therefore, we may assume that (4.2) holds with strict inequality.

Since P_{j-1} is a $\Lambda-$convergent, Theorem 2.1 (iv) and (4.2) imply that

$$|ap_j + p_{j-1} - p| \geq |p_{j-1}|. \tag{4.3}$$

Also, from $pp_j < 0$, $pp_{j-1} > 0$ and we obtain

$$|ap_j + p_{j-1} - p| = |\ |p_{j-1}| - a|p_j| - |p|\ |,$$

and (4.3) implies

$$|\ |p_{j-1}| - a|p_j| - |p|\ | \geq |p_{j-1}|. \tag{4.4}$$

Therefore, $|p_{j-1}| - a|p_j| - |p| < 0$, and from (4.4) we have

$$a|p_j| + |p| - |p_{j-1}| \geq |p_{j-1}|;$$

$$|p| \geq |p_{j-1}| + (|p_{j-1}| - a|p_j|);$$

$$|p| > |p_{j-1}| + |p_{j+1}|,$$

the last inequality following from $|p_{j-1}| = a_j|p_j| + |p_{j+1}|$ and $a < a_j$. This completes the proof of the result.

Proposition 4.3. *Suppose $P = (p,q) \in \Lambda^*$, $q > 0$, is not a $\Lambda-$convergent, and let the integer j be such that $q_{j-1} < q < q_{j+1}$ and $pp_j < 0$.*
(i) If $|pq| \leq 4\mu(\Lambda)$, then either P is a restricted intermediate $\Lambda-$convergent or $P = 2P_{j-1}$ and $|pq| = 4\mu(\Lambda)$.
(ii) If $|pq| \leq 2\mu(\Lambda)$, then $|pq| = 2\mu(\Lambda)$ and $P = P_j + P_{j-1}$. Moreover, $a_j = 2$ and $|p_j|q_j = |p_{j-1}|q_{j-1} = \mu$ with $q_j = (1+\sqrt{2})q_{j-1}$.

Proof. Let $\mu = \mu(\Lambda)$. Since $pp_{j-1} > 0$ and $q > q_{j-1}$, then Proposition 4.1 implies that without loss of generality we may assume that $|p| \leq |p_{j-1}|$. Therefore, by Proposition 4.2 there exists an integer a , $1 \leq a < a_j$, such that $P = aP_j + P_{j-1}$.

For $i = 0, \pm 1$, we set $\gamma_i = \left|\frac{p_{j+i}}{p}\right|$, $\delta_i = \frac{q_{j+i}}{q}$, and then

$$\frac{\mu}{|p|q} \leq \frac{|p_{j+i}|q_{j+i}}{|p|q} = \gamma_i\delta_i.$$

Also, $P = aP_j + P_{j-1}$ and $P_{j+1} = a_jP_j + P_{j-1}$ imply

$$|p| = |p_{j-1}| - a|p_j| \ ; \ q = aq_j + q_{j-1};$$

$$|p_{j+1}| = |p_{j-1}| - a_j|p_j| \ ; \ q_{j+1} = a_jq_j + q_{j-1},$$

from which we obtain

(4.5) $$1 = \gamma_{-1} - a\gamma_0 \ ; \ 1 = a\delta_0 + \delta_{-1};$$

(4.6) $$\gamma_1 = \gamma_{-1} - a_j\gamma_0 \ ; \ \delta_1 = a_j\delta_0 + \delta_{-1}.$$

Using (4.5) in (4.6),

$$\gamma_1 = 1 + (a - a_j)\gamma_0 \ ; \ \delta_1 = (a_j - a)\delta_0 + 1,$$

and (4.5) can be rewritten as

$$\gamma_{-1} = 1 + a\gamma_0 \ ; \ \delta_{-1} = -a\delta_0 + 1.$$

Therefore, for the two choices of $(\gamma,\delta) = (\gamma_1,\delta_1)$ or $(\gamma_{-1},\delta_{-1})$, there exists a nonzero integer D ($D = a - a_j$ or $D = a$, respectively) such that

$$\gamma = 1 + D\gamma_0 \; ; \; \delta = -D\delta_0 + 1. \tag{4.7}$$

For part (i) it suffices to prove that $|D| = 1$ holds for at least one of these D. Using the facts that all $\gamma_i\delta_i \geq \frac{1}{4}$ and (4.7), for each choice of D,

$$\frac{1}{4} \leq \gamma\delta = 1 + D(\gamma_0 - \delta_0) - D^2\gamma_0\delta_0 \leq 1 + D(\gamma_0 - \delta_0) - \frac{D^2}{4};$$

$\frac{D^2-3}{4} \leq D(\gamma_0 - \delta_0)$ holds for both choices of D. Choosing $D = a - a_j < 0$ when $\gamma_0 \geq \delta_0$ and $D = a > 0$ when $\gamma_0 \leq \delta_0$, we thus obtain a choice of D for which $\frac{D^2-3}{4} \leq 0$; that is, $|D| = 1$. This proves part (*i*).

When the stronger inequality $|p|q \leq 2\mu$ holds, we have that all $\gamma_i\delta_i \geq \frac{1}{2}$. Returning to (4.7), for both choices of D,

$$\frac{1}{2} \leq \gamma\delta = (1 + D\gamma_0)(1 - D\delta_0) \leq 1 - D^2\frac{1}{2} + D(\gamma_0 - \delta_0); \tag{4.8}$$

$$\frac{D^2 - 1}{2} \leq D(\gamma_0 - \delta_0); \tag{4.9}$$

$$|D| - \frac{1}{|D|} = \frac{D^2 - 1}{|D|} \leq 2|\gamma_0 - \delta_0|. \tag{4.10}$$

From $0 < \gamma_0 = \frac{|p_j|}{|p|}, \delta_0 = \frac{q_j}{q} < 1$ and $\gamma_0\delta_0 \geq \frac{1}{2}$, we obtain $|\gamma_0 - \delta_0| < \frac{1}{2}$. Hence, (4.10) implies that $|D| = 1$ holds for both choices of $D = a, a - a_j$. These have different signs, and the two values in (4.9) yield both $0 \leq \gamma_0 - \delta_0$ and $0 \leq \delta_0 - \gamma_0$. Therefore, $\gamma_0 - \delta_0 = 0$, and equality holds throughout (4.8); that is, all $\gamma_i\delta_i = \frac{1}{2}$ and $|p|q = 2\mu$. Hence, $P = aP_j + P_{j-1} = P_j + P_{j-1}$ gives

$$2\mu = |p|q = (|p_{j-1}| - |p_j|)(q_{j-1} + q_j) = \mu\Big(\frac{q_j}{q_{j-1}} - \frac{q_{j-1}}{q_j}\Big),$$

the last following from $|p_j|q_j = \mu = |p_{j-1}|q_{j-1}$. Therefore, $\frac{q_j}{q_{j-1}} > 1$ implies $q_j = (1 + \sqrt{2})q_{j-1}$, completing the proof of (*ii*).

Theorem 4.4. *Let Λ be a lattice for which $\mu = \inf\{|pq| > 0 \; : \; (p,q) \in \Lambda\} \neq 0$. Let $P = (p,q) \in \Lambda^*$ with $q > 0$.*
(i) If $|pq| \leq 4\mu$, then either P is a $\Lambda-$convergent, or P is a restricted intermediate $\Lambda-$convergent, or $P = 2P_j$ for some j and $|pq| = 4\mu$.
(ii) If $|pq| \leq 2\mu$, then either P is a $\Lambda-$convergent, or $P = P_j + P_{j-1}$ and

$|pq| = 2\mu$. Moreover, in the second alternative we have $|p_j|q_j = |p_{j-1}|q_{j-1} = \mu$ and $q_j = (1+\sqrt{2})q_{j-1}$.

Proof. By Theorem 2.1 (i), (ii) there exists k such that $q_k < q \leq q_{k+1}$. If $pp_k < 0$, then Proposition 4.3 implies the conclusion. On the other hand, if $pp_k > 0$, then from Theorem 2.1(v) we obtain that $pp_{k+1} < 0$. From $q_k < q \leq q_{k+2}$, the result again follows from Proposition 4.3.

We note that Theorem 4.4 does not hold for arbitrary full lattices: If Λ is the lattice generated by $\{(1,1),(0,\frac{5}{2})\}$, then $P_0 = (-2,\frac{1}{2})$, $P_1 = (1,1)$, $P_2 = (0,\frac{5}{2})$ are consecutive Λ−convergents, and $\mu = 1$. However, $(-1,\frac{3}{2})$ is not a Λ−convergent.

Theorem 4.5. *Let f be a form for which $M(f) = \mu \neq 0$. Let m,n be integers.*
(i) If $|f(m,n)| \leq 4\mu$, then (m,n) is either a convergent or a restricted intermediate convergent of f or twice a convergent of f.
(ii) If $|f(m,n)| < 2\mu$, then (m,n) is a convergent of f.
(iii) If $|f(m,n)| = 2\mu$, then either (m,n) is a convergent of f or f must be equivalent to $\pm\mu(x^2 - 2xy - y^2)$ and (m,n) is a restricted intermediate convergent of f.

Proof. Let $f(x,y) = (ax+by)(cx+dy)$. If Λ is the lattice generated by (a,c) and (b,d), then $\inf\{|pq| > 0 \ : \ (p,q) \in \Lambda\}$ equals $M(f)$, which is nonzero.

We consider integers m,n with $|f(m,n)| \leq 4\mu$ and without loss of generality we assume that (m,n) is not a convergent of f. Then the lattice point $P = m(a,c) + n(b,d)$ is not a Λ−convergent, and parts (*i*) and (*ii*) follow from Theorem 4.4.

When $|f(m,n)| = 2\mu$, Theorem 4.4 (ii) implies that $P = P_j + P_{j-1}$, and (m,n) is a restricted intermediate convergent of f. Since $\{P_j, P_{j-1}\}$ is a basis for Λ, then f is equivalent to $(p_{j-1}x - p_jy)(q_{j-1}x - q_jy)$. From $|p_j|q_j = \mu = |p_{j-1}|q_{j-1}$ and $q_j = (1+\sqrt{2})q_{j-1}$, f is therefore equivalent to

$$p_{j-1}q_{j-1}(x - \frac{p_j}{p_{j-1}}y)(x - \frac{q_j}{q_{j-1}}y) = \pm\mu(x - \frac{p_j}{p_{j-1}}y)(x - (1+\sqrt{2})y).$$

Also, $\frac{|p_j|}{|p_{j-1}|} = \frac{q_{j-1}}{q_j} = 1 - \sqrt{2}$ and $p_jp_{j-1} < 0$ imply that f is equivalent to

$$\pm\mu(x - (1-\sqrt{2})y)(x - (1+\sqrt{2})y) = \pm\mu(x^2 - 2xy - y^2),$$

completing the proof of the theorem.

5.

Theorem 4.5 (ii) is stronger than Theorem 1.2 (ii) (and likewise Theorem 4.5 (i) is stronger than Theorem 2.4) exactly when $4M(f) \geq \sqrt{D(f)}$; that is, when the Markoff value of the form is not greater than 4. In particular, this holds for the elements of the Markoff chain [4,5]; namely, the set of forms whose Markoff value is less than 3.

The first element of the Markoff chain is the class of all forms equivalent to $\mu(x^2+xy-y^2)$, which has roots $[1;\overline{1}\,];[0;\overline{1}\,]$. These forms have no intermediate convergents and the value at any convergent equals μ. Theorem 4.5 thus implies that these forms assume no value in the open interval $(\mu, 4\mu)$.

From Theorem 4.5 (ii), (iii) we have that $|f(m,n)| \leq 2\mu$ holds for non-convergent (m,n) exactly when f is equivalent to $\pm\mu(x^2-2xy-y^2)$, which is the second element of the Markoff chain. Its roots are $[2;\overline{2}\,];[0;\overline{2}\,]$; every intermediate convergent is restricted; the value at every convergent is μ and at every intermediate convergent is 2μ. Therefore, from Theorem 4.5 we obtain that there is no value of the form in $(\mu, 2\mu) \cup (2\mu, 4\mu)$. This pattern of successive isolation up the Markoff chain seems to continue. For example, if the second Markoff forms are disregarded, then the upper bound of 2μ in Theorem 4.5(ii) can be extended to 2.2μ and the third Markoff form is the only new exception to continuing further.

How good is the bound in (ii)? Using terminology from [4,5], let m be a Markoff number with associated Markoff form, f_m. (We then have that $M(f_m)$ equals m and is attained.) Then $(0,-1),(1,0)$ are consecutive convergents of f_m, and $(1,-1)$ is therefore a restricted intermediate convergent to f. From the Markoff theory we know that for any Markoff number m there exist integers u, v with $u^2+1=vm$, $0 \leq u$, $v < m$ such that

$$f_m(x,y) = mx^2 + (3m-2u)xy - (3u-v)y^2;$$

$$\frac{f_m(1,-1)}{m} = -2 + \frac{v-u}{m} = -2 - \frac{u}{m} + \frac{u^2+1}{m^2}.$$

Therefore,

$$\frac{|f_m(1,-1)|}{m} = 2 - \frac{1}{m^2} + \frac{u}{m} - \frac{u^2}{m^2} = \frac{9}{4} - \frac{1}{m^2} - \left(\frac{u}{m} - \frac{1}{2}\right)^2 < \frac{9}{4}.$$

Moreover, for every n, there is a Markoff number m_n such that

$$\frac{u_n}{m_n} = [0;2,(1)_{2n}2] > [0;2,(1)_\infty] = \frac{3-\sqrt{5}}{2}.$$

Therefore, there exist infinitely many forms f for which $(1,-1)$ is an intermediate convergent of f and $|f(1,-1)|$ lies in the open interval $(2,\sqrt{5})$.

References

[1] E.S. Barnes, *The minimum of the product of two values of a quadratic form (I)*, J.L.M.S. **1** (1951), 257-283.

[2] A. M. Legendre, *Essai sur la théorie des nombres, (Second edition)*, 1808, pp. 77.

[3] A. M. Legendre, *Théorie des Nombres, (Third edition), §XII*, 1830, pp. 120-129.

[4] A. Markoff, *Sur les formes quadratiques binaires indéfinies*, Math. Ann. **15** (1879), 381-406.

[5] A. Markoff, *Sur les formes quadratiques binaires indéfinies, II*, Math. Ann. **17** (1880), 379-399.

[6] A. Ostrowski, *Bemerkungen zur Theorie der Diophantischen Approximationen*, Abh. Math. Sem. Hamburg. Univ. (1921), 77-98.

[7] A. Rockett and P. Szüsz, *Continued Fractions*, 1992.

M. E. Flahive
Department of Mathematics
Oregon State University
Corvalis, OR
flahive@math.orst.edu

A. C. Woods
Department of Mathematics
Ohio State University
Columbus, OH 43210
acwoods@math.ohio-state.edu

Bombieri's sieve

John Friedlander and Henryk Iwaniec

Dedicated to Heini Halberstam

1. Introduction

Let $a(n)$ denote a sequence of non-negative reals and $A(x)$ the counting function

$$A(x) = \sum_{n \le x} a(n). \tag{1.1}$$

We are interested in estimating the contribution to $A(x)$ from those integers having few prime factors and, in particular, we shall study for given k the sum

$$S(x) = \sum_{n \le x}{}^{\flat} \, a(n)\Lambda_k(n) \tag{1.2}$$

where the flat "$\flat$" restricts the summation to squarefree numbers. Here the generalized von Mangoldt function Λ_k is defined as the Dirichlet convolution

$$\Lambda_k = \mu * L^k, \quad \text{with } L = \log, \tag{1.3}$$

which has support on positive integers with at most k distinct prime factors. In the case $k = 1$ we have $\Lambda_k = \Lambda$ and

$$S(x) = \sum_{p \le x} a(p) \log p.$$

One should not expect to have too much success studying the case $k = 1$ with sieve methods and we shall be concerned with the cases $k \ge 2$. Special cases of the sum for $k = 2$ were considered by Selberg [S] in connection with the elementary proof of the prime number theorem and in the case of general sequences and general k, this was forged into an elegant and versatile tool in the works [B1, B2] of Bombieri. His results offer a great deal of insight into the parity problem of sieve theory.

The present occasion offers an appropriate opportunity for this paper. The authors first (together) met with Professor Halberstam when he visited Pisa in early 1977 shortly after they had completed their first joint work [FI] which was on this very topic. The current work is, to some extent, an expression of these ideas in the spirit of which the subject of sieve methods has since developed, in no small measure due to the influence of the monograph [HR] of Halberstam and Richert.

Acknowledgement. The authors would like to thank the organizers of this stimulating conference. H.I. thanks the University of Toronto for hospitality during part of the work on this paper. J.F. was supported in part by NSERC grant A5123 and H.I. by NSF grant DMS-9500797.

2. Hypotheses

As has normally been the case (since Eratosthenes) in the use of a sieve we deduce results concerning the sum $S(x)$ from estimates for the sums

$$A_d(x) = \sum_{\substack{n\leq x \\ n\equiv 0 (\bmod d)}} a(n).$$

We consider an approximation,

$$A_d(x) = g(d)A(x) + r_d(x),$$

where $r_d(x)$ is considered as an error term and define the "total remainder"

$$R(x, D) = \sum_{d<D}{}^{\flat} |r_d(x)|.$$

We choose $z = x^{\delta}$ satisfying

$$2 \leq z \leq \min\left(x^{\frac{1}{20}}, x^{\frac{1}{2k}}\right). \tag{2.1}$$

We make three groups of hypotheses. The most stringent of these provides us with an estimate for the total remainder term,

$$R(x, x/z) \leq A(x)(\log x)^{-k-1}. \tag{H$_1$}$$

Remark. In the special case, treated in [B1], when $a(n) = \Lambda(n-2h)$, (H_1), with δ an arbitrarily small positive constant, is essentially the Elliott-Halberstam conjecture but for a number of other important sequences it is known to hold. We mention some examples in the final section of the paper.

The second group of hypotheses gives easy estimates for the counting sums,

$$A(x/z) \leq A(x)(\log x)^{-2}, \tag{H$_2$}$$

$$\sum_{p\geq z} A_{p^2}(x) \log p \leq \delta A(x). \tag{H$_3$}$$

The third group concerns the function g which features some degree of flexibility. This function is not uniquely determined by the sequence $\mathcal{A} = \{a(n)\}$, but in practice g must be well chosen to make the error terms $r_d(x)$ relatively small, at least on average. We can allow g to depend on the range of summation $1 \leq n \leq x$ since x does not change throughout. We regard $g(d)$ as a density and assume that

$$(\mathrm{H}_4) \qquad g \text{ is multiplicative with } 0 \leq g(p) < 1.$$

Some control over the values $g(p)$ will be required. In order to obtain readily applicable results we begin by assuming a crude, one-sided estimate

$$(\mathrm{H}_5) \qquad \sum_{w \leq p < u} \frac{g(p)}{1 - g(p)} \log p \leq \log \frac{cu}{w}$$

for all $u > w \geq 2$, where $c \geq 2$ is a constant. The point is that the densities $g(p)$ may depend on various parameters associated with the sequence $\mathcal{A}$ and the range x, at a few places quite badly, but only in a way which does not affect the constant c. Usually the parameters in question distort g by reducing its values so (H_5) still holds.

With the above hypotheses at hand we shall be able to establish an approximate formula for $S(x)$ in which the connection with the sequence $\mathcal{A}$ reduces simply to the counting sum $A(x)$, see the Proposition in Section 14. However the resulting expression for the main term remains to be refined; as it is, it involves a factor $G(x, z)$ which is a sum of the multiplicative function g and this is not very appealing, see (13.1). To refine the main term we need a considerable regularity in the distribution of $g(p)$ but only at primes $p \geq z$. We express this property by postulating the existence of another function, say f, which is close to g at these primes and well behaving over all integers. Precisely, we assume the following three hypotheses:

$$(\mathrm{H}_6) \qquad f \text{ is completely multiplicative with } 0 \leq f(p) < 1,$$

$$(\mathrm{H}_7) \qquad \Delta(z) = \sum_{p \geq z} |f(p) - g(p)| \leq \delta^3,$$

$$(\mathrm{H}_8) \qquad \sum_{n \leq t} f(n) = a \log t + b + O\left(\left(\log t\right)^{-2}\right),$$

for all $t \geq 2$, where $a > 0$ and b are constants.

We shall see that these hypotheses about f are slightly stronger than (H_5) (they imply (H_5) with a constant c depending on f) and are still easy to verify in practice. The above transition from g to f is convenient to secure uniformity of the final asymptotic formula with respect to the original function

g. Usually g is bad at prime factors of a certain number, say N, and the required modification is considerable, yet it still satisfies

$$|f(p) - g(p)| \le cp^{-1} \quad \text{for every } p|N. \tag{2.2}$$

Moreover, even if not absolutely required, it can be interesting to change g slightly at the other primes as well to make a nicer function f, say we change by less than

$$|f(p) - g(p)| \le cp^{-2} \quad \text{if } p \nmid N. \tag{2.3}$$

For such f, g the variation is bounded by

$$\Delta(z) \le cz^{-1}(\omega(N) + 1) < cz^{-1} \log 4N \tag{2.4}$$

where $\omega(N)$ denotes the number of distinct prime factors of N. Hence (H_7) holds true if $z > c(\log x)^3 \log 4N$.

3. The main result

Theorem. *Let $k \ge 2$ and suppose (H_1)- (H_5) hold true for some $z = x^\delta$ satisfying (2.1). Suppose there is an f which satisfies (H_6)-(H_8). Then*

$$\sum_{n \le x}{}^{b} a(n)\Lambda_k(n) = kHA(x)(\log x)^{k-1}\left\{1 + O\left(\delta(\log \delta)^4\right)\right\} \tag{3.1}$$

where H is a positive constant given by the convergent infinite product

$$H = \prod_p (1 - g(p))\left(1 - \tfrac{1}{p}\right)^{-1} \tag{3.2}$$

and the implied constant depends on c, f and k only.

Note that $g(d)$ is not required to behave nicely below z. However under the only one-sided estimate (H_5) we cannot determine the right order of magnitude of the product H. This can be quite large if $g(p)$ is small at small primes (in terms of x when g depends on x); at any rate it follows quickly from (3.4) that $H > c^{-1} \log 2$.

Note that (H_5) implies

$$g(p) \le \frac{\log c}{\log cp} \tag{3.3}$$

and by partial summation

$$\sum_{w \le p \le u} g(p)\left(1 - g(p)\right)^{-1} \le \log \frac{\log u}{\log w} + \frac{\log c}{\log w}\,.$$

Employing the inequality $-\log(1-\alpha) \le \alpha(1-\alpha)^{-1}$ we deduce that

$$\prod_{w \le p \le u} \left(1 - g(p)\right)^{-1} \le c \frac{\log u}{\log w}. \tag{3.4}$$

Note that we have included the (possibly vacuous) term $p = u$ as we may, since the right-hand sides are continuous functions of u.

4. Properties of $\Lambda_k(n,x)$

Let us recall some properties of the generalized von Mangoldt function Λ_k. We have $L^k = 1 * \Lambda_k$ and $\Lambda_{k+1} = L\Lambda_k + \Lambda * \Lambda_k$, whence $0 \le \Lambda_k \le L^k$ and $\Lambda_j \le L^{j-k}\Lambda_k$ if $0 \le j \le k$. Moreover it follows that $\Lambda_k(n) = 0$ if $\omega(n) > k$, that is $\Lambda_k(n)$ is supported by numbers which have at most k distinct prime factors. We also have the multiplicativity relation

$$\Lambda_k(mn) = \sum_{0 \le j \le k} \binom{k}{j} \Lambda_j(m)\Lambda_{k-j}(n) \quad \text{if } (m,n) = 1.$$

Define

$$\Lambda_k(n,x) = \sum_{d|n} \mu(d) \left(\log \frac{x}{d}\right)^k \tag{4.1}$$

for any $k \ge 0$, $n \ge 1$, and $x > 0$. Note that if $x = n$ this coincides with $\Lambda_k(n)$ and, for any x, $\Lambda_k(n,x) = \Lambda_k(n^*,x)$ where n^* is the product of all distinct prime factors of n, although $\Lambda_k(n) = \Lambda_k(n^*)$ is false. In general, $\Lambda_k(n,x)$ can be expressed in terms of $\Lambda_j(n)$ as follows:

$$\Lambda_k(n,x) = \sum_{0 \le j \le k} \binom{k}{j} \Lambda_j(n) \left(\log \frac{x}{n}\right)^{k-j}. \tag{4.2}$$

Hence,

$$\Lambda_k(n,x) = 0 \qquad \text{if } \omega(n) > k, \tag{4.3}$$

i.e. $\Lambda_k(n,x)$ also is supported by numbers which have at most k distinct prime factors. We derive

$$\Lambda_k(n,x)\left(\frac{\log n}{\log x}\right)^k \le \Lambda_k(n) \le \Lambda_k(n,x) \quad \text{if } n \le x, \tag{4.4}$$

since the terms in (4.2) are non-negative. The lower bound is obtained by inserting $\Lambda_j \le L^{j-k}\Lambda_k$ and the upper bound by taking only $j = k$. These inequalities imply

$$0 \le \Lambda_k(n,x) \le (\log x)^k \quad \text{if } n \le x. \tag{4.5}$$

Moreover we derive

$$\sum_{m|n} \Lambda_k(m,x) \le (\log nx)^k \quad \text{if } n \le x \tag{4.6}$$

by the identity

$$\sum_{m|n} \Lambda_k(m,mx) = (\log nx)^k \tag{4.7}$$

since $\Lambda_k(n,x)$ is non-decreasing in $x \ge n$. Suppose $(m,n)=1$. Any divisor of mn can be written uniquely as a product $d = ab$ with $a \mid m$, $b \mid n$, and accordingly we can write $\log \frac{x}{d} = \log \frac{x}{an} + \log \frac{n}{b}$ showing that

$$\Lambda_k(mn,x) = \sum_{0\le j\le k} \binom{k}{j} \Lambda_j\left(m, \frac{x}{n}\right) \Lambda_{k-j}(n). \tag{4.8}$$

Here we have $j<k$ if $n>1$ because $\Lambda_0(n)=0$. Applying the inequalities $\binom{k}{j} \le k\binom{k-1}{j}$ and $\Lambda_{k-j}(n) \le (\log n)^{k-j}$ we derive from (4.8) that

$$\Lambda_k(mn,x) \le k\Lambda_{k-1}(m,x)\log n \tag{4.9}$$

if $(m,n)=1$, $mn\le x$, $n>1$, $k\ge 1$. Hence in particular,

$$\Lambda_k(mp^\alpha,x) \le k\Lambda_{k-1}(m,x)\log p \tag{4.10}$$

if $k\ge 1$, $\alpha>0$, $p\nmid m$, $mp^\alpha \le x$. By iterated application of this inequality we get

$$\Lambda_k\left(p_1^{\alpha_1}\cdots p_r^{\alpha_r},x\right) \le r!\binom{k}{r}(\log x)^{k-r}\log p_1\cdots\log p_r. \tag{4.11}$$

Hence for a multiplicative function $g\ge 0$ we get

$$\begin{aligned}\sum_{n\le x} g(n)\Lambda_k(n,x) &\le \sum_{0\le r\le k}\binom{k}{r}(\log x)^{k-r}\Big(\sum_{\ell\le x} g(\ell)\Lambda(\ell)\Big)^r\\ &= \Big(\log x + \sum_{\ell\le x} g(\ell)\Lambda(\ell)\Big)^k.\end{aligned} \tag{4.12}$$

This is an equality for $k=0,1$.

5. Fundamental lemma of the sieve method

Let $M \geq 2$. There exist two sequences $\{\lambda_m^+\}$, $\{\lambda_m^-\}$ which have the following properties:

$$\lambda_1^+ = \lambda_1^- = 1 \tag{5.1}$$

$$|\lambda_m^+|,\ |\lambda_m^-| \leq 1 \tag{5.2}$$

$$\lambda_m^+ = \lambda_m^- = 0 \quad \text{if } m \geq M. \tag{5.3}$$

For any $n > 1$

$$\sum_{m|n} \lambda_m^- \leq 0 \leq \sum_{m|n} \lambda_m^+. \tag{5.4}$$

For any multiplicative function g satisfying $0 \leq g(p) < 1$ and (3.4) we have

$$\sum_{m|P(z)} \lambda_m^+ g(m) \leq (1 + \gamma e^{-s}) V(z) \tag{5.5}$$

$$\sum_{m|P(z)} \lambda_m^- g(m) \geq (1 - \gamma e^{-s}) V(z) \tag{5.6}$$

for any $z \leq M$ where

$$P(z) = \prod_{p<z} p, \qquad V(z) = \prod_{p<z} (1 - g(p)) \tag{5.7}$$

$s = \log M / \log z$ and $\gamma \geq 2$ is a number which depends only on the constant c in (3.4). These estimates are very useful for sifting out numbers having small prime factors. Actual constructions of suitable sequences λ_m^+, λ_m^- and proofs can be found in [FI] or [I]; see also [HR].

6. Smoothed sums

We shall consider

$$T(x) = \sum_{n \leq x}^{b} a(n) \Lambda_k(n, x) \tag{6.1}$$

in place of $S(x)$. By (4.4) we have $S(x) \leq T(x)$ and

$$S(x) \geq T(x) \left(\frac{\log y}{\log x} \right)^k - A(y)(\log x)^k$$

for any $y \le x$. For $y = xz^{-1} = x^{1-\delta}$ we get by (H_2)

$$S(x) \ge (1 - \delta k)T(x) - A(x)(\log x)^{k-2}. \tag{6.2}$$

7. Dissecting $T(x)$

We write

$$T(x) = T(x,z) - T'(x,z) + T''(x,z)$$

where

$$T(x,z) = \sum_{\substack{n \le x \\ (n,P)=1}} a(n)\Lambda_k(n,x),$$

$$T'(x,z) = \sideset{}{^\sharp}\sum_{\substack{n \le x \\ (n,P)=1}} a(n)\Lambda_k(n,x),$$

$$T''(x,z) = \sideset{}{^\flat}\sum_{\substack{n \le x \\ (n,P)>1}} a(n)\Lambda_k(n,x),$$

the summation in $\sum^\sharp$ being restricted to numbers which are not squarefree, the summation in $\sum^\flat$ being restricted to numbers which are squarefree, and $P = P(z)$ as in (5.7).

8. Estimate for $T'(x,z)$

From (4.10), (4.5) and (H_3) we infer the following estimates:

$$\begin{aligned} T'(x,z) &\le \sum_{\substack{p \ge z \\ \alpha \ge 2}} \sum_{\substack{m \le xp^{-\alpha} \\ p \nmid m}} a(mp^\alpha)\Lambda_k(mp^\alpha, x) \\ &\le k(\log x)^{k-1} \sum_{p \ge z} A_{p^2}(x) \log p \le \delta k A(x)(\log x)^{k-1}. \end{aligned}$$

9. Estimate for $T''(x,z)$

Since n has at most k prime factors there exists $p \mid n$ with $p \ge n^{1/k}$, thus $n = mp$ and $m \le n^{1-1/k} \le x^{1-1/k}$. For $n > xz^{-1}$ this implies $p \ge z$ so the condition $(n,P)>1$ transfers to that for m. Hence

$$T''(x,z) \le A(x/z)(\log x)^k + \sideset{}{^\flat}\sum_{\substack{m\le x^{1-1/k}\\(m,P)>1}} \sum_{\substack{x/mz<p\le x/m\\ p\nmid m}} a(mp)\Lambda_k(mp,x)$$

$$\le A(x)(\log x)^{k-2} + k(\log x) \sideset{}{^\flat}\sum_{\substack{m\le x^{1-1/k}\\(m,P)>1}} \Lambda_{k-1}(m,x) \sum_{\substack{x/mz<p\le x/m\\ p\nmid m}} a(mp).$$

To the last sum apply an upper-bound sieve of level $D=x/mz$ to relax the primality of p. After having done this ignore the condition $p>D$ getting

$$\sum_p a(mp) \le \sum_{\substack{d|P(D)\\(d,m)=1}} \lambda_d^+ A_{dm}(x) \le g(m)\sigma_m(D)A(x) + \sum_{\substack{d|P(D)\\(d,m)=1}} \lambda_d^+ r_{dm}(x)$$

where

$$\sigma_m(D) = \sum_{\substack{d|P(D)\\(d,m)=1}} \lambda_d^+ g(d) \le (1+\gamma) \prod_{\substack{q<D\\(q,m)=1}} \big(1-g(q)\big)$$

$$\le (1+\gamma)c^{k-1}V(D) \le (1+\gamma)2kc^kV(x)$$

using $\big(1-g(p)\big)^{-1} \le \frac{\log cp}{\log p} \le c$, by (3.3), (3.4), and the fact that m has fewer than k prime factors. Summing over m we see that the resulting total remainder term is bounded by

$$k(\log x)\sideset{}{^\flat}\sum_{dm<x/z} \mid r_{dm}(x)\mid \Lambda_{k-1}(m,x) \le 3(\log x)^{2k-1}R(x,x/z).$$

The resulting sum over m in the main term is bounded by

$$\sideset{}{^\flat}\sum_{m\le x} g(m)\Lambda_{k-1}(m,x) \le k\Big(\log x + \sum_{p\le x} g(p)\log p\Big)^{k-2} \sum_{p<z} g(p)\log p$$

$$\le k(\log cx^2)^{k-2}\log cz$$

by writing $n=mp$ with $p\nmid m$, $p<z$ and using (4.10)–(4.12) and (H_5).

Gathering the above estimates we conclude that

$$T''(x,z) \ll \delta A(x)V(x)(\log x)^k + A(x)(\log x)^{k-2}$$

where the last term is obtained by applying (H_1) and (H_2). Actually the last term is absorbed by the preceding one because

$$V(x) \geq \frac{\log 2}{c \log x} \tag{9.1}$$

by (3.4) and (2.1). Therefore

$$T''(x,z) \ll \delta A(x) V(x) (\log x)^k. \tag{9.2}$$

10. Splitting $T(x,z)$

Put

$$\Lambda_k(n,x,y) = \sum_{\substack{d|n \\ d\leq y}} \mu(d) \left(\log \frac{x}{d}\right)^k \tag{10.1}$$

and let $\Lambda_k^\spadesuit(n,x,y)$ be the complementary sum over $d > y$. Accordingly

$$T(x,y,z) = \sum_{\substack{n\leq x \\ (n,P)=1}} a(n) \Lambda_k(n,x,y)$$

and $T^\spadesuit(x,y,z)$ is the complementary sum, so

$$T(x,z) = T(x,y,z) + T^\spadesuit(x,y,z). \tag{10.2}$$

Changing the order of summation in $T(x,y,z)$ we arrange

$$T(x,y,z) = \sum_{\substack{d\leq y \\ (d,P)=1}} \mu(d) A_d(x,z) \left(\log \frac{x}{d}\right)^k \tag{10.3}$$

where

$$A_d(x,z) = \sum_{\substack{n\leq x,\ n\equiv 0 (\text{mod} d) \\ (n,P)=1}} a(n). \tag{10.4}$$

We estimate the complementary sum in terms of $A_d(x,z)$ by switching to the complementary divisors n/d of n. By (H_3) we get

$$|T^\spadesuit(x,y,z)| \leq \sum_{\substack{d<x/y \\ (d,P)=1}}^{\flat} A_d(x,z) \left(\log \frac{x}{y}\right)^k + O\left(\delta A(x) V(z) \left(\log \frac{x}{y}\right)^k\right). \tag{10.5}$$

11. Evaluation of $A_d(x,z)$

Recall that $P = P(z)$ is the product of all primes $p < z$. We relax the condition $(n,P) = 1$ by applying sieves, say $\{\lambda_m^+\}$ and $\{\lambda_m^-\}$, of level $M = z^s$ with $s \geq 1$ to be chosen later. For any $(d,P) = 1$ we get

$$A_d(x,z) \leq (1 + \gamma e^{-s})V(z)A(x)g(d) + \sum_{m|P} \lambda_m^+ r_{dm}(x)$$

and

$$A_d(x,z) \geq (1 - \gamma e^{-s})V(z)A(x)g(d) + \sum_{m|P} \lambda_m^- r_{dm}(x)$$

by (5.5) and (5.6) respectively. Comparing the two estimates we infer

$$|A_d(x,z) - g(d)V(z)A(x)| \leq \gamma e^{-s} g(d)V(z)A(x) + \sideset{}{^\flat}\sum_{\substack{m<M \\ m|P}} |r_{dm}(x)|. \tag{11.1}$$

12. Estimation of $T^\spadesuit(x,y,z)$

By (10.5) and (11.1) we get

$$|T^\spadesuit(x,y,z)| \ll W\left(\frac{x}{y}, z\right) V(z)A(x)\left(\log \frac{x}{y}\right)^k + R\left(x, M\frac{x}{y}\right)(\log x)^k$$

where

$$W(D,z) = \sideset{}{^\flat}\sum_{\substack{d\leq D \\ (d,P)=1}} g(d) \leq \prod_{z\leq p\leq D} (1 + g(p)) \leq c\frac{\log D}{\log z}. \tag{12.1}$$

To simplify the resulting estimate for $T^\spadesuit(x,y,z)$ we assume that $2\delta(s+1) \leq 1$ and we choose

$$y = xz^{-s-1} = x^{1-\delta(s+1)} \geq x^{1/2},$$

so $Mx/y \leq My = xz^{-1}$ and $R(x, Mx/y) \leq A(x)(\log x)^{-2}$ by (H_1). For this choice

$$|T^\spadesuit(x,y,z)| \ll (s+1)^{k+1}V(z)A(x)(\log z)^k + A(x)(\log x)^{k-2}.$$

Since $k \geq 2$ this yields

$$T^\spadesuit(x,y,z) \ll \delta^2 s^3 V(z)A(x)(\log x)^k. \tag{12.2}$$

13. Evaluation of $T(x, y, z)$

By (10.3) and (11.1) we get

$$\begin{aligned}|T(x,y,z) - G(x,y,z)V(z)A(x)| \\ &\leq \gamma e^{-s} W(y,z)V(z)A(x)(\log x)^k + R(x,My)(\log x)^k \\ &\leq \gamma c \delta^{-1} e^{-s} V(z)A(x)(\log x)^k + A(x)(\log x)^{k-2}\end{aligned}$$

where we have put

$$G(x,y,z) = \sum_{\substack{d \leq y \\ (d,P)=1}} \mu(d) g(d) \left(\log \frac{x}{d}\right)^k .$$

This is well approximated by the complete sum

$$G(x,z) = \sum_{\substack{d < x \\ (d,P(z))=1}} \mu(d) g(d) \left(\log \frac{x}{d}\right)^k . \tag{13.1}$$

The difference is bounded by

$$|G(x,z) - G(x,y,z)| \leq \big(W(x,z) - W(y,z)\big) \left(\log \frac{x}{y}\right)^k .$$

We have

$$\begin{aligned}W(x,z) - W(y,z) &= \sideset{}{^\flat}\sum_{\substack{y<d\leq x \\ (d,P)=1}} g(d) \leq (\log y)^{-1} \sideset{}{^\flat}\sum_{\substack{y<d\leq x \\ (d,P)=1}} g(d) \log d \\ &\leq (\log y)^{-1} \sideset{}{^\flat}\sum_{\substack{m \leq x \\ (m,P)=1}} g(m) \sum_{y/m < p \leq x/m} g(p) \log p \\ &\leq (\log y)^{-1} W(x,z) \log(cx/y) \ll (\log z)^{-1} \log(x/y)\end{aligned}$$

by (H_5) and (12.1). Hence we conclude that

$$T(x,y,z) - G(x,z)V(z)A(x) \ll \left(\delta^2 s^3 + \delta^{-1} e^{-s}\right) V(z)A(x)(\log x)^k . \tag{13.2}$$

On the right side we may insert $V(z) \leq c\delta^{-1} V(x)$ by (3.4).

14. An auxiliary formula

Collecting the results of Sections 7–13 we obtain

$$T(x) - G(x,z)V(z)A(x) \ll \left(\delta s^3 + \delta^{-2}e^{-s}\right) V(x)A(x)(\log x)^k$$

where s (≥ 1) is any number such that $2\delta(s+1) \leq 1$. We choose $s = 3\log(1/\delta)$ getting

Proposition. *Assuming* (H_1)–(H_5) *we have*

$$T(x) = G(x,z)V(z)A(x) + O\left(\delta(\log\delta)^4 V(x)A(x)(\log x)^k\right) \tag{14.1}$$

where $z = x^\delta$, $V(z)$ *is given by (5.7) and* $G(x,z)$ *by (13.1). The implied constant depends on* c *and* k *only.*

15. Comparing $G(x,z)$ with $F(x,z)$

Until now we have not used the approximation (H_7) of g by f. Now we shall exploit this to replace g in $G(x,z)$ by the nicer function f. Put

$$F(x,z) = \sum_{\substack{d<x \\ (d,P(z))=1}} \mu(d)f(d)\left(\log\frac{x}{d}\right)^k. \tag{15.1}$$

We have

$$|F(x,z) - G(x,z)| \leq U(|f-g|)(\log x)^k$$

where

$$U(u) = \sum_{\substack{d<x \\ (d,P)=1}}^{\flat} u(d).$$

Note the inequality

$$|f(d) - g(d)| \leq \sum_{p|d} |f(p) - g(p)| \max\left(f\left(\tfrac{d}{p}\right), g\left(\tfrac{d}{p}\right)\right).$$

To see this apply the identity

$$f(p_1\dots p_r) - g(p_1\dots p_r) = \sum_{1\leq j\leq r} f(p_1\dots p_{j-1})\left(f(p_j) - g(p_j)\right) g(p_{j+1}\dots p_r).$$

If we place the primes p_i with $f(p_i) \le g(p_i)$ before those with $f(p_i) > g(p_i)$, then for each j we have

$$f(p_1 \dots p_{j-1}) g(p_{j+1} \dots p_r) \le \max\left(f\left(\tfrac{d}{p_j}\right), g\left(\tfrac{d}{p_j}\right)\right)$$

leading to the asserted inequality. This inequality gives

$$U(|f-g|) \le \Delta \sum_{\substack{d<x \\ (d,P)=1}} \max\Big(f(d), g(d)\Big).$$

Since $\max(f,g) \le |f-g| + g$ we obtain (provided $\Delta = \Delta(z) \le \frac{1}{2}$)

$$U(|f-g|) \le 2\Delta U(g).$$

Moreover we have $U(g) = W(x,z) \le c\delta^{-1}$ (see (12.1)). Gathering the above estimates we arrive at

$$|F(x,z) - G(x,z)| \le 2c\delta^{-1}\Delta(z)(\log x)^k.$$

Multiply by $V(z) \le c\delta^{-1}V(x)$ and use (H_7) getting

$$G(x,z)V(z) = F(x,z)V(z) + O\left(\delta V(x)(\log x)^k\right). \tag{15.2}$$

16. Basic estimates for sums in f

Before evaluating $F(x,z)$ we shall derive from (H_8) a few elementary estimates for sums in f twisted by Λ_k and μ. Observe that (H_8) implies

$$f(p) \ll (\log p)^{-2}. \tag{16.1}$$

From the fundamental lemma and (H_8) one can derive the bound

$$\sum_{x<p\le 2x} f(p)\log p \ll 1,$$

and then from (16.1) deduce that

$$\sum_{x<n\le 2x} f(n)\Lambda(n) \ll 1. \tag{16.2}$$

This is rather crude. Now we derive a sharper formula by Tchebyshev's method, namely

$$P_1(x) = \sum_{n\le x} f(n)\Lambda(n) = \log x + O(1). \tag{16.3}$$

On one hand we derive from (H_8) by partial summation

$$\sum_{n\le x} f(n)n = ax + O\left(x(\log x)^{-2}\right),$$

$$\sum_{n\le x} f(n)n\log n = ax\log x - ax + O\left(x(\log x)^{-1}\right).$$

On the other hand the latter sum is equal to

$$\begin{aligned}\sum_{mn\le x} f(mn)mn\Lambda(n) &= \sum_{n\le x} f(n)n\Lambda(n)\left\{\frac{ax}{n} + O\left(\frac{x}{n}\left(\log\frac{x}{n}\right)^{-2}\right)\right\}\\ &= ax\,P_1(x) + O(x)\end{aligned}$$

where for the error term we used (16.2). Comparing the results we get (16.3).

Next we generalize to Λ_k. By the recurrence formula $\Lambda_k = L\Lambda_{k-1} + \Lambda * \Lambda_{k-1}$ we deduce that

$$\begin{aligned}P_k(x) &= \sum_{n\le x} f(n)\Lambda_k(n)\\ &= \sum_{n\le x} f(n)\Lambda_{k-1}(n)\left(\log n + P_1\left(\tfrac{x}{n}\right)\right)\\ &= P_{k-1}(x)\left(\log x + O(1)\right).\end{aligned}$$

Hence for any $k\ge 0$

$$P_k(x) = (\log x)^k + O\left((\log x)^{k-1}\right). \tag{16.4}$$

We generalize (16.4) further by partial summation getting

$$\begin{aligned}P_k^{\ell}(x) &= \sum_{n\le x} f(n)\Lambda_k(n)\left(\log\frac{x}{n}\right)^{\ell}\\ &= \frac{k!\,\ell!}{(k+\ell)!}(\log x)^{k+\ell} + O\left((\log x)^{k+\ell-1}\right)\end{aligned} \tag{16.5}$$

for $k,\ell\ge 0$. Hence we shall derive a formula for

$$Q_k(x) = \sum_{n\le x} f(n)\Lambda_k(n,x). \tag{16.6}$$

Since, by (4.2),

$$Q_k(x) = \sum_{0\le j\le k}\binom{k}{j}P_j^{k-j}(x)$$

we get by (16.5) that for any $k \geq 0$

$$Q_k(x) = (k+1)(\log x)^k + O\left((\log x)^{k-1}\right). \tag{16.7}$$

Finally we consider the sum

$$F_k(x) = \sum_{m \leq x} \mu(m) f(m) \left(\log \frac{x}{m}\right)^k. \tag{16.8}$$

Having quite limited information about f we can only succeed for $k \geq 2$ but this is just what we need (another instance where Bombieri's idea features!). We have

$$\begin{aligned} Q_{k-1}(x) &= \sum_{m \leq x} \mu(m) f(m) \left(\log \frac{x}{m}\right)^{k-1} \sum_{n \leq x/m} f(n) \\ &= a\, F_k(x) + b\, F_{k-1}(x) + O\left((\log x)^{k-2} \log\log x\right) \end{aligned}$$

by (H_8). Here the factor $\log\log x$ is not needed if $k \geq 3$. In the same way we get

$$Q_{k-2}(x) = a\, F_{k-1}(x) + O\left((\log x)^{k-2} \log\log x\right).$$

Combining these two formulae with (16.7) we obtain for $k \geq 2$

$$a\, F_k(x) = k(\log x)^{k-1} + O\left((\log x)^{k-2} \log\log x\right). \tag{16.9}$$

17. The constant a_f

The leading coefficient $a = a_f$ in (H_8) can be expressed as an infinite product. Indeed, for $s > 1$ we have

$$\begin{aligned} \prod_p \left(1 - p^{-s}\right) \left(1 - f(p) p^{1-s}\right)^{-1} &= \zeta(s)^{-1} \sum_n f(n) n^{1-s} \\ &= a + \zeta(s)^{-1} \sum_n \left(f(n) n - a\right) n^{-s} \\ &= a + O\left(\zeta(s)^{-1}\right) \end{aligned}$$

since the series is uniformly bounded. Hence letting $s \to 1$ we get

$$a_f = \prod_p \left(1 - p^{-1}\right) \left(1 - f(p)\right)^{-1}. \tag{17.1}$$

18. Evaluation of $F(x, z)$

By (15.1) and (16.8) we have (Eratosthenes sieve)

$$F(x,z) = \sum_{\substack{d \mid P(z)^\infty \\ d \le x}} f(d) F_k\left(\tfrac{x}{d}\right),$$

where $d \mid P^\infty$ means that every prime factor of d divides P. We have

$$a_f F_k\left(\tfrac{x}{d}\right) = k(\log x)^{k-1} + O\left((\log x)^{k-2} \log(d \log x)\right) \tag{18.1}$$

by (16.9). Denote

$$V_f(z) = \prod_{p<z} (1 - f(p)), \tag{18.2}$$

so we have

$$\sum_{d \mid P(z)^\infty} f(d) = V_f(z)^{-1}$$

and

$$\begin{aligned} \sum_{d \mid P(z)^\infty} f(d) \log d &= \sum_{d \mid P(z)^\infty} f(d) \sum_{\ell \mid d} \Lambda(\ell) = V_f(z)^{-1} \sum_{\ell \mid P(z)^\infty} f(\ell)\Lambda(\ell) \\ &= V_f(z)^{-1} \sum_{p<z} \frac{f(p)}{1-f(p)} \log p \ll V_f(z)^{-1} \log z \end{aligned} \tag{18.3}$$

by (16.1) and (16.2). Hence

$$\sum_{\substack{d \mid P(z)^\infty \\ d > x}} f(d) \le \sum_{d \mid P(z)^\infty} f(d) \frac{\log d}{\log x} \ll V_f(z)^{-1} \frac{\log z}{\log x}. \tag{18.4}$$

From the estimates (18.1), (18.3), (18.4), we conclude that for $k \ge 2$

$$a_f F(x,z) V_f(z) = k(\log x)^{k-1} + O\left((\log x)^{k-2} \log(z \log x)\right). \tag{18.5}$$

Observe that the main term does not depend on z.

19. Proof of Theorem

First we compare $a_f V_f(z)$ to $H^{-1}V(z)$. Recall that a_f, H are given by the infinite products (17.1), (3.2) respectively and $V_f(z)$, $V(z)$ by the finite products (18.2), (5.7) respectively. Hence we get

$$a_f V_f(z) H V(z)^{-1} = \prod_{p \ge z} \frac{1-g(p)}{1-f(p)} = 1 + O(\Delta(z)). \tag{19.1}$$

Inserting (19.1) into (18.5) we get

$$F(x,z)V(z) = kH(\log x)^{k-1}\left\{1 + O\left(\Delta(z) + \frac{\log(z\log x)}{\log x}\right)\right\}. \tag{19.2}$$

Notice that for any x

$$H = V(x)\prod_{p<x}\left(1-\tfrac{1}{p}\right)^{-1}\prod_{p\ge x}(1-f(p))\left(1-\tfrac{1}{p}\right)^{-1}\prod_{p\ge x}\frac{1-g(p)}{1-f(p)}.$$

Hence by (16.3) and (H_7) we have $H \asymp V(x)\log x$, so we can replace $V(x)$ by $H(\log x)^{-1}$ in the former error terms. Inserting (19.2) in (15.2) we get

$$G(x,z)V(z) = kH(\log x)^{k-1}\left\{1 + O\left(\delta + \Delta(z) + \frac{\log\log x}{\log x}\right)\right\}. \tag{19.3}$$

Inserting (19.3) in (14.1) we get (note that $\delta \ge \log 2/\log x$)

$$T(x) = kHA(x)(\log x)^{k-1}\left\{1 + O\left(\delta(\log\delta)^4\right)\right\}. \tag{19.4}$$

Finally replace $T(x)$ by $S(x)$ using the inequalities from Section 6. This completes the proof of our theorem.

20. Some applications

We conclude by illustrating some instances where our theorem can be applied. We have already mentioned Bombieri's example, the case of the twin prime problem. The analogous result for the Goldbach problem is the estimate

$$\sum_{m+n=N} \Lambda_k(m)\Lambda_\ell(n) \sim k\ell H N(\log N)^{k+\ell-2}$$

where $H = \mathfrak{S}(N)$ is the usual Goldbach constant

$$\mathfrak{S}(N) = 2\prod_{p>2}\left(1-\frac{1}{(p-1)^2}\right)\prod_{p\mid N}\left(1-\frac{1}{(p-1)}\right)^{-1}.$$

This is valid for N even and $k \geq 2$ provided that we assume the Elliott-Halberstam conjecture in order to have (H_1). Here we take $g(d) = 1/\phi(d)$ for $(d, N) = 1$ and $g(d) = 0$ elsewhere and $f(d) = 1/d$ for all d. All assumptions other than (H_1) are easily verified.

One may consider ternary analogues of the previous example and get unconditional results for the sum

$$\sum_{\ell+m+n=N} \alpha_\ell \beta_m \Lambda_k(n)$$

for quite general sequences α, β, with one well-distributed over arithmetic progressions to small moduli as in the Siegel-Walfisz theorem. In this case the hypothesis (H_1) follows from theorems of Barban-Davenport-Halberstam type. However, in the case of this example, other techniques are more successful.

Even for the simpler case where the sequence is an arithmetic progression, one gets results which do not seem to follow by L-function techniques. Thus for $k \geq 2$ we have, see [F],

$$\sum_{\substack{n \leq x \\ n \equiv a (\bmod q)}} \Lambda_k(n) \sim \frac{kx}{\phi(q)} (\log x)^{k-1}$$

if $(a, q) = 1$ and $\log q = o(\log x)$ as $x \to \infty$. In this case we have $g(d) = 1/d$ for $(d, q) = 1$ and 0 otherwise so we can take $f(d) = 1/d$ for all d.

An interesting example is given in [FoIw] where the hypotheses of Bombieri's sieve are verified for the sequence

$$a(n) = \sum_{\ell^2+m^2=n} \lambda_\ell$$

with fairly general weights $\lambda_\ell \geq 0$. For example it follows by combining both works that

$$\sum_{\ell^2+m^2 \leq x} \Lambda(\ell) \Lambda_k(\ell^2 + m^2) \sim kHx(\log x)^{k-1}$$

provided $k \geq 2$ where

$$H = 2 \prod_p \left(1 - \frac{\chi(p)}{(p-1)(p-\chi(p))}\right)$$

with χ the non-principal character to modulus 4.

References

[B1] E. Bombieri, *On twin almost primes*, Acta Arith **28** (1975), 177–193; ibid. **28** (1976), 457–461.

[B2] ———, *The asymptotic sieve*, Mem. Accad. Naz. dei XL **1/2** (1976), 243–269.

[FoIw] E. Fouvry and H. Iwaniec, *Gaussian primes*, to appear.

[F] J. Friedlander, *Selberg's formula and Siegel's zero*, Recent Progress in Analytic Number Theory. ed. H. Halberstam and C. Hooley, **I**, pp 15–23, Academic Press, London. 1981.

[FI] J. Friedlander and H. Iwaniec, *On Bombieri's asymptotic sieve*, Ann. Sc. Norm Sup (Pisa) **V** (1978), 719–756.

[HR] H. Halberstam and H.-E. Richert, *Sieve Methods*, Academic Press, London, 1974.

[I] H. Iwaniec, *Rosser's sieve*, Acta Arith **36** (1980), 171–202.

[S] A. Selberg, *An elementary proof of the prime number theorem*, Ann. Math. **50** (1949), 305–313.

John Friedlander
Department of Mathematics
University of Toronto
Scarborough College
Toronto, Ontario
CANADA M1C 1A4
frdlndr@math.toronto.edu

Henryk Iwaniec
Department of Mathematics
Rutgers University
New Brunswick, NJ 08903-2101
iwaniec@math.toronto.edu

$\mathbf{B_h}$ sequences

S. W. Graham[1]

Dedicated to Heini Halberstam on the occasion of his retirement

Abstract. Let $\mathcal{A}$ be a set (finite or infinite) of natural numbers, and let a_i denote a generic element of $\mathcal{A}$. We say that $\mathcal{A}$ is a B_h sequence if the sums of the form $a_1 + \ldots + a_h \quad (a_1 \leq \ldots \leq a_h)$ are all distinct. We give an expository account of upper bounds for B_h sequences. We also give a new upper bound for B_3 sequences.

1. Introduction

Let $\mathcal{A}$ be a set (finite or infinite) of natural numbers, and let a_i denote a generic element of $\mathcal{A}$. We say that $\mathcal{A}$ is a B_h sequence if the sums of the form

$$a_1 + \ldots + a_h \qquad (a_1 \leq \ldots \leq a_h) \tag{1}$$

are all distinct. S. Sidon [18] was the first to consider B_2 sequences, and, consequently, B_2 sequences are sometimes called *Sidon sequences.* Sidon was motivated by questions on Fourier series. For example, suppose that $\mathcal{A} = \{\lambda_1, \lambda_2, \ldots\}$ is an infinite B_2 sequence and that $\sum_n (\epsilon'^2_n + \epsilon''^2_n) < \infty$. Sidon proved that there is a continuous function f with Fourier series $\sum_{n=1}^{\infty}(a_n \cos(nx) + b_n \sin(nx))$ and $a_{\lambda_n} = \epsilon'_n, b_{\lambda_n} = \epsilon''_n$.

In 1941, P. Erdős and P. Turán [4] (see also [8], pp. 86–87) gave the first upper bound for finite B_2 sequences. They showed that if $\mathcal{A}$ is a B_2 sequence with $\mathcal{A} \subseteq [1, N]$, then

$$|\mathcal{A}| \leq N^{1/2} + O(N^{1/4}). \tag{2}$$

Their proof used moving averages and the Cauchy-Schwarz inequality. The same ideas are used in the Weyl-van der Corput Lemma of exponential sum theory, and B. Lindström [16] pointed out that this lemma can be used to shorten the proof of (2). With a little care, one can prove the following more precise result.

[1]The author was supported in part by a grant from the National Security Agency.

Theorem 1. *If $\mathcal{A} \subseteq [1, N]$ is a B_2 sequence, then*

$$|\mathcal{A}| \leq N^{1/2} + N^{1/4} + 1.$$

Proof. We first state the Weyl-van der Corput Lemma in the form that we need. This states that if $\xi(n)$ is a real valued function supported on the interval $[1, N]$ and if H is a positive integer, then

$$\left|\sum_n \xi(n)\right|^2 \leq \frac{N+H-1}{H} \sum_{|h|<H} \left(1 - \frac{|h|}{H}\right) \sum_n \xi(n)\xi(n+h). \tag{3}$$

A proof of this can be found in [7], Lemma 2.5. (Note that in the second and third displayed equations of this proof, one can actually write $|I| + H - 1$ instead of $|I| + H$.)

Now we apply Weyl-van der Corput with ξ equal to the characteristic function of $\mathcal{A}$. Because $\mathcal{A}$ is B_2, the inner sum $\sum_n \xi(n)\xi(n+h)$ is $|\mathcal{A}|$ if $h = 0$ and ≤ 1 otherwise. Thus

$$|\mathcal{A}|^2 \leq \frac{N+H-1}{H}(|\mathcal{A}| + H - 1).$$

We take $H = [N^{3/4}] + 1$, so that $H - 1 \leq N^{3/4}$ and $1/H < N^{-3/4}$. This gives $|\mathcal{A}|^2 - (N^{1/4}+1)|\mathcal{A}| \leq N + N^{3/4}$. Using completion of the square and the fact that $N \geq 1$, we deduce that

$$\left(|\mathcal{A}| - \frac{1}{2}(N^{1/4}+1)\right)^2 \leq \left(N^{1/2} + \frac{1}{2}N^{1/4} + \frac{1}{2}\right)^2,$$

and the Theorem follows.

The bound of Theorem 1 was first proved by Lindström [15], who used a combinatorial argument.

In 1939, J. Singer [19] showed that if q is a prime and $N = q^2 + q + 1$, then there is a set $\{a_1, \ldots, a_{q+1}\}$ such that the differences $a_i - a_j (i \neq j)$ form a complete set of non-zero residues $\pmod N$. It is easy to see that this set is also a B_2 sequence; consequently, there is a B_2 sequence $\mathcal{A} \subseteq [1, N]$ with $q+1 > N^{1/2}$ elements. Therefore, the Erdős-Turán bound (2) is asymptotically best possible. Erdős conjectures that the bound $|\mathcal{A}| \leq N^{1/2} + O(1)$ is true, but this appears to be very difficult.

2. Infinite B_2 Sequences

Now suppose that $\mathcal{A}$ is an infinite B_2 sequence, and let $A(n)$ denote the number of elements a_i in $\mathcal{A}$ with $a_i \leq n$. What upper bounds can we give for $A(n)$? Of course, we can consider the finite set $\mathcal{A}\cap[1,n]$ and use the Erdős-Turán bound. However, the fact that $\mathcal{A}\cap[1,n]$ is embedded in an infinite B_2 sequence allows us to say more. The following result is due to Erdős; he communicated it to Stöhr for inclusion in his 1955 survey [2]. An account also appears in the book of Halberstam and Roth [8, pages 89–90].

Theorem 2. *If $\mathcal{A}$ is an infinite B_2 sequence, then*

$$\liminf_{n\to\infty} \frac{A(n)}{n^{1/2}(\log n)^{-1/2}} < \infty.$$

Proof. For $l = 1,\dots,N$, let $\mathfrak{I}_l = ((l-1)N, N]$ and $\mathcal{A}_l = \mathcal{A}\cap\mathfrak{I}_l$. Let $A_l = |\mathcal{A}_l|$, and let ξ_l be the characteristic function of $\mathcal{A}_l$. By inequality (3) with $H = N$,

$$A_l^2 \le 2\sum_{|h|<N}\left(1-\frac{|h|}{N}\right)\sum_n \xi_l(n)\xi(n+h).$$

Summing over l, we get

$$\sum_l A_l^2 \le 2\sum_{|h|<N}\left(1-\frac{|h|}{N}\right)\sum_n \xi(n)\xi(n+h)$$

The inner sum is at most $A(N^2) \ll N$ when $h = 0$ and at most 1 for other values of h; therefore

$$\sum_l A_l^2 \ll N. \tag{4}$$

By Cauchy's inequality,

$$\sum_l A_l^2 \ge \left(\sum_l \frac{A_l}{l^{1/2}}\right)^2\left(\sum_l \frac{1}{l}\right)^{-1}. \tag{5}$$

Now

$$\sum_l \frac{A_l}{l^{1/2}} = \sum_l \frac{A(lN)-A((l-1)N)}{l^{1/2}} \ge \sum_l A(lN)\left(\frac{1}{l^{1/2}}-\frac{1}{(l+1)^{1/2}}\right).$$

Now let

$$\tau = \min_{1\le l\le N}\frac{A(lN)}{(lN)^{1/2}}.$$

Then

$$\begin{aligned}\sum_l \frac{A_l}{l^{1/2}} &\geq \tau N^{1/2} \sum_{l=1}^{N} l^{1/2} \left(\frac{1}{l^{1/2}} - \frac{1}{(l+1)^{1/2}} \right) \\ &\geq \tau N^{1/2} \sum_{l=1}^{N} l^{1/2} \int_l^{l+1} \frac{1}{2} u^{-3/2}\, du \\ &\geq \frac{1}{2} \tau N^{1/2} \sum_{l=1}^{N} l^{-1} \\ &\geq \frac{1}{2} \tau N^{1/2} \log N.\end{aligned}$$

We combine this with (4) and (5) to get

$$N \gg \sum_l A_l^2 \gg \tau^2 N \log N.$$

We therefore conclude that

$$\tau \ll (\log N)^{-1/2},$$

and this implies the desired result.

3. Uniform Distribution

The theorem in the previous section depends on a simple yet very useful observation. This observation is the fact that

$$\sum_l \xi_l(n) \leq \xi(n),$$

where ξ is the characteristic function of $\mathcal{A}$ and ξ_l is the characteristic function of $\mathcal{A} \cap ((l-1)N, lN]$. Erdős and R. Freud [5] also applied this fact in the context of finite B_2 sequences. More specifically, they used it to prove that a maximal B_2 sequence is uniformly distributed.

Theorem 3. (Erdős-Freud) *Suppose $\mathcal{A}$ is a B_2 sequence with $\mathcal{A} \subseteq [1, N]$, and let A denote the number of elements in $\mathcal{A}$. Suppose also that $A = N^{1/2} + O(N^{1/4})$. Let*

$$0 = N_0 < N_1 < \ldots < N_r = N$$

be a partition of the interval $[0, N]$, and let $\mathfrak{I}_j = (N_{j-1}, N_j]$. Let α_j and β_j be defined by the relations

$$|\mathfrak{I}_j| = \alpha_j N \text{ and } |\mathcal{A} \cap \mathfrak{I}_j| = \beta_j A.$$

Then

$$\beta_j = \alpha_j + O\left(N^{-1/8}\alpha_j^{1/2}(\sum_i \alpha_i^{-1})^{1/2}\right).$$

Proof. Let ξ be the characteristic function of $\mathcal{A}$, and let ξ_i be the characteristic function of $\mathcal{A} \cap \mathfrak{I}_i$. By inequality (3),

$$\begin{aligned}\frac{(\beta_i A)^2}{\alpha_i N + H} &\leq \frac{1}{H} \sum_{|h|<H} \left(1 - \frac{|h|}{H}\right) \sum_n \xi_i(n)\xi_i(n+h) \\ &\leq \frac{1}{H} \sum_{|h|<H} \left(1 - \frac{|h|}{H}\right) \sum_n \xi_i(n)\xi(n+h).\end{aligned}$$

Summing this last inequality over i, we get

$$\sum_i \frac{(\beta_i A)^2}{\alpha_i N + H} \leq \frac{1}{H} \sum_{|h|<H} \left(1 - \frac{|h|}{H}\right) \sum_n \xi(n)\xi(n+h). \tag{6}$$

On the right–hand side of (6), the term with $h = 0$ contributes $A \ll N^{1/2}$. The other terms contribute at most 1. Take $H = [N^{3/4}]$; then

$$\alpha_i N + H = \alpha_i N \cdot \{1 + O(\alpha_i^{-1} N^{-1/4})\}.$$

We shall momentarily assume that $\alpha_i > 2N^{-1/4}$ for all i; therefore,

$$\{1 + O(\alpha_i^{-1} N^{-1/4})\}^{-1} = \{1 + O(\alpha_i^{-1} N^{-1/4})\}.$$

We may then rewrite (6) as

$$\frac{A^2}{N} \sum_i \frac{\beta_i^2}{\alpha_i} \{1 + O(\alpha_i^{-1} N^{-1/4})\} \leq 1 + O(N^{-1/4}).$$

Since $A^2/N = 1 + O(N^{-1/4})$, this gives

$$\sum_i \frac{\beta_i^2}{\alpha_i} \leq 1 + O(N^{-1/4} \sum_i \beta_i^2 \alpha_i^{-2}). \tag{7}$$

Now $\mathcal{A} \cap \mathfrak{I}_i$ is a B_2 sequence in an interval of length $\alpha_i N$, so

$$\beta_i A = |\mathcal{A} \cap \mathfrak{I}_i| \ll (\alpha_i N)^{1/2},$$

and thus $\beta_i \ll \alpha^{1/2}$. Combining this observation with (7), we see that

$$\sum_i \frac{\beta_i^2}{\alpha_i} \le 1 + O(N^{-1/4} \sum_i \alpha_i^{-1}). \tag{8}$$

Now we write $\beta_i = \alpha_i + \epsilon_i$, so that

$$\beta_i^2 \alpha_i^{-1} = \alpha_i + 2\epsilon_i + \epsilon_i^2 \alpha_i^{-1}.$$

Since $\sum \alpha_i = \sum \beta_i = 1$, we see that $\sum \epsilon_i = 0$ and

$$\sum_i \epsilon_i^2 \alpha_i^{-1} \ll N^{-1/4} \sum_i \alpha_i^{-1}.$$

If we ignore all the terms on the left except $i = j$, we see that

$$\epsilon_j^2 \ll N^{-1/4} \alpha_j \sum_i \alpha_i^{-1}.$$

The theorem now follows in the case where $\alpha_i > 2N^{-1/4}$ for all i. If, on the other hand, there is some i such that $\alpha_i \le 2N^{-1/4}$, then the error term in the theorem is $\gg \alpha_j^{1/2}$, and the theorem follows trivially from Theorem 1.

4. B_h sequences for $h > 2$: bounds for even h

Now suppose that $\mathcal{A} \subseteq [1, N]$ is a B_h sequence for some $h > 2$. For comparison purposes, it is convenient to note the following simple upper bound for $|\mathcal{A}|$. Consider the set of sums

$$h\mathcal{A} = \{a_1 + a_2 + \ldots + a_h : a_i \in \mathcal{A}, a_1 \le a_2 \le \ldots a_h\}.$$

By definition, the sums in $h\mathcal{A}$ are all distinct. On the other hand, these sums are all contained in the interval $[1, hN]$. Therefore

$$\frac{|\mathcal{A}|^h}{h!} \le \binom{|\mathcal{A}| + h - 1}{h} \le hN,$$

so

$$|\mathcal{A}| \le (h \cdot h!)^{1/h} N^{1/h}. \tag{9}$$

In the case $h = 4$, Lindström [16] proved the stronger bound

$$|\mathcal{A}| \le (8N)^{1/4} + O(N^{1/8}). \tag{10}$$

Lindström proved this by analyzing the set of differences

$$\{a_1 - a_2 : a_1, a_2 \in \mathcal{A}, a_1 \neq a_2\}.$$

Alternatively, one can prove (10) by analyzing $2\mathcal{A}$. Note that $2\mathcal{A} \subseteq [1, 2N]$, so if $2\mathcal{A}$ were a B_2 sequence, then we would have

$$|2\mathcal{A}| \leq (2N)^{1/2} + O(N^{1/4}), \tag{11}$$

and (10) would follow. Although it is not true that $2\mathcal{A}$ is a B_2 sequence (for example, $(a_1+a_2)+(a_3+a_4) = (a_1+a_3)+(a_2+a_4) = (a_1+a_4)+(a_2+a_3)$), we can still use the Weyl-van der Corput inequality to prove (11) and consequently (10).

Xing–De Jia [12] generalized Lindström's result to arbitrary even h, and we present his theorem here.

Theorem 4. (Jia) *Suppose $\mathcal{A}$ is a B_{2r} sequence and $\mathcal{A} \subseteq [1, N]$. Then*

$$|\mathcal{A}| \leq r^{1/2r}(r!)^{1/r} N^{1/2r} + O_r(N^{1/4r}). \tag{12}$$

Proof. As before, let A denote the number of elements in $\mathcal{A}$. Let

$$\mathcal{B} = r\mathcal{A} = \{a_1 + a_2 + \ldots + a_r : a_i \in \mathcal{A}, a_1 \leq a_2 \leq \ldots \leq a_r\},$$

and set $B = |\mathcal{B}|$. Since a B_{2r} sequence is *a priori* a B_r sequence, we see that

$$B = \binom{A + r - 1}{r} \geq \frac{A^r}{r!}, \tag{13}$$

and we can derive an upper bound for A from an upper bound for B.

Apply the Weyl-van der Corput inequality (3) with N replaced by rN and ξ the characteristic function of $\mathcal{B}$. Then

$$B^2 \leq \frac{rN + H}{H} \sum_{|h|<H} \left(1 - \frac{|h|}{H}\right) S_h,$$

where H is an integer to be chosen later, and

$$S_h = \sum_n \xi(n)\xi(n-h).$$

Clearly $S_0 = B$. Assume that $h \neq 0$. Then S_h counts the number of pairs $b, b' \in \mathcal{B}$ with $b - b' = h$. Let

$$b = a_1 + \ldots + a_r, \tag{16}$$

$$b' = a'_1 + \ldots + a'_r \tag{17}$$

be representations of b, b' as sums of r elements of $\mathcal{A}$. Assume that the a_i's and a_i's are given in increasing order; thus, the representations in (16) and (17) are unique. We now write

$$S_h = S'_h + S''_h,$$

where S'_h counts the pairs b, b' in which the representations in (16) and (17) have a common summand, and S''_h counts the remaining pairs. Clearly

$$\sum_{h\neq 0} S'_h \ll_r A^{2r-1} \ll_r N^{(2r-1)/2r}$$

by (9). Now consider S''_h. If $b - b' = b'' - b''' = h$, then $b + b''' = b' + b''$; i.e.,

$$a_1 + \ldots + a_r + a'''_1 + \ldots + a'''_r = a'_1 + \ldots + a'_r + a''_1 + \ldots + a''_r.$$

Since $\mathcal{A}$ is a B_{2r} sequence, this can happen only if

$$\{a_1, \ldots, a_r\} = \{a''_1, \ldots, a''_r\} \text{ and } \{a'_1, \ldots, a'_r\} = \{a'''_1, \ldots, a'''_r\}.$$

The ordering conditions force us to have $a_1 = a''_1, a_2 = a''_2$, *et cetera*. Thus $S''_h \leq 1$, and so

$$B^2 \leq \left(\frac{rN + H}{H}\right)(B + H + O_r(N^{1-1/2r})). \tag{18}$$

Now we first get a crude upper bound for B. Since $\mathcal{B} \subseteq [1, rN]$, we have $B \leq rN$. Using this on the right-hand side of (18) and taking $H = N$ gives $B \ll_r N^{1/2} \ll_r N^{1-1/2r}$. Putting this crude bound back into (18) gives

$$B^2 \leq rN + O_r(H + N^{2-1/2r}H^{-1} + N^{1-1/2r}).$$

We take $H = [N^{1-1/4r}]$ to get

$$B^2 \leq rN + O_r(N^{1-1/4r}).$$

We then combine this with (13) and obtain

$$A^{2r} \leq (r!)^2 rN \cdot \{1 + O_r(N^{-1/4r})\}.$$

Taking $2r$-th roots gives the result.

5. B_h sequences for $h > 2$: bounds for odd h

The key idea in Jia's proof is to consider $r\mathcal{A}$, where r is half of h. But what if h is odd? For example, what is "half" of 3? In this context, it turns out that "half" of 3 is 2; more generally, "half" of $2k-1$ is k.

The following result was discovered independently by S. Chen [2] and S. W. Graham (unpublished).

Theorem 5. *If $\mathcal{A}$ is a B_{2k-1} sequence and $\mathcal{A} \subseteq [1, N]$, then*

$$|\mathcal{A}| \leq (k!)^{2/(2k-1)} N^{1/(2k-1)} \cdot \{1 + O(N^{-1/(4k-2)})\}.$$

We have not attempted to make the estimate uniform in k. If it were desired, however, one could easily obtain such uniformity.

For the proof, we suppose that $\mathcal{A}$ is a B_{2k-1} sequence with $\mathcal{A} \subseteq [1, N]$. We use the symbol $[\mathcal{A}]^k$ to denote the set of unordered k-tuples of $\mathcal{A}$; i.e.,

$$[\mathcal{A}]^k = \{(a_1, a_2, \ldots, a_k) : a_1 < a_2 < \ldots < a_k, a_1, a_2, \ldots, a_k \in \mathcal{A}\}.$$

This notation is commonly used in Ramsey Theory (cf. [6], p. xi). For $\mathbf{a} = (a_1, \ldots, a_k) \in [\mathcal{A}]^k$, we define $\sigma(\mathbf{a}) = a_1 + \ldots + a_k$.

We now set

$$\mathcal{B} = \{\sigma(\mathbf{a}) : \mathbf{a} \in [\mathcal{A}]^k\} = \{a_1 + a_2 + \ldots + a_k : a_1 < a_2 < \ldots < a_k, a_i \in \mathcal{A}\},$$

and $B = |\mathcal{B}|$. Then $\mathcal{B} \subseteq [1, kN]$ and

$$B = \binom{A}{k} = \frac{1}{k!} A^k + O(A^{k-1}). \tag{19}$$

Now let ξ be the characteristic function of $\mathcal{B}$; i.e., $\xi(n) = 1$ if $n \in \mathcal{B}$ and $\xi(n) = 0$ otherwise. Let H be a positive integer to be chosen later. By the Weyl-van der Corput inequality (3), we get

$$\begin{aligned} B^2 &\leq \frac{kN+H}{H} \sum_{|h|<H} \left(1 - \frac{|h|}{H}\right) \sum_n \xi(n)\xi(n-h) \\ &\leq \frac{kN+H}{H} \sum_{|h|<H} \left(1 - \frac{|h|}{H}\right) E(h), \end{aligned} \tag{21}$$

where $E(h) = |\mathcal{E}(h)|$ and $\mathcal{E}(h) = \{n : n \in \mathcal{B} \text{ and } n+h \in \mathcal{B}\}$. The set $\mathcal{E}(h)$ may also be written as

$$\mathcal{E}(h) = \{(a_1, \ldots, a_{2k}) : (a_1, \ldots, a_k) \in [\mathcal{A}]^k, (a_{k+1}, \ldots, a_{2k}) \in [\mathcal{A}]^k,$$
$$a_1 + \ldots + a_k - a_{k+1} - \ldots - a_{2k} = h\}.$$

Note that $E(0) = B$. For $h \neq 0$, we write $\mathcal{E}(h) = \mathcal{E}'(h) \cup \mathcal{E}''(h)$, where $\mathcal{E}'(h)$ consists of those $2k$-tuples in $\mathcal{E}(h)$ with all of $a_1, \ldots, a_{2k}$ distinct, and $\mathcal{E}''(h)$ consists of the remaining $2k$-tuples. In accordance with our notational conventions, we use $E'(h)$ and $E''(h)$ to denote the cardinalities of $\mathcal{E}'(h)$ and $\mathcal{E}''(h)$ respectively.

The contribution of E'' can be estimated immediately, for $\sum_{|h|<H} E''(h)$ is bounded by the number of $2k$-tuples $(a_1, \ldots, a_{2k})$ with at least two of the a_i's equal. The latter is clearly $\ll A^{2k-1}$, so

$$\sum_{1\le|h|\le H} \left(1 - \frac{|h|}{H}\right) E''(h) \ll A^{2k-1}. \tag{22}$$

Now we consider the contribution of the E' terms. Let a be a fixed element of $\mathcal{A}$, and let $E'(h,a)$ be the number of $2k$-tuples $(a_1, \ldots, a_{2k})$ in $\mathcal{E}'(h)$ with a occurring in one of the first k coordinates. Observe that

$$\sum_{a\in\mathcal{A}} E'(h,a) = kE'(h). \tag{23}$$

We claim that for every $h \neq 0$ and every $a \in \mathcal{A}$, $E'(h,a) \le 1$. Suppose that $\mathbf{a} = (a_1, \ldots, a_{2k})$ and $\mathbf{b} = (b_1, \ldots, b_{2k})$ are $2k$-tuples counted in $E'(h,a)$; then there exist i, j with $1 \le i, j \le k$ and $a = a_i = b_j$. Thus

$$\sum_{\substack{\nu=1\\ \nu\neq i}}^{k} a_\nu - \sum_{\nu=k+1}^{2k} a_\nu = \sum_{\substack{\mu=1\\ \mu\neq j}}^{k} b_\mu - \sum_{\mu=k+1}^{2k} b_\mu = h,$$

so

$$\sum_{\substack{\nu=1\\ \nu\neq i}}^{k} a_\nu + \sum_{\mu=k+1}^{2k} b_\mu = \sum_{\substack{\mu=1\\ \mu\neq j}}^{k} b_\mu + \sum_{\nu=k+1}^{2k} a_\nu.$$

Now $\mathcal{A}$ is a B_{2k-1} sequence, so we conclude that

$$\{a_1, \ldots, a_{i-1}, a_{i+1}, \ldots, a_k, b_{k+1}, \ldots, b_{2k}\} = \{b_1, \ldots, b_{j-1}, b_{j+1}, \ldots, b_k, a_{k+1}, \ldots, a_{2k}\}.$$

From our hypotheses, we know that all the a_ν's are distinct, as are all the b_μ's. Therefore, we must have

$$\{a_1, \ldots, a_{i-1}, a_{i+1}, \ldots, a_k\} = \{b_1, \ldots, b_{j-1}, b_{j+1}, \ldots, b_k\}$$

and

$$\{a_{k+1}, \ldots, a_{2k}\} = \{b_{k+1}, \ldots, b_{2k}\}.$$

From the definition of $\mathcal{B}$, we also have the ordering conditions $a_1 < \ldots < a_k$, etc. From these conditions, we conclude that $a_\nu = b_\nu$ for $1 \le \nu \le 2k$ and that $i = j$. This completes the proof of our claim.

The above together with (23) yields

$$E'(h) \le \frac{1}{k} A. \tag{24}$$

Combining (21), (22), and (24), we find that

$$B^2 \le \frac{kN + H}{H}\left(B + \frac{H}{k}A + O(A^{2k-1})\right). \tag{25}$$

Now we combine (19) and (25) and obtain

$$A^{2k} \le (k!)^2 NA + O(HA + NH^{-1}A^{2k-1} + A^{2k-1}).$$

Setting $H = [N^{1/2}A^{k-1}]$ and using the trivial bound $A \ll N^{1/(2k-1)}$ gives

$$A^{2k-1} \le (k!)^2 N \cdot \{1 + O(N^{-1/(4k-2)})\}.$$

The proof is completed by taking $(2k-1)$st roots.

6. Upper bounds for B_3 sequences

When $\mathcal{A}$ is a B_3 sequence, Theorem 5 gives $|\mathcal{A}| \le 4^{1/3}N^{1/3} + O(N^{1/6})$, but this is weaker than a result of An Ping Li [14], who proved that

$$|\mathcal{A}| \le \left(4 - \frac{2\log^2 2}{3\log^2 N}\right)^{1/3} N^{1/3} + O(1).$$

In this paper, we establish the following sharper result.

Theorem 6. *Let $c = 3.99561029143$. If $\mathcal{A}$ is a B_3 sequence and $\mathcal{A} \subseteq [1, N]$, then*

$$|\mathcal{A}| \le c^{1/3}N^{1/3} + O(1).$$

As we shall see in the proof, c may in fact be replaced by an algebraic number c_0 with $c - 10^{-11} < c_0 < c$. In terms of rational approximations, note that

$$4 - \frac{1}{227} < c < 4 - \frac{1}{228}.$$

The improvement in the constant is small, but the point of the theorem is that 4 is not the limit of Li's method.

To get some insight into Li's proof, consider the following variation of the trivial bound (9). Assume that $\mathcal{A}$ is a B_3 sequence, and let

$$\mathcal{B} = \{x_1 + x_2 - x_3 : x_i \in \mathcal{A}, x_1 < x_2, x_1 \neq x_3, x_2 \neq x_3\}.$$

We claim that there is no duplication among the elements of $\mathcal{B}$. For suppose that

$$x_1 + x_2 - x_3 = x_1' + x_2' - x_3'.$$

Then $x_1 + x_2 + x_3' = x_1' + x_2' + x_3$. By the B_3 condition, $\{x_1, x_2, x_3'\} = \{x_1', x_2', x_3\}$. Since x_3 is not equal to either x_1 or x_2, we must have $x_3 = x_3'$. Moreover, the ordering conditions $x_1 < x_2$ and $x_1' < x_2'$ force us to have $x_1 = x_1'$ and $x_2 = x_2'$.

This lack of duplication means that $\mathcal{B}$ has $A(A-1)(A-2)/2$ elements. Since $\mathcal{B} \subseteq (-N, 2N)$, we conclude that $A(A-1)(A-2)/2 < 3N$; i.e.,

$$|\mathcal{A}| \leq (6N)^{1/3} + O(1). \tag{26}$$

By comparison, (9) gives the upper bound $(18N)^{1/3}$.

In his proof, Li considers the set $\mathcal{P} = \mathcal{B} \cap (0, N)$, and he shows that the set $\mathcal{P}$ has more than $A^3/4$ elements. In other words, he shows that the middle third of the interval $(-N, 2N)$ contains more than half of the elements of $\mathcal{B}$.

Now we begin the proof. We may assume that $N/2 \notin \mathcal{A}$; for, in the contrary case, we may delete $N/2$ from $\mathcal{A}$ and absorb it into the "O"–term. For $\mathbf{x} = (x_1, x_2, x_3) \in [\mathcal{A}]^3$, we define

$$f_1(\mathbf{x}) = -x_1 + x_2 + x_3, f_2(\mathbf{x}) = x_1 - x_2 + x_3, \text{ and } f_3(\mathbf{x}) = x_1 + x_2 - x_3.$$

For $i = 1, 2,$ or 3, we define

$$\mathcal{P}_i = \left\{f_i(\mathbf{x}) : \mathbf{x} \in [\mathcal{A}]^3 \text{ and } 0 < f_i(\mathbf{x}) < N\right\}.$$

Then the set $\mathcal{P}$ defined above may be written as $\mathcal{P} = \mathcal{P}_1 \cup \mathcal{P}_2 \cup \mathcal{P}_3$. The argument in the paragraph before (26) shows that $\mathcal{P}_1, \mathcal{P}_2$, and $\mathcal{P}_3$ are mutually disjoint. All of these sets lie in $(0, N)$, so

$$|\mathcal{P}_1| + |\mathcal{P}_2| + |\mathcal{P}_3| < N. \tag{27}$$

Now let $\mathcal{B}_i = \left\{\mathbf{x} \in [\mathcal{A}]^3 : f_i(\mathbf{x}) \in \mathcal{P}_i\right\}$ for $i = 1, 2, 3$. Note that $|\mathcal{B}_i| = |\mathcal{P}_i|$, but $\mathcal{B}_1, \mathcal{B}_2, \mathcal{B}_3$ are not necessarily disjoint. If $\mathbf{x} \in [\mathcal{A}]^3$, then

$$0 < f_2(\mathbf{x}) = x_1 - x_2 + x_3 < N$$

by virtue of the inequalities $0 < x_1 < x_2 < x_3 \le N$. Therefore,

$$|\mathcal{B}_2| = |[\mathcal{A}]^3| = \tfrac{1}{6}A^3 + O(A^2). \tag{28}$$

Similarly, $f_1(\mathbf{x}) > 0$ and $f_3(\mathbf{x}) < N$. Therefore,

$$\mathcal{B}_1 = \left\{\mathbf{x} \in [\mathcal{A}]^3 : f_1(\mathbf{x}) < N\right\} \text{ and } \mathcal{B}_3 = \left\{\mathbf{x} \in [\mathcal{A}]^3 : f_3(\mathbf{x}) > 0\right\}.$$

From (27) and (28), we see that

$$|\mathcal{B}_1| + |\mathcal{B}_3| + \tfrac{1}{6}A^3 + O(A^2) < N.$$

For our purposes, it is convenient to rewrite this as

$$|\mathcal{B}_1 \cup \mathcal{B}_3| + |\mathcal{B}_1 \cap \mathcal{B}_3| + \tfrac{1}{6}A^3 + O(A^2) < N. \tag{29}$$

We can now obtain upper bounds for A by giving lower bounds for $|\mathcal{B}_1 \cup \mathcal{B}_3|$ and $|\mathcal{B}_1 \cap \mathcal{B}_3|$. Note that the trivial lower bounds of 0 give $A \le (6N)^{1/3} + O(1)$, which is the same as (9).

7. Bound for $h = 3$: lower bound for $|\mathcal{B}_1 \cup \mathcal{B}_3|$

Let ξ be the characteristic function of $\mathcal{A}$. We observe that

$$\sum_{a<N/2} \left(\xi(a) + \xi(N-a)\right)\Big(\sum_{b<a}\left(\xi(b) - \xi(N-b)\right)\Big)^2 \ge 0.$$

After expanding this sum, we deduce that

$$S \ge T, \tag{30}$$

where

$$S = \sum_{\substack{a<N/2\\ b,c<a}} \Big\{\xi(a)\xi(b)\xi(c) + \xi(a)\xi(N-b)\xi(N-c) +$$
$$\xi(N-a)\xi(b)\xi(c) + \xi(N-a)\xi(N-b)\xi(N-c)\Big\},$$

and

$$T = 2\sum_{\substack{a<N/2\\ b,c<a}} \Big\{\xi(a)\xi(b)\xi(N-c) + \xi(N-a)\xi(b)\xi(N-c)\Big\}.$$

By using the symmetry in the variables b and c, we find that

$$S = 2 \sum_{c<b<a<N/2} \Big\{ \xi(a)\xi(b)\xi(c) + \xi(a)\xi(N-b)\xi(N-c) + \xi(N-a)\xi(b)\xi(c) + \xi(N-a)\xi(N-b)\xi(N-c) \Big\}$$

$$+ \sum_{b<a<N/2} \Big\{ \xi(a)\xi(b) + \xi(a)\xi(N-b) + \xi(N-a)\xi(b) + \xi(N-a)\xi(N-b) \Big\}$$

$$= 2\big(|\mathcal{C}_1| + |\mathcal{C}_2| + |\mathcal{C}_3| + |\mathcal{C}_4|\big) + |\mathcal{E}|,$$

say, where

$$\begin{aligned}
\mathcal{C}_1 &= \left\{\mathbf{x} \in \lfloor\mathcal{A}\rfloor^3 : x_1 < x_2 < x_3 < N/2\right\}, \\
\mathcal{C}_2 &= \left\{\mathbf{x} \in \lfloor\mathcal{A}\rfloor^3 : x_1 < N/2 < x_2 < x_3, x_1 + x_2 > N\right\}, \\
\mathcal{C}_3 &= \left\{\mathbf{x} \in \lfloor\mathcal{A}\rfloor^3 : x_1 < x_2 < N/2 < x_3, x_2 + x_3 < N\right\}, \\
\mathcal{C}_4 &= \left\{\mathbf{x} \in \lfloor\mathcal{A}\rfloor^3 : N/2 < x_1 < x_2 < x_3\right\}, \text{ and} \\
\mathcal{E} &= \{(x_1, x_2) : x_1 \in \mathcal{A}, x_2 \in \mathcal{A}, x_1 < x_2, x_1 + x_2 \neq N\}.
\end{aligned}$$

If $\mathbf{x} \in \mathcal{C}_1 \cup \mathcal{C}_3$, then $-x_1 + x_2 + x_3 < N$, so $\mathbf{x} \in \mathcal{B}_1$. We infer that $\mathcal{C}_1 \cup \mathcal{C}_3 \subseteq \mathcal{B}_1$. A similar argument gives $\mathcal{C}_2 \cup \mathcal{C}_4 \subseteq \mathcal{B}_3$. After noting that $|\mathcal{E}| \ll A^2$, we get

$$S = 2\sum_{i=1}^{4} |\mathcal{C}_i| + O(A^2) \leq 2|\mathcal{B}_1 \cup \mathcal{B}_3| + O(A^2). \tag{31}$$

A similar analysis of T shows that

$$T = 2(|\mathcal{C}_5| + |\mathcal{C}_6|), \tag{32}$$

where

$$\begin{aligned}
\mathcal{C}_5 &= \left\{\mathbf{x} \in \lfloor\mathcal{A}\rfloor^3 : x_1 < x_2 < N/2 < x_3, x_2 + x_3 > N\right\}, \text{ and} \\
\mathcal{C}_6 &= \left\{\mathbf{x} \in \lfloor\mathcal{A}\rfloor^3 : x_1 < N/2 < x_2 < x_3, x_2 + x_3 < N\right\}.
\end{aligned}$$

Recall that $N/2 \notin \mathcal{A}$, so

$$\lfloor\mathcal{A}\rfloor^3 = \bigcup_{i=1}^{7} \mathcal{C}_i, \tag{33}$$

where

$$\mathcal{C}_7 = \{\mathbf{x} \in \lfloor\mathcal{A}\rfloor^3 : x_1 + x_2 = N \text{ or } x_2 + x_3 = N\}.$$

Now a B_3 sequence is *a priori* a B_2 sequence, so $x_1 + x_2 = N$ has at most one solution with $x_1 < x_2$. We deduce that $|\mathcal{C}_7| \ll A$. Using this with (31), (32), and (33), we obtain

$$\tfrac{1}{3}A^3 + O(A^2) = 2|[\mathcal{A}]^3| = 2\sum_{i=1}^{7} |\mathcal{C}_i| = S + T + O(A^2).$$

From (30) and (31), we infer that

$$S + T \leq 2S \leq 4|\mathcal{B}_1 \cup \mathcal{B}_3| + O(A^2).$$

Combining the last two statements gives the bound

$$|\mathcal{B}_1 \cup \mathcal{B}_3| \geq \tfrac{1}{12}A^3 + O(A^2). \tag{34}$$

If we use the trivial lower bound $|\mathcal{B}_1 \cap \mathcal{B}_3| \geq 0$ together with (29) and (34), we get

$$A \leq (4N)^{1/3} + O(1). \tag{35}$$

We shall use this estimate in the next section. Note that apart from the smaller error term, (35) is the same as Theorem 5.

8. Bound for $h = 3$: conclusion

To get a lower bound for $|\mathcal{B}_1 \cap \mathcal{B}_3|$, we first need the following lemma.

Lemma 1. *Suppose that X is a positive real number with $X \leq N/2$, and suppose that $\mathcal{I}$ is a subinterval of $[1, N]$ with either (i) $\mathcal{I} = (X/2, X]$ or (ii) $\mathcal{I} = (N - X, N - X/2]$. Then $[\mathcal{A} \cap \mathcal{I}]^3 \subseteq \mathcal{B}_1 \cap \mathcal{B}_3$.*

Proof. Suppose that (i) is true, and let $(x_1, x_2, x_3) \in [\mathcal{A} \cap \mathcal{I}]^3$. Then $X/2 < x_1 < x_2 < x_3 \leq X$, so $x_1 + x_2 - x_3 > 2(X/2) - X > 0$ and $-x_1 + x_2 + x_3 < 2X \leq N$. The proof for case (ii) is similar. ∎

Now let k be a positive integer to be chosen later, and define

$$\mathcal{G}_1 = \left(0, 2^{-k-1}N\right], \ \mathcal{G}_2 = \left(2^{-k-1}N, (1 - 2^{-k-1})N\right], \ \mathcal{G}_3 = \left((1 - 2^{-k-1})N, N\right].$$

Write $\mathcal{A} = \mathcal{F}_1 \cup \mathcal{F}_2 \cup \mathcal{F}_3$, where $\mathcal{F}_i = \mathcal{A} \cap \mathcal{G}_i$. Now each $\mathcal{F}_i$ is a B_3-set in a smaller interval. We use (35) to bound $\mathcal{F}_1$ and $\mathcal{F}_3$, and we thereby obtain

$$|\mathcal{F}_1 \cup \mathcal{F}_3| \leq 2\left(4 \cdot 2^{-k-1}N\right)^{\frac{1}{3}} + O(1) = 4^{\frac{1}{3}} 2^{\frac{2-k}{3}} N^{\frac{1}{3}} + O(1). \tag{36}$$

Now assume that b is a real number such that

$$A \geq b^{1/3} N^{1/3}. \tag{37}$$

Then

$$|\mathcal{F}_2| \geq \left(b^{\frac{1}{3}} - 4^{\frac{1}{3}} 2^{\frac{2-k}{3}} \right) N^{1/3} + O(1);$$

this is non–trivial as long as $b \geq 2^{4-k}$. Split the interval $(2^{-k-1}N, (1-2^{-k-1})N]$ into subintervals defined by the relations

$$\mathcal{I}_j = (2^{-j-1}N, 2^{-j}N] \text{ and } \mathcal{I}_{k+j} = ((1-2^{-j})N, (1-2^{-j-1})N]$$

for $j = 1, \dots, k$, and define $\mathcal{A}_j = \mathcal{A} \cap \mathcal{I}_j$. By Hölder's inequality,

$$|\mathcal{F}_2|^3 = \Big\{ \sum_{j=1}^{2k} |\mathcal{A}_j| \Big\}^3 \leq \Big(\sum_{j=1}^{2k} 1 \Big)^2 \sum_{j=1}^{2k} |\mathcal{A}_j|^3.$$

Observe that $|\mathcal{A}_j|^3 = 6|[\mathcal{A}_j]^3| + O(|[\mathcal{A}_j]^2|)$. Using this and Lemma 1, we conclude that

$$\left(b^{\frac{1}{3}} - 4^{\frac{1}{3}} 2^{\frac{2-k}{3}} \right)^3 N \leq 6(2k)^2 |\mathcal{B}_1 \cap \mathcal{B}_3| + O(N^{2/3}).$$

Combining this with (29) and (34), we deduce that

$$N \geq \frac{1}{4} A^3 + \frac{1}{24k^2} \left(b^{\frac{1}{3}} - 4^{\frac{1}{3}} 2^{\frac{2-k}{3}} \right)^3 N + O(N^{2/3}).$$

We incorporate the assumption (37) to get

$$1 \geq \frac{1}{4} b - \frac{1}{24k^2} \left(b^{\frac{1}{3}} - 4^{\frac{1}{3}} 2^{\frac{2-k}{3}} \right)^3 + O(N^{-1/3}).$$

We now conclude that $A \leq b_1^{1/3} N^{1/3} + O(1)$, where b_1 is the largest real root of the equation

$$24k^2 - 6k^2 b - \left(b^{\frac{1}{3}} - 4^{\frac{1}{3}} 2^{\frac{2-k}{3}} \right)^3 = 0. \tag{38}$$

Computations show that the optimal choice for k is $k = 8$; with this choice,

$$b_1 = 3.99561189585308\dots$$

is the largest real root of (38).

Now the preceding calculations may be repeated with 4 replaced by b_1 in (36). In this way, we get a sequence of inequalities of the form $A \le b_j^{1/3} N^{1/3} + O(1)$, where $b_0 = 4$, b_{j+1} is the largest real root of the equation

$$24k_j^2 - 6k_j^2 b - \left(b^{\frac{1}{3}} - (b_j 2^{2-k_j})^{\frac{1}{3}}\right)^3 = 0,$$

and k_j is a positive integer to be chosen at each stage. A modest amount of calculation shows that the optimal choices are $k_0 = k_1 = k_2 = 8$, and these choices yield

$$b_1 = 3.99561189585308\ldots,$$
$$b_2 = 3.99561029201457\ldots, \text{ and}$$
$$b_3 = 3.99561029142809\ldots.$$

(Further iterations give no improvement in the first 11 decimal places.) Now $b_3 < c$, where c was defined in the statement of Theorem 6. This completes the proof.

In conclusion, it is now appropriate to give some details on the relation between our proof and Li's proof. Except for notation and organizational details, we have followed Li in Sections 6 and 7. The improvement comes from the treatment of $\mathcal{B}_1 \cap \mathcal{B}_3$. Li divides the interval $[1, N]$ into $K = 2(\log N/\log 2) + O(1)$ subintervals satisfying the hypotheses of Lemma 1, and he uses Hölder's inequality to deduce that

$$A^3 \le 6K^2\left(|\mathcal{B}_1 \cap \mathcal{B}_3| + O(A^2)\right).$$

In our argument, we restricted the use of Hölder's inequality to $\mathcal{F}_2$. This is more efficient because only a bounded number of subintervals are required in the application of Lemma 1.

9. Infinite B_h sequences

We close with a brief mention of the known results about infinite B_h sequences when $h > 2$. Nash [17] combined Lindström's finite B_4 argument and Erdős' infinite B_2 argument to prove that if $\mathcal{A}$ is an infinite B_4 sequence, then

$$\liminf_{n\to\infty} \frac{A(n)}{n^{1/4}(\log n)^{-1/4}} < \infty.$$

A natural conjecture is that if $\mathcal{A}$ is an infinite B_h sequence, then

$$\liminf_{n\to\infty} \frac{A(n)}{n^{1/h}(\log n)^{-1/h}} < \infty. \tag{39}$$

A straightforward application of the arguments used in proving Theorems 4 and 5 does not work because the bounds for the exceptional cases are insufficiently sharp. However, S. Chen[3] has proved (39) when h is even. This superseded earlier partial results of Jia [13] and M. Helm [9], who independently proved (39) in the even case with the additional hypothesis that $A(n^2) \ll A(n)^2$. Little is known about the corresponding problem when h is odd. However, Helm [11] has shown that if $A(n) \sim \alpha n^{1/3}$ for some α, then $\mathcal{A}$ is not a B_3 sequence.

References

[1] R.C. Bose and S. Chowla, *Theorems in the additive theory of numbers*, Comment. math. helvet. **37** (1962-3), 141–147.

[2] S. Chen, *On the size of finite Sidon sequences*, Proc. Amer. Math. Soc. **121** (1994), 353–356.

[3] S. Chen, *On Sidon sequences of even orders*, Acta Arith **64** (1993), 325–330.

[4] P. Erdős and P. Turán, *On a problem of Sidon in additive number theory, and on some related problems*, J. London Math. Soc. **16** (1941), 212–215; *Addendum*, (by P. Erdős), ibid. **19** (1944), 208.

[5] P. Erdős and R. Freud, *On Sums of a Sidon–Sequence*, J. Number Theory **38** (1991), 196–205.

[6] R.L. Graham, B.L. Rothschild, and J.H. Spencer, *Ramsey Theory*, 2nd Edition, Wiley, New York, 1990.

[7] S.W. Graham and G. Kolesnik, *Van der Corput's method of exponential sums*, Lond. Math. Soc. Lecture Note Series 126, Cambridge University Press, Cambridge, 1991.

[8] H. Halberstam and K.F. Roth, *Sequences*, 2nd Edition, Springer-Verlag, New York, 1983.

[9] M. Helm, *On B_{2k} sequences*, J. No. Theory **48** (1994), 183–196.

[10] M. Helm, *A remark on B_{2k} sequences*, Acta Arithmetica **63** (1994), 367–371.

[11] M. Helm, *On the distribution of B_3 sequences*, J. No. Theory (to appear).

[12] Xing-De Jia, *On finite Sidon sequences*, J. No. Theory **44**, 84–92.

[13] Xing-De Jia, *On B_{2k} sequences*, J. No. Theory **48** (1994), 183–196.

[14] An Ping Li, *On B_3 sequences*, (Chinese), Acta Math. Sinica **34** (1991), 67–71.

[15] B. Lindström, *An inequality for B_2 sequences*, Jour. Comb. Theory **6** (1969), 211–212.

[16] B. Lindström, *A remark on B_4 sequences*, Jour. Comb. Theory **7** (1969), 276–277.

[17] J.C.M. Nash, *On B_4 sequences*, Canad. Math. Bull. **32** (1989), 446–449.

[18] S. Sidon, *Ein Satz über trigonometrische Polynome und seine Anwendung*

in der Theorie der Fourier–Reihen, Math. Ann. **106** (1932), 536–539.

[19] J. Singer, *A theorem in finite projective geometry and some applications to number theory*, Trans. Am. Math. Soc. **43** (1938), 377–385.

[20] A. Stöhr, *Gelöste und ungelöste Fragen über Basen der natürlichen Zahlenreihe, II*, J. Reine Angew. Math. **194** (1955), 111–140.

S. W. Graham
Michigan Technological University,
Houghton, MI 49931 USA
swgraham@mtu.edu

Current address:
National Science Foundation,
Arlington, VA 22230, USA
swgraham@nsf.gov